CONTINGENCY OPERATIONS SERIES

THE U.S. MILITARY INTERVENTION IN PANAMA: OPERATION JUST CAUSE

DECEMBER 1989–JANUARY 1990

by

Lawrence A. Yates

CENTER OF MILITARY HISTORY
UNITED STATES ARMY
WASHINGTON, D.C., 2014

Library of Congress Cataloging-in-Publication Data

Yates, Lawrence A., 1945–
The U.S. military intervention in Panama: Operation Just Cause, December 1989–January 1990 / by Lawrence A. Yates.
pages cm. — (Contingency operations series)
Includes bibliographical references and index.
1. Panama—History—American Invasion, 1989—Causes. 2. Military planning—United States. 3. Unified operations (Military science)—Case studies. 4. Panama—Military relations—United States. 5. United States—Military relations—Panama. I. Title.
F1567.4.Y37 2013
972.8705'3—dc23

2013031795

First Printing

CMH Pub 55–3–1

Contents

☆ ☆ ☆

Page

TABLES

No.

CHARTS

MAPS

Illustrations

Page

Illustrations courtesy of the following sources: p. 4, Jimmy Carter Library; pp. 5, 7, 9 (top), 53, 75, Corbis; pp. 9 (bottom), 98, Department of the Army; p. 17, Stephen Buckley; p. 23, Ron Haviv, VII Photo; p. 39, XVIII Airborne Corps Historian; p. 42, U.S. Southern Command; pp. 56, 110, 413, U.S. Air Force; pp. 86, 156, 160, 166, 167, 171, 216, 223, 262, 267, 293, 296, 344, 360, 428, 451, U.S. Army Center of Military History; pp. 93, 134, 229, 235, 291, 401, 421, Special Operations Command, South; p. 95 (top), Panama Defense Forces; pp. 95 (bottom), 102, 107, U.S. Special Operations Command reconnaissance efforts; p. 124, Maxwell Thurman Photograph Collection, U.S. Army Military History Institute; p. 138, Associated Press; pp. 150, 152 (bottom), History and Museums Division, Headquarters, U.S. Marine Corps; pp. 186, 187, 192, 197, Task Force Wildcat battle book; pp. 253, 269, Army Art Collection; p. 311, U.S. Navy; p. 327, Margaret Liss; p. 396, Getty Images; and pp. 426, 446, U.S. Army psychological operations. All other illustrations are from the files of the Department of Defense.

Foreword

☆ ☆ ☆

On 20 December 1989, the United States launched Operation Just Cause, the invasion of Panama. With only slight exaggeration, spokespersons for the president and the Pentagon declared that, at or near H-hour, twenty-seven thousand U.S. soldiers, marines, sailors, and airmen attacked twenty-seven targets nearly simultaneously, a triumph of planning and execution. Over the course of the next few days, these forces handily defeated the Panamanian military, toppled the brutal and corrupt dictatorship of Manuel Antonio Noriega, and helped return democracy and stability to the troubled isthmus. In the process, they brought down the curtain on a two-year crisis in U.S.-Panamanian relations.

During that lengthy confrontation, the U.S. Southern Command and its service components were still based in Panama, and two American presidents, Ronald W. Reagan and George H. W. Bush, relied on this military presence to help manage the situation and to protect and further U.S. interests in the country. An analytical narrative of how the United States dealt with that simmering crisis can be found in Dr. Lawrence A. Yates' *The U.S. Military Intervention in Panama: Origins, Planning, and Crisis Management, June 1987–December 1989*, published by the U.S. Army Center of Military History in 2008. In *The U.S. Military Intervention in Panama: Operation Just Cause, December 1989–January 1990*, Yates finishes the story as he reviews U.S. contingency planning for the possible use of armed force in Panama and then recounts the execution of those plans.

This study offers a detailed account of the major combat operations that occurred during Just Cause and of the broader themes and issues at play. While it was a short operation, relatively speaking, it was also an incredibly complex one. The challenges to the U.S. military were clear. The planners had to synchronize the preparation and deployment of diverse U.S. units taking part in the invasion, some stationed just minutes from their targets, others having to fly to Panama from bases scattered across the United States. The forces had to strike nearly simultaneously in order to preserve a measure of operational surprise and keep casualties low. And, even though the Army provided the vast majority of combat troops, it was a joint operation, which further complicated the planning. The Goldwater-Nichols Act of 1986 virtually mandated jointness in any significant military undertaking and had strengthened regional combatant commands to better conduct those operations. Southern Command was a joint command and had to determine the best way to task-organize the service units involved and assign each one a mission that would best employ its assets. Also a factor to consider was the nature of the endeavor. The Noriega regime was the enemy, not the Panamanian people, so damage to the country and injury to its populace had to be minimized. The United States had to structure carefully the rules of engagement that would allow soldiers and marines to defend themselves and accomplish their missions, while still protecting Panamanian civilians and, after it was over, assisting them in reforming their democratic government. All planners of military operations should consider the desired political results of any conflict, a lesson that continually seems to need relearning.

The Panama crisis and its military resolution are integral parts of America's military heritage. The insights this case study affords U.S. officers and policy makers today merit its examination a quarter of a century later and well into the future.

Washington, D.C.
31 July 2014

RICHARD W. STEWART
Chief Historian

THE AUTHOR

☆ ☆ ☆

Lawrence A. Yates grew up in Kansas City, Missouri. He received his Ph.D. in history from the University of Kansas in 1981, the same year he joined the Combat Studies Institute at Fort Leavenworth, Kansas. In a 24-year career with the institute, he specialized in teaching and writing about U.S. interventions, contingency and stability operations, and unconventional warfare. In December 1989, he was in Panama during Operation JUST CAUSE, the topic of this volume. Among his other publications, he is the author of Leavenworth Paper 15, *Power Pack: U.S. Intervention in the Dominican Republic, 1965–1966*; Occasional Paper 15, *The US Military's Experience in Stability Operations, 1789–2005*; and the companion piece to this study, *The U.S. Military Intervention in Panama: Origins, Planning, and Crisis Management, June 1987–December 1989*. He is also a coauthor of *My Clan Against the World: U.S. and Coalition Forces in Somalia, 1992–1994*, and coeditor of—and a contributor to—*Block by Block: The Challenges of Urban Operations*. In September 2005, Yates retired from government service. He is currently working on a book assessing the U.S. Army's role in Somalia from December 1992 to March 1994.

PREFACE

☆ ☆ ☆

This book is the final installment of a two-volume set covering the politico-military crisis between the United States government and the Panamanian dictatorship of Manuel Antonio Noriega. The first volume examines various crisis-related activities of the U.S. military from the onset of tensions between the two countries in June 1987 to the decision of President George H. W. Bush to resolve the conflict by military force in December 1989. Included in that book's narrative is an analysis of several issues that placed American armed forces in what several officers stationed in Panama dubbed the Twilight Zone: a limbo between war and peace in which political considerations had their impact not just on military decisions at the strategic level, but on actions at the operational and tactical levels as well. Moreover, the "battlefield" was nonlinear and occupied overwhelmingly by civilians, most of whom were not hostile to the United States. Given these unorthodox circumstances, U.S. combat units stationed in Panama or deployed there had to demonstrate restraint, not their martial prowess, as they operated under highly restrictive rules of engagement that some officers believed would yield the initiative to their adversary in any sudden recourse to open violence. In Washington, two American presidents indicated their desire to see Noriega deposed but failed to devise a timetable or a well-defined, integrated, and effective strategy for achieving that goal, thus contributing to the troops' perception that the crisis could go on indefinitely. Only after a group of Panamanian officers tried and failed to overthrow the dictator in early October 1989 did U.S. officials conclude that the outcome they sought would almost certainly require the use of American military power.

This second and concluding book in the set examines how that power was employed in Operation JUST CAUSE. As covered in Chapters 1 and 2, planning for possible combat in Panama began in early 1988, nearly two years before the invasion that began on 20 December 1989. As the reader will see, just months before JUST CAUSE, changes in the principal contingency plan had returned many traditional-thinking officers to their comfort zone, although they were still left to grapple with such difficult questions as what units to select for the projected joint force and how those units could best be organized, equipped, and utilized to accomplish the plan's objectives. Even more problematic was the fact that much of the fighting would take place in cities and towns, a prospect for which neither the U.S. Army nor Marine Corps was adequately prepared, having years before deemphasized the attention given to urban operations in both doctrine and training. Finally, while the plans discussed the inevitable transition from combat to stability operations and nation building, the procedures for a smooth changeover were not finalized during the planning process. The U.S. military's involvement would be essential in helping to establish a stable, effective, and democratic government to fill the vacuum left by Noriega's defeat. A failure to effect the transition to this phase of the operation risked having the combat successes of American forces go for naught.

Chapters 3 through 12 examine combat operations and how twenty-seven thousand U.S. troops won a decisive victory over their Panamanian opponents. That victory, of course, was never in doubt, but the cost of it was. Fortunately, the professionalism of the joint U.S. forces, the sensible rules of engagement given to them for combat operations, the poor quality of the enemy they faced, and, at times, some very good luck resulted in the accomplishment of the mission with relatively low U.S. and Panamanian casualties and only limited damage to Panama's physical infrastructure. Chapter 13 describes the rough transition to stability operations and nation building and the toll it took on troops who were not prepared to relinquish overnight their status as warriors in order to take on the duties of policemen, social workers, diplomats, political officials, and administrators. Some units performed these functions for a few days, some for a few weeks. In after action reports, many officers expressed the need for training programs to cover such activities. This and other lessons to come out of JUST CAUSE are examined in Chapter 14. Many of them are still highly relevant for U.S. military personnel today, even with the approach this year of the twenty-fifth anniversary of the invasion of Panama.

With regard to the sources used in writing the two volumes, for the first volume, the material I drew on consisted almost exclusively of primary sources, especially interviews and the reports and message traffic that circulated among U.S. headquarters in Panama and the United States. While primary sources constitute the bulk of the citations in this volume, I relied more on interviews and after action reports than on message traffic. I had access to the logs of several units and organizations that participated in JUST CAUSE, but the reliably of some of those logs was later called into question, often by the personnel who kept them. For example, one officer informed me that much of the message traffic received at his post during the first two days of the operation was not recorded formally until hours after the missives were received, a process that often involved guessing as to the timing and the order of their arrival. Partially compensating for this—but only partially—was the fact that many after action reports and oral history interviews reflected a headquarters' or unit's exhaustive efforts to record its activities accurately.

There is also a point I would like to make about the convention I have employed in both volumes with respect to Spanish words. Normally, when a Spanish word requiring an accent mark is used in this book, the accent mark is included, thus informing the reader as to what syllable should be emphasized. (Accent marks in Spanish do not change a word's pronunciation, only the syllabic emphasis.) I have made an exception for accented Spanish words used regularly in English without an accent. For example, the word *río* (river) in Spanish is spelled "rio" in this text, as in Rio Hato. Colón, on the other hand, has an accent mark emphasizing the second syllable. This Spanish word is rarely spoken in English, so a non-Spanish–speaking reader might assume that, absent the accent, it would be pronounced like the English word for a grammatical symbol or an anatomical part.

As I indicated in the first volume, I started the research on the U.S.-Panamanian confrontation in early 1989 in response to a request made by the U.S. Army, South/JTF-Panama commander, Brig. Gen. Bernard Loeffke, and his operations officer, Col. John A. "Jay" Cope. When General Loeffke left Panama for another assignment that summer, his successor, Brig. Gen. (soon promoted to Maj. Gen.) Marc A. Cisneros allowed me to continue my work and, during the first hours of JUST CAUSE, secured my access to the JTF-South operations center at Fort Clayton. I remain exceptionally grateful to these three men, as well as to all those who, serving with them, assisted me during my first six trips to Panama. In the early 1990s, the U.S. Army Center of Military History expanded my assignment by asking me to write the Army's official account of JUST CAUSE. Originally, only the first chapter in the projected volume was to cover the crisis leading up to the invasion; the remaining chapters would address the invasion itself. As I began writing that manuscript, however, I concluded that the U.S. military's role in the two-year crisis deserved its own volume, and the Center agreed. At the time, I was spending most of my working hours performing unrelated duties at Fort Leavenworth's Combat Studies Institute. Thanks to three of my supervisors, however, I was granted significant chunks of time to work on the Panama project. I finished a draft of the first volume on New Year's Day 2001, and, after various reviews, revisions, and editing, it was published in 2008 as the first book in the Center's Contingency Operations Series. By that time, I had retired from the institute and was under contract to the Center to write this, the second volume. I finished the first draft in 2010.

During nearly two and a half decades working on the Panama case study, I have become indebted and grateful to many people and institutions, all of whom I would like to thank and several of whom I need to recognize specifically. In addition to the three officers I have already mentioned, I would also like to acknowledge the official historians at the Southern Command and U.S. Army, South, especially Janet Len-Rios, who helped orient me to the situation in Panama in 1989 and guided me to various documents, individuals, and facilities. At the U.S. Special Operations Command at MacDill Air Force Base, Florida, John Partin, the command's historian, and archivist Gaea Lenders made essential material and their expertise available to me. Similar help was received from historians, archivists, and staff members at the Combined Arms Research Library at Fort Leavenworth; the Joint History Office in the Pentagon; the U.S. Army Military History Institute at Carlisle Barracks, Pennsylvania; and the U.S. Marine Corps History Division, now in Quantico, Virginia. Also, two officers in the XVIII Airborne Corps spent a good deal of time with me, both during and after JUST CAUSE. During the second and third nights of the invasion, Lt. Col. Timothy L. McMahon, one of the corps planners, told me in detail how the contingency plan for combat operations in Panama had evolved and how it was being executed as we spoke. Maj. David Huntoon Jr. also provided insights immediately after JUST CAUSE; then, three years later, as a lieutenant colonel, he spent two days locating and explaining to me critical material kept in the corps' files at Fort Bragg, North Carolina.

Since the early 1990s, the organization furnishing me with the most support in this endeavor has been the aforementioned U.S. Army Center of Military

History, now located at Fort McNair, Washington, D.C. As chief historian, Jeffrey J. Clarke oversaw my work on most of the first volume and was the chief of military history during much of the work on this, the second. Chief Historian Richard W. Stewart arranged the contract for this volume in 2006, and, once I signed it, he became my boss. I could not have asked for a better one. He has opened doors for me, intervened on my behalf with other organizations and individuals, and, most of all, has been extremely understanding and supportive. As for documents regarding Just Cause housed at the Center, a number of people have assisted my collection efforts. One requires special mention: Robert K. Wright Jr., then chief of the Historical Resources Branch. Also serving as both the civilian historian and then an activated reserve officer for the XVIII Airborne Corps, Wright gathered information and interviewed Just Cause participants, material that he readily made available to me, along with his expertise.

In 2011, a panel was convened at the Center to review my manuscript. Richard Stewart chaired the meeting, while General Cisneros, John T. Fishel, and Andrew J. Birtle, chief of the Military Operations Branch in the Center's Histories Division, submitted their critiques. I would like to thank them for their input, which helped in further revising my work before turning it over for formal editing. Fishel, I need to note, served as a reserve lieutenant colonel on active duty during the crisis, participated in almost all phases of the conflict leading up to Just Cause, and was a key player in the nation-building effort that began simultaneously with U.S. combat operations. He is the reigning expert on that phase of the U.S. military's effort and a valuable source of information.

Also on the panel as the Center's editorial adviser was Diane Sedore Arms, who later edited the manuscript's chapters. We had worked together on the first volume, and I was delighted when she was picked to edit the second. Her professionalism, her expertise, and, far from least, her patience has made this a more readable and accurate book. I cannot thank her enough. Others at the Center who were involved in the book's production process were cartographer S. L. Dowdy, who prepared the excellent maps to complement the text, and Michael R. Gill, who collected most of the photographs and prepared the charts and tables and who did the highly professional layout of the text and the cover design. Contractor Kate Mertes prepared the index. Readers will quickly become aware of their talents, and I thank them all, together with Beth F. MacKenzie, chief of the Historical Products Branch in the Center's Histories Division, who managed the final phases of this project.

Kelly Fent assisted me in proofreading the manuscript; I found her help to be indispensable. In conclusion, despite the support I received from everyone I have acknowledged here and from many, many others, I, of course, alone bear complete responsibility for all interpretations and conclusions, as well as for any factual errors found herein.

Overland Park, Kansas
31 July 2014

LAWRENCE A. YATES

The U.S. Military Intervention in Panama: Operation Just Cause

December 1989–January 1990

1

The Panama Crisis

June 1987–December 1989

Early in the afternoon of Sunday, 17 December 1989, President George H. W. Bush met with his national security advisers to discuss events in Panama the previous night. Fewer than twenty-four hours before, members of the Panama Defense Forces (PDF) had shot and killed an American Marine lieutenant, a passenger in a car that had sped away from a checkpoint near the PDF's main headquarters in the capital, Panama City. Two witnesses to the incident, an American Navy lieutenant and his wife, had been taken into police custody. In the ordeal that followed, the officer's tormentors placed a pistol to his head several times and repeatedly hit and kicked him; they also fondled his spouse, slammed her against a wall, and threatened her with rape.[1] Angered by the excessive and unprovoked violence and concerned over the safety of other American citizens living in Panama, the president ordered the execution of Op-

President Bush speaks at an Armed Forces Review and Award Ceremony, with Secretary Cheney at the left and General Powell at the right.

[1] Separate handwritten statements by the U.S. Navy lieutenant and his wife to U.S. Naval Investigative Service, Rodman Naval Station, Panama, 18 Dec 1989.

The Panama Canal's Miraflores Locks

eration Just Cause, the U.S. military invasion of Panama to topple the dictatorship of General Manuel Antonio Noriega Moreno.

Origins of the Crisis

The assault on Panama, which began in the wee hours of 20 December, constituted the climactic episode in a crisis between the two countries that had festered for almost two years.[2] The historical roots of the confrontation could be traced even further back in time to the early years of the twentieth century when President Theodore Roosevelt deployed American naval vessels to the southwest Caribbean in a successful effort to help Panamanians secure their independence from Colombia. The quid pro quo Roosevelt exacted for the show of U.S. force took the form of the Hay–Bunau-Varilla Treaty of 1903 that granted the United States the right to build, operate, and defend a canal across the isthmus of Panama. In addition, the treaty allowed for a strip of land ten miles wide along the length of the canal route (and three nautical miles into the sea on each end) to serve as an operating area within which the American government would exercise, in perpetuity, "all the rights, power, and authority . . . which the United States would possess and exercise if it were the sovereign of the territory."[3] That area became the Canal Zone, bisecting Panama and, together with a provision

[2] The background to the crisis is covered in more detail in ch. 1 of Lawrence A. Yates, *The U.S. Military Intervention in Panama: Origins, Planning, and Crisis Management, June 1987–December 1989* (Washington, D.C.: U.S. Army Center of Military History, 2008).

[3] Quote from the text of the Hay–Bunau-Varilla Treaty, 18 November 1903. The U.S. Senate advised ratification of the treaty on 23 February 1904; President Roosevelt ratified the treaty on 25 February 1904. Interestingly, no Panamanian signed the treaty. Philippe Bunau-Varilla was a French engineer who, having failed in earlier attempts to build a canal in Panama, convinced the United States to try. Instrumental in planning Panama's independence from Colombia, he

in the Panamanian constitution of 1904 that relegated the country to the status of a U.S. protectorate, leaving little doubt of the dominant role Washington intended to play on the isthmus for some time to come.

From the outset, the stationing of U.S. troops in the Canal Zone and the influx of thousands of American citizens living there exempt from Panamanian jurisdiction fueled nationalistic resentments within the fledgling republic. After three decades, the Hull-Alfaro Treaty of 1936 (ratified by the United States in 1939) sought to mitigate these feelings by formally ending Panama's protectorate status and declaring that the Canal Zone was the "territory of the Republic of Panama under the jurisdiction of the United States of America."[4] The concessions did not go far enough, as the pent-up discontent generated by the presence of an unconstrained foreign power in virtual control of the country persisted, sporadically finding release in public demonstrations and isolated acts of defiance. The worst of these occurred in January 1964, after American and Panamanian students fought when the latter attempted to fly their country's flag alongside the U.S. flag in front of a high school in the zone. The initial scuffling quickly escalated into three days of full-scale rioting in which four U.S. servicemen and over twenty Panamanians died and several hundred more demonstrators were injured. In the aftermath of the violence, a number of high-ranking policy makers serving under President Lyndon B. Johnson concluded that continued adherence to the status quo would further undermine America's already tarnished reputation in Latin America and call into question the credibility of Washington's anti-imperialistic and prodemocratic Cold War rhetoric. Accordingly, the president agreed to open talks on a new canal treaty.

Bilateral discussions to this end dragged on for over a decade before the negotiating parties produced an agreement in the form of two documents. Under the first, the Panama Canal Treaty, Panama would assume immediate jurisdiction over the Canal Zone but would allow a new U.S. government organization, the Panama Canal Commission, which was "supervised by a Board composed of nine members, five of whom shall be nationals of the United States of America, and four of whom shall be Panamanian nationals proposed by the Republic of Panama for appointment to such positions by the United States of America in a timely manner," to operate, maintain, and manage the canal. At noon on 31 December 1999, the treaty would be terminated and Panama would assume full control of the waterway. The treaty also mandated that both countries "commit themselves to protect and defend the Panama Canal," with the United States having the "primary responsibility" to do so for the duration of the treaty. In keeping with this provision, the U.S. military would retain control of certain installations—specified in an implementation agreement to the treaty—until the canal passed completely into Panama's hands. American and Panamanian forces would occupy some other installations jointly, and still other sites would gradually be turned over to Panama

---

immediately named himself the new republic's foreign minister and negotiated the canal treaty with American Secretary of State John Hay.

[4] Quote from Walter LaFeber, *The Panama Canal: The Crisis in Historical Perspective* (New York: Oxford University Press, 1978), p. 89.

President Carter and General Torrijos sign the Panama Canal Treaty.

before the treaty expired.[5] The second document, the Neutrality Treaty, committed the United States and Panama to guarantee the neutrality of the canal. Unlike the first document, the Neutrality Treaty had no termination date. On 7 September 1977, President Jimmy Carter of the United States and General Omar Torrijos Herrera, Panama's chief of state, signed the documents.

The following year, the charismatic and flamboyant Torrijos stepped down as head of government but, continuing in his post as commander of Panama's National Guard, maintained his grip on political power. In the United States, the authoritarian nature of the general's military-dominated regime jeopardized ratification of the new treaties, as many Americans who opposed losing control of the canal charged that the waterway's security should not be left to a "tin-horn dictator" with friendly ties to President Fidel Castro in Cuba and the Communist Sandinista movement in Nicaragua. There was also opposition to the treaties in Panama, where traditional anti-American feelings complicated Torrijos' efforts to secure popular support for an agreement that did not relinquish foreign control over the canal immediately. Yet, despite the rancorous rhetoric in both countries, Panamanian voters ratified the treaties via a plebiscite held in October 1977, while the U.S. Senate advised and consented to ratification in two separate and identical votes of 68 to 32. The treaties entered into force on 1 October 1979. At that time, the Canal Zone was formally abolished, and Panama took direct control of about 64 percent of the territory that had been contained therein. Another 18 percent of the former zone became the canal operating area, controlled by the Panama Canal Commission. The remainder was left under the U.S. military.[6]

[5] Quotes from the text of the Panama Canal Treaty. The text of both treaties can be found in app. B of Sandra W. Meditz and Dennis M. Hanratty, eds., *Panama: A Country Study* (Washington, D.C.: Headquarters, Department of the Army, 1989), pp. 273–93.

[6] Ibid., p. 53.

In July 1981, Torrijos died in a plane crash. In the turmoil and uncertainty that followed his death, one feature of Panamanian politics remained constant: the influence of the national guard. In August 1983, Manuel Noriega took over as the organization's new commander.[7] An introvert whose pockmarked face garnered him the sobriquet "*La Piña*" ("the pineapple"), Noriega had received his first military commission in 1967. Torrijos, who became the national guard's commander a year later, looked favorably upon the young officer and made him his protégé, assigning him sensitive duties and, even more critical, covering for his aberrant excesses that, besides drunkenness, included rape and other acts of sexual brutality. As Noriega advanced through the ranks, he developed skills in military intelligence and psychological operations, enabling him to set up a sophisticated domestic espionage network with agents planted inside a variety of Panamanian groups. The information he collected and his willingness to use it to blackmail and intimidate people in positions of power made him an increasingly dangerous man to cross. In 1969, Noriega's timely assistance saved Torrijos from a coup attempt plotted by rival officers in the guard. Torrijos rewarded Noriega by making him chief of military intelligence in Panama and promoting him to lieutenant colonel.

After Torrijos' death, Noriega skillfully outmaneuvered his fellow officers within the national guard's inner circle to emerge in August 1983 as Panama's new strongman. Soon thereafter, he secured passage of Law 20 that created the Panama Defense Forces. The national guard retained its identity but only as one among several entities—including ground, air, and naval forces and the police—constituting the PDF. As commander in chief of the new organization with the rank of *general de fuerzas*, Noriega transformed the PDF into a much more militarized institution than its predecessor. He also acquired under Law 20 the responsibility for implementing "measures needed to guarantee the security of inhabitants and their property and the preservation of public order and social peace," the only limitation being largely cosmetic in that he had to keep the

General Noriega

[7] For biographical details on Noriega up to 1985, see Frederick Kempe, *Divorcing the Dictator: America's Bungled Affair with Noriega* (New York: G. P. Putnam's Sons, 1990), pp. 38–71. This author has also incorporated observations made by Gabriel Marcella and General Frederick F. Woerner Jr., U.S. Army (Ret.), The Road to War: The U.S.-Panamanian Crisis, 1987–1989, draft article, 6 May 1991.

president of Panama informed of national security developments.[8] The military's preeminence within the government was brought home in 1984 when Noriega forced the resignation of Panama's civilian president and, later in the year, helped engineer the fraudulent election that placed the PDF-backed (and U.S.-supported) candidate, Nicolás Ardito Barletta, in the presidential palace.

Law 20 also expanded the role of the military in economic matters by making Panama's ports, airports, immigration, and other revenue-generating enterprises part of the PDF. Also, many officers in advancing their fortunes looked beyond their erstwhile roles as symbolic board members of various companies to become themselves the owners of lucrative firms. Under Torrijos, a formidable coalition of guardsmen, students, rural *campesinos*, urban workers, and the middle class had challenged the political power of Panama's white oligarchy; now, the Panama Defense Forces under Noriega began to encroach—through formal and informal means and through the use of legal and illegal enterprises—upon the elite's economic livelihood.

Noriega's actions generated opposition, but his growing reputation as a ruthless, even sadistic, man silenced most critics. In Panama, there would be no "dirty war," no death squads roaming the country, and no long lists of missing people. Rather, opponents of the regime would be targeted selectively, individually or in small groups; harassed and intimidated; incarcerated if need be; sometimes tortured and sexually abused; but killed only in the most extreme cases. Word of these activities would spread, invariably embellished by the hyperbole of rumor and gossip. The resulting terror was almost always more psychological than physical. The important thing was the effect: acquiescence. One could silence opponents without killing them.

How this ruthless behavior might affect the regime's relationship with the United States seemed of no great concern to Noriega. As a collaborator with the Central Intelligence Agency (CIA), the Drug Enforcement Administration, and U.S. military intelligence and as a confederate in several U.S. foreign policy ventures in Latin America, the general believed that his ties to the administration of President Ronald W. Reagan were strong enough that Washington would either overlook his indiscretions in Panama or, at worst, give him a perfunctory slap on the wrist. It thus came as an unpleasant surprise when, in the spring of 1986, he became the target of two investigations by the U.S. Senate. To make matters worse, he found himself the subject of a front-page exposé in the *New York Times* detailing his alleged involvement in drug trafficking, money laundering, arms shipments to Communist guerrillas, the selling of restricted U.S. technology to Cuba, and the murder of Dr. Hugo Spadafora, a prominent critic of the regime who could not be silenced by intimidation. When President Barletta hinted that he would investigate the Spadafora affair, Noriega replaced him with Vice President Eric Arturo Delvalle, a move that upset some officials in Washington with whom Barletta was very popular.[9] To Noriega's relief, the tempest that followed proved short-lived. When the White House subsequently held a policy review on Panama, even the dictator's detractors supported the decision to put the issue of his

[8] Quote from Meditz and Hanratty, eds., *Panama,* p. 225.

[9] *New York Times*, 12 Jun 1986.

objectionable behavior "on the shelf" until the Reagan administration's efforts to overthrow the Communist Sandinista government in Nicaragua bore fruit.[10]

Events in Panama dashed the administration's hopes of downplaying its Noriega problem. On 1 June 1987, the general cashiered an ambitious rival, Col. Roberto Díaz Herrera, the PDF's chief of staff. Díaz Herrera swiftly retaliated, publicly accusing Noriega of fraud, corruption, and drug trafficking. He also told reporters that his former boss was responsible for the murder of Spadafora and, even more damning, had engineered the plane crash that killed Torrijos. Although evidence to support the last charge was sketchy at best, these allegations, coming from such a high-ranking member of the PDF leadership, rocked the country. Spontaneous demonstrations erupted throughout Panama City and in Colón at the northern end of the canal. Crowds waved white handkerchiefs and banged pots, two gestures of defiance that almost overnight became symbols of the internal opposition to Noriega. That opposition acquired organizational status on 10 June after leading Panamanian businessmen—some motivated by principle, others resentful of PDF encroachments on their economic turf—formed the National Civic Crusade. Several diverse groups including students, teachers, workers, medical organizations, and religious leaders proclaimed their support. When the crusade began promoting mass demonstrations, a national strike, and car caravans through the capital as part of its protest, the regime responded with antiriot police and counterdemonstrations. Panama suddenly found itself in the throes of a serious political crisis.

Many of the public protests in Panama City took place within a few miles of where the U.S. Southern Command (SOUTHCOM) and its service components were located (*Map 1*). Headquartered at Quarry Heights on Ancon Hill overlooking the capital, the unified command managed U.S. military programs

Demonstrators bang pots in protest against the Noriega regime.

[10] Kempe, *Divorcing the Dictator*, pp. 169–79.

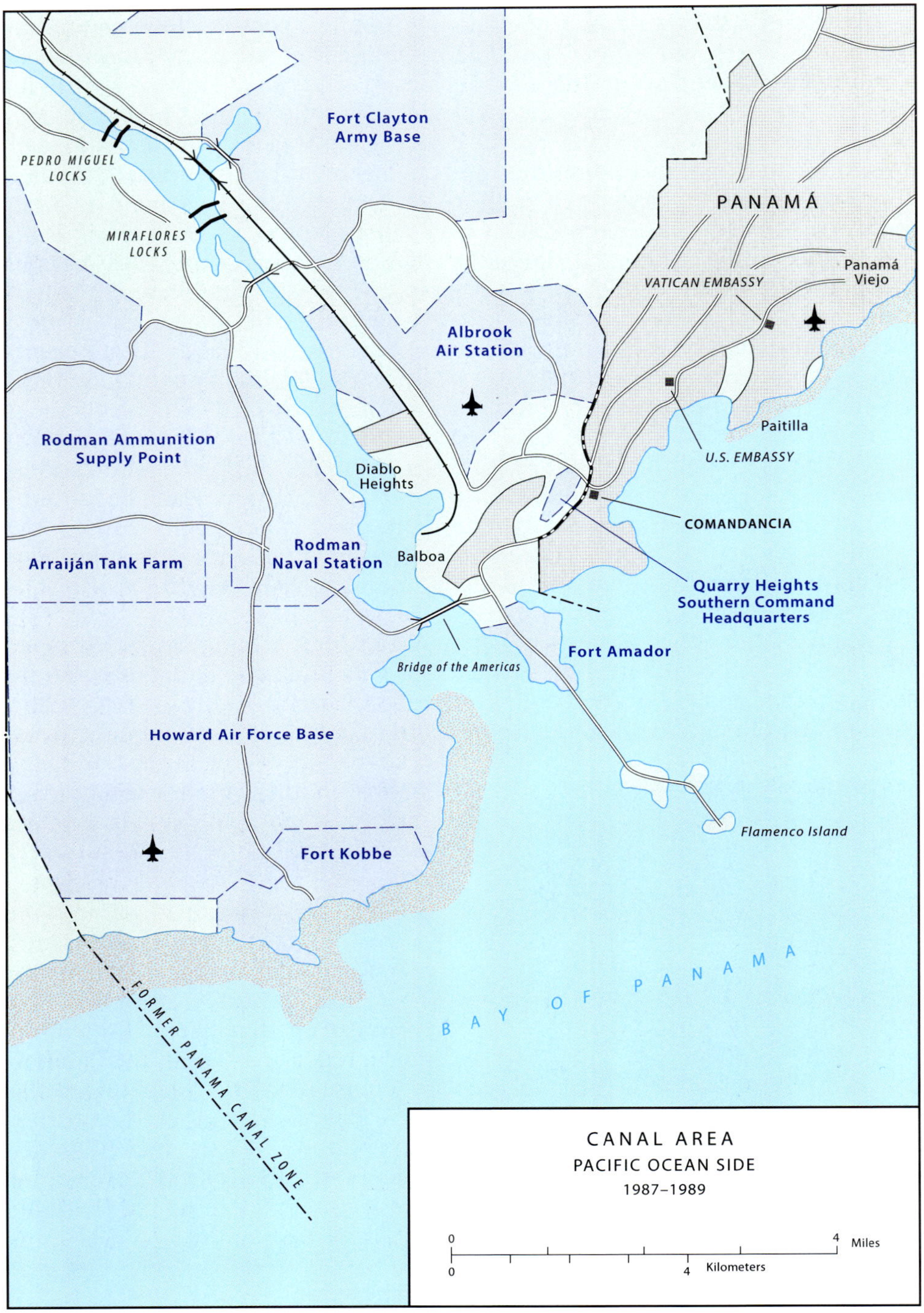

Fort Clayton Army Base
PEDRO MIGUEL LOCKS
MIRAFLORES LOCKS
PANAMÁ
Panamá Viejo
VATICAN EMBASSY
Albrook Air Station
Paitilla
U.S. EMBASSY
Rodman Ammunition Supply Point
Diablo Heights
COMANDANCIA
Rodman Naval Station
Arraiján Tank Farm
Balboa
Quarry Heights Southern Command Headquarters
Fort Amador
Bridge of the Americas
Howard Air Force Base
Flamenco Island
Fort Kobbe
BAY OF PANAMA
FORMER PANAMA CANAL ZONE
CANAL AREA
PACIFIC OCEAN SIDE
1987–1989
0
4
Miles
0
4
Kilometers

Map 1

A Panama Defense Forces (PDF) antiriot unit uses a water cannon to put down a street demonstration.

General Woerner

and activities in Central and South America. With such a broad and wide-ranging mission, SOUTHCOM's new commander in chief (CINCSO), General Frederick F. Woerner Jr., was concerned that the crisis in Panama might draw Washington's attention—and his command's resources—away from more important issues affecting U.S. interests in the region.[11] Worse, should the United States somehow become involved in Panama's domestic unrest, the fallout could have a negative impact on America's image in Latin America and disrupt SOUTHCOM's long-term relationship with the Panama Defense Forces, which under the treaties was a partner in ensuring the security of the canal. Woerner had no respect for Noriega, whom he considered a thug, but the PDF was another matter. In his opin-

[11] At the time of the U.S.-Panama crisis and Operation JUST CAUSE, the U.S. Army was the predominant service in the Southern Command. Thus, the service of U.S. military personnel identified in this volume will be listed only if it is not the Army.

ion, the organization possessed the potential for becoming, if not a completely apolitical institution, a more professional one, willing to support some form of legitimate, nondictatorial government in Panama. Integral to this transformation was the example of military professionalism U.S. forces in the country provided while interacting daily with their Panamanian counterparts. The positive impact of that face-to-face relationship would likely be placed in jeopardy should the United States and Panama end up at loggerheads because of the current situation.

There was another crisis-related issue that weighed heavily on the general: his responsibility to protect the people under his command from the violence in Panama. There were some fifty thousand American citizens living on Panamanian territory, especially in and around the capital, and among them were several thousand U.S. service personnel and their dependents. They already risked being innocent bystanders in the demonstrations rocking the country's largest cities. Should the United States openly criticize or disavow the current regime, Noriega might respond by mounting a campaign of intentional intimidation and harassment against them. For these and other reasons, Woerner repeatedly emphasized his position that Panama's internal problems should not be allowed to escalate into an unwanted and unnecessary U.S.-Panamanian confrontation.

The general's hopes that the crisis could be confined to Panama were not to be realized. As anti-Noriega protests continued throughout the remainder of 1987, the dictator's countermeasures grew increasingly harsh. In addition to declaring a state of emergency, the regime unleashed two police companies with antiriot capabilities, the Dobermans and Centurions, against the demonstrators, one result being the televised spectacle of unarmed civilians falling victim to truncheons, water cannon, and birdshot. At the same time, as part of the effort to shore up the substantial, if diminishing, political support he did enjoy in Panama, Noriega resorted to increasingly shrill rhetoric mixing appeals to Panamanian nationalism with strong doses of anti-American bombast.

The latter tactic was nothing new in Panamanian politics, and, during past occurrences, U.S. officials had generally accepted it for what it was, political theater performed primarily for the domestic audience. But this time Noriega overplayed his hand. Four days after the U.S. Senate passed a resolution critical of his regime, progovernment demonstrators outside the U.S. Embassy in Panama City threw rocks and paint-filled balloons at the building, broke windows, and damaged cars in the area. American Ambassador Arthur H. Davis Jr. angrily denounced the attack and recommended the suspension of economic and military assistance to Panama. The State Department agreed, arguing that economic pressure could be used to rid Panama of an unsavory dictator and the United States of a troublesome "friend." Over the objections of the Central Intelligence Agency and the Pentagon, President Reagan adopted this position, signing legislation in December that restricted U.S. assistance to Panama until that country's government could provide evidence of a meaningful transition to democracy. The sanctions hurt the Noriega dictatorship but also had an adverse effect on businessmen, many of whom formed the core of the opposition movement, and on the Panamanian people.

In early 1988, relations between the Reagan administration and the government in Panama took an even more dramatic turn for the worse. On 5 February, two federal grand juries in Florida, one in Tampa, the other in Miami, announced their separate indictments against Noriega on charges of drug trafficking, racketeering, and money laundering. The general denounced the action as illegal, condemned what he called U.S. aggression in the region, and had the Panama Defense Forces take even stronger measures against the regime's domestic opposition. He also threatened to expel U.S. service personnel and their families from Panama. When President Delvalle, under pressure from friends among the opposition, tried to assert his independence and relieve the PDF leader of command, he, like his predecessor, was pushed aside by the same assembly that had elevated him. The upshot was that the indictments and Noriega's reaction to them had the effect of transforming the crisis along the very lines Woerner had hoped to avoid, from a largely internal affair between the Panamanian dictator and the National Civic Crusade, with some external pressure from Washington, into a full-blown political confrontation between the governments of Panama and the United States.

The Southern Command Reacts

The escalation of the crisis did not lead immediately to any regime-sponsored campaign to harass U.S. citizens in Panama or to threaten American property and the canal.[12] When minor incidents and treaty violations did occur, the Southern Command through its Center for Treaty Affairs issued formal, if largely ineffectual, protests. Initially, Woerner and many of his subordinates regarded the marked deterioration in U.S.-Panamanian relations as a political storm that might soon abate without leaving in its wake irrevocable damage to SOUTHCOM-PDF relations. In the meantime, articles in the command's newspaper, broadcasts over the SOUTHCOM television station, and a series of town hall meetings in which high-ranking officers answered questions regarding the crisis sought to allay the fears and concerns of American military personnel and their dependents, while helping them avoid dangerous locations where demonstrations were scheduled to occur. As another means of keeping these people out of harm's way, the command put into effect a series of Personnel Movement Limitations (PMLs), ranging from Alpha, which allowed normal travel by military personnel and their dependents, to Echo, which would curtail all travel. The daily PML level was then adjusted to the threat posed by crisis-related developments in the streets. Yet, even as these measures sought to exert a calming effect, Woerner could not ignore the possibility of the crisis becoming more inflamed, perhaps degenerating to a point at which U.S. troops would find themselves facing off against anti-American demonstrators or, worse, the Panama Defense Forces. Despite the

[12] For the Southern Command's (SOUTHCOM's) reaction to developments in the crisis and the command's early contingency planning, see Yates, *U.S. Military Intervention in Panama, June 1987–December 1989,* chs. 1 and 2.

seeming unlikelihood of either eventuality, the general found it only prudent in early 1988 to have his staff review and update its "security posture." Among other things, this meant planning for a variety of military contingencies that might arise as a result of the crisis.

On 28 February, the U.S. Joint Chiefs of Staff (JCS) at Woerner's request authorized such planning in the "crisis action mode," and on 4 March the Southern Command's Operations Directorate, or J–3, produced its first draft of CINCSO ELABORATE MAZE, an operation order (OPORD) that covered a wide range of possible U.S. military initiatives against Panamanian forces. Woerner and the Joint Chiefs examined the document and mandated some essential changes. The result was a plan consisting of five phases—the first three covering defensive operations and a troop buildup; the fourth, offensive combat operations; and the last, civil-military operations that would necessarily occur during and after the combat phase. Given the length and complexity of the plan, the Joint Chiefs in April directed that it be broken down into four separate operation orders: ELDER STATESMAN (later POST TIME) for defensive operations and, if necessary, a buildup of U.S. troops; BLUE SPOON for offensive operations (the invasion plan); and KRYSTAL BALL (later BLIND LOGIC) for the civil-military phase. As a precautionary measure, the fourth operation order, KLONDIKE KEY, was added to cover the evacuation of American noncombatants from Panama. Together, these contingency plans constituted the PRAYER BOOK. Over the next year and a half, they would undergo extensive and continuous revision.[13]

The CINCSO version of ELABORATE MAZE (and the PRAYER BOOK operation orders derived from it) provided only strategic guidance for conducting U.S. military operations against Panamanian forces and rebuilding the country afterward. Transforming these general guidelines into highly detailed blueprints for specific actions called for the writing of supporting plans, the initial group of which would be developed for headquarters one echelon below the Southern Command. Those headquarters would have responsibility for executing the PRAYER BOOK at the operational level while exercising control over units carrying out tactical missions in the field. In the case of the CINCSO plan for rebuilding Panama, for example, a Civil-Military Operations Task Force (CMOTF) would serve as the subordinate headquarters, although it would not be activated unless hostilities were imminent or just under way. Once set up, it would answer directly to the SOUTHCOM commander. In the meantime, writing that headquarters' supporting plan fell to a handful of civil affairs officers in the Southern Command J–5 directorate, a few of whom had worked on the strategic, or CINCSO, version. In their efforts, they were assisted by reservists coming to Panama on 31-day tours. The document the officers produced, ultimately referred to as the CMOTF BLIND LOGIC operation order, contained a systematic and comprehensive approach to nation building

[13] ELABORATE MAZE and later the PRAYER BOOK plans were drafted as operation orders not operation plans (OPLANs). The main reason for this was because in 1988 the plans division was in SOUTHCOM's J–3 shop. There was also the prospect, although considered highly unlikely by Woerner, that the plans would be executed soon after they were drafted. John T. Fishel, *The Fog of Peace: Planning and Executing the Restoration of Panama* (Carlisle Barracks, Pa.: Strategic Studies Institute, 1992), pp. 14–15.

in Panama following open hostilities in which U.S. forces would have successfully mounted offensive combat operations to topple the Noriega regime.

The Southern Command's contingency plan for offensive combat operations in Panama required two supporting plans, one for U.S. Special Operations Forces (SOF), the other for conventional U.S. forces.[14] Woerner's guidance for both documents emphasized that hostilities should be conducted in such a way as to minimize damage to Panama's infrastructure, economy, and military organizations. To this end, the preferred scenario envisaged a series of surgical strikes aimed only at key PDF command, control, and communications nodes and a few troop bases in and near Panama City and Colón. The objective would be to bring down the Noriega regime in such a way as to "decapitate" the Panama Defense Forces but not destroy it as an institution. Special operations units deploying from the United States would be responsible for most of the surgical strikes, which estimates suggested would last no more than a few hours. At the same time, conventional forces stationed in Panama or deployed from outside the country would be conducting a series of largely defensive operations aimed at protecting U.S. installations, property, and citizens. Once the strikes ended, the special operators would turn control of their targets over to the conventional units, who would then play a crucial role in stability operations and the reconstruction of Panama.

Having developed this concept of operations, Woerner initially directed one of his components, the Special Operations Command, South (SOCSOUTH), to write the supporting plan for the use of Special Operations Forces. When Col. Chuck Fry, the SOCSOUTH commander, informed him that the headquarters was too overextended throughout the region to do so, the general requested assistance from the U.S. Special Operations Command at MacDill Air Force Base, Florida. The commander there, General James J. Lindsay, deployed a small planning team to Panama in mid-March under Col. Joseph S. Stringham. By the end of the month, the team had completed its assignment, and on 24 March Woerner formally activated the Joint Special Operations Task Force (JSOTF) that would revise, update, and, if necessary, execute the SOF supporting plan for what soon would be code-named Blue Spoon. For the remainder of the crisis, this operational-level headquarters answerable to the SOUTHCOM commander would work out of the special operations headquarters at Fort Bragg, North Carolina.

Conventional U.S. forces engaged in any combat operations in Panama would come under the operational control of another headquarters, Joint Task Force-Panama (JTF-Panama), which like its special operations counterpart would work directly for the SOUTHCOM commander. Long-standing plans for the defense of the Panama Canal called for this headquarters to be activated in a crisis and placed under the commanding general of U.S. Army, South (USARSO), the Southern Command's Army component. In late

[14] In the original combat planning, staff officers envisaged a third task force, a U.S. Navy carrier battle group that would be subordinate to the SOUTHCOM commander. Because of interservice rivalry—despite the emphasis on jointness contained in the Goldwater-Nichols Department of Defense Reorganization Act of 1986—the plan was basically stillborn, as is discussed in a bit more detail in Yates, *U.S. Military Intervention in Panama, June 1987–December 1989,* pp. 91, 155.

March and early April, as the first round of supporting plans for Elaborate Maze were being written, Woerner proved reluctant to set up the organization. Unlike the Joint Special Operations Task Force at Fort Bragg, JTF-Panama would be run out of the headquarters building at Fort Clayton, Panama, just a ten-minute drive from Quarry Heights. Since Clayton was located adjacent to Panama City, Noriega and his people would likely interpret the presence of the joint task force as an escalation in the crisis, a signal Woerner did not want to send. The general's hesitation, however, adversely affected contingency planning in that a handful of Army staff officers at Fort Clayton, almost all of them lacking the joint experience and resources needed for their task, had to write the conventional forces' supporting plan with scant help from the other services. Despite these difficulties, the officers' efforts bore fruit on 21 March with the completion of Commander, JTF-Panama, OPORD 1–88, Elaborate Maze. Soon thereafter, in early April, Woerner decided he could no longer put off activating JTF-Panama under Maj. Gen. Bernard Loeffke, the USARSO commanding general. As a joint task force commander, Loeffke now had operational control of Army, Air Force, Marine, and Special Operations components, on which he could draw for both planning and operational assistance (*Chart 1*). As with the special operations task force, JTF-Panama was charged with updating its supporting plans as needed and with executing them if required. To ensure a safe and efficient interaction between Special Operations Forces and conventional units should hostilities occur, both joint task forces coordinated the invasion plan with one another daily in the months that followed. In addition, JTF-Panama also received the task of revising and updating other supporting plans initially generated by the USARSO staff, namely those Prayer Book operation orders calling for defensive actions and the possible evacuation of civilians.[15]

More than the demands of contingency planning compelled Woerner to activate JTF-Panama. In March and April 1988, the crisis had heated up, requiring the kind of continuous oversight and hands-on involvement that the Southern Command, with its regional responsibilities and strategic orientation, was not configured or expected to perform. JTF-Panama, on the other hand, was an operational-level headquarters exercising control over a variety of tactical units and other multiservice assets spread throughout the canal area. Once activated, it would be in an ideal position to provide the sort of day-to-day crisis management that the proximity of events demanded. To be sure, General Woerner and his staff would continue to monitor the crisis and provide JTF-Panama with guidance they deemed appropriate (even if the joint task force staff often derided much of this guidance as micromanagement from on high). For his part, General Loeffke undertook few important initiatives without first consulting his four-star superior a few miles away.

Among its crisis management responsibilities, JTF-Panama had to absorb newly arrived "security enhancement" forces that had been deployed to Panama in March and April. This first buildup of American troops in the crisis signaled not a prelude to war but a response to several concerns: the grow-

[15] From this point on in the text, the code names used will be those given for the operation orders in the Prayer Book.

Chart 1—Command Relationships: Joint Task Force-Panama (as of May 1988)

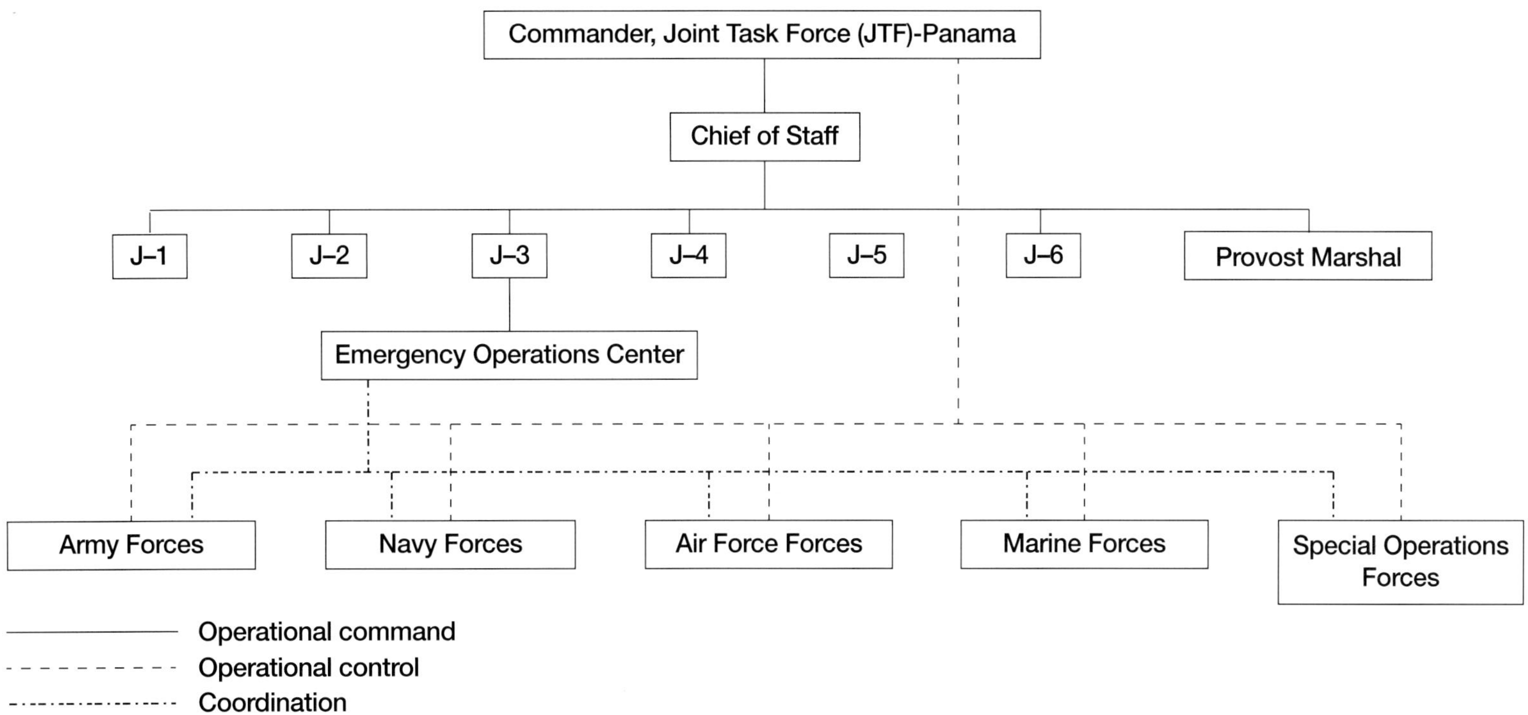

ing number of burglaries and intrusions on U.S. bases and other installations triggered in part by a downturn in the Panamanian economy; the increasing incidents of low-level but daily harassment of American military personnel by members of the Panama Defense Forces, especially the *tránsitos,* or traffic police; and, in Woerner's words, the "fluid, volatile, and unpredictable" situation in Panama, especially after a handful of PDF officers had tried to depose Noriega on 16 March. Until a week or so before the failed coup d'état, the Southern Command had sought to rely on its service components, primarily USARSO, to cope with the mounting threats and uncertainty, but the expanded patrols and guard duty this entailed interfered with the ability of the affected units to carry out their normal duties. As a result, in mid-March the command requested and the Pentagon approved the deployment of additional forces to Panama, including a military police (MP) battalion from Fort Bragg, to improve security. Later that month, after the Panama Defense Forces roughed up and arrested a number of journalists, including some Americans, staying at Panama City's Marriott Hotel, even more troops and equipment were deployed, consisting of additional military police, an infantry company of marines, and a package of aviation assets, soon called Task Force Hawk, from the 7th Infantry Division (Light) at Fort Ord, California. Almost all of the deploying forces would go under the operational control of the newly activated JTF-Panama.[16]

Just as the security enhancement buildup was nearing its end, the crisis escalated dramatically. When the Marine infantry company arrived in early April, its commander, Capt. Joseph P. Valore, learned that the bulk of his unit would take up positions guarding the Arraiján Tank Farm on the west side of the canal. A fuel depot with dense jungle on three sides and the Inter-American Highway on the fourth, the facility seemed particularly vulnerable to PDF intrusions and sabotage. Tragically, on the company's first night patrolling the

[16] As suggested above, the security enhancement buildup occurred in two phases. On 12 March, the secretary of defense authorized the Joint Chiefs to direct the commander in chief, U.S. Atlantic Command (USCINCLANT), to deploy a Marine Fleet Anti-Terrorist Security Team; the commander in chief, U.S. Forces Command (CINCFOR), to deploy an Army military police battalion from Fort Bragg, an Army Arrival/Departure Airfield Control Group, and an Army movement control team; the chief of staff of the Air Force to deploy an Air Force Airbase Ground Defense Flight and an Air Force Dog Flight; the chief of staff of the Army to deploy a counterintelligence detachment (Spanish language qualified); and the commander in chief, U.S. Special Operations Command (USCINCSOC), to deploy an Army signal battalion (minus). The commander in chief, Military Airlift Command (CINCMAC), would provide airlift support, and the commander in chief, Strategic Air Command (CINCSAC), would provide air refueling support. On 1 April, the Defense Department announced the second troop augmentation. Specifically, the units that deployed to Panama over the course of the following week included the 519th Military Police Battalion; the Headquarters and Headquarters Company, 16th Military Police Brigade; an Army reconnaissance troop; three Air Force base ground defense units; two Air Force security police patrol dog squads; a security police squadron patrol dog unit; a Marine infantry company; and a combat aviation company from the 7th Infantry Division (Light), which consisted of four OH–58 Kiowas, fifteen UH–60 Black Hawks, and seven AH–1 Cobras. Quote from Msg, Commander in Chief, Southern Command (CINCSO), to Joint Chiefs of Staff (JCS), 161815Z Mar 1988, sub: Panama Security Enhancement.

A Marine patrol on the Arraiján Tank Farm

tank farm, one of its men was killed by friendly fire. The very next night, an intense firefight erupted between the marines and armed intruders. While the defenders suffered no casualties, they claimed to have seen several of their opponents fall. Loeffke removed the unit from the facility and replaced it with a battalion from USARSO's 193d Infantry Brigade. The marines were debriefed and, soon thereafter, gave their account of events at a SOUTHCOM press conference. To their surprise and dismay, the command's public affairs officer cast doubts on what had actually happened, particularly on whether there had been any intruders, much less an exchange of gunfire. The ramifications of the tank farm episode persisted throughout the remainder of the crisis, with senior officers in the Southern Command and JTF-Panama headquarters expressing concerns about marines being generally "trigger happy," and with the marines objecting both to this lack of trust and to the subsequent promulgation of highly restrictive rules of engagement (ROE) regarding the use of deadly force at the facility. So strict were the rules, they asserted, that they no longer enjoyed the inherent right of self-defense accorded all military personnel.[17]

Within a week of the Arraiján Tank Farm firefight, another shooting occurred, this time in the jungle west of Howard Air Force Base, also on the west side of the canal, just south of the fuel depot. On the night of 20 April, three U.S. Army Special Forces keeping watch for armed intruders in the area detonated a claymore mine and emptied their M16 rifles at a group they believed to be hostile. After daybreak, a rescue column extracted the three men who, during their debriefing, readily admitted that the group they had encountered

[17] For additional information on the tank farm firefight, see Yates, *U.S. Military Intervention in Panama, June 1987–December 1989,* pp. 66–77; Lt. Col. Nicholas E. Reynolds, *Just Cause: Marine Operations in Panama, 1988–1990* (Washington, D.C.: U.S. Marine Corps History and Museums Division, 1996), pp. 6–10. The "trigger happy" quote came from several informal conversations between the author and JTF-Panama officers and enlisted men.

had not fired on them. Their commander, Colonel Fry, commended them for their honesty, but the incident quickly raised issues that went beyond personal character.

For one, prior to sending the three-man team into the jungle, Fry's staff had asked the battalion commander of a nearby USARSO infantry unit to be ready to send in troops to help out if there was trouble. When the battalion commander tried to get details on the team's mission and position, he was told he was not cleared to know, causing him to question the propriety of putting his own men at risk without understanding more about the Special Forces operation. Fry's staff countered that disclosure of additional information would compromise operations security, thus placing a small and already vulnerable team in greater jeopardy. Since both sides were voicing deep-seated concerns about the safety of their personnel, neither headquarters felt inclined to yield. Friction between special operations and conventional forces was hardly uncommon within the military. In Panama, an exchange of liaison officers and an effort to achieve more openness at the tactical level reduced but did not eliminate the discord between the two groups after the 20 April episode.

A second issue raised by the incident concerned command and control. Once the shooting broke out west of Howard, Loeffke joined Fry in the latter's tactical operations center at Howard. In Loeffke's opinion, Fry's men were engaged in finding intruders, an undertaking related directly to the crisis in Panama. That being the case, the team involved should have come under the operational control of JTF-Panama. Fry vehemently disagreed, arguing that his men answered only to him and that, as head of SOCSOUTH, a subunified command, he answered only to the commander of the unified command above him, meaning Woerner. Resolution of the dispute rested with Woerner, who sided with JTF-Panama. Yet, despite the decision and the expressed desire by both Loeffke and Fry to remove the friction between their headquarters, the working relationship between SOCSOUTH and JTF-Panama remained contentious throughout most of the crisis.[18]

Routine

From early February into late April 1988, the increased tensions and violent confrontations in Panama led some observers to conclude that a U.S. invasion of the country was inevitable and imminent. Woerner did not think so, and he proved correct. Although an effort by the Reagan administration to negotiate Noriega's departure from power broke down during this period, the unproductive talks did not lead to an escalation of provocative moves by either side. The Panamanian regime had no desire to provoke the United States into a war, and Reagan had other more pressing regional and international issues with which to grapple. Accordingly, in May, the crisis settled into something of a routine, a status it would retain for another year. U.S.-Panamanian relations remained strained and some form of low-grade confrontation between

---

[18] For the shooting west of Howard Air Force Base and the issues raised by it, see Yates, *U.S. Military Intervention in Panama, June 1987–December 1989,* pp. 77–89.

the PDF and U.S. military personnel occurred on a near daily basis. In short, the hostility persisted but generally at levels deemed tolerable by all involved.

Occasionally, there would be a spike in the routine. For example, armed intruders continued to infiltrate facilities such as the Arraiján Tank Farm, presumably on PDF training exercises, occasionally exchanging shots with the marines or soldiers on guard duty. Only once, on Halloween night, did the shooting on the fuel depot approach the intensity of the April firefight. Again, there were no American casualties, but the incident served as a reminder of how rapidly the crisis might spiral out of control, despite the wishes of the two governments to keep it in check. During this period, the Panama Defense Forces also continued what had become a persistent campaign of harassment generally conducted at low to moderate levels. This, for the most part, included such acts as detaining U.S. military personnel, interfering with freedom-of-movement rights guaranteed U.S. officials under the canal treaty, demanding bribes for minor or fabricated traffic violations, and, occasionally, roughing someone up. But here, too, matters could occasionally escalate beyond the norm, as when several Panamanian policemen abducted a U.S. warrant officer, took him out on an isolated road, beat him, put a gun near his head, and pulled the trigger, causing the man to pass out from fear. The worst incident occurred in early March 1989, when Panama Defense Forces stopped several buses carrying American children to school. As if this was not bad enough, the Panamanians had their guns drawn and were threatening to tow the buses away with the children still on board. The vehicles, to be sure, lacked licenses required by the regime in what SOUTHCOM regarded as another harassment tactic. But frightening American children crossed the line of good judgment and raised tensions in the crisis to a higher level. The subsequent headline in the *Army Times* read "Terror in Panama." The Southern Command's Center for Treaty Affairs protested the school bus and other incidents, but to little avail. Through Brig. Gen. Marc Cisneros, the command's director of operations, Woerner did send the regime a clear message: if the PDF killed an American, the United States would respond with force. Whether Washington would make good on this threat, though, was problematic. For the remainder of Woerner's tour of duty, this explicit criterion for testing America's resolve did not materialize.[19]

As the crisis ebbed and flowed over the course of a year, Woerner remained optimistic that the situation could be resolved without bloodshed. Noriega, he realized, would not likely give up power voluntarily (in return for a fashionable life in some country from which the United States could not extradite him), and U.S. economic sanctions were having little negative impact on the regime itself. But, Woerner reasoned, a more comprehensive program of American pressure, one that played upon the regime's vulnerabilities and the existing "fissures" within the Panama Defense Forces, might prove decisive. Such a program, however, required the various U.S. government departments and agencies with a stake in the crisis to cooperate and coordinate at the interagency level. In promoting such an integrated approach, Woerner frequently sent Admiral William Crowe Jr., the chairman of the Joint Chiefs, a list of precise

[19] For a lengthier list of the more abusive incidents, see ibid., pp. 139–50; *Army Times*, 20 Mar 1989.

recommendations for what needed to be done. For the proposals to work, the general always insisted, the full range of his recommendations needed to be accepted and orchestrated as a coherent campaign plan. It would not be effective, he explained by way of illustration, to take his list of actions ranging from *A* to *Z* and implement only *A*, *L*, *O*, and *T*. Yet, this piecemeal approach was invariably the response he received to his démarche, despite the strong support he received from Crowe in interagency forums. The result, in Woerner's opinion, was a disjointed and ineffectual approach to the crisis, not a well-thought-out and coherent strategy. The impasse dismayed him but did not prevent him from resubmitting his proposals whenever the occasion presented itself.[20]

At some point in mid-1988, observers of the situation in Panama correctly surmised that the Reagan administration was not contemplating any far-reaching initiatives to resolve the crisis. Vice President George H. W. Bush was gearing up for a run at the presidency, which resulted in Woerner and his staff receiving direct and unambiguous instructions to "keep the situation as quiet as possible" through the election. In the mid-1970s, Bush had been director of the Central Intelligence Agency and, as such, had had dealings with Noriega. Any escalation of the situation in Panama, the vice president's campaign managers worried, would make those dealings front-page news, hurting his chances at the polls in November. Woerner's closest military and civilian advisers regarded this directive as the flagrant and improper intrusion of partisan politics into the daily business of a military command, even though they, too, for their own reasons, desired no intensification of the crisis.[21] Given these circumstances, any full-scale review of possible U.S. courses of action in Panama would have to await a new administration. In November, Bush beat his Democratic challenger, Michael Dukakis, to become the president-elect.

As the presidential campaign played itself out in the United States, planners at Quarry Heights, Fort Clayton, and Fort Bragg continued revising and updating the PRAYER BOOK operation orders. Early in the planning process, Cisneros had developed doubts about whether JTF-Panama had the staff officers needed to work on the BLUE SPOON contingency plan for a U.S. invasion, especially because the plan had begun to evolve in such a way as to give conventional forces under the joint task force a greater role in offensive combat operations. Cisneros confided his concerns to Woerner, and both generals explored the possibility of having the XVIII Airborne Corps, the country's contingency corps, take over executive responsibility for further work on the conventional version of the BLUE SPOON supporting plan. There were, however, at least two obstacles to such an arrangement. First, JTF-Panama, or at least its largest element, U.S. Army, South, would surely object to being replaced as the principal headquarters for planning and executing conventional operations. Second, while Woerner had no objection to having the XVIII Airborne Corps oversee BLUE SPOON planning, he was not enthusiastic about having an

---

[20] For Woerner's "fissures" approach to resolving the crisis, see Yates, *U.S. Military Intervention in Panama, June 1987–December 1989,* pp. 97–101.

[21] For the impact of domestic politics in the United States on Woerner's handling of the crisis in Panama, see ibid., pp. 100–101. Quote from Thomas Donnelly, Margaret Roth, and Caleb Baker, *Operation Just Cause: The Storming of Panama* (New York: Lexington Books, 1991), p. 36.

outside command geared to "warfighting" being given the responsibility for actually running combat operations in Panama. The use of force, if it became necessary, needed to be very selective and exercised with restraint, he believed. The Southern Command's future relations with the Panama Defense Forces and the Panamanian people, not to mention America's standing in Latin America, made such an approach imperative. In the end, Woerner finessed the matter, getting the Joint Staff in the Pentagon to assign the XVIII Airborne Corps as the executive agency for Blue Spoon planning, while keeping the corps guessing as to the exact point at which it would be brought in to replace JTF-Panama and to direct combat operations should the plan be executed. As expected, the USARSO staff officers in their JTF-Panama hats raised strenuous objections to the new arrangement, while the corps demanded to be placed in charge of any combat operations from the outset. Neither headquarters received satisfaction. In February 1989, in a meeting at Fort Bragg, JTF-Panama planners formally handed responsibility for the conventional Blue Spoon supporting plan over to the corps, which was still uncertain about the role it would play in the event of an invasion.[22]

At the time of the meeting at Fort Bragg, the Bush administration had been in office less than a month, and the new president was having difficulty filling all the positions on his national security team. One effect of the delay was to postpone a high-level review of future U.S. options vis-à-vis the Noriega regime. During a speech Woerner made that February, the general mentioned the deferred debate, citing a "policy vacuum in Washington" as the reason. The word *vacuum* came in a sentence referring directly to the State Department's need to fill the position of assistant secretary of state for inter-American affairs, vacant since January. But as a provocative headline in the *New York Times* implied, the general seemed to be criticizing the new president. Woerner protested to Crowe that this was not his intention, and he received assurances that key administration officials did not hold the remarks against him. Later, he would have reason to question the veracity of those assurances.[23]

Soon after Woerner's speech, the president and his advisers did begin the long-awaited reassessment of what needed to be done about Noriega. Again, overtures were made to see if the dictator could be persuaded to step down, but the diplomatic initiative fared no better than it had the year before. That caused the administration to focus on what it needed to do to prepare for the next milestone in the crisis, 7 May, the date Panamanians were scheduled to go to the polls in their country's presidential elections. In Washington and Panama, U.S. officials took for granted that Noriega would "steal" the election by rigging the vote. Once the dictator committed this high-handed offense, President Bush would have a pretext for shifting to a much tougher policy, a course of action he had concluded was necessary. Woerner concurred in this assessment, using the proposed policy change to reiterate for his new bosses

[22] On bringing the XVIII Airborne Corps into the crisis, see Ronald H. Cole, *Operation Just Cause: The Planning and Execution of Joint Operations in Panama, February 1988–January 1990* (Washington, D.C.: Joint History Office, Office of the Chairman of the Joint Chiefs of Staff, 1995), pp. 9–10. The word *warfighting* is a U.S. Army doctrinal term.

[23] *New York Times*, 24 Feb 1989; Yates, *U.S. Military Intervention in Panama, June 1987–December 1989,* pp. 137–38.

his theretofore futile appeals for a comprehensive interagency strategy. He also emphasized in his messages to Admiral Crowe that a tougher policy ran the risk of intensifying the crisis, perhaps to the point where the United States would have to employ military force. That was a risk, he advised, from which U.S. policy makers should not shrink.

Although Woerner still desired a peaceful settlement of the crisis or at least a settlement that did not involve a U.S. invasion of Panama, he had never ruled out the use of military force, arguing only that it should be employed as a last resort. His tough talk during the strategy discussions in the two months before the Panamanian elections did not contradict his conviction that the best outcome to the crisis would involve a solution devised by the Panamanians themselves. But, failing that outcome, he recognized that U.S. forces in Panama and elsewhere had to be prepared to execute a wide range of options from all-out intervention to a series of less violent but nevertheless assertive measures. Among the latter considered in the administration's policy review were initiatives to put pressure on the Panamanian regime and the Panama Defense Forces to take action against Noriega on their own.

Indeed, the application of U.S. military pressure, not the exercise of armed force, became the centerpiece of the "get tough" strategy the president planned to implement after the Panamanian elections. Once Noriega had canceled or rigged the voting—again, a foregone conclusion in the White House—Bush would announce the augmentation of U.S. forces in Panama, as well as an accelerated movement of American military dependents living on Panamanian territory either onto U.S. bases or out of the country. While the troop buildup would not constitute a formal execution of the Prayer Book's Post Time operation order, several of the conventional forces scheduled to deploy could be found on that plan's troop list. Once the units arrived in Panama and were placed under General Loeffke's joint task force, they would begin a series of actions intended to reassert U.S. rights under the canal treaties. These and other operations would be carried out in such a way as to intimidate Panamanian forces while keeping them off-balance and confused. As the PDF's morale plummeted, its officers, acting for the good of the institution, would feel compelled to depose their commander. Helping to apply this pressure, other U.S. agencies would initiate a variety of nonmilitary measures aimed at the regime, although, to Woerner's chagrin, the piecemeal way in which they would do so once again fell well short of the comprehensive interagency program he had long advocated.[24]

On Sunday, 7 May, Panamanians went to the polls to elect a president and two vice presidents. A smattering of violence and a number of irregularities marred the voting process but could not detract from the central fact that the overwhelming majority of ballots had been cast for the regime's opponents. Surprised by this outcome, Noriega annulled the election on Wednesday, but not until he had unleashed his so-called Dignity Battalions—namely, PDF members (out of uniform), regime employees and supporters, and various

[24] For the Bush administration's review of options in the Panama crisis, as well as Woerner's role in that review, see Yates, *U.S. Military Intervention in Panama, June 1987–December 1989,* pp. 150–56.

A Dignity Battalion member attacks Guillermo "Billy" Ford (*left*).

thugs—against the three opposition candidates as they and their supporters held a victory parade. In the melee, all three candidates were injured, with photographs of vice presidential contender Guillermo "Billy" Ford, his tropical shirt soaked in blood, appearing in newspapers and news magazines around the world. (The blood was not Ford's, it turned out, but that of his bodyguard, who had been fatally shot in the head, a deliberate target of assassination.) The injured parties took refuge where they could, while the regime arrested and incarcerated an estimated three hundred opposition members, some to be tortured as an object lesson for the others.

Operation Nimrod Dancer

On 10 May, the day following the postelection violence, President Bush, according to plan, recalled Ambassador Davis and announced that he was pulling American military dependents in Panama onto U.S. bases or out of the country—the operation was code-named Blade Jewel—and sending in additional U.S. troops as part of Operation Nimrod Dancer.[25] The main conventional units to deploy over the next two weeks were a brigade headquarters and a battalion from the 7th Infantry Division at Fort Ord, another Marine infantry company (this one equipped with light armored vehicles, or LAVs), and a battalion from the 5th Infantry Division (Mechanized) based at Fort Polk, Louisiana. (Some special operations personnel also flew into Panama on highly classified missions.) To beef up the "light fighters" from the 7th Infantry Division, a second battalion from Fort Ord deployed for the ostensible

[25] For Operation Nimrod Dancer, the U.S. troop deployments, and the operations to reassert U.S. treaty rights in Panama, see ibid., chs. 6 and 7.

purpose of doing a rotation in the Jungle Operations Training Center (JOTC) on the Atlantic side of the country.

As these units came under the operational control of JTF-Panama, Loeffke and his staff simplified command and control arrangements by creating three subordinate task forces. Task Force Atlantic, under the command of Col. David R. E. Hale, the 7th Infantry Division brigade commander, operated out of Fort Sherman and assumed responsibility for crisis-related U.S. military activities on the Atlantic, or northern, side of the canal area. On the Pacific Ocean side, Task Force Bayonet, led by the commander of USARSO's 193d Infantry Brigade, was responsible for the east bank of the canal area. The west bank fell under Task Force Semper Fi, commanded by a Marine colonel and numbering among its forces the two Marine companies that had been sent to Panama because of the crisis.

The arrival of the NIMROD DANCER units was monitored carefully by the international press. The uppermost question in the minds of reporters, as well as the American and Panamanian people, was whether the deployment of these troops was a prelude to an invasion and war. The Bush administration took immediate steps to dampen such speculation, mounting a public relations campaign to emphasize the troops' primary mission spelled out in the president's 10 May statement: to protect American property and personnel. As Bush told reporters in an impromptu press briefing, it was up to the Panamanian people and the Panama Defense Forces to deal with Noriega. After a few days passed without violence, the news media became convinced that hostilities, in fact, were not imminent. At that point, other trouble spots and weightier issues once again captured the headlines and dominated network broadcasts.

Overlooked in the turmoil surrounding the beginning of Operation NIMROD DANCER was a simple sentence in the 10 May presidential message stating that the United States would "assert its treaty rights in Panama."[26] A perusal of the leading newspapers at the time would suggest that no one even bothered to ask, much less investigate, what Bush meant by these words or what actions he contemplated. Ironically, as events unfolded, just as the media furor over NIMROD DANCER was beginning to subside, the role of the U.S. military in Panama was about to become much more confrontational.

Throughout the crisis, the United States had charged the Panamanian regime with violating various clauses of the canal treaty as part of Noriega's harassment campaign. Of these violations, one involved interfering with the legal movement of U.S. government vehicles on official business. The interference would often begin with the Panama Defense Forces stopping individual vehicles at a single location. Once several had been pulled over, they were designated a convoy, which under the terms of the treaty required a formal Panamanian escort. The contrivance was disruptive to official business, a nuisance, a clear example of harassment, and a treaty violation. For most of 1988, when Woerner was under instructions from Washington to keep the crisis from intensifying, the Southern Command could only protest such behavior. Once

[26] The full text of the president's statement is in *Department of State Bulletin* 89 (July 1989): 70–71.

the Bush administration came in and seriously began to review its options in Panama, however, Woerner and his advisers were able to make the case that U.S. forces should challenge the regime on the freedom-of-movement issue. Bush's national security people agreed.

The plan was to run daily convoys with an armed U.S. escort the length of the transisthmian highway. In accordance with the treaty, PDF officials would be notified of the convoy movement, but only an hour or two before it began. If they did not arrive with an escort, the convoy would set out anyway. If there was any attempt to interfere with its freedom of movement, the convoy's escorts would resist. Just what form that resistance would take became the object of detailed discussions, both between the Southern Command and Washington, and among the troop units involved. Existing rules of engagement were reviewed and revised when necessary, and what Colonel Hale called rules of confrontation, which would cover actions short of the use of deadly force, were developed for all anticipated contingencies. What the Bush administration did not want was to have U.S. forces engaged in a shootout. Indeed, Admiral Crowe had informed Woerner that the troops under his command via JTF-Panama were to be forceful but were not to start a war. (Woerner conceded that, at that point in the crisis, this seemingly ambivalent advice made perfect sense to him.) The goal of the operation was to reassert American treaty rights, put pressure on the Panamanian armed forces, keep them guessing as to ultimate U.S. intentions, undermine their morale, and just possibly induce them to oust Noriega in order to ensure the PDF's survival as an institution.

The first freedom-of-movement operation took place in late May, with two convoys running simultaneously, one north to south from Fort Sherman, the other south to north from Fort Clayton. According to the plan, each had a small unit of armed escorts, and each had a formidable air armada covering it, to include two layers of attack helicopters, two layers of fighter aircraft, and an AC–130 Spectre gunship. To no one's surprise, the Panama Defense Forces reacted as expected: they chose not to interfere with the movement of the vehicles. In the days and weeks that followed, JTF-Panama, on Woerner's instructions, altered the frequency and composition of the convoys to keep the Panamanians off guard; Loeffke also reduced the convoys' air cover to a minimum as a means of saving on fuel and maintenance costs. For their part, Panamanian troops and police coming into contact with the vehicles rarely deviated from their initial reaction of noninterference.

In addition to the convoys, JTF-Panama mounted several other operations to reassert American treaty rights, such as staging a show of force to reclaim the Fort Espinar officers' club, a U.S. facility on the Atlantic side that had been rented to the Panama Defense Forces, only to have them default on the payments. There were also a host of training exercises, all legal under the treaty, designed to intimidate the Panamanian military with demonstrations of American strength and firepower. During May and June, the effect of these various undertakings was mixed. Intelligence reports indicated that the new, tougher policy was taking a psychological toll on the PDF rank and file, although not necessarily to the degree that Washington had desired. Moreover, inherent in the U.S. initiatives was the risk that something could go wrong or that there could be a flash point in which one side or the other would open fire, thus

General Cisneros

igniting the general hostilities that neither government wanted. In the first six weeks of Nimrod Dancer alone, there had been a half-dozen near misses. In one, the PDF commander at Fort Espinar had not received accurate information from his chain of command about the number of U.S. forces that would arrive to take over the officers' club. When a company from the 7th Infantry Division showed up at the front gate, he reacted strongly, setting up mortars and having his troops man other firing positions. U.S. forces responded in kind. After several hours, both sides stood down, with lessons concerning the "fog and friction" and unintended consequences of military operations firmly lodged in their minds.[27]

Despite the risks involved, the feeling at Quarry Heights, Fort Clayton, and in Washington was that U.S. troops needed to apply even more pressure—things needed to be *ratcheted up*, according to a widely used term at the time. Within the Southern Command and JTF-Panama, officers held brainstorming sessions to determine what additional measures might achieve this end. The lists they compiled, once vetted by Woerner, followed the chain of command to Washington, where Bush's national security advisers weighed the ramifications of each recommendation. Following these deliberations, the president on 22 July approved a national security directive, NSD-17, that set forth new guidelines for U.S. military forces in Panama to follow in their efforts to assert treaty rights and intimidate Noriega and his followers. The directive detailed four categories of operations differentiated from one another by the level of risk and visibility involved. Low-risk, low-visibility activities fell under Category I. As examples, NSD-17 listed publicizing the evacuation of U.S. government dependents, expanding media and psychological campaigns against the regime, and placing PDF members under formal escort whenever they entered a U.S. base. Low-risk, high-visibility activities constituted Category II, which included increasing military police patrols between U.S. installations, bringing AH–64 Apache helicopter crews and battalion-size units to Panama for training, and practicing amphibious and combat operations at night. Category III consisted of medium-risk, high-visibility measures, such as the increase of armed patrols and reconnaissance in the vicinity of key PDF installations. The most sensitive

[27] For more details on the half-dozen near misses, see Yates, *U.S. Military Intervention in Panama, June 1987–December 1989,* chs. 7 and 8.

General Thurman

and dangerous category, Category IV, encompassed high-risk, high-visibility operations and exercises. Examples included reestablishing U.S. access to a causeway at Fort Amador and reasserting U.S. control over certain facilities at Fort Amador, Fort Espinar, and Quarry Heights. Because of the risks involved, all Category III and IV proposals had to be submitted to the Joint Chiefs of Staff for approval at the interagency level or higher.[28]

At the time President Bush signed NSD-17, few of the top-ranking U.S. military officers in Panama who for months had dealt with the crisis on a daily basis still remained. In June, most had either retired or moved to other assignments. At Quarry Heights, both Woerner's director of intelligence, Brig. Gen. John F. Stewart Jr., and the special operations commander, Colonel Fry, had left the country; at JTF-Panama, General Loeffke and his operations director, Col. John A. Cope Jr., had taken new assignments in Washington, and the chief of staff, Col. Arnold T. Rossi, was preparing to retire. As a result of these and other personnel changes, the institutional memory with respect to the crisis would, at the highest military levels, reside with Woerner, who had one more year to go in his tour as the SOUTHCOM commander, and with General Cisneros, who would advance from the Southern Command's operations directorate to take over from Loeffke as both commanding general, U.S. Army, South, and commander, JTF-Panama. (In October, Cisneros would be promoted to major general.)

On 6 July 1989, this small remnant of top-level continuity was halved when Woerner received word that he was being relieved of his command. Secretary of Defense Richard Cheney refused to give Woerner a reason for this drastic action, although Secretary of State James A. Baker indicated in his memoirs that the administration did not consider the four-star general to be a team player. Woerner's defenders correctly countered that, as the SOUTHCOM commander, the general had done everything asked of him throughout the crisis. Regardless, General Maxwell R. Thurman, who was just stepping down as the commanding general of the Army's Training and Doctrine Command, was tapped as Woerner's successor. As Thurman recalled, he received the nod not because he was a war fighter or a Latin American specialist, but because he knew how Washington worked. He

[28] On "ratcheting up" and NSD-17, see ibid., ch. 8; Cole, *Operation Just Cause*, pp. 11–12.

had been vice chief of staff of the Army and had served in other Pentagon positions in which he had spent much time testifying before Congress. Soon after he agreed to take over from Woerner, he received a three-hour lecture from Admiral Crowe on the need to review U.S. strategy not just for Panama but for all of Latin America within the Southern Command's purview and then to help thrash out a consensus in Washington for any required changes. To prepare for these new responsibilities, the general had only about two months. The change of command ceremony was scheduled for 30 September.[29]

Woerner's departure would leave Cisneros as the general officer on the scene most knowledgeable about the inner workings of the crisis. Cisneros had been a loyal subordinate to Woerner and supported his boss' belief that an invasion of Panama would not be in the best interests of the United States. But Cisneros also believed that, short of war, U.S. military forces in Panama could be much more aggressive in their actions to compel the regime to bring an end to the crisis. He thus welcomed and helped execute the wider range of operations—many that Woerner, too, had recommended—covered in NSD-17 and subsequent directives from Washington. As for the troops under his command and operational control, Cisneros continued the joint training events Loeffke had initiated under JTF-Panama, while adding a few twists of his own. The new measures included Sand Fleas (low-key exercises and operations designed, like their namesake, to irritate the PDF) and contingency readiness exercises, which would allow in-country U.S. forces to rehearse portions of highly classified Blue Spoon battle plans without realizing that was what they were doing.

The Road to Just Cause

The SOUTHCOM change of command ceremony took place in Panama as scheduled. The next evening, the newly ensconced General Thurman received a disturbing report: a Panamanian officer, Maj. Moisés Giroldi Vega, commander of the PDF 4th Infantry Company, was planning a coup d'état against Noriega, slated to take place within twelve hours. The general was skeptical, thinking Giroldi might be part of a regime plot to plant rumors and then embarrass him when he overreacted. Reinforcing his doubts were reports he received that Noriega and the major were close associates, the latter having helped put down the March 1988 coup attempt against the dictator. Giroldi, Thurman learned, was also known to have conducted entrapment operations in the past. Cisneros was less wary and argued for a stepped-up level of readi-

[29] For Woerner's relief and the reactions to it, see Yates, *U.S. Military Intervention in Panama, June 1987–December 1989,* pp. 223–24; James A. Baker III with Thomas M. DeFrank, *The Politics of Diplomacy: Revolution, War, and Peace, 1989–1992* (New York: G. P. Putnam's Sons, 1995), pp. 183–84 (including the quote); Bob Woodward, *The Commanders* (New York: Simon & Schuster, 1991), pp. 96–98. For Thurman's account of his reasons for being chosen as Woerner's successor and of his meeting with Crowe, see Briefing, Gen Maxwell R. Thurman at the School for Advanced Military Studies (SAMS) (hereafter cited as Thurman SAMS briefing), 23 Sep 1992.

ness for JTF-Panama, but Thurman approved only minimal measures. He also informed President Bush's key national security advisers about these sudden and unexpected developments.

As the hours passed Monday morning, the purported time for the coup came and passed without incident. Thurman and, thanks to him, the White House became further convinced that nothing was going happen. When word came from Giroldi's wife that her husband had delayed action until the next day, the source of the information alone generated even more skepticism. Consequently, on Tuesday morning, 3 October, Bush met only briefly with his advisers to discuss possible developments in Panama. The president then left the Oval Office to keep his schedule for the day. Unknown to him, the coup attempt in Panama was already under way, as Giroldi's men accosted Noriega at his headquarters, the Comandancia. By the time Bush and his advisers could meet several hours later to discuss the situation, units loyal to Noriega had moved against the major, freed their commander, and forced the rebels to surrender. Giroldi and several other plotters paid with their lives; the remaining conspirators were incarcerated, some after being tortured.

The failure of the Bush administration and the Southern Command to take action that would have ensured the success of the coup became a subject of heated debate in Washington. To some extent, the criticism of the president reflected partisan politics, although some Republicans in Congress openly questioned his handling of the situation. It had been, the critics charged, a missed opportunity. Summoned to Washington, Thurman briefed the president and his advisers on 16 October and then testified before Congress. The gist of the general's assessment became a mantra in the succeeding weeks and months: the coup attempt was poorly planned, poorly organized, and poorly led; Giroldi, had he succeeded, would have been little improvement over Noriega; and for the Southern Command to have taken action in support of Giroldi and the "coupmongers" would have meant reacting to fluid and unpredictable events rather than having the United States itself determine the conditions under which it would intervene militarily. Not all who listened to Thurman's explanation found it convincing. As for Bush, he had to endure once again the epithet hurled at him during his presidential campaign that he was a "wimp" in foreign policy, incapable of acting decisively. After talking with the president, Thurman thought otherwise. Upon leaving Washington, the general was convinced that the next major disruption in the Panama crisis would cause Bush to order U.S. intervention, especially if harm came to any American serviceman or civilian. Besides, few other options remained. The administration and the Southern Command had tried for months to goad the Panama Defense Forces into ousting Noriega on its own. Now, some Panamanian officers had tried to do just that—albeit for reasons not directly related to U.S. pressure—and failed. Given Giroldi's brutal fate, few analysts at Quarry Heights thought it likely that any other coterie within the PDF would find the courage to move against the dictator anytime soon.[30]

---

[30] For the 3 October 1989 abortive coup and its aftermath, see Yates, *U.S. Military Intervention in Panama, June 1987–December 1989,* pp. 248–63; Kempe, *Divorcing the Dictator,*

Once the abortive coup of 3 October increased the likelihood that American armed force would be used to resolve the crisis with Panama's dictator, U.S. officers involved in planning such a contingency had to devote more time to reviewing and updating the pertinent operation orders. Helping to focus the planners' attention, Thurman had intimated even before he became the SOUTHCOM commander that he intended to make significant changes to the existing invasion plan.[31] In early August, as CINCSO-designate, the general had visited Fort Bragg and received briefings on the Blue Spoon supporting plans. On that occasion, officers from special operations and the XVIII Airborne Corps revealed to him some of their serious reservations concerning the transition to combat operations, should Blue Spoon be executed. Under Woerner's long-standing guidance, U.S. forces would not initiate hostilities against the Noriega regime until the troop buildup called for in the Post Time operation order had been completed. True, a number of those units had arrived in Panama during Operation Nimrod Dancer, but there was still a sizable force yet to be deployed. The problem was that transporting the remaining units into the country could take up to two weeks. This was a deliberate calculation on Woerner's part. His strategy was to sacrifice one of the principles of war, surprise, in order to emphasize another, mass. In Woerner's assessment, a steady and public flow of thousands of additional U.S. forces into the country, this time for the expressed purpose of toppling the regime, might finally compel Noriega to step down or the Panamanian military to remove him. The planners at Fort Bragg disagreed. In their view, a gradual buildup of the remaining Post Time forces would leave the initiative for action with Panamanian units, which might take hostages or commit other hostile acts before the incoming U.S. troops were ready to launch combat operations. A better Blue Spoon strategy, they told Thurman, would be for American forces to launch a massive surprise attack against their selected targets. Thurman agreed, telling those present that, after the change of command at Quarry Heights, he would authorize a new plan along those lines.

In fact, the Southern Command was already working on an alternative to the current Blue Spoon. The directive to do so had come from the Joint Chiefs, but the main impetus behind it had been Lt. Gen. Thomas Kelly, the Joint Staff's director of operations, an officer who shared the same doubts about Woerner's

---

pp. 369–97; Woodward, *The Commanders,* pp. 119–25. For Thurman's account of his post-coup trip to Washington, see Thurman SAMS briefing, 23 Sep 1992. Thurman used the term *coupmongers*.

[31] For U.S. military planning from August into December 1988, see Yates, *U.S. Military Intervention in Panama, June 1987–December 1989,* pp. 263–70; Cole, *Operation Just Cause,* pp. 14, 17–24; Tom Clancy with Gen Carl Stiner (Ret.) and Tony Koltz, *Shadow Warriors: Inside the Special Forces* (New York: G. P. Putnam's Sons, 2002), pp. 307–27; Interv, author with Brig Gen William Hartzog, U.S. Army, and Col Thomas A. Braaten, U.S. Marine Corps, 29 Jun 1990, Quarry Heights, Panama; Telephone Interv, author with Maj David Huntoon Jr., U.S. Army, 27 Jun 1990; Intervs, Robert K. Wright Jr. with Lt Gen Carl W. Stiner, U.S. Army, 2, 7, 27 Mar and 11 Jun 1990, Fort Bragg, N.C.; and with G–3 Plans Section, XVIII Airborne Corps, 30 Mar 1990, Fort Bragg. The Wright-with-Stiner interview is compiled as one interview, though it was conducted on four days.

concept of operations as the planners at Fort Bragg. When the Joint Chiefs' directive reached Quarry Heights, Woerner stayed aloof from the enterprise, allowing his new operations officer, Brig. Gen. William Hartzog, a great deal of latitude in formulating a response. During July and August, Hartzog drafted a plan code-named Banner Savior, the contents of which sought to correct perceived Blue Spoon shortcomings such as the gradual buildup of U.S. forces; the sacrifice of secrecy for mass; and the command and control arrangements that had two separate joint task forces, one for conventional units, one for Special Operations Forces, reporting separately and directly to the SOUTHCOM commander. As Hartzog fleshed out and continuously revised Banner Savior, he briefed Kelly twice and consulted with Thurman soon after the four-star's visit to Fort Bragg. Upon learning that Woerner's designated successor favored a new concept of operations and that the planners at Bragg knew this, Hartzog began to coordinate his efforts with the XVIII Airborne Corps commander, Lt. Gen. Carl W. Stiner, and his staff.

This coordination between Hartzog and the corps was essential, mainly because Stiner, after Thurman's visit, had directed two of his planners, Lt. Col. Timothy L. McMahon and Maj. David Huntoon Jr., to write a new supporting plan based on the incoming SOUTHCOM commander's stated intentions. By the time Thurman assumed command at Quarry Heights on 30 September, the two officers had completed their draft, which the corps titled Joint Task Force-South (JTF-South) Operation Plan (OPLAN) 90–1. Much of the plan's content duplicated what was contained in the original JTF-Panama Blue Spoon, including many of the listed units and targets, the projection of 0100 as the preferred H-hour, and the restrictive rules of engagement designed to minimize damage to Panama and its people. The major changes were conceptual, as the Woerner strategy entailing a gradual buildup of U.S. forces prior to offensive combat operations was dropped in favor of a sudden, swift, and all-out assault on the assigned targets. The new plan thus locked in place the strategic combination of surprise and mass preferred by the airborne corps, special operators, the Joint Staff, and Thurman. General Stiner was optimistic. If, in the event of hostilities, U.S. combat operations were launched at 0100 as planned, the heaviest of the fighting, he predicted, would be over by dawn.

Shortly after the coup attempt on 3 October, Thurman convened a meeting in Panama so that he and his staff could discuss the status of the pertinent Prayer Book contingency plans. Stiner and key members of his staff attended, as did the JSOTF commander, Maj. Gen. Gary Luck, members of the JSOTF staff, and representatives from JTF-Panama. At this and a subsequent planning session, the SOUTHCOM commander reviewed the corps' JTF-South OPLAN 90–1 and indicated the changes he wanted made in light of the failed coup. To begin with, the target list would have to be expanded to include PDF units that, while based on the periphery of the capital and even farther away, had managed to intervene on 3 October to compel Giroldi's surrender. Revisions to the operation plan also had to reflect the relocation of certain PDF units and the shakeup in the organization's top leadership that Noriega mandated after nearly being deposed. Thurman further directed that 90–1 be updated to specify what U.S. forces would do in the event that disgruntled PDF officers would try again to overthrow their commander. This unlikely possibility raised another issue. As Thurman told Hartzog, Stiner, and Luck, they needed to look into three different scenarios: which U.S. forces would be available and what they could accomplish if the war plan had to be executed within 2 hours, 20 hours, or 48 hours of a trigger event, such as another coup attempt.[32]

Besides revising the current contingency plans, Thurman used the first post-coup planning session at Quarry Heights to remove the ambiguity Woerner had deliberately created concerning the XVIII Airborne Corps' operational role should hostilities break out in Panama. To the assembled group, Thurman formally made known that, should the president decide to use force, the corps' commander and key staff officers would deploy to Panama immediately—not at some unspecified future date—and set up JTF-South as a war-fighting headquarters working directly under the SOUTHCOM commander. Once JTF-South was activated, it would absorb JTF-Panama's staff directorates and assume operational control over U.S. forces currently under JTF-Panama as well as those deploying from the United States, including the Joint Special Operations Task Force. By putting the JSOTF under JTF-South, Thurman resolved the unity of command issue critics of Woerner's plan had raised. General Luck did not object to the arrangement. Given Stiner's extensive background in special operations, the JSOTF commander had no reason to believe that JTF-South would assign his troops missions that would be inappropriate and unsuited to their unique capabilities. As for Stiner, Thurman had informed him in August of the role he envisaged the XVIII Airborne Corps playing during BLUE SPOON. Now, in early October, the assignment was official.

During the remainder of the month, SOUTHCOM, XVIII Airborne Corps, and JSOTF staff officers worked together to revise their respective combat plans. After Thurman signed the SOUTHCOM-level version on 30 October, Hartzog, Stiner, and Luck flew to Washington to brief the final drafts to Thurman and Kelly, then to the new chairman of the Joint Chiefs of Staff, General Colin Powell, and finally to the Joint Chiefs as a group. Because Army units constituted the overwhelming majority of the projected combat force, General Carl E. Vuono and his vice chief requested and received a separate briefing. On 3 November, the plans were approved as CINCSO OPORD 1–90 (BLUE SPOON), the Southern Command's conceptual plan; JTF-South OPLAN 90–2, the XVIII Airborne Corps' revised operational-level plan; and the JSOTF BLUE SPOON plan. For the rest of November and into December, the staff officers working on each plan continued to refine and embroider their handiwork, holding periodic meetings in Panama and at Fort Bragg. There were also a series of briefings for the commanders of the U.S. Atlantic Command, the Military Airlift Command, and the Forces Command, all of whom had vital roles to play under BLUE SPOON. In the case of the Military Airlift Command, the Air Force component of the Transportation Command, a one-

[32] Once Thurman made the XVIII Airborne Corps' 90–1 the basis for discussion and further planning, Hartzog's work on BANNER SAVIOR came to an end. Instead, the SOUTHCOM operations officer began revising the existing CINCSO version of BLUE SPOON to reflect both Thurman's strategic guidance and the operational input of Stiner's staff.

week conference among the various planning staffs determined airlift requirements for the contingency operation.

During this time, one of Stiner's highest priorities was to enhance the weapons available to any invasion force by pre-positioning M551 Sheridan light tanks and Apache helicopters in Panama. Accordingly, in mid-November, four Sheridans from the 82d Airborne Division were flown into Howard Air Force Base in a secret night flight and then moved to a "motor pool" nearby, where they were kept concealed in tents during daylight. At the same time, six Apaches and three OH–58 Kiowa scout helicopters also arrived via C–5 Galaxy cargo planes, with the Apaches being kept under cover in hangars at Howard. As Stiner saw it, the Sheridans could be used in a direct-fire role against a heavy target such as the PDF's Comandancia complex. As for the Apaches, they boasted a greater night-vision capability than did the AH–1 Cobra attack helicopters already in Panama; they could also fire laser-guided Hellfire missiles as surgical-strike weapons.[33]

While the Southern Command orchestrated the planning process and intensified preparations for possible hostilities with the Panama Defense Forces, JTF-Panama continued its day-to-day management of the crisis. One issue with which it grappled in the fall of 1989 was ensuring that the units rotating in to replace those deployed as a continuation of Operation Nimrod Dancer and the 1988 security enhancement received a thorough orientation as well as continuous updates on the situation they faced. Complicating the process, the fall rotations, when combined with routine personnel moves, resulted in all three task forces JTF-Panama had established at the outset of Nimrod Dancer receiving new commanders: Col. Michael G. Snell took over Task Force Bayonet; Marine Col. Charles E. Richardson, Task Force Semper Fi; and Col. Keith Kellogg of the 7th Infantry Division, Task Force Atlantic.

While becoming acclimated to an environment in which neither peace nor war prevailed—the closest doctrinal term the Army had for describing the crisis in Panama was *low-intensity conflict*—the arriving troops found themselves immediately involved in a series of major operations, joint training events, contingency readiness exercises, Purple Storms, and Sand Fleas.[34] These activities allowed JTF-Panama to keep pressure on the PDF; to acquaint newly arrived units with the urban terrain on which they might have to fight; and, as Cisneros intended, use the repetition and frequency of the exercises to lull the PDF into a false sense of security. As before, some of these measures were designed to rehearse aspects of the highly classified Blue Spoon supporting plan without the participants knowing it. Cisneros, the JTF-Panama commander, made sure the operations and exercises were adjusted to reflect the latest changes to the plan coming out of Quarry Heights, Fort Bragg, and his own operations directorate at Fort Clayton. Furthermore, by varying the rehearsals just slightly from the actual plan, Cisneros was having the units under his control

---

[33] Interv, Wright with Stiner, 2, 7, 27 Mar and 11 Jun 1990; Donnelly, Roth, and Baker, *Operation Just Cause*, p. 91; Capt. Kevin J. Hammond and Capt. Frank Sherman, "Sheridans in Panama," *Armor* 99 (March-April 1990): 8.

[34] For the doctrinal aspects of the crisis in Panama, see Yates, *U.S. Military Intervention in Panama, June 1987–December 1989,* pp. 34–35, 112–13.

establish a false footprint that might deceive the PDF into expecting one U.S. maneuver only to encounter something else should Blue Spoon be executed.

Around 18 November, the Saturday before Thanksgiving, the Southern Command learned through an informant that Colombian drug cartels intended to detonate car bombs on and near U.S. installations in Panama, presumably in retaliation for certain of the Bush administration's antidrug activities. By coincidence, the bomb scare occurred at a time when Stiner and his staff were scheduled to be in Panama for a planning conference. Soon after being apprised of the alleged Colombian threat, Thurman summoned Stiner and told him that the potential danger and the risk of war that went with it had to be taken seriously. Thurman therefore decided to activate JTF-South as a precaution. He later acknowledged that the XVIII Airborne Corps did not "belong to me" and that he should have asked permission to activate a joint task force through his chain of command, beginning with General Powell. But, as he told a group of officers at Fort Leavenworth, he "did not, repeat, did not ask authority about that from the JCS." "I'm not going to ask the Department of Defense, or the Chairman what he thinks about that because I'm liable to get the wrong answer, which is, 'No.'" Powell and Secretary of Defense Cheney were not pleased with Thurman's fait accompli but, given the situation, allowed it to stand. In other protective measures, Thurman ordered that concertina wire and concrete blockades be placed around U.S. bases and facilities and that individuals and vehicles entering these locations be fully checked by security guards. In the end, the informant was found to be lying—Thurman would later refer to the whole affair as the "bomb scam." Still, the episode had allowed the XVIII Airborne Corps to acquire some valuable command and control experience as a joint task force headquarters operating in Panama under the Southern Command. One postscript to the hoax was that, to the inconvenience and annoyance of many U.S. military personnel, dependents, and employees, the concrete and wire obstacles set up during the bomb scare remained in place.[35]

In mid-December, as the U.S. forces included on Blue Spoon troop lists looked forward to the holiday season, they were at the same time well prepared for any escalation in the crisis. In Panama, as noted, JTF-Panama had put units under its control through quasi-rehearsals of the plan, while in the United States elements of the 82d Airborne Division, the 7th Infantry Division, U.S. Rangers, and Special Operations Forces had conducted even more elaborate rehearsals and field exercises geared to Blue Spoon. But now a lull in all this activity seemed likely. Veterans of the crisis related to newcomers the axiom that, in Panama, troublesome situations were always put on hold during the holiday season. To be sure, SOUTHCOM prognosticators referred to Noriega's more erratic behavior in public as a sign that sooner or later the dictator would precipitate an incident that would trigger a U.S. invasion. But, again, the consensus was that such a provocation was not likely to occur before January 1990. Thurman and Cisneros did not let down their guard, but neither

[35] On the Thanksgiving bomb scare and the activation of JTF-South, see ibid., pp. 271–72; Woodward, *The Commanders,* pp. 142–44; Clancy with Stiner and Koltz, *Shadow Warriors*, pp. 327–28; Telephone Interv, author with Huntoon, 27 Jun 1990. Quotes from Thurman SAMS briefing, 23 Sep 1992.

did they discourage the host of festivities and formal parties that began to intrude on daily routine.

On Friday, 15 December, Noriega dampened the holiday spirit somewhat when he gave a major televised speech before the Panamanian national assembly, declaring himself head of government and proclaiming that Panama was in a state of war because of North American aggression. While the reference to war in the speech was reported in the United States with hyperbolic rhetoric and a sense of urgency, most observers at the Southern Command and JTF-Panama recognized it for what it was—more political theater. Still, the speech created a mood that could not help but exacerbate the crisis. It was in this highly charged atmosphere that, on the following night, the shooting incident at the PDF checkpoint occurred, which left Marine 1st Lt. Robert Paz dead and the Navy lieutenant and his wife brutalized. Reportedly, the latter incident angered President Bush more than the shooting. The following day, he met with his national security team for a briefing on the situation. That session, which began at 1130, followed a flurry of Sunday morning phone calls between General Powell and key military players involved with Blue Spoon, beginning with Thurman. In essence, the briefing took the form of presenting the invasion plan to the president and recommending that he approve its execution. Bush reviewed the plan and its ramifications carefully, asking detailed questions throughout. Then, convinced that the situation in Panama was only going to worsen and that the time had come to take decisive action, he concluded simply, "Okay, let's go. We're going to go."[36] The two-year crisis in U.S.-Panamanian relations was about to be resolved by force of arms.

[36] On the meeting in which President Bush authorized execution of Blue Spoon, see Woodward, *The Commanders,* pp. 167–71.

2

Preparations and Adjustments

17–19 December 1989

☆ ☆ ☆

Sunday, 17 December

Among U.S. service personnel in Panama, there was a good deal of speculation Sunday morning and afternoon about whether the Bush administration would respond militarily to the killing of Lieutenant Paz the night before. A number of officers, unaware of the brutal treatment meted out to the Navy lieutenant and his wife, ventured that nothing would come of the episode save another official and ineffective protest from the Southern Command. It had been "a bad night" in Panama City, as one U.S. officer lamented, but not a *casus belli*.

Maj. Gen. Marc A. Cisneros was not so sure. Meeting with JTF-Panama's commanders and staff at noon that Sunday, he admitted that he had no indication of what Washington would do but that he could not imagine "nothing being done on this one." To be prudent, the joint task force had to make final preparations for a war that might be just a few days away. Cisneros went around the table asking his commanders and staff officers if there were any "war stoppers." Only some minor issues arose. He then expressed to those assembled in the classified forum his two major concerns: the potential for friendly fire and the reliability of communications in the field. Quickly, several officers assured him that the means existed to avoid the former while maximizing the latter. At one point, the general also directed that, while preparing for possible combat, the troops under JTF-Panama should outwardly appear to be conducting business as usual. To enhance that perception, he stated that someone needed to contact PDF Maj. Moisés Cortizo, one of Noriega's confidants, and tell him that "we want to settle this," thus helping to create the overall impression that "we're trying to defuse this thing." In sum, every effort needed to be made to maintain operations security so that, in the event of hostilities, U.S. forces would enjoy the element of tactical surprise.[1]

In Washington that afternoon, President Bush himself emphasized secrecy as he decided in favor of military action, scheduled to begin in three days on Wednesday, 20 December. Speaking forcefully to his advisers, the

[1] Author's notes, JTF-Panama meeting, 17 Dec 1989, Fort Clayton, Panama.

General Powell

General Kelly

president directed that only people essential to the success of the operation were to be informed of his decision and, then, only when it became absolutely necessary. This injunction remained in the thoughts of General Powell and General Kelly as they rode from the White House back to the Pentagon. Both believed that, realistically, operations security surrounding the invasion could be maintained through Monday, with any leaks on Tuesday coming too late to jeopardize the undertaking. As one measure to limit the number of people aware of the impending operation, Powell told Kelly that the vice chairman of the Joint Chiefs and the director of the Joint Staff need not be told of the president's decision until the next day. When Kelly got back to his office, he took an additional precautionary step by sending home the bulk of his staff, keeping behind only his deputy and four junior officers, to whom he revealed, "We're going to execute Blue Spoon."[2]

Between 1700 and 1730 that afternoon, the two generals notified the commanders and operations officers whose responsibilities did require immediate knowledge of the president's decision. Powell personally telephoned General Thurman; General Lindsay at the U.S. Special Operations Command; General Edwin H. Burba Jr., commander in chief, Forces Command; and Air Force General Hansford T. Johnson, commander of both the U.S. Transportation Command and the Military Airlift Command. Among the officers Kelly called was General Stiner. After summarizing

[2] Quote from Woodward, *The Commanders,* pp. 171–73. Cole, *Operation Just Cause*, pp. 28, 30.

General Stiner

the White House meeting for the airborne corps commander, Kelly asked him to relay the information to Maj. Gen. Wayne A. Downing, who had just taken over from General Luck as the commander of special operations forces in the plan. Among the responsibilities Downing inherited was command of the Joint Special Operations Task Force that would be responsible for executing many of the H-hour surgical strikes called for in the invasion plan.[3]

H-hour was among the critical pieces of information Powell and Kelly discussed with the officers they contacted on Sunday. It would remain 0100, the same time chosen nearly two years earlier in the first versions of the Elaborate Maze and Blue Spoon contingency plans. The early hour would allow U.S. forces to bring their superior night-fighting capabilities into play; it would also find the international airport in the capital virtually closed for the night and the volume of vehicular traffic on the streets of the country's two largest cities, Panama City and Colón, much reduced. As for D-day, Stiner, for one, was pleased to hear that, in keeping with a recommendation he had made to his superiors, the president had agreed to Wednesday. This would give the airborne commander sufficient opportunity "to pull the air package together" so that the military could "go with the total force—not piecemeal." The time the Air Force would need to prepare had already been addressed early Sunday morning in a discussion between General Johnson at the Military Airlift Command and General Thurman at Quarry Heights. Johnson's director of operations had calculated it would take sixty hours to marshal the crews and aircraft required for transporting units from the United States to Panama. When Thurman heard the estimate, what he later described as a "donnybrook" ensued, with him arguing for a shorter, 48-hour window. But, as he would later relate, "the JCS settled the hash and said 60." For General Downing, the debate was largely irrelevant. Waiting until Wednesday worked well for his JSOTF forces, but, as he had told Thurman, his troops could be ready to go on Tuesday, if necessary. Some would be ready sooner, namely units already in Panama under the Special Operations Command, South, and certain other special

[3] Woodward, *The Commanders,* p. 172; Interv, Robert K. Wright Jr. with Lt Gen Carl W. Stiner, U.S. Army, 2, 7, 27 Mar and 11 Jun 1990, Fort Bragg, N.C. The Stiner interview is compiled as one interview, though it was conducted on four days.

Secretary Cheney

operations personnel who would begin deploying from the United States Sunday night, as soon as Secretary of Defense Cheney authorized their departures.[4]

After talking with Kelly and phoning Downing, Stiner thought it best to attend a scheduled Christmas party that Sunday evening to give the appearance of business-as-usual. Afterward, though, around 2000, he went to the plans area at corps headquarters where he met with key staff officers and commanders who needed to know immediately about the decision to launch combat operations. The group included the corps' deputy commander, its chief of staff, and a select group of planners; Downing and his operations officer; Maj. Gen. James H. Johnson Jr., commanding general of the 82d Airborne Division, accompanied by the division's operations officer; and the commanders of the 4th Psychological Operations Group and the 96th Civil Affairs Battalion. Conveniently, all of those present came from headquarters located on Fort Bragg.

As was the case with officials in Washington, an immediate and urgent concern of the group gathered around Stiner was operations security. One question surfaced immediately: what cover story could be devised to explain the flurry of activity, inherent in any major combat deployment, that would soon engulf Fort Bragg and adjacent Pope Air Force Base? The 82d commander suggested that corps spokespersons describe the imminent increase in troop and aircraft movements as an emergency deployment

[4] The first quotes on air package and the total force are from Interv, Wright with Stiner, 2, 7, 27 Mar and 11 Jun 1990. Interv, Lt Col Steven P. Bucci with Gen Carl W. Stiner, U.S. Army (Ret.), 1999, Senior Officer Oral History Program, U.S. Army War College (USAWC)/ U.S. Army Military History Institute (USAMHI), Carlisle Barracks, Pa. Second and third quotes from Interv, John Partin with Gen Maxwell R. Thurman, U.S. Army, 12 Apr 1990, Fort Benning, Ga. See also Cole, *Operation Just Cause*, pp. 28, 30; Woodward, *The Commanders*, p. 172; Military Airlift Command (MAC), *Anything, Anywhere, Anytime: An Illustrated History of the Military Airlift Command, 1941–1991* (Scott Air Force Base, Ill.: Office of History, Military Airlift Command, 1991), p. 195. Panama and the East Coast of the United States are in the same time zone, and, in December, the United States is on standard, not daylight savings, time. Thus, the time in Washington in December 1989 was the same as in Panama. Official messages often use Greenwich Mean Time. In such cases, 0100 (H-hour) in Panama often would have been expressed as 0600Z, the Z time zone being five hours ahead of Panama's time zone, R.

readiness exercise. Under this cover, the prescribed steps for preparing the units involved—referred to in military jargon as the notification-hour, or N-hour, deployment sequence—could commence the next day, 18 December, at 1300. Beginning at that time, additional personnel would be notified of the operation, the troops participating in BLUE SPOON would be marshaled and isolated, and all involved would begin initiating standard operating procedures for the eighteen-hour countdown until the first planes carrying the 82d's paratroopers took off for their objectives in Panama. One advantage of starting the N-hour sequence at 1300 was that the time fit well with the half-day holiday work schedule set to go into effect Monday. In the afternoon, fewer employees would be on post to witness the increased activity. Another benefit, given the checklists and timetables to be followed, was that the loading of many transport planes would not begin until after dark on the eighteenth.

Several officers present at the meeting with Stiner were part of the "group of twenty" that the general had created back in mid-October. He had told them then to be ready to deploy to Panama on a moment's notice; now, they would be among an advance party that would precede him into the area of operations. Consequently, part of the discussion Sunday night touched on what they would do after their arrival in Panama. According to Stiner, "They were to get down there and get the operations center fully operational, staffed up, prepared to receive the rest of us, and prepared to fulfill its role in executing the operation." After the general adjourned the meeting, members of the advance group grabbed their bags, many of which had been packed for weeks, and, led by the corps' chief of staff, Brig. Gen. Edson Scholes, departed shortly before midnight for Howard Air Force Base. Meanwhile, Stiner's Panama crisis action team at Bragg established tactical satellite communications with the Southern Command and JTF-Panama.[5]

In Panama that night, both of these headquarters were in the process of reviewing their contingency plans and refining essential communications arrangements. At Fort Clayton, Cisneros had not yet received formal notification of the decision to execute JTF-South Operation Plan 90–2. Fortuitously, his top planner, Lt. Col. Robert Pantier, had gone to Quarry Heights around 2030 to help a SOUTHCOM staff officer assess the command and control arrangements contained in BLIND LOGIC, the civil-military operation plan for a postinvasion Panama. As Pantier read over the plan, he concluded that it did not provide for a smooth transition from combat to stability operations. Instead, he later said, it represented a "disaster waiting to happen." After suggesting possible ways to salvage parts of the plan, he prepared to leave, only to be asked to sit in on a

[5] Details of the meeting Sunday night at XVIII Airborne Corps headquarters can be found in After Action Report, Operation JUST CAUSE, 20 Dec 1989–12 Jan 1990, Joint Task Force-South, 19 Mar 1990 [hereafter cited as JTF-South AAR], p. 3. Quote from Interv, Wright with Stiner, 2, 7, 27 Mar and 11 Jun 1990. Intervs, Wright with Maj Gen James H. Johnson Jr., U.S. Army, 5 Mar 1990, Fort Bragg, N.C.; author with Lt Col Tim McMahon, U.S. Army, 21 Dec 1989, Fort Clayton, Panama; and with Lt Col David Huntoon Jr., U.S. Army, 7 Apr 1992, Fort Bragg, N.C.

U.S. Southern Command headquarters

briefing for General Thurman that was about to begin, the topic being the reconnaissance and surveillance plan contained in the special operations version of Blue Spoon. Pantier believed that the briefing was simply part of the ongoing planning process of which he had been a participant since arriving at U.S. Army, South, the previous year. He therefore sensed nothing out of the ordinary when General Hartzog left the room to take a call from Stiner. When the meeting ended, Hartzog was still on the phone. From what Pantier could hear, the conversation seemed to concern one of the semimonthly planning sessions the Southern Command had hosted since October, most likely the one scheduled well in advance to begin the next day. Because he was responsible for arranging transportation for the officers arriving from Fort Bragg, Pantier waited for Hartzog to finish and then asked the general when Stiner and his people could be expected. Hartzog replied that Stiner would not come until Monday night, but that an advance party from the corps would land at Howard early in the morning. Pantier knew that a routine planning session would not necessitate the dispatch of an advance detachment from Fort Bragg. Execution of the JTF-South Operation Plan 90–2, however, would. Hartzog's comments thus "sent up a flag" for the lieutenant colonel. What remaining doubts he had about what was transpiring quickly vanished when he was directed to tell Cisneros to report for a meeting with Thurman early in the morning so that the JTF-Panama commander could be apprised of late-breaking developments in the crisis.[6]

As of late Sunday night, notifications of key military personnel concerning President Bush's decision that afternoon had been made and preliminary steps had been taken within the U.S. military establishment to transform the

[6] JTF-South AAR, p. 3. Quotes from Interv, author with Lt Col Robert Pantier, U.S. Army, 13 Jun 1990, Fort Clayton, Panama.

complexities of a written contingency plan into the reality of a major combat operation. In two days, American troops would be fighting in Panama.

Monday, 18 December

Judged by the content of newspapers and television news broadcasts in the United States Monday morning, the secrecy surrounding the impending invasion seemed to be holding. On NBC, for example, the lead story repeated details of Noriega's "declaration of war" on Friday, and the killing of Lieutenant Paz and the mistreatment of the Navy lieutenant and his wife Saturday night. Yet, the report concluded, the administration seemed to be "playing it cautious," concerned perhaps that American public opinion would not support a strong U.S. military response, especially since some official pronouncements had suggested that the occupants of the car in which Paz had been a passenger may not have behaved prudently at the PDF checkpoint. The Southern Command's own newspaper, *Tropic Times*, did its part to project a sense of calm. While containing detailed accounts of the weekend's events and while citing Secretary Cheney's statement that Noriega had "encouraged this kind of lawlessness," it also quoted a U.S. official as saying the incident was only "coincidental" to Noriega's "state of war" speech and that to "draw conclusions from the rhetoric of the last few days would be misleading at this time." Echoing this sentiment, Air Force Col. Ronald T. Sconyers, the Southern Command's public affairs officer, told a CNN audience that "most of the indications point that this is an isolated incident, but certainly a very tragic and needless one."[7]

How long the secrecy surrounding the president's decision would survive once U.S.-based units slated to take part in Blue Spoon began their N-hour deployment sequences on Monday remained to be seen. Further complicating the matter was news reaching Fort Bragg that Air Force C–141 Starlifter transports, slated to be loaded with heavy equipment, were going to begin arriving at Pope Air Force Base around noon, much sooner than Stiner had anticipated. The Military Airlift Command had begun assembling the required air armada Sunday night, right after receiving word from Powell about the Oval Office meeting. The time of year worked to expedite the process, in that most of the required aircraft were readily available thanks to reduced transport commitments during the holiday season. Stiner, after digesting this and other information, decided at a 0730 briefing to move N-hour forward from 1300 to 0900. Reinforcing the decision was a weather forecast that called for a storm front to move through North Carolina later in the day. If the worst predictions concerning the severity of the storm proved true, there would almost certainly be significant delays in launching any aircraft from Pope. Since the C–141s were going to arrive early, the anticipated delays could be offset somewhat by beginning the rigging

---

[7] Author's notes, 18 Dec 1989, Fort Clayton, Panama. Quotes from *Tropic Times*, 18 Dec 1989.

C–141 Starlifter

and loading of the transports soon after they arrived, even if that meant performing these tasks in daylight.[8]

One scenario concerning the incoming transports, both those for heavy equipment and those for personnel, worried Stiner; he feared they would all arrive about the same time, further jeopardizing the secrecy surrounding the airlift while overtaxing landing and loading facilities at Pope. To ameliorate this possible effect, the general later recounted,

> we came up with a plan whereby we would bring in the heavy-drop and CDS [container delivery system] birds and load them, and launch them to Charleston [Air Force Base, South Carolina]. They would be brought in by non-airdrop qualified crews; would go to Charleston where the airdrop qualified crews were in [crew] rest and were receiving their briefings, and the planes would be topped off on fuel, and they would launch from there. At the required time, the twenty personnel birds would come into Pope, pick up the jumpers, and launch. The birds at Charleston would join them en route in order to make the TOT [time on target].

With the loaded heavy-drop transports already at Charleston, Stiner believed that, regardless of the weather, at least ten of the twenty projected troop-carrying planes would be able to take off from Pope in the initial launch scheduled for the night of the nineteenth. Even after taking the possibility of such a major adjustment into account, he still felt confident that enough

[8] Intervs, Wright with Stiner, 2, 7, 27 Mar and 11 Jun 1990, and Johnson, 5 Mar 1990; author with McMahon, 21 Dec 1989; MAC, *Anything, Anywhere, Anytime*, p. 195.

troops would drop into Panama "to do the majority of the critical tasks that we absolutely had to do that night," meaning D-day.[9]

Having personally observed the beginning of the N-hour deployment sequence at Bragg and having helped make adjustments to the airflow plan, Stiner, together with his main party, boarded a C–20 Gulfstream aircraft in midafternoon and flew to Panama. At that time, the holiday work schedule, the cover story about a readiness exercise, and even the bad weather were all helping to maintain operations security. By evening, however, some network news broadcasts in the United States were noting an increased level of activity at both Bragg and Pope, although reporters on the scene were for the moment repeating the official explanation that what they were witnessing was a no-notice training event.[10]

While Stiner had been monitoring the preparations for war from his headquarters Monday morning, key officials and officers in the Pentagon were reviewing the SOUTHCOM and XVIII Airborne Corps versions of BLUE SPOON. All realized the risks entailed in the contemplated operation, and all could remember the *Mayaguez* incident of 1975, the bombing of the Marine barracks in Beirut in 1983, and the U.S. invasion of Grenada just a few days later. All three operations—even the Pentagon's marred success in Grenada—were regarded as failures of the military art. Secretary of Defense Cheney, wanting to avoid any such calamities on his watch, played an active role in the review process—first, to ensure that the BLUE SPOON chain of command (of which he was the first link under the president) was short and efficient; second, to minimize interference from individuals and organizations outside the chain of command; and, third, to be able to answer any Panama-related military questions the president might have. Beginning Monday morning, pursuit of these goals led him to talk with General Powell at length, during which time the secretary posed a series of pointed questions. "It seemed to Powell," reporter Bob Woodward later wrote, "that Cheney wanted to know all the details, right down to the squad level." After the morning sessions, Powell called in Kelly and, in turn, subjected him to a similar interrogation, one purpose of which was to find and eliminate any superfluous requirements remaining in the plan.[11]

Later, both Powell and Kelly briefed Cheney on the execute order formally authorizing the use of military force against Noriega and his regime. The document contained one cosmetic but important revision from its original draft. In a telephone conversation with Kelly Sunday evening, General Lindsay of the U.S. Special Operations Command had asked almost rhetorically, "Do you want your grandchildren to say you were in BLUE SPOON?" Kelly's deputy recommended calling the operation JUST CAUSE instead, and that was the code name used in the order briefed to Cheney. The secretary accepted the change. As will be seen, he also made several last-minute modifications to the part of

---

[9] Quotes from Interv, Wright with Stiner, 2, 7, 27 Mar and 11 Jun 1990. The bracketed material in the block quotation is in the original transcript.

[10] Ibid.; Author's notes, 18 Dec 1989, Fort Clayton, Panama.

[11] Quote from Woodward, *The Commanders*, pp. 175–76. Cole, *Operation Just Cause*, pp. 31–32.

the operation plan that dealt with the employment of the Air Force's state-of-the-art but expensive F–117A Stealth fighters.[12]

At 1825 Eastern Standard Time, the Pentagon transmitted the Just Cause execute order to participating headquarters, departments, and agencies. Upon receiving his copy of the message, the commander in chief, Forces Command, who had the doctrinal responsibility for delegating command and control arrangements for American units deploying from the United States, sent his own message to Stiner, informing him that the 82d Airborne Division, the 7th Infantry Division (Light), and, if necessary, the 24th Infantry Division (Mechanized) would be placed under the operational control of JTF-South, as planned.[13]

Although Stiner and his entourage were still in transit to Panama when the execute order was issued, his advance party had been in the country since early morning. After arriving at Howard Air Force Base, the group had gone to Fort Clayton and begun setting up JTF-South (Forward) in the U.S. Army, South, Emergency Operations Center. With operations security in mind, formal activation of the joint task force would await the actual deployment of combat units the next evening. In the meantime, the group from Fort Bragg worked in the operations center, attended briefings, and scrutinized the Blue Spoon operation checklist.[14]

Around 1300, while one of the meetings to review the checklist was in progress in the Simón Bolívar conference room at Fort Clayton, reports arrived of another shooting incident in Panama City, this one in the Curundu area near Albrook Air Station. Details were sketchy, but the first accounts suggested that an American officer had shot a uniformed Panamanian. As more information became known, the U.S. Defense Department issued a statement to the effect that an American Army lieutenant had left a local laundry and gotten into his car. At that point, he was approached by a PDF traffic policeman who was reaching for his pistol. The Army officer drew his own pistol and fired twice. The Panamanian fell to the ground, then got up and fled the scene. "The extent of his injuries," the statement said, "is unknown at this time." The next day, a story in *Tropic Times* reported the incident but did not repeat the rumors then circulating that there was a direct connection between the shooting and the events of Saturday night. The article, however, did cite a White House statement linking the death of Lieutenant Paz that evening to "a coordinated wave of violence started by Panama's weekend declaration that a state of war existed with the United States." The article also noted that the twelve thousand U.S. troops in Panama were on a high state of alert and that they and their families had been told to avoid public places.[15]

The *Tropic Times* offered no insight as to why the American officer was armed or why one preliminary statement by the Pentagon indicated that the Panamanian

---

[12] Quote from Woodward, *The Commanders*, p. 173, and see also pp. 176–77. Cole, *Operation Just Cause*, pp. 31–32. A more detailed discussion of the F–117A issue can be found in ch. 9 of this volume.

[13] Cole, *Operation Just Cause*, p. 32; Telephone Interv, author with Maj David Huntoon Jr., 27 Jun 1990.

[14] JTF-South AAR, p. 3; Interv, Wright with Stiner, 2, 7, 27 Mar and 11 Jun 1990.

[15] Quotes from *Tropic Times,* 19 Dec 1989. Author's notes, 18 Dec 1989, Fort Clayton, Panama.

actually had his pistol drawn, as opposed to reaching for it. The story also seemed to contradict indications picked up by U.S. military intelligence in Panama that Noriega had told the Panama Defense Forces to avoid confrontations with U.S. service personnel after Saturday night's shooting. Yet, the possibility that the Army lieutenant at the laundry was indeed a new victim in a regime-orchestrated campaign of violence could not be dismissed. To some, however, it seemed more likely that the American officer had gone out seeking to avenge the shooting of Paz, a prospect that had high-ranking officers at the Southern Command and JTF-Panama livid. They had been trying to maintain a business-as-usual demeanor so as not to tip off the regime to any military moves the Bush administration might make. Now the actions of an Army lieutenant—whether justified or inexcusable—placed the plausibility of that façade in jeopardy and risked putting the Panama Defense Forces members on their guard. Worse, the PDF might decide to go "tit for tat," as one JTF-Panama planner put it, thus "screwing up" the invasion timetable by causing U.S. forces to employ their troops prematurely. Musing on the incident, one person recalled President Kennedy's terse remark about the American U–2 pilot who flew over the Soviet Union at the height of the Cuban missile crisis: "There's always some SOB who doesn't get the word."[16]

Aside from the U.S. investigators assigned to the incident, there was little effort after a day or so to follow up on the fate of the unfortunate or foolhardy Army officer (whose interrogation, it was rumored, had been severe). More pressing matters were at hand. On Monday evening, Stiner's plane arrived at Howard. The general and the flag officers with him headed to Quarry Heights to meet with Thurman and others, while staff officers in the group traveled to Fort Clayton. In a practice adopted during the semimonthly planning visits, all were dressed in civilian clothes, another ploy to avoid raising the regime's suspicions.

At Fort Clayton that night, Cisneros scheduled a JTF-Panama meeting. As the word circulated, virtually everyone in the USARSO headquarters building poured into the Bolívar conference room. The general arrived over an hour late, around 2230, and immediately ordered the room vacated, calling back only his component commanders and key staff officers. He then briefed them on the decision to launch Operation JUST CAUSE. Shortly after midnight, Stiner and others joined the group, having wrapped up their meetings at Quarry Heights. The assemblage was significant for at least two reasons. First, it was the first gathering of the war-fighting commanders and most of their staff since the presidential decision on Sunday. Second, to Stiner's relief, the officers who would run JTF-South once it was activated were in Panama and setting up shop *before* hostilities began.

Among the increasing number of American officials and officers aware of the impending invasion, there was a pervasive concern that the Panama Defense Forces, if tipped to what was coming, might preempt the U.S. attack by launching one of its own, before JUST CAUSE forces based outside Panama were able to deploy. Throughout the crisis, the Southern Command, JTF-Panama, and the XVIII Airborne Corps had briefed a no-notice, or reactive, contingency plan,

[16] Author's observations and conversations in Panama, 18–19 Dec 1989; Interv, author with Lt Col Robert Pantier, U.S. Army, 21 May 1991, Fort Leavenworth, Kans.

code-named High Anxiety, that called for in-country U.S. forces to deal with any preemptive PDF attack until the stateside troops listed in the Post Time and Blue Spoon operation orders could arrive.[17] In the October planning sessions, Thurman had also emphasized the need to know how Blue Spoon would be executed in response to another coup attempt or some other trigger event. Now on the eve of taking military action, Stiner stood ready to handle such unwelcome contingencies if need be, although no one relished contending with the increased level of confusion and uncertainty bound to accompany hostilities initiated on the enemy's timetable. To the relief of all, as Monday night gave way to Tuesday morning, the secrecy surrounding Just Cause seemed to be holding. It needed to last for just under another twenty-four hours.

The Plan

When Stiner entered the Simón Bolívar conference room shortly after midnight on 19 December, he listened to the remainder of Cisneros' briefing to JTF-Panama's principal commanders and staff. Then, at 0045, the corps commander began his own hour-long briefing of the execute order, the first substantive sentence of which asserted that the National Command Authority "has authorized the execution of USCINCSO OPORD 1–90 (Blue Spoon)/30 Oct 89." There followed a statement of the military mission.

> USCINCSO will conduct joint offensive operations to neutralize the PDF and other combatants, as required, so as to protect US lives, property, and interests in Panama and to assure the full exercise of rights accorded by international law and the US-Panamanian treaties.

More specifically,

> the purpose of these offensive operations is to ensure: continuing freedom of transit through the Panama Canal, freedom from PDF abuse and harassment, freedom to exercise US treaty rights and responsibilities, the removal of Noriega from power in Panama, the removal of Noriega's cronies and accomplices from office, the creation of a PDF responsive to and supportive of an emergent democratic government of Panama [GOP], and a freely elected GOP which is allowed to govern.[18]

The execute order designated the commander in chief, U.S. Southern Command, as the supported commander. All forces would come under his operational control once they entered his area of responsibility. The list of supporting commanders consisted of the commanders in chief of the Forces Command,

[17] Fragmentary Order (FRAGO) (High Anxiety) to Operation Order (OPORD) 7–88 (Blue Spoon) folder. The folder, located in the Combined Arms Research Library at Fort Leavenworth, Kansas, contains copies of the no-alert fragmentary orders from before and after the 3 October abortive coup.

[18] Msg, Chairman, Joint Chiefs of Staff (CJCS), to CINCSO et al., 182325Z Dec 1989, sub: EXORD [Execute Order]\USCINCSO OPORD 1–90 (Blue Spoon).

Transportation Command, Special Operations Command, Atlantic Command, Pacific Command, Strategic Air Command, Tactical Air Command, and special operations forces. Other supporting organizations included the Defense Intelligence Agency, the Central Intelligence Agency, the National Security Agency, the Defense Logistics Agency, the Defense Mapping Agency, and the Defense Communications Agency.[19]

The longest portion of what was a very brief message covered rules of engagement. In general, commanders were to "conduct all military operations in accordance with the laws and directives governing armed conflict." While every element of the Panama Defense Forces as well as its armed supporters (such as the Dignity Battalions) was subject to attack, only the "minimum force necessary" was to be used to achieve the operation's objectives; when possible, "hostile forces should be warned and given the opportunity to surrender." Furthermore, because the Panamanian people were not the enemy, every effort would be made to prevent "injury to noncombatants," damage to civilian property and public buildings, and any "disruptive influence" on the country's economy. The section on rules of engagement concluded by saying, "The right of self-defense is never denied. These ROE do not restrict the right of every commander or individual to use necessary and appropriate force to defend himself or his unit against hostile intent or hostile act." In employing the necessary level of force, U.S. troops would have their "normal organic weapons," buttressed in some specific cases by M136 AT4 antitank weapons, AC–130 gunships, Sheridan light tanks, Marine light armored vehicles, Apache helicopters with Hellfire missiles, and .50-caliber machine guns. Field artillery could be used in a direct-fire mode, but indirect fire would have to be approved by higher headquarters. Also, with the exception of the bombs scheduled to be dropped by F–117As, there would be no air strikes without Stiner's authorization. In sum, the rules of engagement contained some significant but not excessive limitations and restrictions on the circumstances in which deadly force could be used and on the weapons U.S. forces could employ.[20]

The Pentagon execute order briefed by Stiner offered little or no elaboration on the tasks U.S. troops and the supporting commands and agencies were to perform. That detailed information was contained in the Southern Command's operation order, 1–90; in the JTF-South and Joint Special Operations Task Force supporting plans; and in the lower-echelon tactical blueprints derived from them.[21] These documents had been pieced together over nearly two years under Woerner's and Thurman's guidance, with the input of Generals Powell, Kelly, Stiner, Luck, Downing, Loeffke, and Cisneros. In their final versions, approved by the Joint Chiefs, they reflected the XVIII Airborne Corps' way of doing business. Consequently, Operation JUST CAUSE would employ overwhelming military force and the element of tactical surprise to mount nighttime surgical strikes on over two dozen enemy targets, most of

[19] Ibid.

[20] Quotes from ibid. USAWC/USAMHI Interv, Bucci with Stiner, 1999.

[21] Sanitized versions of CINCSO OPORD 1–90 (BLUE SPOON), 30 Oct 1989, and the JTF-South Operation Plan (OPLAN) 90–2, 3 Nov 1989, have been declassified and provide the basis of much of the discussion that follows.

which would be attacked simultaneously. Once the combat objectives had been achieved, several of the participating units would assist in the transition to stability operations aimed at rebuilding the country. For most of the combat troops airlifted in from the United States, especially those of the Special Operations Forces, the intent was to redeploy them to their home bases as soon as the fighting ended. Despite the magnitude and complexity of the operation, estimated times for actual fighting at most targets were brief, ranging from no more than a few hours to perhaps a day, after which the targeted PDF units would have been isolated or defeated. As for the complete neutralization of the enemy's command, control, and communications capabilities, the JTF-South operation plan predicted that the mission "should be completed within three days following our initial assaults."[22]

To a great extent, these optimistic estimates as well as the plan's only moderately restrictive rules of engagement spoke to the generally low regard in which U.S. officers held the Panama Defense Forces and its leadership. Besides the well-known technological gap favoring the American military, intelligence reports citing low morale and discipline within the PDF's ranks, the substandard training afforded many of its units, and the penchant for self-preservation many Panamanian officers had demonstrated during the failed coup of 3 October further diminished the likelihood of extended hostilities. Noriega, to be sure, had made several changes in the Panama Defense Forces after the coup attempt, removing some commanders of questionable loyalty, promoting others who had stood by him, and relocating various units to make them less vulnerable to subversive influences. But through a variety of open and covert methods, U.S. military intelligence knew about most of these adjustments and was able to update assessments of the Panamanian forces and their order of battle. On the eve of Operation Just Cause, revised figures placed PDF membership at fifteen thousand, of which approximately thirty-five hundred were listed as combat troops organized primarily into two battalions, ten infantry companies, a ceremonial cavalry squadron, a riot control company (the Dobermans), and a special operations command composed of commandos and an elite unit known by its Spanish acronym UESAT. U.S. estimates of the wartime intentions and capabilities of these combat forces varied from unit to unit, with expectations that some would fight with determination while others would surrender to U.S. troops with minimal or no resistance. As for the Panamanian air force, it consisted, in general, of 38 fixed-wing aircraft, 17 helicopters, and some air defense guns; the navy, of 12 vessels and a naval infantry company. The vast majority of PDF members served in nonmilitary positions, including police and investigation officers, customs and conservation officials, and administrative personnel.[23]

---

[22] JTF-South OPLAN 90–2, 3 Nov 1989.

[23] Briefing, 1st Corps Support Command (COSCOM), 22 Feb 1990, Fort Leavenworth, Kans.; Capt. Joseph M. Nemmers et al., *United States Army South Staff Ride: Operation Just Cause, 20 December 1989–31 January 1990*, updated ed. (1991; repr., Fort Clayton, Panama: Historical Office, Headquarters, U.S. Army, South, 1998), p. 9; Donnelly, Roth, and Baker, *Operation Just Cause*, p. 75; Interv, author with Lt Cdr Kenneth D. Pierce, U.S. Navy, 21 May 1998, Fort Leavenworth, Kans. The strength and location of specific PDF units will be cited in the chapters covering the operations in which they took part.

When preparing and disseminating the PDF order of battle, U.S. analysts generally included information on Noriega's paramilitary Dignity Battalions, or, as American military personnel derisively called them, DigBats (a play on the word *dingbat*, meaning, among other things, "an empty-headed or silly person.") Formed in 1988 ostensibly under the regime's minister of public works, the Dignity Battalions dressed in civilian clothes—often in T-shirts with *Batallón Dignidad* stenciled on them—and included PDF members, government workers, regime supporters, the unemployed, and criminals and thugs. Some members joined of their free will, some participated under pressure, some were paid, and some owed their jobs to the dictatorship. Whatever the motives, when called on, this mishmash of people could swell the ranks of a regime-sponsored political rally or demonstration; members could also be used to intimidate or physically abuse Noriega's opponents. In May 1989, the DigBats attracted international attention when they carried out their most publicized act of violence, the postelection attack on the "victory" march staged by the opposition candidates and their supporters. Most of the time, the Dignity Battalions carried sticks, hoses, and pipes with which to beat their victims. Given their close ties to the government, however, several analysts assumed that, in the event of war with the United States, they would likely have access to more lethal weapons. Thus, when the JUST CAUSE execute order of 18 December stated that "all parts" of the Panama Defense Forces could be attacked by U.S. forces, the Dignity Battalions were included among the eligible elements. As their name implied, DigBats presumed that they enjoyed a quasi-military status. Yet their organization and training was never so structured as to provide U.S. intelligence personnel any clear idea of the number of Panamanians involved.[24]

To defeat the Panama Defense Forces and any other armed groups, to capture Noriega, and to begin the transformation of Panama into a true democracy, the U.S. invasion plan called for the employment of just over twenty-seven thousand troops. Nearly half of this force was already in Panama, either because the units were based there, as in the case of the Southern Command's component forces, or because they had rotated into the country as a part of troop augmentations conducted over the course of the crisis, namely, the security enhancement buildup of early 1988 and Operation NIMROD DANCER in May 1989. The remainder of the invasion force would deploy from bases in the United States. The U.S. order of battle identified Operation JUST CAUSE as a joint undertaking, although the Army represented the largest portion of the force, nearly 80 percent. Of the conventional troops involved, 18,587 were soldiers and 2,850 more came from Army units in the Special Operations Forces (*Table 1*). Not that service allegiance mattered that much to Stiner. For him, the important fact was that the Pentagon had given him all the troops he had requested through the Southern Command. There had been no haggling over numbers.[25]

---

[24] Kempe, *Divorcing the Dictator*, pp. 357–58; Msg, CJCS to CINCSO et al., 182325Z Dec 1989, sub: EXORD [Execute Order]\USCINCSO OPORD 1–90 (BLUE SPOON). As for the word *dingbat*, it came readily to mind for U.S. officers because it had been popularized in a 1970s television comedy, *All in the Family*.

[25] The troop numbers are taken from Briefing Slides, SOUTHCOM, Operation JUST CAUSE: "Rebirth of a Nation," n.d. Stiner's account of receiving all the troops he requested (and even

TABLE 1—U.S. MILITARY PERSONNEL IN OPERATION JUST CAUSE

MILITARY COMPONENT	NUMBER OF PERSONNEL
Army	18,587
Navy	521
Air Force	3,133
Marine Corps	690
Special Operations, by component	
Army	2,850
Navy	500
Air Force	800
TOTAL	27,081

Source: Briefing Slides, U.S. Southern Command, Operation JUST CAUSE, 20 Dec 1989–31 Jan 1990, n.d.

If twenty-seven thousand U.S. troops pitted against a much smaller and much less professional and competent enemy force amounted to overkill, it was meant to. The use of disproportionate force was, under Presidents Reagan and Bush, promulgated in the so-called Weinberger-Powell doctrine, a product of the Vietnam syndrome in the United States. The quagmire and eventual American defeat in Southeast Asia, many officers and analysts believed, had resulted to a large extent because U.S. policy makers had committed combat units piecemeal and in numbers inadequate to win the war. In future conflicts, therefore, once the president and key national security advisers had determined America's military objectives in a situation requiring the use of force, the next step was to employ from the beginning the combat power necessary to achieve those objectives. For many advocates of this approach, the phrase *all force necessary* translated easily into overwhelming force. Why leave the outcome to chance?[26]

turning down some additional units offered to him, as he did in a conversation with the Marine Corps commandant) is in Interv, Wright with Stiner, 2, 7, 27 Mar and 11 Jun 1990.

[26] In November 1984, Secretary of Defense Caspar W. Weinberger enumerated several criteria that should govern the commitment of U.S. forces to combat. His list, which became known as the Weinberger Doctrine, included the wholehearted commitment of troops for the purpose of winning. Prior to Operations DESERT SHIELD and DESERT STORM, General Powell reiterated and slightly revised the doctrine, although both versions implied the use of decisive and overwhelming force to attain specific and achievable objectives. In the first few post-Vietnam de-

Dignity Battalion members engage in street violence.

The command and control relationships for such a large invasion force incorporated the guidance Thurman had provided Stiner's people and the JSOTF commander at Fort Bragg in early August and at Quarry Heights in October. Military operations would take place in the SOUTHCOM commander's area of responsibility, and elements of the XVIII Airborne Corps, with its extensive experience in planning and conducting contingency operations, would serve as the war-fighting headquarters—Joint Task Force-South—with Stiner answering directly to Thurman. To ensure unity of command, JTF-South would assume operational control of the Joint Special Operations Task Force commanded by General Downing. As for the commander and most staff directors in JTF-Panama, they would accept assignments as deputies in Stiner's headquarters, while the joint task force itself would not be operative so long as JTF-South remained activated.

The organizational charts depicting this arrangement showed JTF-South at the top of a wiring diagram containing over a dozen task forces and component headquarters (*Chart 2*). Of these, the Joint Special Operations Task Force controlled several of the most elite units in the U.S. military. In contrast to this multiservice organization, the Air Force Forces (AFFOR) and the Naval Forces (NAVFOR) were component headquarters that were service specific. The NAVFOR commander was Rear Adm. Gerald E. Gneckow, and the naval forces under him (including some with special operations capabilities) were the same ones he controlled in his role as both the SOUTHCOM and JTF-Panama

cades, several commentators regarded this as something new in recent American history. Yet, if one examines just the period of the Cold War, on at least two occasions—the U.S. interventions in Lebanon in 1958 and in the Dominican Republic in 1965—the United States deliberately employed more than enough combat power to subdue any military opposition.

CHART 2—OPERATION JUST CAUSE: OPLAN 90–2 COMMAND AND CONTROL

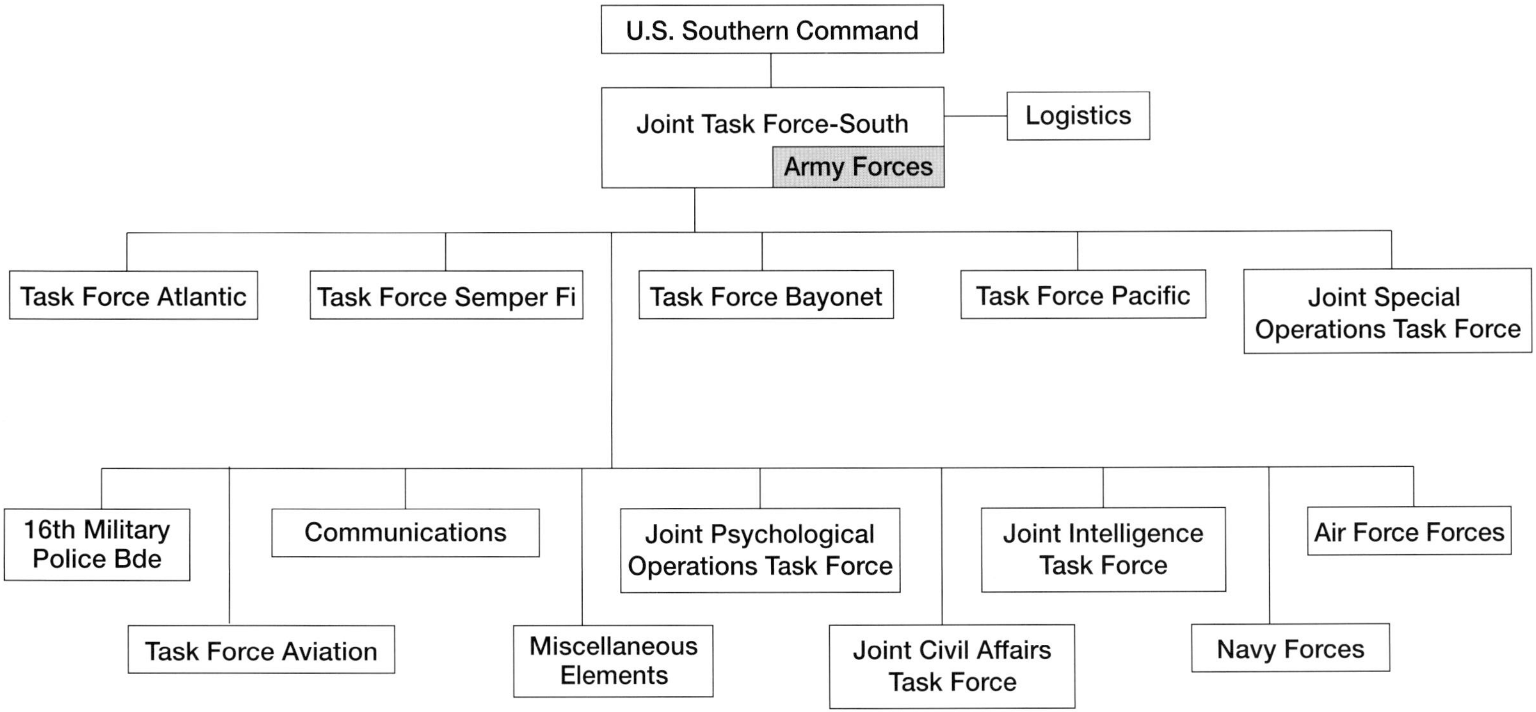

Source: Joint Task Force-South Operation Plan 90–2, 3 Nov 1989.

Navy component commander. They were Special Boat Unit 26, a Navy Sea-Air-Land (SEAL) platoon, a Naval Security Group, the Naval Small Craft Instruction and Technical Training School, and Mine Division 127. That these units did not exactly constitute a significant U.S. naval presence in Panama was unimportant, given the negligible size and poor quality of the PDF navy. In Stiner's judgment, the NAVFOR's assets, especially the special boat unit and fast patrol boats, would be adequate for the security, reconnaissance and surveillance, and patrolling missions they were to conduct in support of conventional and special operations under JTF-South Operation Plan 90–2. Experience was also a factor. According to Stiner, Gneckow's units were "very well trained, and had a lot of expertise operating in that area. They knew the Panama Canal, they knew the surrounding waters . . ., they knew how to deal with the Panama Canal Commission." He could have added that they also had knowledge of Panama's ports and possessed the capability to defend the aquatic approaches to Howard Air Force Base.[27]

In the case of the AFFOR, Maj. Gen. Peter T. Kempf, the Southern Command's Air Force component commander and the commanding general of the Twelfth Air Force at Bergstrom Air Force Base, Texas, coordinated airlift and combat air requirements with Stiner during the planning process. He also designated Brig. Gen. Robin Tornow, commander of the Southern Air Division, to be the JTF-South air component commander. As such, Tornow would also serve as the air defense commander and the joint force air commander with airspace control authority. With respect to aircraft, if the U.S. Navy's assets in Panama were considered just adequate for the impending invasion, the Air Force's did not meet even that standard. Needed were additional AC–130 gunships to supplement the two already in the country; a number of EF–111 Ravens and EC–130 Volant Solos for conducting electronic warfare against PDF command, control, and communications targets; the two F–117As Cheney had authorized; the Military Airlift Command's troop and equipment transports; an Airborne Warning and Control System; a C–130H Hercules to use as an airborne battlefield command and control center, an alternate command post; and various aircraft from the Tactical Air Command. Thanks in part to the lead time between the president's decision and the execution of the operation plan, all of these planes would be available when needed for deployment to Panama. Also on hand, as a result of General Kempf's efforts, were Strategic Air Command tankers, essential for providing in-air refueling for the transports and gunships en route to Panama from the United States. On the eve of Operation JUST CAUSE, then, JTF-South's reinforced Air Force component appeared ready to perform its airlift, air cover, close air support, command and control, and reconnaissance missions.[28]

Under the JTF-South organizational diagram, there was no Marine Forces component per se. During the meetings in October and early November, Col. Charles E. Richardson, the Marine commander under JTF-

[27] JTF-South OPLAN 90–2, 3 Nov 1989. Quote from Interv, Wright with Stiner, 2, 7, 27 Mar and 11 Jun 1990. For a more detailed U.S. order of battle, see An. A to JTF-South OPLAN, 3 Nov 1989.

[28] JTF-South OPLAN 90–2, 3 Nov 1989; Interv, Wright with Stiner, 2, 7, 27 Mar and 11 Jun 1990.

Panama, had tried to persuade Stiner to include more Marine elements, particularly a Marine expeditionary brigade, in the plan. The larger Marine contingent, Richardson contended, was needed to secure the vast area west of the canal and to subjugate the Panamanian forces located there. The 1988 drafts of Elaborate Maze and Blue Spoon had made provisions for a Marine brigade. Should not the current plan do so as well? In the end, Richardson failed to convince Stiner, who turned down not just the brigade request but one for an additional Marine battalion as well. The general did, however, put an Army engineer battalion and a military police company under Richardson's operational control and promised to send the colonel a battalion from the 82d Airborne Division (later switched to one from the 7th Infantry Division) within thirty-six hours of H-hour. On 18 December, when General Powell met with the Joint Chiefs prior to briefing President Bush, the Marine commandant revisited the question of adding more marines to the troop list, asking specifically for a Marine expeditionary unit, but he fared no better than Richardson. In the hours leading up to the commencement of Operation Just Cause, the units under Colonel Richardson did not technically constitute a Marine Force component; rather his task force retained the Task Force Semper Fi designation it had acquired as part of JTF-Panama. Its mission under the 90–2 operation plan was to secure various sites and areas along the western approaches to Panama City.[29]

General Tornow

On the east side of the canal, directly across from Richardson's area of responsibility, Col. Michael G. Snell's Task Force Bayonet faced what many planners considered to be the most difficult D-day missions: moving against the Comandancia, the PDF's main headquarters complex, and other enemy units and headquarters in its vicinity. Snell commanded the 193d Infantry Brigade, the largest Army combat unit based in Panama, and, for Just Cause, he would retain under his control the unit's two infantry battalions, engineer company, and field artillery battery. Task-organizing bolstered his force, as he also received operational control of the platoon of predeployed M551 Sheridans from the 82d Airborne Division; the 519th Military

[29] On Task Force Semper Fi, see Reynolds, *Just Cause*, pp. 17, 19–20. See also JTF-South OPLAN 90–2, 3 Nov 1989; Interv, Wright with Stiner, 2, 7, 27 Mar and 11 Jun 1990.

Police Battalion from Fort Meade, Maryland; a Marine light amphibious infantry platoon; and the 4th Battalion, 6th Infantry, 5th Infantry Division (Mechanized), the last two units being part of the continuing Nimrod Dancer rotation. Around H-hour, the bulk of Task Force Bayonet would break off into three smaller elements: Task Force Gator, charged with the Comandancia mission; Task Force Wildcat, which would seize and secure PDF facilities in the vicinity of Quarry Heights and the Comandancia; and Task Force Black Devil, which would neutralize the PDF 5th Infantry Company at Fort Amador.[30]

The targets assigned to Task Force Bayonet were clustered around Ancon Hill, near the heart of downtown Panama City. Other PDF installations and forces were located farther out on the periphery of or just beyond the capital, but within striking distance of the Comandancia and Snell's other objectives. To prevent these enemy troops from reinforcing the Ancon Hill area or interfering in any other way with the operations of Task Force Bayonet, a brigade-size element from the 82d Airborne Division was assigned to isolate and, if necessary, neutralize or destroy PDF units at Las Tinajitas, Fort Cimarrón, and Panama Viejo. According to the plan, after U.S. Rangers operating under the Joint Special Operations Task Force had seized and secured the Torrijos-Tocumen airport complex in eastern Panama City, the paratroopers would drop onto the site, board helicopters, and launch air assaults on the three targets nearby. Each objective was expected to be secured before dawn.[31]

Bringing in paratroopers to deal with PDF garrisons in and near the capital reflected a change in the Blue Spoon plans caused by the failed coup. Prior to 3 October, a brigade from the 7th Infantry Division was to be the only major conventional unit to deploy from the United States to participate in D-day combat operations on the Pacific side of Panama. An airborne brigade was to be brought in only as a follow-on force. After the critical role outlying PDF garrisons played in putting down the coup attempt, however, Fort Cimarrón, Las Tinajitas, and Panama Viejo went from being secondary concerns to primary targets, the neutralization of which would require air assault operations. So Stiner's planners decided to flip-flop the missions and scheduled arrival of the two stateside brigades. The logic behind the realignment was simple enough. In the same time it would take to land and begin unloading light fighters from the 7th Infantry Division flying into Torrijos-Tocumen, paratroopers from the 82d could airdrop, assemble, board waiting helicopters, and be en route to the three targeted PDF bases. The commander of the light division, Maj. Gen. Carmen Cavezza, was not convinced the change to the plan was necessary, especially given the 7th's extensive experience in the crisis to that date, but he nevertheless acceded to Stiner's reworking of the plan. As for command and control, once the paratroopers arrived in Panama, they would become part of Task Force Pacific, a headquarters that would be set up under a

---

[30] Nemmers, *United States Army South Staff Ride*, pp. 10, 13; JTF-South OPLAN 90–2, 3 Nov 1989.

[31] JTF-South OPLAN 90–2, 3 Nov 1989.

general from the 82d and which, at some point in Operation Just Cause, would absorb Snell's Task Force Bayonet and Richardson's Task Force Semper Fi.[32]

The higher prioritization accorded Fort Cimarrón, Panama Viejo, and Las Tinajitas after the failed coup increased from two to five the number of D-day air assaults called for in the operation plan, raising the question of whether the Army had sufficient aircraft in Panama to carry out the additional missions. As part of the security enhancement buildup in 1988, the assets of the 193d Brigade's 1st Battalion, 228th Aviation, had been augmented by about a third of the 7th Infantry Division's aviation equipment and personnel, designated under JTF-Panama as Task Force Hawk. In Stiner's opinion, however, more was still needed in light of changing war plans. He therefore requested and received authorization for the secret deployment of Apache and Kiowa helicopters to Panama in November; he also reinforced the in-country aviation battalion and Task Force Hawk with UH–60 Black Hawk, UH–1H Huey, and CH–47D Chinook helicopter crews and door gunners from Fort Bragg. These additions of equipment and manpower, in turn, necessitated an organizational change. "In order to bring this whole thing together," one corps planner later noted, "we felt that it would be worthwhile to create yet another subordinate task force, a Task Force Aviation, and designated the commander of the 7th ID(L) Aviation Brigade [Col. Douglas R. Terrell] as the commander." Because it was the last task force added to the plan and because its elements were located in multiple locations, Task Force Aviation had "the biggest problem of getting its command and control worked out prior to the operation." In the meantime, it also had to conduct a "very comprehensive training program to include FARP (Forward Area Rearming and Refueling Point), or refueling operations in the field."[33]

Almost all of the task forces under JTF-South concentrated mainly on enemy targets in or near Panama City on the southern edge of the canal area. One exception was Task Force Atlantic, which had the mission of confronting Panamanian forces in an area extending from Cerro Tigre near Fort Clayton to the northern tip of the canal. As with Task Force Bayonet and Task Force Semper Fi, Task Force Atlantic had been activated as a JTF-Panama component at the beginning of Operation Nimrod Dancer. Headquartered at Fort Sherman, the task force commander since early October had been Col. Keith Kellogg, commander of the 7th Infantry Division's 3d Brigade. In mid-December, the units under Kellogg included the 4th Battalion, 17th Infantry, from Kellogg's brigade; a field artillery battery; an air defense artillery battery; an infantry battalion from the 82d Airborne Division that was going through the Jungle Operations Training Center at Fort Sherman; and several smaller

[32] For the change of roles between the paratroopers and the light fighters, see Interv, author with Maj Harry Tomlin, U.S. Army, 22 Mar 1991, Fort Leavenworth, Kans. Tomlin was a planner for the 7th Infantry Division (Light) who attended the post–3 October planning sessions at Fort Bragg and in Panama. See also Donnelly, Roth, and Baker, *Operation Just Cause*, p. 55; Cole, *Operation Just Cause,* p. 74; Interv, author, Wright, and Joe D. Huddleston with Lt Gen Carmen Cavezza, U.S. Army, 30 Apr 1992, Fort Lewis, Wash.

[33] Interv, Wright with Stiner, 2, 7, 27 Mar and 11 Jun 1990. Quotes from Interv, Wright with G–3 Plans Section, XVIII Abn Corps, 30 Mar 1990, Fort Bragg, N.C.

outfits that provided medical, engineering, maintenance, supply, military police, and psychological operations (PSYOP) capabilities. Responsible for 90–2 operations in well over half of the canal area, Kellogg had a lengthy list of missions and targets, the latter of which included Colón, Panama's second largest city; Fort Espinar, home base of the PDF's 8th Infantry Company; Coco Solo; Renacer Prison; Madden Dam; and Cerro Tigre.[34]

Rounding out the subordinate headquarters under JTF-South was a group of specialized and functional commands: the 16th Military Police Brigade, the 1109th and 35th Signal Brigades, a joint intelligence task force, a joint psychological operations group, and a joint civil affairs task force. The military police brigade was to deploy from Fort Bragg with one of its battalions. Once engaged in Operation JUST CAUSE, it would assume responsibility for collecting and handling prisoners and detainees; provide security for convoys and key facilities; and stand ready to support civil-military operations by helping to reestablish law and order in Panama. Of the signal brigades, the 1109th would install, operate, and maintain JTF-South's communications network, while the 35th would provide personnel and equipment in support of the JTF-South communications plan. The Panama-based 470th Military Intelligence Brigade had the mission of processing incoming information and supplying the product to JTF-South and its components. An officer from the 4th Psychological Operations Group commanded JTF-South's PSYOP component with the mission of supporting U.S. combat operations in the canal area down to the company level through the use of such assets as loudspeaker teams and prerecorded broadcasts. Finally, the commander of the 96th Civil Affairs Battalion, the only active duty civil affairs unit in the U.S. Army, would run the task force whose mission contained such priorities as conducting civil-military operations during the combat phase of JUST CAUSE and supporting "the restoration of law and order, the reestablishment of PANAMANIAN civil infrastructure, and the restoration of civil works and public utilities."[35]

JTF-South Operation Plan 90–2 had, as Powell noted on reading a draft, "a lot of moving parts." It was far too complicated and complex for either Thurman or Stiner to micromanage its execution. Thus, SOUTHCOM and corps planners, having incorporated the appropriate rules of engagement, gave the subordinate commanders "mission-type taskings" in the form of very general printed and verbal guidance. For example, Operation Plan 90–2 might simply state a task force's mission as "neutralize PDF in area of operations" or "seize and secure PDF site." Stiner or the planners might then provide some verbal elaboration on what *neutralize*, *seize*, and *secure* meant to them. After that, subordinate commanders assumed responsibility for working out the details of how to accomplish their missions, determining in the process the composition and organization of the available force, the equipment and weapons to be used, the routes to be taken, the tactics to be employed, and so forth. Once these determinations were made, the lower echelon officers would brief their plans back up the chain of command. Approval at higher

---

[34] JTF-South OPLAN 90–2, 3 Nov 1989. For an account of Task Force Atlantic's role in the pre–JUST CAUSE portion of the crisis, see Yates, *U.S. Military Intervention in Panama, June 1987–December 1986,* chs. 6–9 passim.

[35] JTF-South OPLAN 90–2, 3 Nov 1989.

levels, even if it came from Thurman or Stiner, did not mean that work on a plan stopped; new intelligence, changing circumstances, and ongoing analysis would necessitate revisions, often up to the last minute.[36]

The size and nature of the projected operations necessitated a communications system that units from all services could use, that provided redundancy, and that allowed any one commander to talk to another at any time. The most recent U.S. contingency operation of any significance involving the invasion of a foreign country had been Operation Urgent Fury in Grenada in October 1983. Initially hailed as a near-flawless victory of American armed forces, accounts of the venture gradually emerged disputing that judgment—not the victory, of course, but the ease with which it had been achieved. The joint operation, it turned out, had been plagued from the outset by command and control problems, interservice rivalry, single-service parochialism, and interoperability difficulties. Most of all, there had been abysmal failures in joint communications, with a story making the rounds—later regarded as apocryphal—that an officer from one service had needed to use his own AT&T phone card and a commercial telephone on the island to call in fire support from another service. The problems that had afflicted joint forces on Grenada played a large part in the passage by Congress of the Goldwater-Nichols Department of Defense Reorganization Act in 1986, the main purpose of which was to streamline the military's command, control, and communications arrangements in the joint arena and to promote interservice cooperation, coordination, and interoperability.[37]

Senior and mid-level U.S. commanders and staff involved in the Panama crisis were well aware of the lessons of Urgent Fury, an operation that had been planned in less than a week.[38] They also knew that what transpired in Panama would be evaluated in terms of the emphasis Goldwater-Nichols placed on joint operations and interservice cooperation. Communications would come under special scrutiny, so the XVIII Airborne Corps, once it formally became part of the Blue Spoon planning process, began developing joint communications-electronic operating instructions, looking to improve

[36] Powell quoted words from Donnelly, Roth, and Baker, *Operation Just Cause*, p. 79, and see also p. 80. Interv, Wright with Stiner, 2, 7, 27 Mar and 11 Jun 1990. Taskings quote from Interv, Wright with G–3 Plans Section, XVIII Abn Corps, 30 Mar 1990. In this last oral history interview, one of the staff officers said that JTF-South's reliance on "mission-type taskings" was in contrast to JTF-Panama, whose versions of Blue Spoon, in his opinion, had been highly detailed in the orders given the various subordinate commanders.

[37] Three excellent sources on Grenada and the lessons learned in that operation are Mark Adkin, *Urgent Fury: The Battle for Grenada* (Lexington, Mass.: D. C. Heath, 1989); Ronald H. Cole, *Operation Urgent Fury: Grenada* (Washington, D.C.: Joint History Office, Office of the Chairman of the Joint Chiefs of Staff, 1997); and Edgar F. Raines Jr., *The Rucksack War: U.S. Army Operational Logistics in Grenada, 1983* (Washington, D.C.: U.S. Army Center of Military History, 2010). While the Raines volume focuses on logistics, it also contains an excellent analysis of the planning process and preparations for Urgent Fury.

[38] The XVIII Airborne Corps planners who wrote OPLAN 90–2 for Panama discussed the impact the lessons of Urgent Fury had on them in Interv, Wright with G–3 Plans Section, XVIII Abn Corps, 30 Mar 1990. The head of JTF-Panama's planning division, Lt. Col. Robert Pantier, also told the author that, in working on Blue Spoon, he read official accounts of the intervention in Grenada.

on those the corps had used earlier in the year for a joint training exercise. It took the XVIII Airborne Corps' signal staff 2½ weeks to finish the instructions (an amount of time, they noted, that might not be available in future contingencies). The final document, simplified to "stick in your pocket," as General Stiner put it, "provided the common information necessary for the efficient use of all radio nets used throughout the operation." As such, the instructions became a critical piece in an elaborate communications plan that involved equipment and support from several organizations, ranging from the Joint Chiefs, the Defense Communications Agency, and the National Security Agency at the top, downward to the communications units of the Southern Command; the XVIII Airborne Corps; U.S. Army, South; and the participating military services in general. The methods of communication included tactical satellites, very high frequency–frequency modulated (VHF–FM) and high-frequency (HF) radios, the World Wide Military Command and Control System, secure phone systems, and the automated digital network (AUTODIN) data communications service. JTF-South, according to the plan, would make use of the Panama-based 1109th Signal Brigade's in-country communications facilities to the "maximum extent possible."[39]

The size and complexity of the impending contingency operation also demanded a comprehensive plan for ensuring that the participating troops, headquarters, and units received essential supplies in a timely way. SOUTHCOM's service components, especially U.S. Army, South, had various resources already in the theater, and these, the operations plan emphasized, would "be used to the maximum extent." Units deploying to Panama would also bring with them enough food—mainly meals, ready to eat—for three days; enough petroleum-oil-lubricants packages for five days; and their basic loads of ammunition, minus fragmentation, incendiary, and concussion grenades. Each service component headquarters under JTF-South was responsible for logistical support, primarily through its own service supply channels. As for the Army Forces (ARFOR), which constituted the overwhelming majority of U.S. units under 90–2, the XVIII Airborne Corps' 1st Corps Support Command would meet the component's combat service support and medical needs (especially those exceeding what USARSO could provide). In performing this critical mission, the support command became the executive agent for most of the combat service support in the theater, serving under the JTF-South commander who, also dual-hatted as the ARFOR commander, would "exercise directive logistics authority to preclude unnecessary duplication among supporting forces."[40]

Because the objectives of Operation Just Cause included the removal of the Noriega regime, JTF-South Operation Plan 90–2 was treated as a four-phase campaign plan. Phase I was labeled "predeployment/crisis action" and covered the establishment of the JTF-South command and control element,

---

[39] Stiner's pocket quote and the communications facilities usage quote from Interv, Wright with Stiner, 2, 7, 27 Mar and 11 Jun 1990. Interv, Wright with G–3 Plans Section, XVIII Abn Corps, 30 Mar 1990. Stiner's common information quote from JTF-South AAR, Encls 1, 6; CINCSO OPORD 1–90, 30 Oct 1989, an. K; JTF-South OPLAN 90–2, 3 Nov 1989, an. K.

[40] Quotes from JTF-South OPLAN 90–2, 3 Nov 1989, an. D. Briefing, 1st COSCOM, 22 Feb 1990.

as well as preparation of American forces in Panama and the United States to launch offensive and defensive operations against the Panama Defense Forces and the regime it served. Phase II consisted of those operations, with the objectives of protecting American lives and property, capturing Noriega, and isolating or neutralizing enemy forces and their means of command and control. Phase III involved the transition to stability force operations that would feature civil-military activities and the restructuring of the Panamanian armed forces. Finally, combat units from the United States would redeploy during Phase IV.[41]

Going into Tuesday, 19 December, several Phase I objectives had been accomplished, such as the arrival in Panama of General Stiner, General Downing, and the JTF-South staff officers. Other parts of Phase I were still in progress, such as the loading aboard transport aircraft of the U.S.-based troops and equipment. The transition to Phase II also seemed to be proceeding on schedule. No one, though, expected this promising beginning to continue without disruption. As Stiner had cautioned at the end of one meeting in late October, "It won't go exactly as planned." As H-hour approached, his observation—taken as axiomatic within the military profession—would be borne out.[42]

TUESDAY, 19 DECEMBER

In Washington and in Panama, Tuesday was another day of meetings, decisions, and last-minute tweaking of the operation plan. At Fort Clayton, additional staff arrived, including twelve individuals from the Twelfth Air Force, to fill out the ranks of JTF-South, while the limited space available in the USARSO operations center continued to be transformed into a headquarters for a corps-size command and control element. Around noon, JTF-South issued a detailed JUST CAUSE execution checklist for the subordinate commands, containing the objectives of the operation and delegating responsibilities for achieving them.[43]

In Washington, Powell tried to keep appointments he had scheduled before the president's decision on Sunday in order to maintain the semblance of a routine working day. Then, at 1400, he and Cheney met with Bush to see if the president had any last-minute decisions, questions, or doubts concerning the invasion. Bush approved use of the F–117A fighters, and he indicated he would explain his decision to use armed force in Panama to the press, Congress, and the American public on Wednesday morning. He further stated that he would not postpone or cancel the invasion should the Panamanian opposition leaders who had claimed victory in the May presidential election balk when informed of the role Washington had in mind for them to play in replacing the current regime. At some point, apparently in this session with the secretary of defense

[41] JTF-South AAR.

[42] Stiner's remark is quoted in Donnelly, Roth, and Baker, *Operation Just Cause*, p. 89.

[43] Interv, author with McMahon, 21 Dec 1989.

and the chairman of the Joint Chiefs, the president also approved calling out the Defense Department's National Media Pool and sending it to Panama.[44]

At 1700, Powell assembled the Joint Chiefs in his office. When Cheney dropped in, Marine Commandant General Alfred M. Gray Jr. reassured the secretary that, no matter how negative or confused the initial reports of the fighting that night might appear, the operation would be a success. Cheney was pleased to see the chiefs still involved, though they had no authority to command any of the units taking part in the operation. Powell, according to reporter Bob Woodward, was simply pleased that they "had finally stopped messing with the plan." Once the group broke up, Powell successively instructed his staff to keep Cheney updated, learned to his dismay that some reporters in the news media were speculating about impending U.S. military action, called Thurman to wish him good luck, and went home for dinner. When he returned to the Pentagon, he took a nap and then joined Cheney and Kelly in the Crisis Situation Room at 2352, an hour before Operation JUST CAUSE was scheduled to begin. Two phone lines in the room connected the three men with General Thurman and his staff.[45]

In Panama, about the time Tuesday afternoon Powell and Cheney were wrapping up their meeting with the president, Stiner was in the conference room of Building 95 at Fort Clayton conducting an update briefing for key staff officers and his subordinate commanders. To those assembled, he distributed copies of the execute order and the final version of the joint execution checklist. He also told the commanders that at 1800 they could begin informing those officers under them who had not yet been told about the operation, and at 2100 the troops could be informed and briefed. Stiner was still talking when Thurman entered the room. The

[44] The Department of Defense (DoD) media call out did not begin until Tuesday evening at 1930, which all but ensured that the pool would not arrive in time to cover the initial fighting. Reporters were irate, arguing that the call out should have begun sooner and that the Pentagon should have alerted American reporters who were already in Panama. In January 1990, Pete Williams, the assistant secretary of defense for public affairs, asked his assistant Fred S. Hoffman to "research the facts surrounding the DoD National Media Pool deployment to Panama in conjunction with Operation JUST CAUSE and to provide his findings and recommendations." In March, those findings and recommendations were distributed to various correspondents. In the document, Hoffman attributed the failure to call out the Defense Department's media pool earlier to an "excessive concern for secrecy" and "a secrecy-driven decision" by Cheney. Hoffman was also critical of Williams for not trying "to convince Cheney that the pool had to be launched early enough to reach Panama before the operation kicked off." In the report itself, Cheney did not back away from his decision. Hoffman would later indicate that Powell shared Cheney's view. Thurman apparently did not, as he is quoted by Hoffman as indicating that "we made a mistake by not having some of the press pool in with the 18th Airborne Corps so they could move with the troops." Stiner is cited as saying such an arrangement would have been possible. Fred S. Hoffman, Review of Panama Pool Deployment: December 1989, Mar 1990. A later assessment of the report can be found in Cheryl Arvidson, "Press advocates: Cheney kept a tight rein on media during military conflicts," *Freedom Forum Online*, http://www.freedomforum.org/templates/document.asp?documentID=3869. For a succinct overview of the news media's problems during the Panama invasion, see Pascale M. Combelles, "Operation JUST CAUSE: A Military-Media Fiasco," *Military Review* 75 (May-June 1995): 77–85.

[45] Cole, *Operation JUST CAUSE*, pp. 32–33. Quote from Woodward, *The Commanders*, p. 180, and see also pp. 179, 181.

SOUTHCOM commander advised the group to stick to the plan but know when to adjust; he also revealed to many of them for the first time that one H-hour objective, highly classified to that point, would be the rescue of an American citizen who, since April, had been held prisoner in the Carcel Modelo, a PDF jail across the street from the Comandancia.[46] After the briefing, the commanders headed to their respective headquarters. Thurman would monitor developments from the complex of rooms called the Tunnel at Quarry Heights, while Stiner planned to arrive at the JTF-South operations center at Fort Clayton a couple of hours before the fighting began.

As the SOUTHCOM and JTF-South operations centers gradually filled with the officers who would monitor and direct the invasion, the units taking part were being briefed for the first time on what they would be facing. Meanwhile, transport aircraft in the United States were set to take off with troops or equipment aboard. At 1816, C–130s carrying a ranger battalion from Fort Lewis, Washington, lifted off from Lawson Army Airfield, Georgia; an hour later, a second ranger battalion, also on C–130s, took off from Hunter Army Airfield, Georgia. And an hour or so after that, the last of the C–141s carrying heavy-drop cargo took off from Charleston Air Force Base. In all, the Military Airlift Command contributed over 60 C–141s, over 20 C–130s, and 2 C–5s to the airlift needed for the commencement of Operation JUST CAUSE.[47]

It was during the airlift that the inevitable fog (both literally and figuratively in this case) and friction of war began to insinuate themselves significantly into the transition from Phase I to Phase II of the operation plan. At Fort Bragg and Pope Air Force Base, the worst ice storm in twenty years moved through on the nineteenth. Two battalions from the 82d Airborne Division's Division Ready Brigade had to wait while their transports were being deiced. An hour after the scheduled departure time, only ten of the twenty C–141s carrying the paratroopers were able to get into the air. The remaining ten would not take off for another 4½ hours. After consulting with General Johnson, the division commander, Stiner determined that the first wave would not wait for the second to catch up before parachuting into the Torrijos-Tocumen airport complex in Panama City. Stiner wanted to keep "as closely as possible to the original timetable." As the JTF-South operations officer, Col. Thomas H. Needham, later explained, "We had a very detailed execution checklist. Once you start adjusting times, your execution checklist falls out, and so, I think, that it was pretty much the staff's position and the boss's position—let's stick with the plan."[48]

If nature conspired to disrupt timetables, last-minute but not wholly unanticipated lapses in operations security endangered the element of tactical surprise U.S. forces hoped to achieve. In the United States, NBC reported on its television newscast Tuesday evening that C–141 transports were landing at Howard Air Force Base in Panama every ten minutes. On CBS, newscaster

[46] Interv, author with McMahon, 21 Dec 1989; JTF-South AAR.

[47] Operation JUST CAUSE Significant Activities, n.d. [XVIII Abn Corps chronology of events]; Cole, *Operation Just Cause*, p. 33; MAC, *Anything, Anywhere, Anytime*, p. 195.

[48] Stiner quote from Cole, *Operation Just Cause*, p. 33. Needham quote from Interv, Wright with Col Thomas H. Needham, U.S. Army, 6 Mar 1990, Fort Bragg, N.C.

Dan Rather balanced speculation that a U.S. invasion of Panama might be imminent with references to the Pentagon's cover story that the troop activity at Fort Bragg and Pope Air Force Base was a readiness exercise. The *New York Times*, sources indicated, also had the invasion story, but it would not appear until the paper's Wednesday morning edition, well after JUST CAUSE was under way. In Panama City, there was a palpable feeling of unease, a strong sense that something was about to happen. While Stiner was conducting his commanders' briefing at Fort Clayton, a telephone call came into the USARSO command group. The caller had helped organize a block party for American students who, attending college in the United States, were back in Panama on Christmas vacation to visit their families and friends. The man had heard rumors of military activity and wondered whether or not he should cancel the celebration. Operations security dictated the reply he received from the Army major who answered the phone: there was no need to call off the festivities. The party was held as scheduled, with what turned out to be tragic consequences.[49]

Around 1700, the U.S. intelligence community began to pick up the first hints that some senior PDF officers knew military action against them was imminent, although they clearly did not know what form it would take, with one of Noriega's confidants guessing there would be a snatch operation to seize the Panama strongman. As the evening progressed, H-hour, too, appeared to have been compromised. The text of one message emanating from the Comandancia declared: "They're coming. The ballgame is at 1AM. Report to your units . . . draw your weapons and prepare to fight." The breaches of security troubled Thurman and Stiner, but they found no reason to panic. They had yet to receive any reports suggesting that Panamanian forces were formulating a systematic and coordinated response to what was about to happen. In Stiner's opinion, enemy leaders were too incompetent to present an effective defense on such short notice. Besides that, "They didn't know all the targets," the general later commented, "or what our tactics were."[50]

Still, the fact that several PDF officers knew as much as they did had to be taken seriously. For U.S. forces to lose the element of surprise opened up all sorts of dire scenarios that had haunted BLUE SPOON planners for nearly two years. Panamanian forces had hostage lists, Thurman and Stiner knew, and might use the remaining time before H-hour to start rounding

[49] Cole, *Operation Just Cause*, pp. 33–34; Woodward, *The Commanders*, p. 180; Donnelly, Roth, and Baker, *Operation Just Cause*, pp. 102–03. The author was present when the major at Fort Clayton took the call regarding the Christmas party. See also Maria Len-Rios, "Operation Just Cause: An Eyewitness Account," *Focal Point* (Winter 1990): 10–11, 25–27, for more on the Christmas party.

[50] First quote from Cole, *Operation Just Cause*, p. 34. Woodward, *The Commanders*, p. 181. Second quote from Interv, Wright with Stiner, 2, 7, 27 Mar and 11 Jun 1990. In this interview, Stiner attributed at least part of the compromise of operations security to the speculation on television newscasts in the United States. "The Panamanians watch television," he noted. "Many of the Panamanian officers that were captured said they saw [the U.S. news broadcasts] and called each other on the telephone and discussed in great detail what they should do. I don't know to what extent this influenced their motivation, but it is obvious that some took action to alert their troops."

up Americans on those lists. Meanwhile, other enemy personnel—and it would not require a large number—might place themselves in advantageous locations from which they could use shoulder-fired surface-to-air missiles known to be in the PDF inventory to shoot down transport planes carrying U.S. Rangers and paratroopers. The number of service personnel who would die in the loss of even one troop carrier would forever tarnish the all-but-inevitable American victory. Yet, throughout the planning process, no participant had been able to devise an effective countermeasure to the missile threat, save strategic and tactical surprise, which was now at risk. There was also the possibility that, with prior warning, hostile forces would bring Howard Air Force Base under indirect mortar fire before American countermortar radar and other systems could locate and neutralize the weapons. If the mortar rounds were accurate, the runway at Howard would become unusable to U.S. transport planes. Another alarming prospect concerned guerrilla warfare. If Noriega and enough PDF members did manage a coordinated reaction to the advance warnings they had received, they might simply head into the jungle to mount unconventional operations against U.S. forces and any new Panamanian government. U.S. intelligence collectors had uncovered plans for this course of action but believed it unlikely given what was perceived as the low level of professionalism and the dubious loyalty, dedication, and stamina of the PDF senior leadership. As Colonel Needham remarked of dictators in general, "You get used to the good life for so long that you've got to be pretty tough to go out into the jungle and survive." He clearly did not regard Noriega or his inner military circle as being that tough.[51]

As the time for initiating combat operations neared, Stiner considered whether H-hour should be advanced, at least for U.S. troops based in Panama, given that some enemy forces knew that hostilities were imminent and "the message traffic . . . indicated that they were making preparations at certain installations." At first, he thought about moving H-hour up by thirty minutes, but after talking with Downing, the JSOTF commander, he concluded that the most he could request was a fifteen-minute change. The critical factor influencing his thinking was the need to coordinate Task Force Gator's attack on the PDF's main headquarters complex, the Comandancia, with the assault of a special operations commando team to rescue the American citizen jailed across the street in the Carcel Modelo. One problem was that, before Task Force Gator could move to the Comandancia, located on the east side of the canal, it had to receive the four M551 Sheridans that had been secreted on the west side for the past month. For that to happen, a swing bridge in front of Fort Clayton had to be rotated into position over the canal to allow vehicular traffic, which also meant stopping ships transiting the waterway. JTF-South did not want to open the bridge until the last southbound commercial vessel had cleared the canal's Pacific outlet. By the time that occurred, the earliest the Sheridans would be able to cross to the east side would be around 0030. Informed of this, Stiner, from his command post

[51] Intervs, Wright with Stiner, 2, 7, 27 Mar and 11 Jun 1990, and G–3 Plans Section, XVIII Abn Corps, 30 Mar 1990. Quote from Interv, Wright with Needham, 6 Mar 1990.

at Clayton, telephoned Thurman at Quarry Heights and asked that H-hour be changed to 0045. Thurman approved the request, informed the Pentagon of his decision, and, shortly after midnight, issued the order to Downing to begin the attack on the Comandancia and Carcel Modelo fifteen minutes earlier than planned.[52]

Downing had already sent a message to his forces letting them know that the enemy was aware of the impending attack and warning all special operations elements to be prepared to fight their way to their objectives. Around 0030, the U.S. military in Panama went to Defcon 1, the highest defense readiness condition. Thurman also implemented Personnel Movement Limitation Echo, which meant that, with the exception of the forces involved in the operation, all service personnel and their dependents were not to travel anywhere in Panama. (The need for such restrictive movement would soon be obvious to all.) By this time, a number of U.S. units were already on the move to their targets and sporadic shooting could be heard in the southeastern portions of Panama City. In effect, Operation JUST CAUSE had begun.[53]

[52] Quote from Interv, Wright with Stiner, 2, 7, 27 Mar and 11 Jun 1990. Interv, author with Maj Mark Collier, U.S. Army, 24 Jan 1990, Fort Clayton, Panama; Woodward, *The Commanders*, pp. 181–82; Capt. Kevin J. Hammond and Capt. Frank Sherman, "Sheridans in Panama," *Armor* 99 (March-April 1990): 9.

[53] Memo, HQ, 4th Bn (Mech), 6th Inf, for Commanding General (CG), U.S. Army Forces, et al., 27 Mar 1990, sub: Operation "Just Cause" After Action Report; Significant Activities, n.d. [from G–3 logs]; Woodward, *The Commanders*, p. 183.

3

Special Operations at the Outset of Just Cause

☆ ☆ ☆

Of the 27,000 U.S. military personnel who had specific combat missions during Operation Just Cause, just over 4,100—or 15 percent—were under the operational control of General Downing's Joint Special Operations Task Force. Most of the units to which these troops belonged had been listed in the first Blue Spoon contingency plans written in 1988. At that time, planners had envisaged Special Operations Forces conducting surgical strikes on a limited number of critical PDF command and control nodes, securing those objectives in a matter of hours, and then quickly turning them over to conventional U.S. units. In the ensuing year and a half, particularly after the Nimrod Dancer buildup of U.S. conventional forces in Panama in mid-1989, the four-star change of command at Quarry Heights that September, and Major Giroldi's attempted coup against Noriega just days later, the plans changed significantly, with new targets being added and with conventional units acquiring responsibility for some of the H-hour missions originally assigned to the JSOTF. The targeting shuffle, though, in no way altered the central role of special operators in the event of an invasion; when they lost a mission to conventional troops, it was only to take on another.

As H-hour approached, these missions included locating and apprehending Noriega, preventing him from leaving the country to evade capture, closing off the Atlantic and Pacific entrances to the Panama Canal, neutralizing enemy units at Rio Hato, seizing Panama's Torrijos-Tocumen airfield complex, rescuing an American citizen the regime had imprisoned in April 1989, helping to isolate the battlefield in downtown Panama City, and, when necessary, providing support to conventional forces.

To accomplish these missions, Downing's joint task force was organized into several subordinate task forces, some of which were further subdivided into specific task units. Task Force Red, consisting of the 75th Ranger Regiment, was the largest special operations element by far. Its mission was to seize and secure the Torrijos-Tocumen airfield facilities, while subduing the PDF 2d Infantry Company located there, and to neutralize the PDF 6th and 7th Infantry Companies at Rio Hato. A considerably smaller element, Task Force Black, consisted of the 3d Battalion, 7th Special Forces Group (Airborne), based in Panama, and Company A, 1st Battalion, 7th Special Forces Group, from Fort Bragg, North Carolina. One of its H-hour missions was to prevent the PDF's Battalion 2000 at Fort Cimarrón from reinforcing enemy units in the capital. Navy SEALs, Naval Special Warfare Unit 8, and Special Boat Unit 26 formed Task Force White, responsible for closing off the canal and disabling naval vessels or aircraft that could whisk Noriega out of the country. Two other task forces, Blue and Green, were composed of elite special mission units that would seek to capture Noriega. Task Force Green had the additional assignment of rescuing the incarcerated American citizen. Finally, belonging to no one task force but present to support special operations in general were the 112th Signal Battalion (minus), the 160th Aviation Group (minus), the 528th Support Battalion (minus), the 617th Aviation Detachment, and psychological operations and civil affairs units.[1]

Regime Change

From the beginning of contingency planning in February 1988, commanders and staff involved in the process recognized that a U.S. invasion of Panama would result in the overthrow of the Noriega regime.[2] Whether the strongman fled or was captured or killed in the process mattered little; thus, although compartmentalized plans existed for seizing and bringing him to the United States to stand trial on drug-trafficking charges, early iterations of Blue Spoon did not make his apprehension a specific mission. Instead, military planners focused not on the general's fate but on what would follow him in Panama. The working assumption envisaged a democratic government replacing the toppled dictatorship. Estimates held that a month would be needed for the new regime to be installed, establish its legitimacy, and assert its ability to govern. During the transition, a U.S. military government headed by the SOUTHCOM commander would run the country. For well over a year, this scenario informed the thinking of staff officers writing the Blind Logic operation order for nation building in a post-Noriega Panama.[3]

By late 1989, the basic premise underlying Blind Logic had been overtaken by events. Of these, the most important was the country's presidential election in May 1989, in which Guillermo Endara, the opposition candidate for president, and Ricardo Arias Calderón and Guillermo "Billy"

[1] The missions and organization of the Joint Special Operations Task Force (JSOTF) are discussed in U.S. Special Operations Command, *History, 1987–2007*, 20th anniversary ed. (MacDill Air Force Base, Fla.: U.S. Special Operations Command, n.d.), pp. 32–44. A JSOTF order of battle can be found at the U.S. Army Center of Military History Web site on Operation Just Cause (http://www.army.mil/cmh-pg/documents/panama/taskorg.htm) and in Donnelly, Roth, and Baker, *Operation Just Cause,* pp. 81–82.

[2] For a more detailed account of the planning process for stability operations and nation building in Panama, see the appropriate sections in Yates, *U.S. Military Intervention in Panama, June 1987–December 1989*; Fishel, *The Fog of Peace*.

[3] Early plans were also vague about the fate of the Panama Defense Forces. If reformed, some U.S. officers believed, the organization might support a democratic government. Only well into the planning process did the defeat and destruction of the Panamanian armed forces become an objective of U.S. combat operations.

Ford, his two vice presidential running mates, won a clear victory, and which Noriega had promptly annulled. To a number of U.S. officers surveying the postelection chaos from the perspective of Quarry Heights, the logical step to take in the event of a U.S. invasion would be to forgo the establishment of a SOUTHCOM-led military government and, instead, swear in the men who had a legal, legitimate, and popular claim to enter the presidential palace. Woerner accepted this argument. He also concluded sometime in mid-1989 that the capture of Noriega should be an explicitly stated BLUE SPOON mission. Left at large following the overthrow of his regime, the deposed dictator might foment an insurgency against a U.S.-backed democratic government, although few analysts at the Southern Command thought that likely. A more credible threat was that of Noriega creating political mischief for the Endara administration from some hidden location in Panama or from exile abroad.[4]

This appreciably revised blueprint for what the U.S. military could hope to accomplish politically in Panama through armed force survived the change of command from Woerner to Thurman. Soon thereafter, once the failed coup of 3 October caused the new SOUTHCOM commander to believe that war with the Panama Defense Forces was inevitable, the "Noriega watch" went into effect at Quarry Heights, with "a six-person intelligence team, supplemented by experts from the National Security Agency and the Central Intelligence Agency in Washington" working to keep track of the dictator's whereabouts around-the-clock. At some point, Thurman requested special operations tracking teams to participate in the effort. General Cisneros spoke later of "the room," a clearinghouse for information on the dictator's movements. The data came from satellite imagery, monitored communications, and surveillance teams. A map in the SOUTHCOM operations center provided a visual record of the group's calculation of Noriega's whereabouts at any given moment.[5]

That those keeping tabs on Noriega knew where he was located about 80 percent of the time—Stiner's estimate—testifies to the magnitude and intensity of the endeavor and the skill of all involved. That the dictator could not be found the other 20 percent reflected the uncertainties that infuse any complex enterprise, made more challenging by Noriega's deliberate attempts to elude his trackers and by his consummate skill at deception. Once he learned that there was a concerted U.S. effort to follow him, he used doubles, decoy vehicles and aircraft, false convoys, frequent changes of clothes, recordings of his voice, and other subterfuges to thwart anyone trying to determine his location at those times when he did not want it known. He would also frequently change his whereabouts in an effort to keep his adversaries guessing or simply

[4] Kevin Buckley, *Panama: The Whole Story* (New York: Simon & Schuster, 1991), p. 189. For adding the capture of Noriega as a BLUE SPOON mission, see Yates, *U.S. Military Intervention in Panama, June 1987–December 1989,* p. 222.

[5] Briefing, Gen Maxwell R. Thurman at the School for Advanced Military Studies (SAMS) (hereafter cited as Thurman SAMS briefing), 23 Sep 1992. First and second quotes from Donnelly, Roth, and Baker, *Operation Just Cause*, p. 104, and see also p. 105. Kempe, *Divorcing the Dictator,* p. 14; Edward M. Flanagan Jr., *Battle for Panama: Inside Operation Just Cause* (New York: Brassey's, 1993), p. 214. Third quote from Interv with Lt Gen Marc A. Cisneros, U.S. Army (Ret.), 1998 (for the television program, *Secrets of War*).

to overload the system with information. As President Bush was told at one point, the general was "moving around like a Mexican jumping bean."[6]

On Tuesday, 19 December, just hours before the beginning of Operation JUST CAUSE, the Noriega watch map accurately showed him to be in Colón. Since Sunday, as analysts would later determine, the general had been receiving reports that an invasion was imminent, but he had dismissed them as baseless. Then, while in Colón Tuesday, he learned that Cuba's intelligence service was reporting U.S. troop transports in the air, due to arrive over Panama in a matter of hours. According to some members of Noriega's entourage, this information caused him to leave Colón for Panama City late that afternoon. While a decoy convoy of official vehicles wound its way across the transisthmian highway, Noriega, his driver, assistants, and bodyguards covered the same route in an inconspicuous white van.

Special operations personnel tracking the general followed the convoy, whose route ended at the Comandancia. When a poor choice for a Noriega look-alike emerged from one of the cars, even a staged honor guard ceremony could not maintain a convincing charade. The trackers knew that their prey had won the cat-and-mouse game. "Where was the real one?" General Stiner later asked rhetorically. "We didn't know." The real one, in fact, arrived back in the capital around 2000, after which he and his companions borrowed a Hyundai automobile and went to one of the several small houses he maintained in the city for official use. From there, he made telephone calls to close friends, all of whom told him that the situation looked dire. To the astonishment of the men with him, the general still failed to realize the full extent of his precarious position. The United States, he claimed, was simply engaging in a disinformation campaign to convince him to flee the country. That Noriega was allegedly drinking at the time possibly clouded his judgment. Whatever the case, no one ordered the mobilization of the Panama Defense Forces and Dignity Battalions for the war that was fast approaching. In the words of one author, "PDF units shifted around in a disorganized way. Action seemed to depend on how seriously individual units took the rumors." As for the dictator, he "decided that entertainment was what he needed."

> A sergeant on his staff was ordered to pick up one of Noriega's occasional girlfriends and bring her to La Siesta, a hotel near the [Tocumen] airport that had been converted into a PDF officers' club. It was another of Noriega's haunts. Noriega climbed into the back of a four-door Hyundai for the trip to the club. A Mercedes and a Toyota Land Cruiser filled out the caravan.

Noriega was still at the Ceremi Recreation Center inside La Siesta when, around 0030, he heard explosions. Half an hour later, one of his bodyguards went outside and discovered "a sky full of paratroopers." That shock prompted the general to dress and flee with his men in the Hyundai. After evading U.S.

[6] Clancy with Stiner and Koltz, *Shadow Warriors*, p. 333; Donnelly, Roth, and Baker, *Operation Just Cause*, p. 105. Quote from Kempe, *Divorcing the Dictator*, p. 11, and see also p. 14. Before the "Noriega watch" went into effect, according to Stiner, U.S. officials only knew where the dictator could be located about 50 percent of the time. Interv, Robert K. Wright Jr. with Lt Gen Carl W. Stiner, 2, 7, 27 Mar and 11 Jun 1990, Fort Bragg, N.C.

Rangers at Tocumen, they drove around the capital during the predawn hours, taking refuge in the houses or apartments of friends at four different locations. From each, Noriega would make telephone calls to military officers, friends, and girlfriends, seeking information and giving and receiving advice. He learned that his daughter and her family had sought political asylum and that, within hours of the invasion, the Panama Defense Forces as an organization had practically ceased to exist. Some associates urged him to resist the Americans, with Capt. Asunción Eliécer Gaitán, a close confidant and security escort, convincing the general to record a belligerent speech over the telephone that Gaitán could later broadcast on a radio station still under PDF control. Dawn found Noriega in an apartment, hoping that U.S. troops in the vicinity would not see him. Now a fugitive, he had devised no specific plan to ensure his escape.[7]

For the U.S. commands in Panama and for the Bush administration in Washington, the news that Noriega had not been captured before or soon after H-hour sounded a sour note in what otherwise was playing out as a well-orchestrated military operation. Throughout the night, the U.S. forces charged with capturing the Panamanian strongman intensified their efforts, and analysts intercepted several of the dictator's phone conversations without, however, being able to determine his exact location. Noriega's known haunts were all targeted, but he was not in any of them. In a couple of instances, accurate information led the troops to residences where Noriega had taken refuge, but, when they arrived, they found he had already departed, if only by a few minutes. Also, during the night, one of Noriega's bodyguards told the general that he was going to go on ahead of the group to scout new hideouts. Instead, he surrendered to a group of U.S. soldiers and tried to tell them where his boss was located, only to discover that none in the group spoke Spanish. By the time the bodyguard was brought to General Cisneros' attention, the information was too old to be of value. As Thurman concluded later, "We had 400 calls the first day on the whereabouts of Noriega. We chased him all over Hell's half acre."[8]

The failure to apprehend Noriega produced significant friction, in one instance between the special operations community and U.S. civilian intelligence agencies, and in another, between the Bush administration and the news media. In the first case, both General Lindsay at the Special Operations Command and General Downing believed that information processed by the civilian agencies was not reaching the special operations units quickly

[7] Stiner's quotes and his account of Noriega's departure from Colón are from Clancy with Stiner and Koltz, *Shadow Warriors*, pp. 333–34. According to Stiner, Noriega had been in the official convoy leaving Colón, but the convoy had at some point split in two, with the special operators following the decoy section that went to the Comandancia. The fuller account of Noriega's movements the night of the invasion is from Kempe, *Divorcing the Dictator*, pp. 13–16, and Buckley, *Panama*, pp. 234–38 (from which all other quotes are taken). The narratives by Kempe and Buckley were pieced together from information they received from the men around Noriega on the night of 19–20 December. Please note that in some sources Gaitán is spelled Gaytán.

[8] Woodward, *The Commanders*, pp. 185–86; Donnelly, Roth, and Baker, *Operation Just Cause*, pp. 105, 107. Quote from Interv, John Partin with Gen Maxwell R. Thurman, 12 Apr 1990, Fort Benning, Ga. Television Interv with Cisneros, 1998.

General Downing

enough to be useful. As Downing later lamented, "The intelligence people are not going to tell you what they got until the last minute," presumably in order to protect their sources.[9]

Regarding the administration and the media, the first official announcement of Operation Just Cause came at a White House press conference convened at 0140 on 20 December by Bush's press secretary, M. Marlin Fitzwater. Fitzwater read a statement listing the objectives of the invasion, including the apprehension of Noriega. When a reporter asked, "Has General Noriega been captured yet?" the press secretary replied, "We don't know how long it will take, but that is our ultimate objective. It has not happened at this time." The choice of the word *ultimate* was perhaps unnecessary, unfortunate, and unwise. U.S. forces could overthrow Panama's dictatorship—in fact, at the time Fitzwater was speaking, they already had—and pave the way for democracy in Panama without apprehending Noriega. Fitzwater's adjective, however slightly, hinted otherwise.[10]

At 0720, President Bush addressed the country from the Oval Office, stating his reasons for ordering the invasion. His only reference to the fallen dictator was to say, "General Noriega is in hiding." Immediately after the address, Cheney and Powell conducted a briefing at the Pentagon. In his opening statement, the secretary of defense included the acknowledgment, "At this point, of course General Noriega is still at large. . . . He is, at this point, a fugitive." Powell followed, devoting most of his presentation to the U.S. units participating in Just Cause, the targets they were attacking, and the success of the operation to date. When he turned his attention to Noriega, he asserted, "We have not yet located the General. . . . But, as a practical matter, we have decapitated him from the dictatorship of his country and he is now a fugitive and will be treated as such." Once the formal briefing ended, the two men entertained questions, a number of which requested further information on Noriega's status. In answering, Powell indicated that "the reign of terror is over," and "We're looking for him. He's not running anything."[11]

[9] Donnelly, Roth, and Baker, *Operation Just Cause*, p. 106. Quote from Interv, Partin with Maj Gen Wayne A. Downing, U.S. Army, 6 Jan 1990, Howard Air Force Base, Panama.

[10] Woodward, *The Commanders*, pp. 184–86.

[11] First Powell quote from *New York Times*, 21 Dec 1989. Remaining Powell quotes from Woodward, *The Commanders*, pp. 187–89. See http://bushlibrary.tamu.edu/research/public_

The point that Powell tried to make was that the military operation was achieving its principal objectives, even though Panama's dictator was still at large.[12] In the days that followed, many reporters took a different position, making Noriega's capture the litmus test for the success or failure of Operation JUST CAUSE. That this view betrayed a grossly oversimplified assessment of a complex undertaking seemed to matter little, especially when reinforced by such developments as the broadcast of Noriega's bombastic speech—the one Gaitán had recorded—over a Panamanian radio station on the afternoon of 20 December. So long as the dictator was on the loose, reporters continuously put administration spokespersons on the defensive when discussing events in Panama.

If Noriega could not be readily located, the same could not be said of Panama's new leaders. Advisers to President Bush had told him that the best way to transform Panama from a dictatorship to a democracy was to follow the SOUTHCOM recommendation for replacing the general with the true winners of the May 1989 presidential election—Endara, Arias Calderón, and Ford. On the eve of JUST CAUSE, Bush was convinced, in the words of one reporter, "that the point of no return would be achieved once Endara agreed to be sworn in as president of Panama and to request U.S. intervention." One question for senior U.S. officers and officials in Panama was when to convey this news to Endara and the two men who were to be his vice presidents. To

(*Second from the left to right*) **Billy Ford, Guillermo Endara, and Ricardo Arias Calderón lead a May 1989 election march.**

papers.php?id=1356&year=1989&month=12 for a transcript of President Bush's address to the nation.

[12] Concerning the issue of the relative importance of Noriega's apprehension to the success of JUST CAUSE, the author was standing within earshot of General Stiner during the first hours of the operation and heard the general say into a telephone something to the effect, "No, we don't have him yet, but we've got more important things to do." Author's notes written several days after the beginning of Operation JUST CAUSE.

inform them well in advance of the invasion might compromise operations security, which would place the three Panamanians, not to mention the entire military enterprise, in jeopardy. Better the three be told of the impending invasion just an hour or so before it began. The next question, then, was how to convey the news. Initial suggestions called for U.S. Army Special Forces to go into Panama City and round up the three men shortly before H-hour, but Col. Robert C. "Jake" Jacobelly, commander of Special Operations Command, South, objected to the risks and recommended instead that Thurman or John Bushnell, the U.S. Embassy's relatively new deputy chief of mission and, in the ambassador's absence, the chargé d'affaires in Panama, invite the three to dinner at Howard Air Force Base the night of the invasion. Stiner approved Jacobelly's plan, and, on orders from Secretary of State Baker, Bushnell issued the invitation. Baker also told the chargé to help Endara form a Panamanian government that could be operational within hours of the invasion. In order "to maintain the element of surprise," the secretary added, Bushnell was to undertake his assignment without telling any of his staff at the embassy.[13]

Endara, Arias Calderón, and Ford accepted the invitation. Around 2030, Bushnell, after pledging the three to secrecy, told them that a U.S. invasion was imminent and asked them to consider taking over the government of Panama, a move the United States would support. "They're in a state of shock . . .," Bushnell reported to Baker. "They never expected we'd do this much. They'll do everything they can to make this work." "I felt like a big sledgehammer hit my head," Endara recalled. Although he and his two colleagues welcomed the proposition, they realized that they would be assuming the leadership of a country "under occupation by American forces." The reaction throughout Latin America, they knew, would be negative—the Bush administration also anticipated an adverse regional response—but the feeling was that, after initial protests, most governments in the hemisphere would resume normal diplomatic relations with Panama, few being sorry to see Noriega deposed. For their safety, the three men were flown by helicopter from Howard to Fort Clayton, where they were sworn in. At 0039, Thurman informed Washington that the ceremony had been completed.[14]

In his address to the nation on the morning of 20 December, President Bush declared, "The brave Panamanians elected by the people of Panama in the elections last May, President Guillermo Endara and Vice Presidents Calderon and Ford, have assumed the rightful leadership of their country. . . . The United States today recognizes the democratically elected government of President Endara. I will send our Ambassador back to Panama immediately."

---

[13] First quote from Woodward, *The Commanders*, p. 182, and see also p. 170. Donnelly, Roth, and Baker, *Operation Just Cause*, p. 122. Second quote from Baker with DeFrank, *Politics of Diplomacy*, p. 190.

[14] Third and fourth quotes from Buckley, *Panama*, p. 234. First and second quotes (Bushnell) from Baker with DeFrank, *Politics of Diplomacy*, p. 190. Woodward, *The Commanders*, p. 182. General Stiner, in Clancy with Stiner and Koltz, *Shadow Warriors,* p. 333, states that the swearing-in ceremony took place in Thurman's office at Quarry Heights, not at Fort Clayton, with a Panamanian judge administering the oaths of office. After the ceremony, the three men were then taken to a safe house on Clayton.

During the Pentagon briefing that followed, one reporter asked whether Endara had been sworn in on an American base. Yes, Powell responded. Thus surfaced the controversy that would quickly surround the circumstances under which the new government had been installed, confirming in the process the fears of Endara and the others that, initially at least, they would be regarded as American "puppets."[15]

Task Force White: The Canal, Balboa Harbor, and Paitilla Airport

As elite U.S. troops were trying to locate Noriega, additional special operations units were engaged in various other missions. Of the naval special warfare elements involved, most belonged to Task Force White, whose operations center was located at the Rodman Naval Station on the Pacific side of the canal area. According to a U.S. Special Operations Command history, Task Force White "consisted of five SEAL platoons, three patrol boats, four riverine patrol boats, and two light patrol boats (22-foot Boston Whalers)." This force, in turn, was organized into four task units, each with a separate mission. Task Unit Charlie was to control the waters at the Atlantic (also referred to as the Caribbean) entrance to the Panama Canal, while Task Unit Foxtrot was to do the same at the Pacific entrance. Task Unit Whiskey was to destroy the *Presidente Porras* patrol boat in Balboa Harbor, and Task Unit Papa was to seize Paitilla airfield.[16]

Task Unit Charlie was a last-minute addition to the task force. During a preinvasion briefing on 19 December, Lt. Cdr. Marshall D. Daugherty from Naval Special Warfare Unit 8 raised the possibility of "establishing a task unit on the north coast" at the canal's entrance near Colón. The group with Daugherty deemed the suggestion feasible, and it was strongly supported by Task Force Atlantic, the JTF-South component headquarters in the area. Daugherty received command of the hastily created task unit, which initially consisted of eight SEALs detached from two different SEAL teams, and two riverine patrol boats manned by eleven crew members from Special Boat Unit 26, based at Rodman. The two patrol boats began to transit the canal northward toward Colón at 1700. Once Daugherty reached the Atlantic side, he met with Col. Keith Kellogg, the Task Force Atlantic commander; received from him operational control of two Army landing crafts, mechanized (LCMs), with twelve crew members; went over Task Unit Charlie's missions in direct support of Task Force Atlantic; and coordinated communication frequencies and call signs. The lieutenant commander then set up his headquarters at the Jungle Operations Training Center, close to the piers where the LCMs and patrol boats could be moored. In the short time remaining before H-hour, he briefed his subordinate commanders on

[15] Bush quote from Transcript of President George H. W. Bush's address, 20 Dec 1989. Powell response from *New York Times*, 21 Dec 1989. Buckley, *Panama*, p. 234.

[16] U.S. Special Operations Command, *History, 1987–2007*, p. 38.

the forthcoming action, including the rules of engagement under which they would be operating.[17]

At 0040, ten minutes after the two patrol boats and the two mechanized landing craft had begun movement to their combat positions, shooting could be heard at nearby Coco Solo. By 0050, the four craft were on station at the mouth of Limón Bay (*Map 2*). As part of its mission, Task Unit Charlie was to shut down all commercial and enemy maritime traffic in the vicinity. To that end, Daugherty ordered one of the patrol boat commanders to contact the port authority and broadcast over the marine band channel word that "the Port of Colon and the Panama Canal are closed. This area has been declared a war zone and all maritime traffic in and out of this area is prohibited."[18] The announcement warned that the Americans would be patrolling the proscribed waters but also provided information detailing how the ships still docked in port might contact the U.S. gunboats. The broadcast, with timely updates, was repeated four times an hour.

Closing the canal and stopping all shipping through it was partly intended to keep vessels of nonbelligerent countries out of harm's way. The action would also make it difficult for any Panamanian forces to enter or flee the war zone via the waterway. At 0130, Task Unit Charlie's two riverine patrol boats began patrolling the west side of the shipping channel near Colón. Shortly after the patrols began, Daugherty directed the crews to go after a small craft in the area reported to be carrying PDF reinforcements to Coco Solo. As the patrol boats entered the waters where the enemy vessel had been sited, Daugherty learned that U.S. attack helicopters and an AC–130 gunship maneuvering overhead were requesting permission to open fire on what they believed to be PDF patrol boats they had observed. Fearing that the targets were in fact the two boats belonging to Task Unit Charlie, the lieutenant commander immediately ordered the crews to return to their initial rendezvous point. Through his chain of command, he also alerted the aircraft overhead that "friendly craft were operating in Limón Bay." Two hours later, there was a virtual replay of the episode when the patrol boats received orders to find a PDF craft reported damaged and sinking and to put SEALs aboard to search it. By the time the two boats reached the pier where the enemy vessel was supposed to be, it was not in sight, leading to the conclusion that it had already sunk. Once again, Daugherty learned that friendly troops in the area, this time on land and armed with AT4 rockets, were preparing to engage "enemy" patrol boats they could see from their positions near the water. And, once again, the lieutenant commander ordered his craft out of the area. In his opinion, two potential "blue-on-blue" incidents were two too many, so he informed his superiors that

---

[17] Ibid., p. 31. Quote from Memo, Cdr, Task Unit (TU) Charlie, for Cdr, Task Force (TF) White, 31 Dec 1989, sub: After Action Report Submission. TU-Whiskey Augmentation of TU-Carib: Maritime Interdiction Operations in Support of TF Atlantic, n.d., atch to AAR, TU-Whiskey Operational Summary, n.d., which appears to have been written soon after the events it describes, U.S. Special Operations Command (SOCOM) History Office files. The account in the text of Task Unit Charlie's activities in Operation Just Cause is pieced together from aforementioned sources.

[18] Quote from Memo, Cdr, TU Charlie, for Cdr, TF White, 31 Dec 1989, sub: After Action Report Submission.

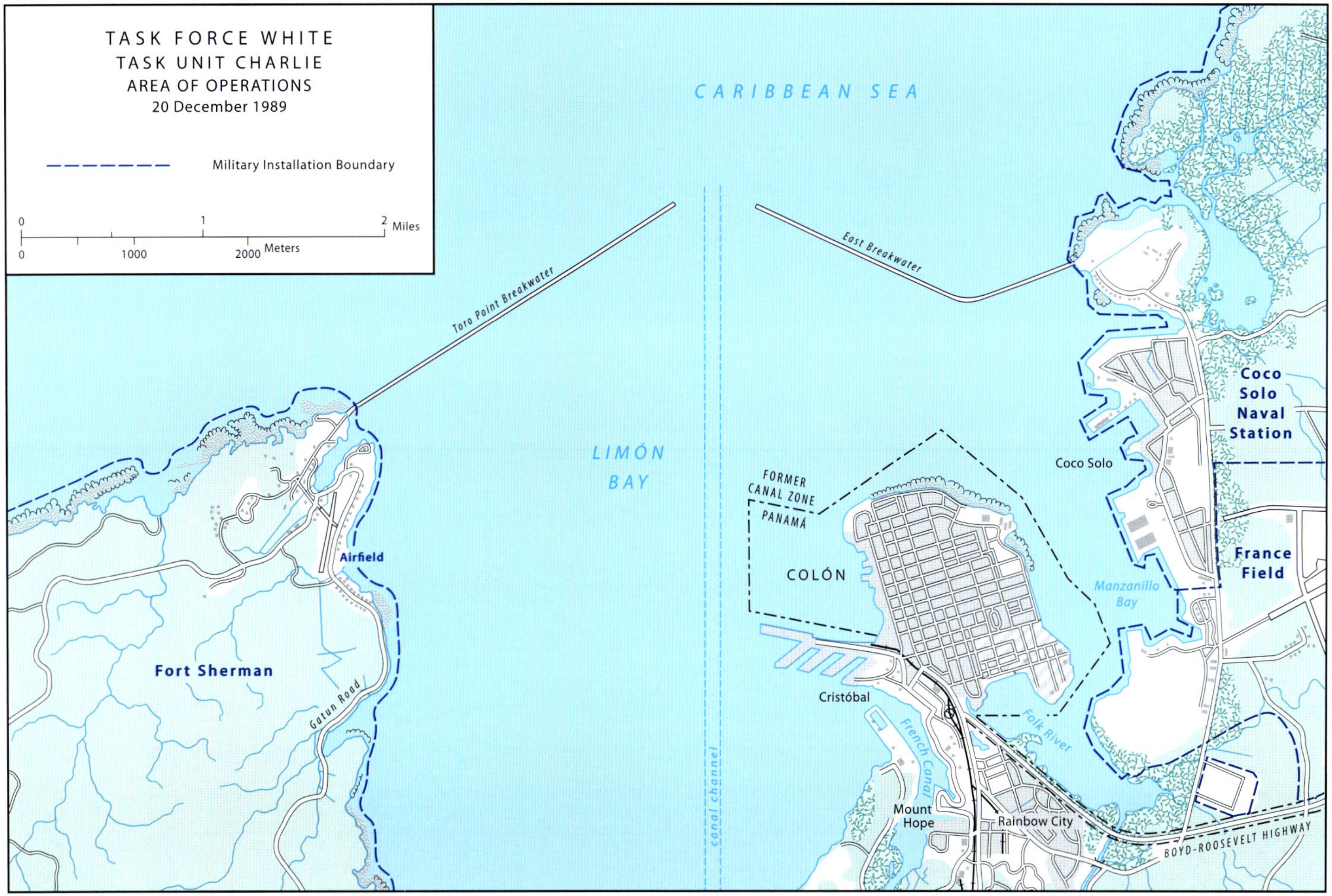
TASK FORCE WHITE
TASK UNIT CHARLIE
AREA OF OPERATIONS
20 December 1989
Military Installation Boundary
0
1
2 Miles
0
1000
2000 Meters
CARIBBEAN SEA
Toro Point Breakwater
East Breakwater
LIMÓN BAY
canal channel
Airfield
Fort Sherman
Gatun Road
FORMER CANAL ZONE
PANAMÁ
COLÓN
Cristóbal
French Canal
Mount Hope
Rainbow City
Folk River
Manzanillo Bay
Coco Solo
Coco Solo Naval Station
France Field
BOYD-ROOSEVELT HIGHWAY

Map 2

"no further patrols would take place within 1000 yards of [the] Coco Solo or Colón waterfront during hours of darkness due to the high risk of taking friendly fire."[19]

After daybreak on the twentieth, Task Unit Charlie turned its attention to a German merchant ship, *Asian Senator*, then moored in the port of Cristóbal. Reports indicated that at least two dozen Panamanian troops had taken refuge aboard. The task unit's two patrol boats and an LCM with Daugherty on board converged on the scene and took up positions that allowed them to observe each side of the boat and part of the pier area. Initial attempts to contact anyone on the *Asian Senator* by radio and loudspeaker failed. Meanwhile, the task unit spied U.S. Army special operations helicopters conducting strafing runs on a building at the end of the pier. When PDF elements in the building shot back at the helicopters, Daugherty received permission to have one of his patrol boats try to suppress the enemy fire, using several of the weapons available to the crew. This, in turn, caused whoever was in the building to open fire on the patrol boat, a fusillade that lasted several seconds. The episode ended when a rocket fired from one of the helicopters made a direct hit on the target.

Just as the threat from the pier had been removed, the Panamanian troops who were holed up on the *Asian Senator* made their first appearance, with thirteen of them fleeing down the gangplank from the ship to the pier. One of the patrol boats and a SEAL detachment fired over the heads of the men, but to no avail. Seven more Panamanians then made a similar exit from the ship, after which Daugherty received orders that those of his crew who could see the gangplank should destroy it with gunfire. The ensuing volleys failed to hit the target, although rounds did strike containers aboard the *Asian Senator*, not to mention the ship's superstructure. Daugherty then called a halt to the shooting, citing the probability of "considerable collateral damage" as his reason. Despite the errant rounds, the fusillade did have at least one positive effect: soon after the bombardment stopped, the ship's captain radioed one of the patrol boats and requested a cease-fire, obviously unaware that, thanks to Daugherty, one was already in effect. The captain went on to say that a number of well-armed Panamanians who had come onto the *Asian Senator* were ready to surrender.[20]

Daugherty told the ship's captain to have the Panamanians move to the top of the gangplank and await further instructions. He then directed one of the LCMs to land five SEALs to round up the prisoners. Once on the pier, the SEALs established security and then approached the gangplank. As they did so, another group of unarmed Panamanians appeared at the end of the pier carrying a white flag. Daugherty tried to contact the SEALs to provide guidance on how to handle each group, but communications failed, and he ended up going ashore himself. As all this was transpiring, the SEALs found themselves coming under increasing small-arms fire from buildings near the pier. They returned fire and watched as a U.S. helicopter tried to neutralize one of the enemy positions, only to be hit by ground fire. (Task Unit Charlie

[19] Ibid.
[20] Ibid.

was later informed that the helicopter had crashed.) At one point, Daugherty requested an air strike against one of the buildings, only to be told that there were no "air assets" available. Later, after the SEALs had placed their prisoners on an LCM and were in the process of loading captured PDF weapons onto it, enemy fire reached an "unacceptable level," and Daugherty again made a request up the chain, this time for additional special operations personnel to help secure the area. He also informed his superiors that containers aboard the *Asian Senator* blocked his patrol boats from providing supporting fire. His request was again denied. Instead, he was told to finish collecting weapons that had been stacked on the pier and to have a SEAL element search the merchant ship for any remaining PDF personnel. Daugherty responded that, "without the availability of [the] requested support, TU Charlie was departing the pier area immediately to turn in the prisoners at Coco Solo." The two patrol boats and LCM made the return voyage around 1430. After discharging thirty-six captured Panamanians, the crews resumed their blockade duties around Colón. The next day, the canal would reopen to shipping but under very strict controls.[21]

While Task Unit Charlie was patrolling the waters on the Atlantic side of the canal, Task Unit Foxtrot was performing similar missions on the Pacific side in the vicinity of Panama City. According to one official source, "SEALs in three patrol boats guarded the waters around Howard Air Force Base, and two riverine patrol boats covered the approaches to the Bridge of the Americas." Of particular concern were three small islands near Howard where intelligence reports speculated that concealed enemy personnel were ideally situated to launch surface-to-air missiles against U.S. aircraft flying into the air base. To determine the accuracy of the reports, a four-man SEAL team launched a canoe from one of the patrol boats and landed on each of the islands. Their search ended "without evidence of enemy activity," so the four returned to the patrol boat.[22]

While this was transpiring, another element from Foxtrot, an assault team aboard three combat rubber raiding crafts, worked its way from the Bay of Panama up the Rio Tapia and linked up with ground troops in the Don Bosco neighborhood near the Torrijos-Tocumen airport complex. The mission was to capture Luis A. Cordoba, a former PDF intelligence chief, or at least to prevent his escape via a water route. At dawn, U.S. forces stormed a compound where Cordoba was thought to be, but the quarry was not present. After the area was cleared and searched, the assault team made its way back to the patrol boats.[23]

As Task Unit Foxtrot conducted maritime patrol and interdiction operations, largely without incident, Task Unit Whiskey was carrying out its reconnaissance and surveillance, maritime interdiction and containment, and direct action missions, the principal one of which was to blow up Noriega's personal patrol boat, the *Presidente Porras*, docked at Pier 18 in Balboa Harbor (*Map 3*). The boat, if manned and at sea, presented a threat to U.S. missions

[21] Ibid.

[22] First quote from U.S. Special Operations Command, *History, 1987–2007*, p. 40. Second quote from Howard AFB [Air Force Base] Security, n.d., SOCOM History Office files.

[23] Cordoba Operation, n.d., SOCOM History Office files.

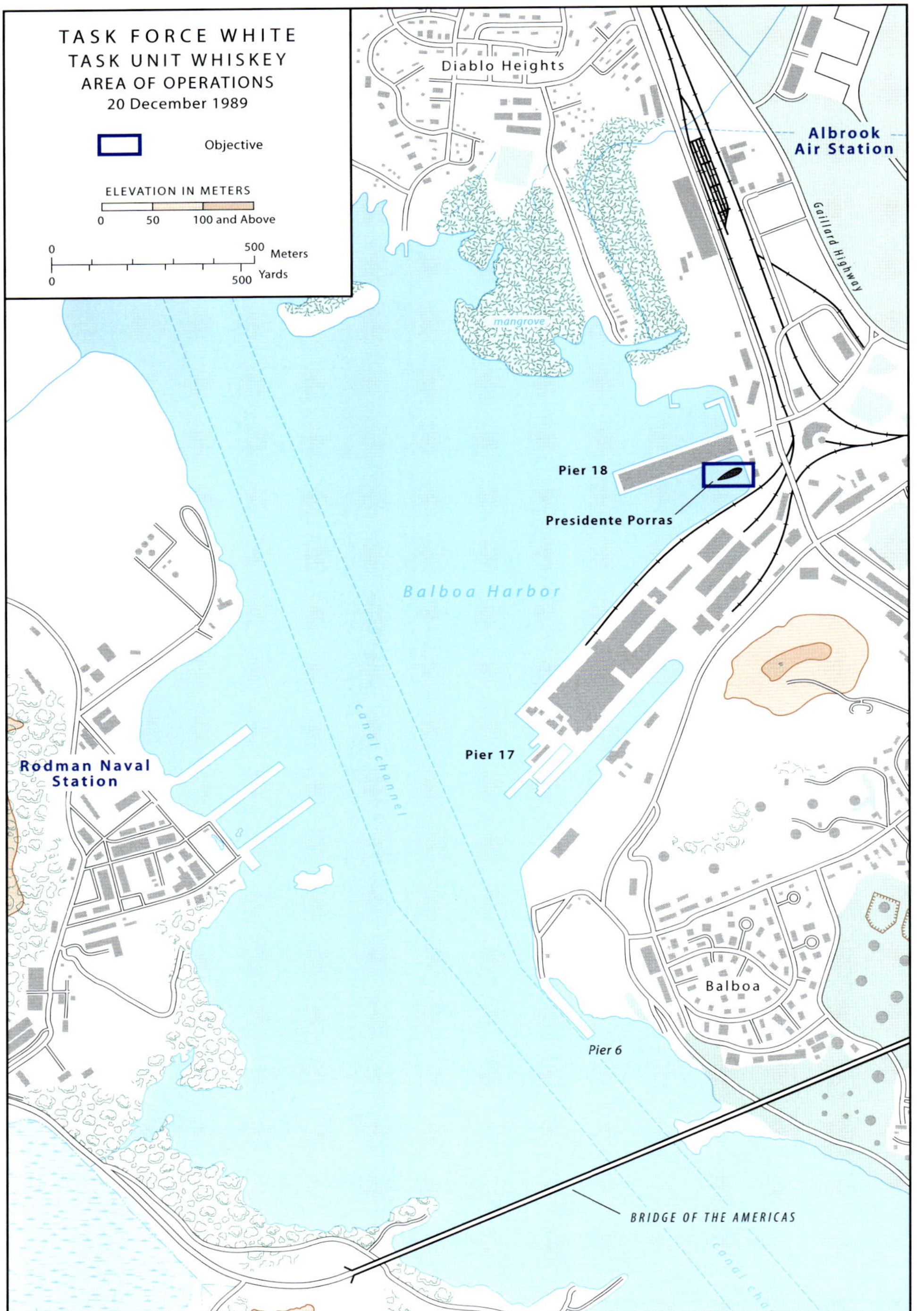
TASK FORCE WHITE
TASK UNIT WHISKEY
AREA OF OPERATIONS
20 December 1989
Objective
ELEVATION IN METERS
0
50
100 and Above
0
500 Meters
0
500 Yards
Diablo Heights
Albrook
Air Station
Gaillard Highway
mangrove
Pier 18
Presidente Porras
Balboa Harbor
canal channel
Pier 17
Rodman Naval
Station
Balboa
Pier 6
BRIDGE OF THE AMERICAS
canal channel

Map 3

Combat rubber raiding craft

in two ways. With its automatic cannon and machine guns, it could disrupt operations against the Panama Defense Forces at Fort Amador, the Bridge of the Americas, and Paitilla airfield. It could also provide Noriega a means of escape. To prevent either possibility, BLUE SPOON planners had earmarked the vessel for destruction.[24]

U.S. naval craft, attack helicopters, and AC–130 gunships available for JUST CAUSE were capable of performing this mission, but they were committed to other targets. Consequently, two pairs of divers from SEAL Team 2, a Virginia-based unit, received the assignment to destroy the *Presidente Porras*. Use of the two pairs ensured redundancy in the event that Panamanian forces should neutralize one of them. The plan called for the four men to conduct a "combat swimmer demolition attack," the first such attempt by Navy divers since World War II to destroy a surface vessel with underwater mines, in this case haversacks filled with C4 plastic explosives. The men would wear Draeger underwater breathing equipment, which was designed to leave no trail of air bubbles. Other elements of Task Unit Whiskey at Rodman Naval Station, armed with machine guns, grenade launchers, and mortars, would provide them fire support if needed. Reinforcing Whiskey were two riverine patrol boats and a SEAL squad from Task Unit Foxtrot, which would take up blocking positions near the east side of the canal.[25]

---

[24] Donnelly, Roth, and Baker, *Operation Just Cause*, pp. 120–21. According to one Navy SEAL, the original plan called for seizing the patrol boat, towing it to Rodman Naval Station, and turning it over to the new government. After General Lindsay at the U.S. Special Operations Command decided that this was too risky, the revised plan called for destroying the boat. Interv, Partin with Cdr Patrick T. Toohey, 18 Nov 1992, Fort Bragg, N.C.

[25] This account of the Balboa Harbor mission is from Donnelly, Roth, and Baker, *Operation Just Cause*, pp. 120–21; Flanagan, *Battle for Panama*, pp. 82–84; TU-Whiskey Operational Summary, n.d., and Foxtrot Plt Support of TU Whiskey Combat Swimmer Operation, n.d.,

Rodman Naval Station

Shortly before the operation got under way, the Task Unit Whiskey headquarters at Rodman received a query from the Joint Special Operations Task Force as to whether the unit could accommodate moving H-hour from 0100 to 0030. That meant detonating explosive charges under the patrol boat thirty minutes earlier than called for in the plan. The timers for the explosives had already been set, though, and the divers were ready to depart. To stop and reset the timers would prevent the men from reaching their objective on time. To reset the timers on the move and in the dark was a highly risky procedure that could result in an accidental explosion and the death of the divers. Given these considerations, the task unit commander, Navy Cdr. Norman Carley, told his superiors that H-hour needed to remain the same. Believing he had made his point, Carley prepared to accompany the SEALs to their insertion site on the other side of the canal.

At 2300, two raiding craft—one with Carley, the other with a communications man, and each carrying two divers, a coxswain, and an M60 machine gunner—departed Rodman. Because of "unanticipated boat traffic in the harbor," including some vessels with searchlights, they took a "circuitous route" to a line of mangrove trees north of Pier 18. En route, the boat with the communications equipment experienced problems with its outboard motor, which quit working entirely once the craft reached the mangroves. While in the tree line, the SEALs spied a PDF patrol boat coming at them, but it sped by

both in SOCOM History Office files. Quote from U.S. Special Operations Command, *History, 1987–2007*, p. 40. Col Frank Akers, comp., TU-Whiskey: Balboa Harbor PDF Patrol Boat Operation, in Operation JUST CAUSE, *The Warriors*, n.d., a series of vignettes released by the U.S. Southern Command; F. K. Butler Jr., "Closed-circuit oxygen diving in the U.S. Navy," *Undersea and Hyperbaric Medicine* 31, no. 1 (2004): 15–17; Television program, *Navy SEALs: Canal Zone*, New Dominion Pictures, 2001. Sea-Air-Land (SEAL) Team 2 was airlifted to Panama from its home at the Naval Amphibious Base at Little Creek, Va.

without incident. At 2330, the raiding craft with Carley aboard maneuvered to an insertion point within 150 yards of Pier 18, and Diver Pair 1 entered the water. Carley then motored back to the disabled boat and towed it to the insertion point, where Diver Pair 2 slipped over the side, five minutes behind schedule. The commander then decided to tow the raiding craft to Rodman, where its motor could be changed. Both of the rubber boats would then return to the preplanned extraction point for the divers. As the two craft began transiting the canal back to the naval station, Carley learned H-hour had in fact been moved up fifteen minutes. He realized immediately that the change would leave his divers trying to reach their extraction point in the middle of intense combat operations, but at that moment he had no way to contact and warn them.[26]

As for the divers, after each pair arrived at Pier 18, they used the piling and overhead docks as cover while they maneuvered independently, "alternating between surface and subsurface movement," to the position where *Presidente Porras* was moored. Diver Pair 2, despite its delay, reached the underside of the patrol boat first. After confirming that it was the target, they attached their haversack carrying the explosives to the port propeller and pulled the pins on the timers. Diver Pair 1 arrived three minutes later and performed the same operation on the starboard propeller. For each pair of divers, the procedure took fewer than two minutes. While Diver Pair 1 was just finishing its work, the boat's engines started, raising the deadly prospect that the propeller blades would start rotating while both SEALs were right next to them. The two swam away in time, however, and both pairs of divers headed separately toward Pier 17. Suddenly, they experienced two intense explosions, presumably from grenades dropped into the water by Panamanian forces above them after the shooting started on shore. The SEALs surfaced and sought safety behind the pier's pilings and then resumed maneuvering underwater beneath Pier 17 in the direction of their extraction point. Four more explosions again forced them to surface and seek cover. The delays caused by the protective maneuvers meant that the two divers were still under the piers when the haversacks they had attached to the patrol boat exploded at exactly 0100. "The boat reared up forward . . . it went straight up—the bow went up," one of them stated later. The vessel sank within two minutes, as crews on other Panamanian vessels in the harbor began to turn on their propellers as "an anti-swimmer measure" in response to the blast.[27]

Once the divers swam from the pier area into the main channel of the canal, they heard overhead what sounded like a deep-draft vessel, perhaps an ocean-going freighter, coming straight at them. "You can't tell under the water exactly where a vessel is," one recounted, "you just hear it getting louder, and louder—it sounds like a freight train coming." To save themselves, they descended from their normal depth of twenty feet to forty-five feet, no routine maneuver. As one study has observed, "The increased toxicity of the pure oxygen in the Draeger system in deeper water was risky." In this instance, the

[26] TU-Whiskey Operational Summary, n.d.

[27] Ibid.

SEALs suffered no harmful effects from their hasty descent and the ten to fifteen minutes they spent at the hazardous depth.[28]

Meanwhile, at 0045, the two raiding craft that had returned to Rodman were back in Balboa Harbor, arriving ten minutes later near the extraction point for the divers at the southern end of Pier 6, 750 yards from the flaming wreckage of the *Presidente Porras*. As Carley and the others waited under the pier for the SEALs to appear, tracer rounds darting over their heads kept them continuously aware of the heavy fighting now taking place on the east side of the canal, including one intermittent firefight right above them. When Carley realized that the divers were well behind schedule, he sent the other rubber boat to the next pier over to search for them, but the craft returned without having made contact. Not until 0200, did Diver Pair 1 arrive, followed five minutes later by Diver Pair 2. Once aboard the raiding craft, the SEALs safely crossed the harbor back to Rodman, their mission a complete success.

Besides the *Presidente Porras*, there was one other means of transportation near Panama City that U.S. planners feared Noriega might use to flee the country: his Learjet maintained at Paitilla airport, a private facility for domestic airlines and small planes several miles up the coastline eastward from the heart of the capital. Another concern about Paitilla was that it might be used as a rallying point for PDF reinforcements flying in from outlying bases. U.S. forces thus needed to disable the facility's 5,000-foot runway, destroy the

Paitilla airfield, with Panama City in the background

[28] Butler, "Closed-circuit oxygen diving in the U.S. Navy," p. 17.

Learjet, and deny use of the airfield to the enemy. The attempt to do so would prove, in terms of U.S. fatalities, the deadliest single action of Operation JUST CAUSE.[29]

The mission fell to Task Unit Papa, the largest component of Task Force White. SEAL Team 4, commanded by Navy Cdr. Thomas McGrath and based at the Naval Amphibious Base, Little Creek, Virginia, detailed three sixteen-man SEAL platoons (Bravo, Delta, and Golf) and supporting units to Task Unit Papa. As with other stateside military units listed in the JTF-South war plans, the SEAL team had rehearsed its mission in mid-December, by chance just days before President Bush's decision to invade Panama. Conducted at Eglin Air Force Base in the Florida panhandle, the rehearsal had not gone well, for while the SEALs disabled the stand-in for Noriega's Learjet and rendered the practice runway unusable by blocking it with other aircraft, they incurred several "casualties" from men playing the defenders and experienced significant communication and coordination difficulties. Afterward, commanders and staff analyzed the problems in what were at times heated exchanges, mainly between McGrath and his immediate superior, Capt. John Sandoz, a Vietnam War veteran who was now in command of Naval Special Warfare Group 2. When they briefed General Lindsay, however, the SOCOM commander came away convinced that the rehearsal had been a success. Consequently, nothing was done to rectify any shortcomings in the plan. The one significant change that did take place occurred largely as a result of the friction between Sandoz and McGrath. Lacking confidence in his subordinate, Sandoz revised some of Task Unit Papa's command, control, and organizational arrangements. McGrath would not lead the assault elements as planned but would remain with two patrol boats and a support group a couple of miles offshore. While he would still command the operation, Lt. Cdr. Patrick Toohey, his executive officer, would be in charge of the ground force. Sandoz, in later explaining his decision to leave the SEAL Team 4 commander at sea, stated somewhat disingenuously, "From that vantage point, McGrath could remain in constant communication with Toohey, the AC–130 Spectre gunship flying above, as well as other SEAL operations going on simultaneously in Panama."[30]

Once the SEAL Team 4 platoons had been airlifted to Panama, the briefings they received prior to the assault indicated that any enemy resistance

[29] This account of the SEALs at Paitilla airfield is from U.S. Special Operations Command, *History, 1987–2007*, pp. 38–39; Donnelly, Roth, and Baker, *Operation Just Cause*, pp. 114–20; Flanagan, *Battle for Panama*, pp. 84–86; Malcolm McConnell, *Just Cause*, paperback ed. (New York: St. Martin's Press, 1991), pp. 51–78; Paitilla Operation, n.d., SOCOM History Office files; The SEAL assault on Patilla [*sic*] Airfield during Operation Just Cause, n.d., http://www.specwarnet.net/miscinfo/patilla.htm; Orr Kelly, *Brave Men, Dark Waters: The Untold Story of the Navy SEALs* (New York: Pocket Books, 1992), pp. 1–4; Interv, Partin with Toohey, 18 Nov 1992.

[30] Quote from Donnelly, Roth, and Baker, *Operation Just Cause*, p. 115. For the rehearsal problems and consequent change in command arrangements, see also Mir Bahmanyar with Chris Osman, *SEALs: The US Navy's Elite Fighting Force* (Oxford, U.K.: Osprey Publishing, 2007), p. 82; Interv, Partin with Toohey, 18 Nov 1992; and Brian W. Reeves, "Navy SEALs: Theory and Reality" (Master's thesis, Naval Postgraduate School, 1997), pp. 77–81. Most of Reeves' discussion of the Paitilla operation is based on the document, Naval Special Warfare Command Lessons Learned, Operation JUST CAUSE, n.d.

at Paitilla would likely come from only six to eight PDF guards who could be persuaded to surrender or, failing that, easily subdued. Other security personnel—rent-a-cops or, as McGrath was reported to have said, "Old men with rusty rifles"—might also be present as guards for private planes, including those used by drug dealers, but these civilians were not expected to pose a threat. "Boy scouts can do this job," McGrath supposedly claimed, much to Toohey's chagrin. Some ranking SEALs asked for an opportunity to reconnoiter the airfield themselves. On the advice of intelligence officers concerned about compromising the operation, Downing turned down the request, but Navy lobbying caused Thurman to reconsider his own aversion to last-minute surveillance forays and override the JSOTF commander. On Tuesday, then, a number of SEALs dressed in civilian clothes observed the target from the second floor of a club at the southern end of the facility and from a highway north of it. What they discovered generated mixed feelings. The three hangars on the west side of the runway, including the one for Noriega's plane, did not seem to be fortified. Nor were there numerous armed guards at the site, possibly because the Learjet was not there. But the SEALs could not see inside the jet's hangar, only the open runway and aprons over which they would have to advance. Armed mainly with rifles, machine guns, AT4 rockets, fragmentation grenades and grenade launchers, and a mortar, the attack force would be exposed and vulnerable to enemy fire coming from inside any of the hangars.[31]

Despite these concerns, the plan remained basically unchanged. The three lightly armed SEAL platoons would land near the southern end of the airport and move northward up Paitilla's runway. Golf would destroy the Learjet; Delta would deal with facilities on the eastern side of the field, across from the hangars; Bravo would bring up the rear, providing covering fire and helping to obstruct the runway by dragging planes and other obstacles onto it. If any of the platoons ran into serious resistance, an AC–130H overhead could be contacted by an Air Force combat control officer with the assault party to knock out any defensive positions the SEALs themselves could not surmount.

Sometime after dark, well in advance of H-hour, the 62-man force of three SEAL platoons and other personnel who would participate in the landing boarded fifteen combat rubber raiding craft. Launching from a beach at Fort Kobbe on the west side of the canal, the task unit rendezvoused with two patrol boats out of Rodman Naval Station, one of which had McGrath aboard, and were towed to a point offshore within two miles of the airport's southern edge. Meanwhile, on the north side of the objective, three two-man reconnaissance and surveillance teams took up positions along a highway running between Paitilla and Panama Viejo, the latter being home to a PDF cavalry squadron. Once Task Unit Papa's rubber craft were in position, two scouts swam ashore ahead of the ground force to reconnoiter the landing site and to place strobe lights at the location. The SEAL platoons did not disembark near the airfield until 0030. By then, Commander McGrath had received a message directing the assault force not to destroy certain aircraft at Paitilla. McGrath sought

[31] Quotes from Bahmanyar with Osman, *SEALs*, p. 85. See also Interv, Partin with Toohey, 18 Nov 1992.

clarification from the Task Force White operations center at Rodman. What did the term *minimal damage* mean? The response he received defined minimal damage as shooting out the planes' tires and cutting their control wires. In what would become a point of controversy after the operation, he concluded that this applied to the Learjet, as well as to other aircraft. He then passed the information to Commander Toohey, who was not pleased to hear it. If his men were supposed to disable rather than destroy the Learjet, they would have to make some last-minute adjustments to their tactics; namely, they would have to get closer to the plane, perhaps even entering its hangar, thus putting themselves at greater risk.[32]

Once ashore, Toohey and the ground force made their way to the southern end of the airfield, which proved to be deserted. From their position, however, they could hear the shooting already taking place in and around Panama City. After establishing a security perimeter, Toohey set up his command post near the task unit's mortar, while the three SEAL platoons began moving up the runway, with Bravo and Golf advancing along the west side toward the hangars, and Delta along the east side toward a control tower and a cantina frequented by the PDF guards (*Map 4*). As one squad was leapfrogging past another, they yelled at civilian watchmen and security guards they met en route to get out of the area. The shouting—together with the combat in Panama City—ensured that Task Unit Papa had lost the element of tactical surprise. Meanwhile, Toohey received another troublesome message, this one indicating, incorrectly, that Noriega had left Colón by helicopter and was heading to Paitilla. Fearing that the general might arrive before the SEALs had accomplished their mission, Toohey ordered the men to pick up the pace of their advance. This put them into their assault positions at approximately 0105. Of the three northernmost hangars at the airfield, the middle one housed the Learjet, which was clearly visible to the Golf Two SEAL squad that lay prone thirty yards opposite the structure. On Golf Two's right and perpendicular to it, another squad, Golf One, provided flank protection. Together, the squads formed an L shape, with Golf One in line to thwart any threat from the north, and Golf Two facing the hangars. The view from both vantage points revealed that the Panamanian guards on the scene stood ready to open fire from behind oil drums, cinder-block walls, and other available cover.

---

[32] Interv, Partin with Cdr Thomas R. Williams, n.d., SOCOM History Office files. Williams was the Task Force White operations officer located at Rodman Naval Station during JUST CAUSE. In the interview, he confirmed the content of the "minimal damage" message that he received from JSOTF headquarters at Hangar 3 at Howard Air Force Base at 2030 and that he relayed it to McGrath. He also said that the guidance applied to only two aircraft, neither of them the Learjet, but that, in seeking clarification, McGrath did not raise the issue of whether the directive for "minimal damage" applied to Noriega's plane. After the operation, General Thurman weighed in on the issue, stating emphatically that "I did not issue an order to change the mission." Donnelly, Roth, and Baker, *Operation Just Cause*, p. 120. Toohey admitted being "upset" when he received McGrath's message. Interv, Partin with Toohey, 18 Nov 1992. Years later, Partin told the author that pre–JUST CAUSE briefings concerning rules of engagement for specific missions contained some vague and contradictory guidance, just enough to generate confusion on such issues as the Learjet. Telecon, author with Partin, 6 Jun 2012.

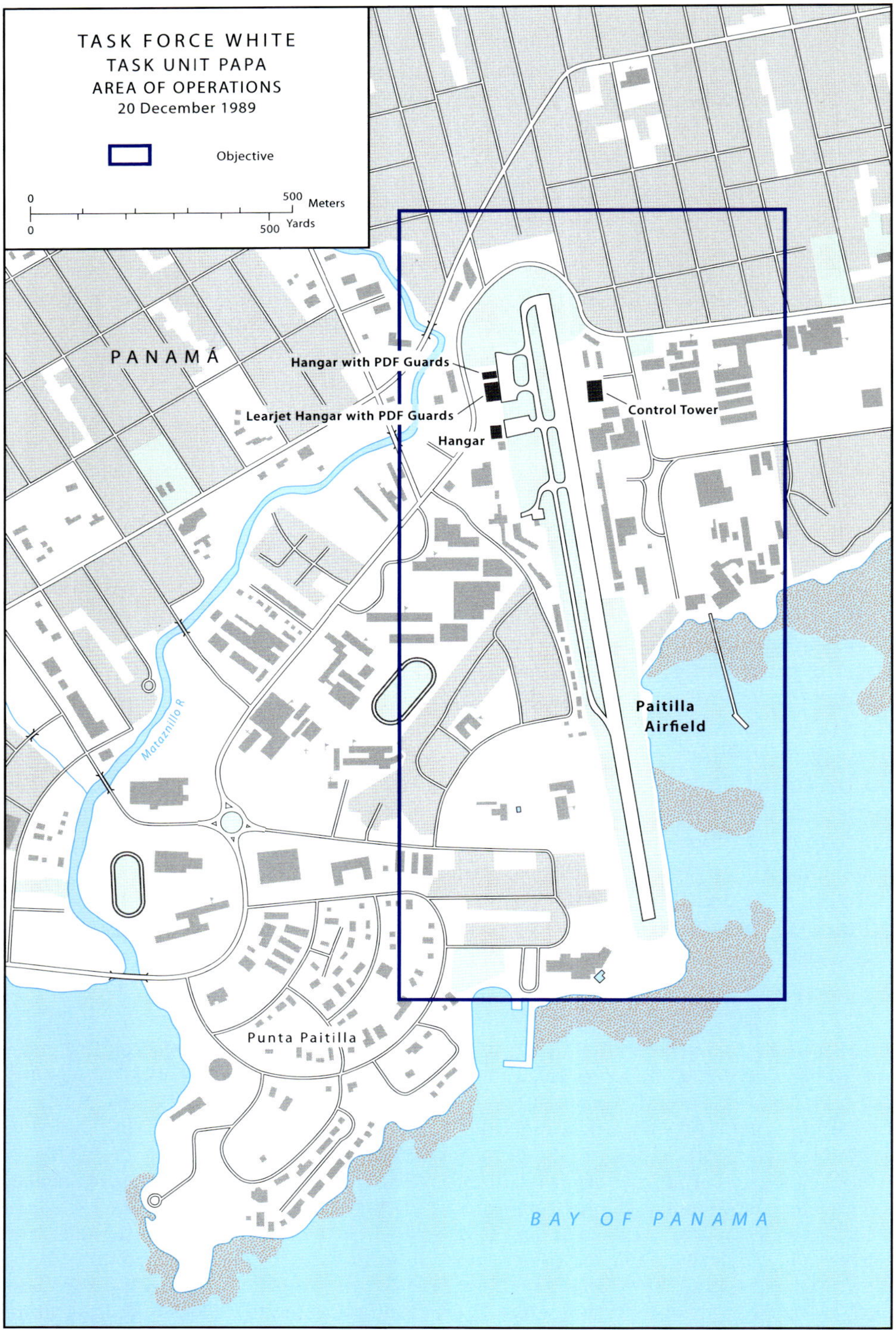
TASK FORCE WHITE
TASK UNIT PAPA
AREA OF OPERATIONS
20 December 1989
Objective
0
500 Meters
0
500 Yards
PANAMÁ
Hangar with PDF Guards
Learjet Hangar with PDF Guards
Hangar
Control Tower
Matasnillo R.
Paitilla
Airfield
Punta Paitilla
BAY OF PANAMA

Map 4

Before the SEALs could launch their assault, Toohey received yet another message on his radio, this one informing him that three enemy V300 Cadillac Gage armored reconnaissance vehicles were heading full speed toward the airport. If true, SEALs would be vulnerable to the vehicles' 90-mm. cannon. Toohey thus had little choice but to order Golf One to move to the thoroughfare and take up an ambush position. What happened next is not entirely clear. As the squad stood up, a SEAL from Bravo platoon, which was closing on the hangars, apparently fired at a Panamanian guard who was taking aim at the line of men now plainly visible in front of him. Immediately, one or more enemy soldiers opened fire. The scene that followed was horrific, as bullets from AK47s raked the nine SEALs of Golf One. Eight men went down within seconds, one killed, one dying, and six more wounded, some from ricochets off the runway.

Seeing the carnage to its right, Golf Two poured small-arms rounds into the middle and northernmost hangars. As the defenders took cover, two SEALs from the squad ran over to help their wounded comrades; in a second enemy volley, one of them was killed. Toohey then ordered his other two platoons to move north and reinforce Golf. Bravo, commanded by Lt. (jg) John Connors, arrived first and, in a 45-second fusillade, directed more fire into the first two hangars. Delta platoon soon joined in, and, in the ensuing firefight, any enemy defenders that had not yet fled were killed or wounded. But they had again exacted a high price, wounding two more SEALs, one critically, and killing Connors after he had exposed himself in an attempt to fire a grenade into the hangar with the Learjet. The battle at Paitilla airfield lasted little more than seven minutes. Three SEALs had been killed outright, another died soon thereafter, and eight others survived moderate to severe wounds. A captured Panamanian defender claimed that the men in the hangar, far from being rent-a-cops, were highly trained PDF special operators. The Americans, he said, had killed three of their number, while eight others had been wounded and evacuated. As for the V300 vehicles heading toward Paitilla, they ended up going past the airfield en route to downtown Panama City.[33]

At no time during the shootout did the AC–130H flying overhead engage the enemy. Soon after the SEALs came ashore at Paitilla, the Air Force officer whose job it was to coordinate with the gunship had set up his communications equipment at Toohey's command post but had not been able to make contact with the plane. The problem continued during most of the battle, with communications finally being established just as the firing was tapering off. By then, the Spectre's powerful and precise weapons were no longer needed. (Later reports speculated that the plane's howitzer, if there had been an attempt to fire it, might not have been in working condition.)[34]

---

[33] The SEALs killed in action at Paitilla airfield were Lt. John Connors, Chief Engineman Donald McFaul, Torpedoman's Mate 2d Class Isaac Rodriguez, and Boatswain's Mate 1st Class Christopher Tilghman. Connors received the Silver Star posthumously. For PDF casualties, see *Los Angeles Times*, 27 Jan 1990.

[34] After the fact, a debate developed over what impact the AC–130H might have had on the battle at Paitilla if the ground element had been able to contact it. Some have argued that the gunship's weapons would have made quick work of the PDF defenses. Others, apparently including the AC–130's fire-control officer, questioned whether the Spectre would have been able

Soon after the first SEALs were hit, Toohey had sent an urgent message requesting medical evacuation (MEDEVAC) helicopters to extract the wounded. The message reached the 214th Medical Detachment at Howard Air Force Base at 0115. There, medics in one of two Black Hawk crews dedicated to the Special Operations Forces in JUST CAUSE took off at 0130, arriving at Paitilla eight miles away around 0200. After ten minutes on the ground, during which time the eight wounded SEALs were squeezed aboard, the helicopter departed and flew at maximum speed to the Joint Casualty Collection Point at Howard, arriving there at 0225. All of the wounded survived the flight.[35]

The fact that fifteen minutes elapsed before the MEDEVAC helicopter departed Howard after receiving Task Unit Papa's urgent request for help became another point of controversy after the operation, as did the fact that it took the pilots almost half an hour to fly the short distance from the Air Force base to Paitilla. In their defense, the pilots later stated that the initial message did not include coordinates for the airfield's location, which they did not know, nor did it apprise them of the radio frequency employed by the task unit's communications equipment. A power outage in the medical detachment's operations section prevented them from acquiring this essential information. Once they boarded their Black Hawk, the dense air traffic at Howard further delayed their departure until they were finally told they could take off "at their own risk." As for the length of the flight to the battle area, several reasons were advanced by way of explanation. These included the crew members' lack of familiarity with the terrain around Paitilla; the lack of communications with Task Unit Papa; the blacked-out conditions under which they had to fly, which increased the risk of a collision with other friendly aircraft; and the need to fly a "circuitous route" over the ocean because of the combat operations taking place at Flamenco Island and Fort Amador. When the helicopter arrived at Paitilla, the pilots "landed next to a burning building and were guided in by a strobe light which they hoped was friendly."[36]

Once the wounded were en route to Howard, the remainder of Task Unit Papa began setting up "a more defendable perimeter on the southeast side of the airfield." A reaction platoon from Rodman arrived sometime

to use its weapons at all, given the proximity of the SEALs to the PDF defenders in the hangars. For an overview of the debate, see Donnelly, Roth, and Baker, *Operation Just Cause*, p. 119, and McConnell, *Just Cause*, pp. 64–65, 72–73. For more information on the problem of communicating with the AC–130, see Interv, Partin with Williams, n.d.; Reeves, "Navy SEALs," pp. 95–96. Reeves, p. 95, indicates that, "to comply with last minute changes to the communications plan," the Spectre had switched frequencies, in the process failing to notify the air controller with Task Unit Papa.

[35] Interv, SOCOM Inspector General Team with Medical Operation—Punta Paitilla Airfield, 14 Feb 1990, Howard Air Force Base, Panama. Being interviewed were the two medical evacuation (MEDEVAC) pilots who flew the Paitilla mission, CWO2 Robert Eugene Lowe and WO1 Michael Paul Premo.

[36] Ibid. For a systematic account of the medical treatment received by the wounded SEALs, see Lt. Cdr. James J. Mucciarone, Col. Craig H. Llewellyn, and Lt. Col. John M. Wightman, "Tactical Combat Casualty Care in the Assault on Punta Paitilla Airfield," *Journal of Special Operations Medicine* 8, no.1 (Winter 2008): 106–10.

Noriega's disabled Learjet

after 0315, and, by 0330, Paitilla was considered secured. At dawn, "a patrol conducted a reconnaissance of the hangars," while, according to plan, "other SEALs dragged airplanes onto the runway to block its use." During the fighting, an AT4 rocket had rendered Noriega's Learjet inoperable, thus accomplishing the ground force's mission, although at a high price. The next day, the SEALs departed Paitilla, to be replaced by a ranger company.[37]

Both at the outset of the planning process and following the invasion, there were officers in the special operations community who argued that the mission to seize Paitilla airfield should have been assigned to a ranger company or battalion. For starters, the Rangers trained for such missions, the SEALs did not. Furthermore, SEALs were used to operating in small groups of from two to sixteen men, not in actions that required the coordinated efforts of three platoons. At one point, a SEAL officer had recommended three alternatives to the larger force, with each alternative involving only two or eight men, but Sandoz had rejected them. As one assessment concluded, "The middle ground was taken in the form of a unit too large for a clandestine strike, but not large enough to handle a straightforward direct action mission." Both General Lindsay and General Stiner disagreed, maintaining that the assignment called for blocking a runway that ran along a waterline, not for seizing an airfield. That being the case, the use of the SEALs was appropriate. Whatever the merits of either side of the argument, no party to the controversy could deny that Task Unit Papa's casualties "made that brief action the deadliest for the SEALs in the nearly thirty years since the Sea, Air, Land teams were formed." They also contributed to the conclusion that among "the dozens of actions

[37] U.S. Special Operations Command, *History, 1987–2007*, p. 39.

of Just Cause, the Paitilla mission may be the most controversial among the military," with everything from planning, leadership, weaponry, and tactics being called into question.[38]

The Rescue of Kurt Muse

If the Paitilla mission generated the most controversy, another special operations undertaking, the rescue of American citizen Kurt Muse from a PDF prison, the Carcel Modelo, became one of the most publicized. The Panama Defense Forces had arrested Muse in April for "subversive" activities and accused him of crimes against state security, charges that, under the Panama Canal Treaty, did not require the Panamanian government to turn him over to the United States. After Muse admitted that, in advance of the May presidential elections in Panama, he had set up a clandestine communications network to jam regime-operated frequencies and broadcast his own messages against Noriega's hand-picked candidates, he was placed in the Carcel Modelo, across the street from the main PDF headquarters, the Comandancia, in Panama City. Although allowed regular visits from an officer, doctor, and lawyer representing U.S. Forces, he had little hope for a quick release. Nor was his experience pleasant. As he languished for months in a crude 8-foot-by-12-foot cell, he witnessed, both visually and audibly, the torture of fellow prisoners, especially after the incarceration of opposition members during the postelection violence in May. He was also informed after the 3 October coup attempt that, should the United States invade Panama or in any way try to rescue him militarily, he would be shot on the spot. A guard stood watch outside his cell, ready to make good on the threat.[39]

Muse's arrest and detention garnered more attention within the Bush administration than would normally have been the case. Muse belonged to a group of conspirators—the others having escaped after hearing of his arrest—that at one point in its underground activities had agreed to accept support from the U.S. Central Intelligence Agency. For this reason, CIA Director

---

[38] First quote from Bahmanyar with Osman, *SEALs*, p. 84. Second quote from Kelly, *Brave Men, Dark Waters*, pp. 2–3. For Lindsay's and Stiner's position, the debate in general, and the third quote, see Donnelly, Roth, and Baker, *Operation Just Cause*, pp. 119–20. See also The SEAL assault on Patilla [*sic*] Airfield during Operation Just Cause, n.d.; *Los Angeles Times*, 27 Jan 1990; James W. Collins, "Blue and Purple: Optimizing the Command and Control of Forward Deployed Naval Special Warfare" (Master's thesis, U.S. Army Command and General Staff College, 1997), pp. 71–74. Besides the issues raised in the text, critics also questioned the initial plans; the last-minute change in command and control arrangements; the tactics employed in which the one SEAL squad stood up in line, thus providing an ideal target for the PDF guards; and the weapons carried by the SEALs, in the sense that the M203 grenade launchers did not deliver enough punch to overcome a well-entrenched enemy.

[39] Muse's account of his clandestine activities and his incarceration are in Kurt Muse and John Gilstrap, *Six Minutes to Freedom* (New York: Citadel Press, 2006). A summary of these events placed into the broader context of the U.S.-Panama crisis is in Yates, *U.S. Military Intervention in Panama, June 1987–December 1989*, pp. 158–61. See also Thomas B. Hunter, "Operation Acid Gambit: The Rescue of Kurt Muse," http://www.specialoperations.com/Operations/Just_Cause/Acid_Gambit/default2.html.

William H. Webster reportedly "pressed [Secretary of Defense] Cheney to have the military draw up a rescue plan for Muse that would be ready for execution on short notice." That the president, himself a former director of central intelligence, also expressed concern about Muse's situation, especially after he received a personal letter from the prisoner smuggled out of the Carcel Modelo

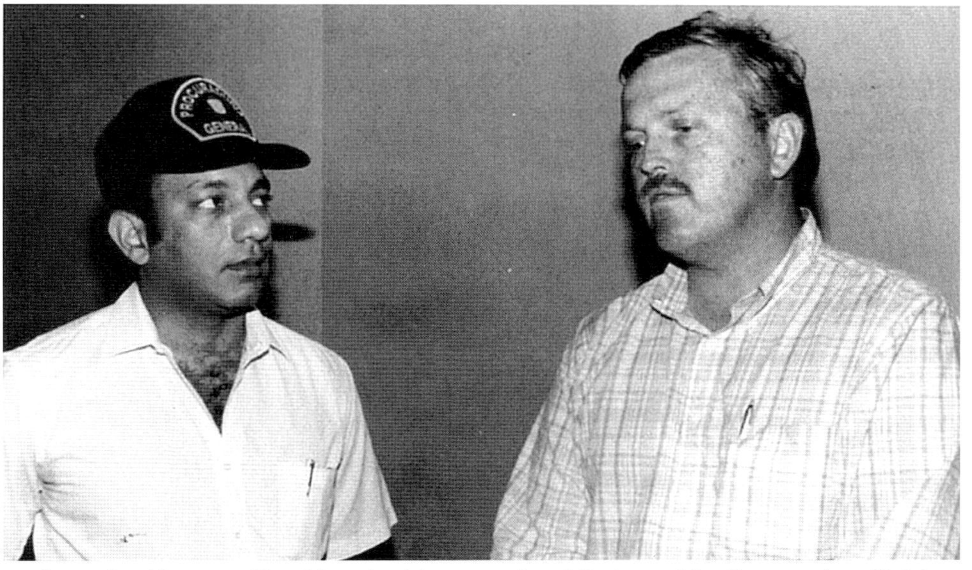

Soon after his arrest, Kurt Muse (*right*) appears in public, here with a Panamanian official.

The Carcel Modelo compound, with the jail at lower left

in a book, lent weight to Webster's entreaties. As a result, staff personnel in special operations began planning a rescue mission that would involve one of the country's most elite counterterrorist units. Code-named Acid Gambit, the plan could be executed if the president judged Muse's safety to be in jeopardy. By the time Thurman convened a Blue Spoon planning session with Stiner and others in Panama after the failed coup, Acid Gambit had acquired the status of an "unwritten but high-priority mission from Washington." On 16 October, Maj. Gen. Gary Luck, then the special operations commander, gave a half-hour briefing to Bush on the contingency plan, using a scale model of the prison to cover every phase of the rescue operation in minute detail. At the time of the briefing, Acid Gambit was still regarded as a contingency that could be executed independently or as a part of Blue Spoon. But Secretary Cheney, among others, drew the logical conclusion that the execution of the rescue mission by itself, even if successful, could result in dire consequences because Noriega might make other American nationals living in Panama pay dearly for the humiliation he would suffer. As for whether the rescue mission should be executed simultaneously with Blue Spoon, Thurman believed that it should, stipulating that freeing Muse would precede the other planned actions. "It became clear to me," Thurman would say, "that he'd be dead if we didn't get him fast." Thus, when Thurman, Stiner, and Luck sought the Joint Chiefs' approval of JTF-South Operation Plan 90–2 in their 3 November briefings at the Pentagon, they presented Acid Gambit as a mission that would most likely be executed as an integral part of Blue Spoon.[40]

At the tactical level, only one of three squadrons within the Army's elite special missions unit would be needed for the operation; which one it would be depended on the timing of a presidential order to execute Blue Spoon. Since no one knew if or when that would occur, all three squadrons visited Panama separately, in part to reconnoiter the prison and its surroundings and in part to meet with the commander and key staff officers of the mechanized battalion over which JTF-Panama had operational control. The coordination was essential because that battalion would form the core of Task Force Gator, a task-organized unit that, under the latest versions of Blue Spoon, would mount an H-hour assault against the Comandancia compound adjacent to the Carcel Modelo (*Map 5*). Of necessity, the special operation to free Muse would have to occur simultaneously with the attack on the PDF headquarters. To have the conventional attack begin before the rescue mission would cost the commandos the element of surprise and, more than likely, Muse his life. To begin Acid Gambit before the conventional attack would leave the commandos vulnerable to the firepower of Panamanian forces defending the Comandancia. For both attacks to begin at the same time and to achieve maximum effect, arrangements needed to be made in advance so that the two forces would not be working at cross purposes. But there was more involved than just staying out of each other's way. If the rescue mission went badly, the

[40] First quote from Woodward, *The Commanders*, pp. 91, 135–39. Second quote from Clancy with Stiner and Koltz, *Shadow Warriors,* p. 314. Third quote from Interv, Partin with Thurman, 12 Apr 1990.

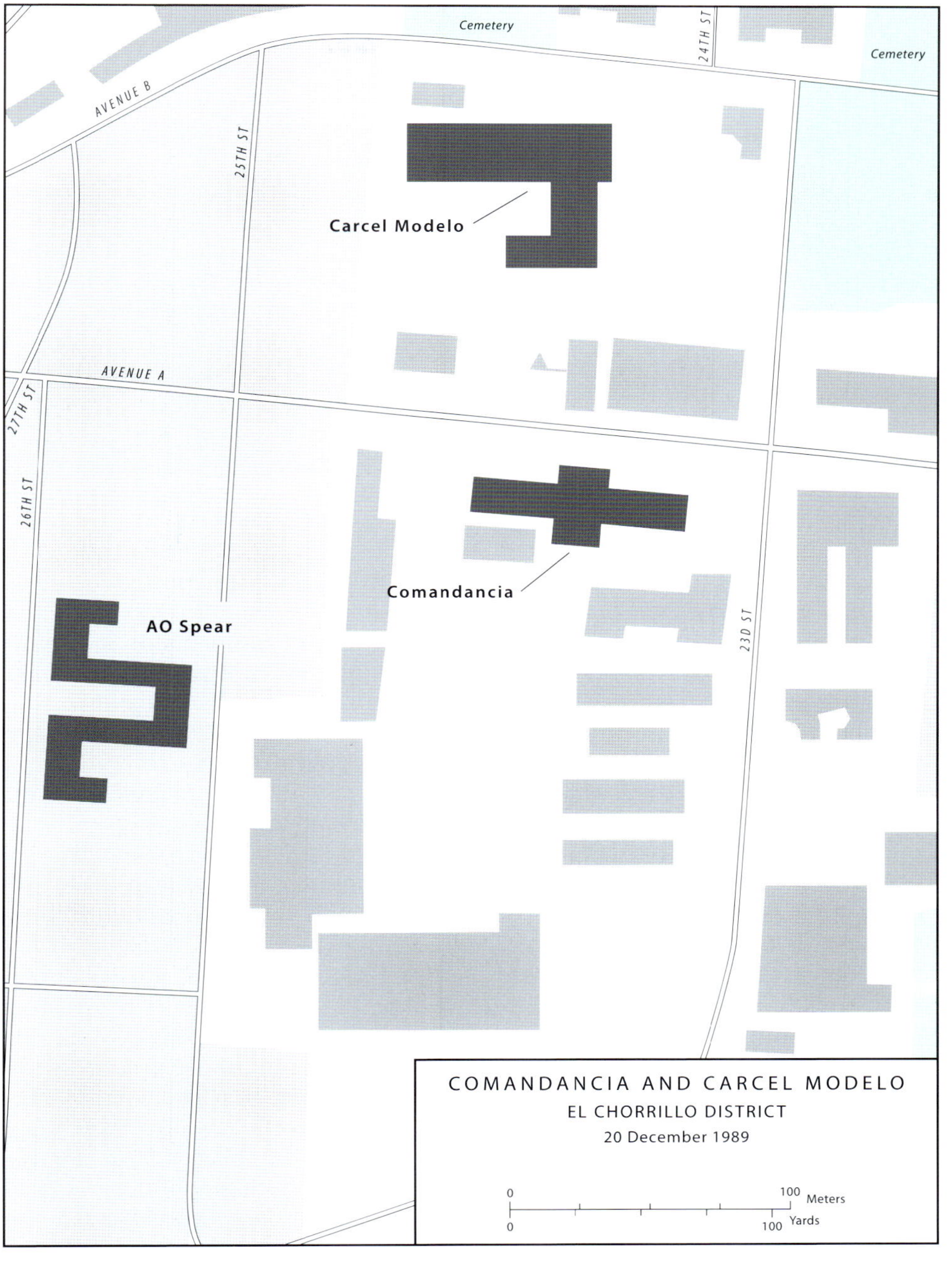

Cemetery
Cemetery
24TH ST
AVENUE B
25TH ST
Carcel Modelo
AVENUE A
27TH ST
26TH ST
Comandancia
AO Spear
23D ST
COMANDANCIA AND CARCEL MODELO
EL CHORRILLO DISTRICT
20 December 1989
0
100 Meters
0
100 Yards

Map 5

MH–6 Little Bird

commandos might require support from Task Force Gator, probably in the form of several M113 armored personnel carriers.[41]

Along with members of the commando squadrons, elements of the 160th Aviation Group, known as the Night Stalkers, arrived in Panama from Fort Campbell, Kentucky, in early October to get a better assessment of the role they would play in a rescue mission. The 160th had a company of MH–6 Little Bird helicopters that could be modified with outboard benches to hold up to three commandos on each side. There was also a company of AH–6 Little Bird gunships that could escort and provide covering fire for the troop carriers. Finally, the unit had Black Hawk helicopters that could be used to ferry troops, furnish protective fires, or conduct armed reconnaissance. Once in Panama, 160th pilots flew training missions, sometimes with the commandos if they happened to be in the country, and gathered all the visual intelligence they could. One means of accomplishing the latter task was to board helicopters making routine medical flights to Gorgas Army Hospital near the Carcel Modelo. As one author has noted, "Medevac helos bringing 'injured' troops to the hospital were often carrying pilots and commandos scribbling furiously on notepads." The deception paid off, with the resulting intelligence revealing, among other things, that a communications antenna on the roof of the Carcel Modelo would prohibit any Black Hawks from landing there. An alternative was to have the commandos fast-rope onto the roof, but that would leave them exposed to gunfire from enemy troops nearby. The responsibility for getting the assault force to the target therefore fell to the unarmed and smaller MH–6s. Once that decision was made, the Little Bird crews began "an intense training

[41] Intervs, Wright with Lt Col James W. Reed, U.S. Army, 6 Jan 1990, Fort Clayton, Panama; with Maj Terry Freeze, U.S. Army, 6 Jan 1990, Fort Clayton, Panama.

regimen" with the commandos, "working long nights assaulting an elementary school at Howard that closely resembled the prison structure."[42]

In early December, the pilots, crews, and support personnel from the 160th returned from Panama to the United States in time to participate with other Special Operations Forces in the full rehearsal of the JSOTF portion of the Blue Spoon plan. As was the case with Task Unit Papa, the Acid Gambit rehearsal took place at Eglin Air Force Base, where a three-story mock-up of the four-story Carcel Modelo had been constructed. (A four-story replica was not essential since the only course of action called for the assault team to move from the roof of the real prison downward to Muse's cell on the second floor.) The specific features of the mock-up, according to one account, were continuously being "updated by reports from those military personnel who were permitted to visit Muse in his cell."[43]

The plan for the rescue mission, though not complex, did require quick action and precise timing. Four MH–6s covered by AH–6s and Black Hawks would fly the commandos to the target, placing them on the roof. The aircraft would then fly to a concrete-covered water reservoir at Quarry Heights and await word to return and pick up the assault team. Meanwhile, as two additional AH–6s, two AC–130 gunships from the Air Force's 1st Special Operations Wing, and SOF snipers engaged enemy forces in the area, one group of commandos would blow open a rooftop door into the prison and proceed downstairs to Muse's cell. In their descent, they would deal with any PDF resistance by whatever means they deemed appropriate. Another group would stay on the roof and knock down the communications antenna, thus making room for a Black Hawk to land if needed. Once Muse was freed, his rescuers would retrace their steps back to the roof, board the returning MH–6s, and spirit their "precious cargo" to a predetermined rendezvous point.[44]

When, just a few days after the rehearsal, Bush ordered military action, the assault force and aviators returned to Panama. Folded up and loaded aboard C–5 Galaxy transport planes at Fort Campbell, the MH–6s, AH–6s, and Black Hawks made the return trip as well. In Panama, on Monday morning, Lt. Col. James W. Reed, who as commander of the mechanized battalion was the Task Force Gator commander as well, met with special operations personnel at Howard Air Force Base to update and coordinate plans for the simultaneous assaults on the Comandancia and Carcel Modelo. As Reed knew, for the duration of the rescue mission, Task Force Gator, composed of conventional units, would come under the operational control of Downing's Joint Special Operations Task Force. Reed's highest priority would still be the battle for the Comandancia, but one consequence of being under Downing was the directive he received to deploy three of his M113 armored personnel carriers—designated Team King after the Army

[42] Michael J. Durant and Steven Hartov with Lt. Col. Robert L. Johnson (Ret.), *The Night Stalkers* (New York: G. P. Putnam's Sons, 2006), pp. 124–25.

[43] Woodward, *The Commanders*, p. 145; Durant and Hartov with Johnson, *The Night Stalkers*, p. 125. Quote from Hunter, "Operation Acid Gambit."

[44] Hunter, "Operation Acid Gambit"; Durant and Hartov with Johnson, *The Night Stalkers*, pp. 124–59 passim.

lieutenant who was in charge of them—near the Carcel Modelo in case they were needed to help evacuate the assault force. With that mission in mind, the M113s were to be positioned with their ramps down. There would only be a driver and vehicle commander in each, and a special operations liaison officer in one of the three.[45]

On Monday and Tuesday, when General Stiner talked with commanders and staff officers cleared to be briefed on the full range of JUST CAUSE operations, many in his audience heard for the first time about the rescue mission. Most paid it little heed because it did not conflict or impinge upon their own operational assignments and responsibilities. Some, though, did wonder about the wisdom of the undertaking. General Cisneros was one. In his view, the decision to attack the Comandancia instead of simply isolating it (as originally planned under General Woerner) was in part a consequence of the requirement to rescue Muse. While Cisneros did not oppose the mission, he lamented that the Comandancia operation would now require troops who could have been used in downtown Panama City to maintain law and order, prevent looting, and possibly locate Noriega. "The battle," Cisneros said at the time—meaning the critical battle that would determine the long-term success of JUST CAUSE—was "going to be in the city." Furthermore, he reasoned, focusing on the Comandancia "because we had one American in jail" ignored that "we have also a lot of Americans that live downtown," who, without the proper number of forces to protect them, would find their lives placed in jeopardy. "In fact, one of them was kidnapped and one was killed," he later related, and "it was because we had nobody downtown."[46]

For the Carcel Modelo assault force and the special operations pilots and crews who would transport and protect them, the two days leading up to JUST CAUSE afforded them a last chance to go over the plans, receive updated intelligence briefings, and, in the case of pre-positioned snipers at Quarry Heights as well as several commandos walking the area in civilian clothes, directly observe the prison and headquarters compound adjacent to it. The units also reviewed their execution checklists, procedures, rules of engagement, and communications. As for Kurt Muse, the central character in all this, on the morning of 19 December, around 1100, he received a visit from Lt. Col. Robert Perry, the Southern Command's treaty affairs officer and a man Muse had not seen since being arrested nine months earlier. The meeting took place in the visitors' area of the Carcel Modelo, and several journalists, prison guards, and PDF officers were in close proximity. Soon after Muse and Perry began talking, the noise of U.S. helicopters circling the prison filled the room. Then, raising his voice to be heard, Perry declared dramatically, "I want you to know that if anything happens to you," alluding to Noriega's public threats to execute Muse, "no one will walk out of this prison alive." The lieutenant

---

[45] For the C–5A airlift of the 160th's helicopters, see Durant and Hartov with Johnson, *The Night Stalkers*, p. 126. For Reed's meetings with special operations personnel at Howard, see Intervs, Wright with Reed, 6 Jan 1990; with Freeze, 6 Jan 1990; and Chapter 6 of this volume.

[46] Interv, Partin with Maj Gen Marc A. Cisneros, U.S. Army, 2 Mar 1990, Fort Clayton, Panama.

colonel then rose, shook Muse's hand, and left the room. The message he had delivered, according to Muse, had not been lost on its intended audience.[47]

That night, shortly before H-hour, many of the aircraft supporting U.S. special operations in JUST CAUSE began assembling "into their various assault elements on the blacked-out runways of Howard." When word reached them that H-hour had been advanced by fifteen minutes, the commandos strapped themselves onto the outboard benches of the MH–6s, and the air flotilla took off for the Carcel Modelo. The formation for the ten-minute flight included the four troop carriers and four AH–6s, two of which peeled off as the target came into sight and headed for their attack positions above the Comandancia. According to plan, the first two MH–6s landed on the prison roof, off-loaded their passengers, and took off. As the second pair of Little Birds prepared to land, the commandos sitting on the benches opened fire on Panamanian soldiers crossing a walkway between the Comandancia and the Carcel Modelo. The sniper team positioned at Quarry Heights also picked off targets, so that bullets from both sources resulted in several PDF guards being killed or wounded. By this time, the helicopters were under fire and the attack on the PDF headquarters was about to begin. Once all members of the assault team were on the roof, they blew open the door as planned, and several of them descended the stairs, making their way to Muse's second-floor cell. On the way down, they encountered one PDF guard, who quickly surrendered, was flex-cuffed, and left behind on the floor. When the commandos reached Muse, the guard outside his cell had fled and locked himself inside an office close by. As two of the rescue team blew the door's lock and killed the guard, two others, one with a twelve-gauge shotgun, shot the lock off Muse's cell. Once they had Muse in their possession, they adorned him with a Kevlar helmet and body armor, hustled him up to the roof, and called for the MH–6s to return. When the first Little Bird touched down for the pickup, Muse was stuffed inside the tiny craft, while six of his rescuers settled onto the benches. All this had been accomplished in just six minutes following the team's landing on the prison.[48]

So far, the operation had proceeded exactly as planned and with no U.S. casualties. Muse had been rescued and, surrounded by commandos, was on a helicopter ready to take off from the prison's roof. Another MH–6 with six commandos was also preparing to take off, and two more Little Birds were seconds away from landing and extracting the remainder of the assault force. At that precise moment, when all seemed to be going so well, the fog and friction of war intruded. Pre–JUST CAUSE intelligence briefings indicated there were no power lines near the roof of the prison. That information turned out to be wrong. As the Little Bird with Muse took off under intense fire, one of the two pilots noticed power lines right in front of him. The other pilot

[47] Hunter, "Operation Acid Gambit"; Durant and Hartov with Johnson, *The Night Stalkers,* pp. 126–27. For Muse's account of the meeting with Perry and Perry's quote, see Muse and Gilstrap, *Six Minutes to Freedom*, p. 266, and see also pp. 262–65, 267.

[48] The account of the rescue mission thus far is based on the following sources: Hunter, "Operation Acid Gambit." Quote from Durant and Hartov with Johnson, *The Night Stalkers*, p. 128, and see also pp. 129–31, 148–49. Muse and Gilstrap, *Six Minutes to Freedom*, pp. 277–89. Durant and Hartov with Johnson provide a detailed account of the helicopter operations, while Muse and Gilstrap do the same for the rescue team's actions.

The Little Bird that crashed during the Muse rescue mission

pulled up over the wires, but, because of the stress the maneuver placed on the overloaded craft, the helicopter quickly lost altitude. Enemy bullets did the rest, as the MH–6 took direct hits. Despite the pilot's efforts to keep the Little Bird airborne, it crashed into the street about a block and a half from the prison. At almost the same time, as the second pair of MH–6s approached the Carcel Modelo's roof, an enemy round pierced the window of one, severely wounding one of the pilots. The helicopter still managed to land, but, when it took off, only five, not six, commandos were on the benches. That left seven team members for the last Little Bird to extract, but they proved too much of a load. Three of the seven, including the commander of the team, jumped back onto the roof and, while the helicopter departed, continued to shoot at enemy positions around them. Fortunately for the stranded men, the rescue plan called for a Black Hawk to be on-call in the area, and the pilot, once apprised of the situation, headed for the prison. With the communications antenna having been removed, he was able to land the helicopter, its gunners firing the whole time, pick up the last three commandos, and head back to Howard.[49]

Reports that the helicopter with Muse aboard had crashed quickly made their way up the chain of command. When word reached Downing's headquarters at Howard, "everybody just kind of sucked in a breath," according to the general. "We were all just sick to our stomachs." Several

[49] Durant and Hartov with Johnson, *The Night Stalkers*, pp. 150–53; Hunter, "Operation Acid Gambit"; Muse and Gilstrap, *Six Minutes to Freedom*, pp. 292–94. In the account by Muse and Gilstrap, the Little Bird carrying Muse crashed solely as the result of enemy fire. In the versions presented by Hunter and by Durant and Hartov with Johnson, the unanticipated presence of the utility wire was a contributing factor to the crash. The authors of both accounts talked with the commandos and air crews who participated in the mission.

minutes later, Downing received an index card containing the names of those aboard the downed MH–6. "I just thought I had read the death list," he revealed afterward. From Howard, the bad news worked its way up to Stiner's and Thurman's respective headquarters, with the latter informing General Powell in Washington. According to Bob Woodward, "Disappointment was written all over Powell's face. Muse was the guy they were going to rescue for the President, and now his helicopter had gone down." The assumption in Washington was the same as in Panama: Muse was most likely dead.[50]

To the contrary, there had been no fatalities at the crash site, although everyone, with the exception of Muse, had been injured by the impact, wounded by enemy fire, or both, some severely. With difficulty, the commandos who were mobile moved Muse to the safety of an apartment building, established a defensive perimeter around an abandoned jeep, and waited. So that friendly forces could locate the position, the team activated an infrared strobe light. A Black Hawk pilot flying overhead saw the light, signaled the commandos by "wiggling" his fuel pods, and transmitted the information back to headquarters. Team King, with its three M113 armored personnel carriers, was waiting in the vicinity for just such a development. The team leader had seen the Little Bird go down and, once informed of its location, moved out to extract its crew and passengers. Arriving about fifteen minutes after the crash, he got the mangled troops and Muse aboard his vehicles and headed toward an intermediate rendezvous point between two schools in the Balboa district near the Comandancia. From there, Muse and the more severely wounded commandos were taken by helicopter to Howard.[51]

As with reports of the crash, the news that Muse was alive and safe made its way up each link of the chain from Downing in Hangar 3 at Howard to Washington. "God, did we feel good," Downing related. The report also changed the mood in the Pentagon, where Powell received the good news.[52] Yet, exhilarating as it was, Muse's rescue could not compensate for the SEAL fatalities at Paitilla. On balance, however, the H-hour missions performed by Special Operations Forces had all been accomplished, though not without cost. There was still much to do, however. By the time the canal had been shut down, the airfield at Paitilla secured, Noriega's patrol boat destroyed, and Muse rescued, Operation JUST CAUSE was still in its initial stages. As the commandos who rescued Muse left the Carcel Modelo behind them, the night-long battle for the Comandancia across the street, the centerpiece of the invasion, was still only minutes old.

---

[50] First, second, and third quotes from Interv, Partin with Downing, 6 Jan 1990. Interv, Wright with Stiner, 2, 7, 27 Mar and 11 Jun 1990. Last quote from Woodward, *The Commanders*, p. 184.

[51] Platoon/Headquarters and Headquarters Company (HHC), "Team King," in Nemmers, *United States Army South Staff Ride,* pp. 57–58; Hunter, "Operation Acid Gambit"; Durant and Hartov with Johnson, *The Night Stalkers*, pp. 153–54; Muse and Gilstrap, *Six Minutes to Freedom*, pp. 294–307.

[52] Quote from Interv, Partin with Downing, 6 Jan 1990. Woodward, *The Commanders,* p. 185.

4

Task Force Gator and the Comandancia

☆ ☆ ☆

The detonation of the entry charge that allowed the assault force on the roof of the Carcel Modelo access to Kurt Muse also served as a signal to launch the U.S. attack on the Comandancia. One participant called the battle that followed "the toughest mission in Panama." It was certainly one of the most complex and involved some of the fiercest fighting.[1]

Plans and Preparations

Back in March and April 1988, a key sentence in the first contingency plans for U.S. offensive operations in Panama targeted "key PDF leadership, command and control facilities, military and civilian airfields, and other critical nodes within the Republic of Panama."[2] As the "Pentagon" of the Panama Defense Forces, the Comandancia clearly met the first criterion, and, as was the case with most offensive operations planned early in the crisis, the mission for isolating the facility and, if necessary, neutralizing enemy troops located there was assigned to Special Operations Forces, specifically U.S. Army Rangers. In the concept of operations that evolved over the succeeding months, a ranger battalion under the operational control of the Joint Special Operations Task Force would fly from the United States to Howard Air Force Base, board waiting helicopters, and launch an air assault on the Comandancia complex. Once the mission had been accomplished, the Rangers would turn over security of the site to conventional units under JTF-Panama. In terms of command and control, both the Joint Special Operations Task Force and JTF-Panama would answer directly to the commander in chief, U.S. Southern Command.[3]

Plans for a ranger air assault on the Comandancia appeared in successive drafts of Blue Spoon supporting plans well into 1989. Then, following the

---

[1] Donnelly, Roth, and Baker, *Operation Just Cause*, p. 134. Quote from Flanagan, *Battle for Panama*, p. 101.

[2] Msg, CINCSO to JCS, 042255Z Mar 1988.

[3] Interv, author with Lt Col Robert Pantier, U.S. Army, 13 Jun 1990, Fort Clayton, Panama. For most of 1988 and all of 1989, Pantier was the senior planner for U.S. Army, South, and JTF-Panama.

failed coup d'état against Noriega on 3 October, a number of considerations caused U.S. planners to reexamine portions of the operation order that pertained to the Pacific Ocean side of the country. To begin with, the role that certain Panamanian forces had played in rescuing Noriega from his captors prompted a new look at what measures needed to be taken against units such as the 6th and 7th Infantry Companies and Battalion 2000. Also in need of evaluation were changes Noriega made within the Panama Defense Forces after his narrow escape. Some units, such as the UESAT, were relocated, while some of the command and control functions formerly centralized at the Comandancia were dispersed to other headquarters. As a clearer picture of the post–3 October Panamanian military began to emerge, certain BLUE SPOON objectives theretofore regarded as peripheral—Rio Hato, Las Tinajitas, and Fort Cimarrón, for example—acquired a much higher priority. The Comandancia retained its status as a priority-one target, but questions arose as to which U.S. force could best be employed to subdue the facility. Some officers argued that the ranger battalion originally designated for the mission could be used more productively elsewhere, while the Comandancia assignment could be entrusted to conventional units.

That idea was a logical extension of the BLUE SPOON fragmentary order, code-named HIGH ANXIETY, drafted earlier in the crisis to cover a no-alert, or reactive, scenario in which the Panama Defense Forces moved without warning against U.S. units and sites. In this contingency, as stated in the 20 June 1989 version of the order, the situation would "warrant an immediate execution" of BLUE SPOON before forces from the United States had arrived in Panama. The burden of holding the line would fall on JTF-Panama, which would conduct "economy-of-force joint offensive operations with in-place forces to neutralize and ultimately disarm the PDF and any other combatants within the Panama City and Colón areas." These operations, in turn, would give "sufficient time for augmenting forces from [the United States] to close in order to continue possible extended operations." Under this and a subsequent version of HIGH ANXIETY, U.S. Air Force elements under JTF-Panama had responsibility for any military action against the Comandancia. Following the coup attempt of 3 October, however, the no-alert Comandancia mission was transferred from the Air Force to Task Force Bayonet, a grouping of ground forces task-organized under JTF-Panama and led by the commander of U.S. Army, South's 193d Infantry Brigade, headquartered at Fort Clayton. According to an 11 October draft of HIGH ANXIETY, Task Force Bayonet shouldered the responsibility for isolating the PDF headquarters. If deemed necessary, the troops could be ordered to seize and secure the facility as well.[4]

Once Task Force Bayonet acquired the Comandancia mission under the no-notice fragmentary order, JSOTF planners saw no reason why that conventional headquarters could not take action against the same target should BLUE SPOON be executed in the way the Pentagon and the Southern Command preferred: as a "deliberate" undertaking in which the U.S. military, in responding with force to a regime provocation, would have ample time to

[4] Ibid.; FRAGOs (HIGH ANXIETY) to JTF-Panama OPORD 7–88 (BLUE SPOON), 20 Jun 1989, 20 Aug 1989, and 11 Oct 1989; Briefing Slides, BLUE SPOON: The Evolution of the Plan for JUST CAUSE, n.d.

marshal the necessary troops and supplies before launching a coordinated attack at a time determined by the White House, not the enemy. After JTF-Panama and the XVIII Airborne Corps agreed to revising the unit assignments, the substitution of Task Force Bayonet for Special Operations Forces at the Comandancia appeared in the late-October revisions of the CINCSO and JTF-South versions of the BLUE SPOON operation orders. The ranger battalion

The Comandancia compound, with the PDF headquarters in the foreground

affected by the change quickly picked up responsibility for operations against Panamanian units at Rio Hato.

The switch meant that Task Force Bayonet, under either the reactive or deliberate contingency, would have to isolate the Comandancia by occupying a number of defensive positions in the target's vicinity. The task force might also have to launch a ground attack to accomplish what the Rangers had planned to do through an air assault: neutralize the Panamanian forces in and around a compound that enclosed the PDF headquarters and over a dozen other buildings, including troop barracks for two enemy companies. This more decisive action, in turn, would require a robust force in which technology would have to overcome what planners calculated to be a friendly-to-enemy troop ratio of one to one—approximately 350 U.S. soldiers to an estimated 390 Panamanian soldiers—at the target, far short of the traditional and doctrinally sound three-to-one ratio of attacker to defender (or the higher four-to-one, even six-to-one, ratio desired in urban operations). In bringing together units that would maximize the assault force's capabilities, Colonel Snell, the 193d Brigade commander, regarded the mechanized battalion placed under Task Force Bayonet during NIMROD DANCER as the logical unit to spearhead an operation against a fortified position in a built-up urban area.

At the time Snell picked up the Comandancia mission, the mechanized battalion in Panama was the 4th Battalion, 6th Infantry, from the parent 5th Infantry Division (Mechanized) at Fort Polk, Louisiana. The unit had rotated into the country on 19 September, replacing the original NIMROD DANCER mechanized force, the 5th Battalion, 6th Infantry, also from the 5th Division. The 4th Battalion deployed with its four line companies and only six dozen or so M113 armored personnel carriers, a shortfall that resulted in each maneuver platoon having only three of the vehicles instead of four. The battalion's fifth company, an antitank unit, did not deploy but provided the Panama-bound force with three tube-launched, optically tracked, wire command-link-guided (TOW) missile vehicles and crews, which were attached to the battalion's scout platoons. Once in Panama, the battalion headquarters, support units, and one of its companies moved to Camp Rousseau (later nicknamed Camp Gator) on the west side of the canal next to Rodman Naval Station. (The camp would also serve as the secret "motor pool" for the M551 Sheridans flown into Panama in November.) Another mechanized company from the 4th was located just north of Rousseau on the Army's Empire Range, a training area. A third, which served as JTF-Panama's quick reaction force, went to the east side at Corozal, and the fourth stood in support near Albrook Air Station. About once a week, the four companies changed these locations with one another.

That the 4th Battalion, as well as the battalion that had preceded it in the NIMROD DANCER rotation, arrived in Panama with M113s instead of newer Bradley fighting vehicles was no accident. The 5th Division was not scheduled to replace the older armored personnel carriers for another year, a timetable that planners for both POST TIME and BLUE SPOON took into account when drawing up their respective troop lists. The M113 was smaller than the Bradley but could hold more people, thus making it a more practical vehicle for negotiating the very narrow streets of Panama City and for evacuating civilians or taking troops into combat. So strongly did the planners favor deployment of the M113s that urgings from the Pentagon during NIMROD DANCER to have Bradley-equipped mechanized battalions sent to Panama met with no success.[5]

The suitability of the M113s to urban operations created a demand for mechanized units at targets other than the Comandancia. Consequently, only two of the 4th Battalion's four line companies, B and D, would mount the attack on the PDF headquarters. Snell needed the battalion's Company A and its vehicles for another mission, while Company C (minus) would serve as Task Force Bayonet's reserve force. As partial compensation for withholding these two companies from a main attack, Snell cross-attached Company C from his Panama-based brigade's 1st Battalion, 508th Infantry (Airborne), to the Comandancia assault force. Company D also received a platoon from the 4th's Company C, as well as a number of engineers for specific missions involving explosives. From the 82d Airborne Division, the force also acquired operational control over the platoon of four M551A1 Sheridan armored reconnaissance vehicles under wraps at Camp Rousseau. When married up with a platoon of

[5] Intervs, Robert K. Wright Jr. with Lt Col James W. Reed, U.S. Army, 6 Jan 1990, Fort Clayton, Panama; with Maj Terry Freeze, U.S. Army, 6 Jan 1990, Fort Clayton, Panama; author with Pantier, 13 Jun 1990; Telephone Intervs, author with 1st Lt Douglas L. Rubin, U.S. Army, 5, 6, 11 Apr 1990.

the Marines' light armored vehicles—LAV–25s—from Task Force Semper Fi's light amphibious infantry company, the Sheridans became part of "Team Armor," commanded by the airborne company commander (and later his replacement) who had flown to Panama as part of the M551 command and control element. Rounding out the ground forces that would take part in the Comandancia operation were two platoons of military police, a platoon of engineers, medics, and some other personnel. Organized as Task Force Gator and led by the mechanized battalion commander, the major conventional units listed for offensive operations at the Comandancia appeared as follows on the eve of Operation JUST CAUSE: Company B, 4th Battalion, 6th Infantry, 5th Infantry Division; Company D (plus), 4th Battalion, 6th Infantry, 5th Infantry Division; Company C, 1st Battalion, 508th Infantry, 193d Infantry Brigade; the M551 platoon, Company C, 3d Battalion, 73d Armor, 82d Airborne Division; 2d Platoon, Company D, 2d Light Amphibious Infantry Battalion, U.S. Marine Corps; Headquarters and Headquarters Company (minus), 4th Battalion, 6th Infantry, 5th Infantry Division; 1st and 3d Platoons, 988th Military Police Company; and 1st Platoon, Company A, 7th Engineer Battalion.[6]

Between mid-October and mid-December, as the order of battle for Task Force Gator was being finalized, commanders and planners were also developing a concept of operations to replace the ranger air assault at the Comandancia. To facilitate the process, the mechanized battalion's Company B and D commanders, Capt. Joseph D. Goss and Capt. Matthew R. Ethridge, respectively, were issued interim top secret clearances that would allow them access to the plans, and, soon thereafter, their platoon leaders received special permission to be brought into the deliberations as well. The same arrangement held true for Capt. Timothy John Flynn, the new commander of Company C, 1st Battalion, 508th Infantry, and his platoon leaders. Throughout the process, officers at all levels made observations and offered suggestions, although, in general, the tactical commanders were restricted to selecting which of their platoons and squads would carry out the specific tasks contained in the evolving plan. The basic blueprint for the assault was in place by late October, with planners continuing to revise it over the next two months. In the midst of all this, Lt. Col. James Reed took command of the mechanized battalion on 1 December, making him Task Force Gator's new commander as well.

Reed was quickly subjected to a ten-day barrage of briefings from his staff and company commanders to acquaint him with the complexities of the plan. In general terms, the newly developed concept of operations called for Task Force Gator, either in response to hostile actions by the Panama Defense Forces or as part of the deliberate execution of the BLUE SPOON operation order, to isolate the Comandancia compound and the adjacent Carcel Modelo in Panama City. If Panamanian forces in the area ignored a broadcasted

---

[6] Nemmers, *United States Army South Staff Ride*, p. 47. A more detailed order of battle for Task Force Gator is in Memo, Lt Col James W. Reed for CG, U.S. Army Forces Command (FORSCOM), and Cdr, Center for Army Lessons Learned (CALL), 27 Mar 1990. Attached to the memo is the JUST CAUSE After Action Report of the 4th Battalion (Mechanized), 6th Infantry. The U.S. Army, South, staff ride cited above provides a chronological timeline for each of the principal units in Task Force Gator, as well as a summary of the operations.

entreaty to surrender, Reed was to order his forces to open fire. The Americans' advantage in firepower, the planners assumed, would inflict a heavy toll on Panamanian defenses and defenders. Once resistance had subsided, U.S. troops were to enter each of the two compounds and clear the buildings therein.[7]

Command and control arrangements would depend on the circumstances under which the operation was executed. If Colonel Reed conducted the attack as a reactive measure under the revised no-alert High Anxiety fragmentary order, Task Force Gator would remain attached to Colonel Snell's Task Force Bayonet, which, in turn, would remain under JTF-Panama's operational control.[8] If, on the other hand, the attack occurred under the deliberate Blue Spoon plan and in conjunction with the special operations raid on the Carcel Modelo to rescue Kurt Muse, Gator would go under the operational control of the Joint Special Operations Task Force for an indeterminate time. This

AC–130 Spectre gunship

second possibility, considered the most desirable and the most likely scenario, meant that AH–6 attack helicopters and at least one AC–130 gunship would be employed over the Comandancia, thus increasing tremendously the firepower that U.S. commanders and planners sought to concentrate on the target.

The deliberate scenario incorporating the Comandancia and Carcel Modelo missions also necessitated coordination between the mechanized battalion and the squadron-size special operations assault force. As noted

---

[7] Intervs, Wright with Col Michael G. Snell, U.S. Army, 1 Jan 1990, Fort Clayton, Panama; with Capt Joseph D. Goss, U.S. Army, 13 Jan 1990, Corozal Barracks Building, Panama (draft transcript); with Capt Matthew R. Ethridge, U.S. Army, 10 Jan 1990, Fort Clayton, Panama; with Freeze, 6 Jan 1990; with Reed, 6 Jan 1990; author with Capt William R. Reagan, Capt Robert Zebrowski, and Capt Timothy John Flynn, all U.S. Army, 21 Jun 1990, Fort Clayton, Panama.

[8] Draft, FRAGO (High Anxiety) to JTF-Panama OPORD 7–88 (Blue Spoon), 11 Oct 1989.

in the previous chapter, the rescue of Muse would fall to one of the three commando squadrons eligible to execute it. Prudence therefore dictated that officers from all three travel at various times to Panama where, working out of Hangar 3 at Howard Air Force Base, they could meet with Task Force Gator personnel—usually the commander and his operations, intelligence, and fire support officers—for the purpose of examining and adjusting the two sets of highly sensitive operation orders. During these sessions, the special operations–conventional interaction was professional but not free from disagreement. As the mechanized battalion's executive officer later explained, the revisions the special operations personnel made to the plan at each meeting often caused the battalion staff to play "a twenty questions game": "Well, why did we do this, how come we did that, . . . we've had this for three months, why are we just now changing it?"[9]

When Reed assumed command of Task Force Gator, he was reluctant to urge major changes to a plan he knew had already received General Stiner's blessings and perhaps General Thurman's as well. Yet, with respect to one provision inserted by the special operations people, the lieutenant colonel listened to the objections of his subordinates and then strongly suggested an adjustment be made. The issue concerned the JSOTF's insistence that, during the attack, a platoon from the mechanized force crash through the southeast gate of the Comandancia in order to set up an overwatch position, designated Point E, inside the compound itself. The position would certainly give the platoon a clearer field of fire at any vehicular movement in or out of the compound or on a street nearby. The problem was that an estimated one hundred Panamanian defenders would likely be occupying a building located right next to the platoon. To Captain Ethridge and the battalion staff, "It was kind of putting our guys out on a limb." Reed agreed and recommended that the overwatch position be established outside the compound, arguing that the .50-caliber machine gun on an M113 was capable of firing over the wall. The recommendation was accepted up the chain of command.[10]

There was a second issue related to Point E on which Reed again asked for changes to the plan. The agreement to relocate the overwatch position outside the compound placed the mechanized platoon farther from the building in question but still within firing range of its Panamanian occupants. Given this ongoing albeit diminished threat, Reed's staff was unhappy that the fire support plan it received had removed those enemy troops as targets for the AC–130s because of a fuel storage point located near the building. According to Reed's executive officer, the planners "were afraid they would hit the fuel tanks on the ground there and set off a big fireball explosion, and do more collateral damage than they were led to believe was necessary." Reed, on the other hand, wanted his men protected and requested that the building be reinstated as a target. The result was a compromise, in which the Air Force agreed to have the AC–130 use its .40-mm. guns rather than its powerful 105-mm. howitzer to fire on the structure.[11]

---

[9] Interv, Wright with Freeze, 6 Jan 1990.

[10] Interv, Wright with Reed, 6 Jan 1990. Quote from Interv, Wright with Freeze, 6 Jan 1990. Interv, Wright with Ethridge, 10 Jan 1990.

[11] Interv, Wright with Freeze, 6 Jan 1990.

Between mid-October and mid-December, while Reed and his predecessor were working with the special operators to formulate and refine Task Force Gator's role in the event of hostilities, the ground units that would actually mount an attack on the Comandancia were preparing and rehearsing for the mission as best they could. Because the task force was already in Panama, it could not construct without detection the elaborate mock-ups available to Blue Spoon forces in the United States. Working in Gator's favor, though, were the maneuvers and exercises that U.S. forces had conducted as a part of Operation Nimrod Dancer to assert American treaty rights in the country. More specifically, in the two months following the October coup attempt, JTF-Panama adapted its Sand Fleas, contingency readiness exercises, and Purple Storms to the evolving plan as it pertained to the Comandancia. Mechanized units in M113s passed closer to the headquarters, in the process becoming familiar with the streets in the vicinity and conducting various timing and communications tests. There were limits on just how close to the PDF complex the armored personnel carriers could get, with Fourth of July Avenue, a main street a few blocks west of the compound—the Panamanians knew it as the Avenue of the Martyrs—representing a boundary that U.S. troops could not cross as part of their movements. To offset this restriction, the Company B and Company D commanders, Captains Goss and Ethridge, donned civilian clothes and, in their own cars, reconnoitered the streets off-limits to U.S. military vehicles. During their forays, they looked for routes whose use by their mounted soldiers would be permitted during routine exercises.[12]

The nonmechanized units in Task Force Gator did not enjoy this kind of proximity to the Comandancia. In the case of Company C, 1st Battalion, 508th Infantry, located adjacent to Howard Air Force Base at Fort Kobbe, Captain Flynn was not able to alter the routes of his Sand Fleas and readiness exercises without jeopardizing operations security. Prior to 3 October, his company's Blue Spoon target had been Fort Amador, and the unit had conducted its day-to-day maneuvers in that area. To change the truck routes suddenly and drastically would almost certainly raise PDF suspicions. The best Flynn could do was to have his platoons and squads assume the order of march he intended to employ in the event there was an attack on the Comandancia. Even farther removed from the PDF headquarters than Flynn were the Marine light armored vehicles that would be attached to Gator if Blue Spoon were executed. Located with Task Force Semper Fi on the west side of the canal, they had no credible reason for passing anywhere near the Comandancia in the course of a routine Sand Flea. At a greater disadvantage, the four M551 Sheridans, the other half of Gator's Team Armor, were not only on the same side of the canal as the LAVs, but their presence still remained a closely held secret. Consequently, the only movement training allowed the Sheridan crews was once a week between 2200 and 0200, when they took the "vehicles out of their concealed locations and drove them around the motor pool." During daylight, the crews compensated for these security restrictions by riding in high mobility, multipurpose wheeled vehicles (HMMWVs) through the streets of Panama

[12] Intervs, Wright with Snell, 1 Jan 1990; with Goss, 13 Jan 1990; with Ethridge, 10 Jan 1990; with Freeze, 6 Jan 1990; with Reed, 6 Jan 1990; author with Reagan, Zebrowski, and Flynn, 21 Jun 1990.

City, acquainting themselves with the various routes that the M551s might use in wartime. Although a new team commander took over on 8 December, both the outgoing and incoming captains, together with the Sheridan platoon leader and the LAV platoon leader, "conducted reconnaissance of the area of operation and gathered intelligence" and coordinated "routes and plans for link-up, frequencies and call signs."[13]

In further preparation for the Comandancia mission, some Gator units sought to hone their skills in military operations on urbanized terrain (MOUT). In August, prior to deploying to Panama, each company in the 4th Battalion (Mechanized) had already finished a week of training in the urban operations layout at Fort Hood, Texas. The units had also conducted roadblock drills, Ethridge later related, in anticipation of "the missions that we saw mostly likely that we'd be performing down here." Once in Panama, where they acquired the Comandancia assignment after 3 October, the battalion's Companies B and D were able to rehearse the assault on a handful of occasions, employing a demolished housing area at Rodman Naval Station that in many respects resembled the layout of the real target. At the platoon and squad levels, the battalion also conducted live-fire exercises and "mini-mechanized gunnery" on the Army's Empire Range. These activities allowed the soldiers to fire their weapons under control of a squad leader, even though the confined area of the range imposed limitations on what the units could do.[14]

Mixed in with the troop exercises and training were a slew of other measures designed to enhance a unit's familiarity with the plan. Each Gator infantry company prepared a battle book to have ready for subordinate units in the event of war. The books contained the pertinent operation order, specific guidance for subordinate units, critical intelligence, and relevant maps and photographs. (While the maps received high praise, there was, according to some users, a dearth of photographs of the target area.) Officers also conducted jeep exercises, map exercises, and "oral rehearsals" of the plan. All of this created a frenetic operational tempo that left many of the troops frustrated. Flynn later observed that official prohibitions on various off-duty social activities, combined with the feeling there was no end to the crisis in sight, also contributed to a decline in morale as the holidays approached. The mechanized battalion's executive officer, Maj. Terry Freeze, added that levels of frustration grew when a company rotating into the role of JTF-Panama's quick reaction force was too often being sent into the field as the result of false alarms and unconfirmed reports. Over time, some staff officers started

---

[13] Interv, author with Reagan, Zebrowski, and Flynn, 21 Jun 1990; Col. Frank Akers, comp., "Discipline on Ancon Hill," Operation JUST CAUSE, *The Warriors*. Quotes from Capt. Kevin J. Hammond and Capt. Frank Sherman, "Sheridans in Panama," *Armor* 99 (March-April 1990): 8.

[14] Intervs, Wright with Reed, 6 Jan 1990; with Freeze, 6 Jan 1990. Quote from Interv, Wright with Ethridge, 10 Jan 1990. Interv, Wright with Goss, 13 Jan 1990. Captain Flynn, whose Company C, 1st Battalion, 508th Infantry, also engaged in live-fire training, noted that, at the urging of General Cisneros and Colonel Snell, "no-shoot" targets were added to the training to approximate better the urban environment in which combat operations might be conducted. Interv, author with Reagan, Zebrowski, and Flynn, 21 Jun 1990.

referring to "reverse Sand Fleas," Freeze indicated, because the PDF was "harassing us more than we were harassing the PDF, we thought."[15]

Perhaps the holidays would provide a respite. Reinforcing the conventional wisdom that, in Panama, crises did not heat up during festive occasions were intelligence reports that Panamanian forces were "taking off" for Christmas and New Year's Day. Aware of this assessment, Reed decided that, once his Companies B and D had run a "confidence building" exercise on 14–15 December, he would "reduce the OPTEMPO a little bit over about a week-long period." At that point in mid-December, he and his staff had nailed down the details of the Comandancia plan, having repeatedly discussed and run through them with the special operations people at Howard. Yet, as the lieutenant colonel later acknowledged, "Just the overall complexity of the plan was of some concern to me." Also worrisome was the fact that, despite the flurry of small-unit exercises and operations by U.S. forces in Panama after the 3 October coup attempt, the demands of the crisis, the dictates of operations security, and the absence of large U.S. training areas in-country combined to prevent Task Force Gator from coming together as a whole to rehearse its Blue Spoon mission. Still, Reed believed the troops had earned a respite, no matter how brief. Before they could get it, however, the shooting at the Comandancia checkpoint and the death of Lieutenant Paz changed everything.[16]

News of the incident spread quickly through the chain of command, with most of the ranking U.S. officers in Panama being notified while attending unit and organizational Christmas parties that Saturday night. The festivities ended abruptly, as commanders and staff officers departed for their respective headquarters. For several hours, the Southern Command and JTF-Panama gathered as much information as they could. Air reconnaissance platforms were launched to determine if there were any Panamanian troop movements, if there was another coup attempt in progress, or if the shooting was part of some broader threat. At 2150, less than an hour after receiving an order to muster, units under Task Force Bayonet were directed to mount a contingency readiness exercise that, among other things, called for the two mechanized companies located on the west side of the canal to cross over the swing bridge to the east side, thus placing them in a more advantageous position should they have to execute their High Anxiety or Blue Spoon contingency plans. Colonel Snell also issued a directive for the units under his operational control to begin task organizing in accordance with JTF-South Operation Plan 90–2.[17]

Of the three infantry companies constituting the core of Task Force Gator, Sunday morning found the soldiers and M113s of Company D relocated from

[15] Intervs, Wright with Snell, 1 Jan 1990; with Reed, 6 Jan 1990; with Ethridge, 10 Jan 1990; with Goss, 13 Jan 1990; author with Reagan, Zebrowski, and Flynn, 21 Jun 1990. Quote from Interv, Wright with Freeze, 6 Jan 1990.

[16] Quotes from Interv, Wright with Reed, 6 Jan 1990. Interv, Wright with Freeze, 6 Jan 1990. One platoon leader in Company D noted after Just Cause the difficulties created during the operation by the fact that a variety of units—airborne, Rangers, marines, mechanized forces, and light infantry—had not had a chance to work together prior to the invasion. Telephone Interv, author with Rubin, 5 Apr 1990.

[17] A brief chronology of the unit movements on the night of 16–17 December is in Nemmers, *United States Army South Staff Ride,* pp. 47, 52; and Memo, Reed for CG, FORSCOM, and Cdr, CALL, 27 Mar 1990.

Camp Rousseau across the canal to the Fort Clayton gymnasium. As for Company B, its rotation status as the quick reaction force on one-hour alert meant that it was already located on the east side of the canal at Corozal. After deploying to Fort Amador and then to Quarry Heights during the readiness exercise, the company returned to Corozal for the remainder of the day. In the case of Company C, 1st Battalion, 508th Infantry, Captain Flynn had initially moved it from its barracks at Fort Kobbe across the canal into the Balboa area on the east bank. It was the first time an exercise had placed his unit near the Comandancia, prompting some of the men to speculate that their objective in combat might not be Fort Amador, as previous exercises had suggested. Shortly after his troops reached Balboa, Flynn received word to take the company to Fort Clayton. At some point, he also learned that, in accordance with Snell's task-organizing directive, the company had gone under the operational control of the mechanized battalion. Flynn made contact with Captain Ethridge's Company D at Clayton but later Sunday afternoon returned with his men to Fort Kobbe, a part of General Cisneros' desire to give the appearance of business as usual. Even so, deception had its limits. Colonel Reed, for example, learned at a JTF-Panama meeting that the mechanized companies that had crossed the swing bridge were not to return to their bases but were to remain on the east bank of the canal indefinitely. The measure was highly visible but also prudent and precautionary, given the possibility of war.[18]

The next day, well before dawn, Reed and his operations officer, Maj. James M. Donivan, were summoned to Howard Air Force Base to meet with the special operations personnel who would conduct the Muse rescue mission. Reed later recalled that this was the first indication he had that BLUE SPOON would, in fact, be executed. For the remainder of the day, the group updated plans for the assault on the Carcel Modelo and the deliberate attack on the Comandancia. Reed also received from the Joint Special Operations Task Force liaison personnel for the three M113s constituting Team King, a communications team including a satellite communications operator, and a combat coordination team. Meanwhile, rumors began circulating within the battalion that war was imminent. For Task Force Gator's company commanders who, unlike their counterparts assigned to other BLUE SPOON missions, had been involved in the planning process since mid-October, the rumors were confirmed Monday evening, almost twenty-four hours before company commanders in other task forces would be told by their superiors.[19]

For Reed and his staff, the anxiety generated by impending combat was aggravated by the ambiguity surrounding command and control relationships in the hours leading up to hostilities. According to the deliberate attack scenario in BLUE SPOON, Task Force Gator would be placed under the operational control of the special operations commander on the scene until the Muse rescue mission and possibly certain other SOF objectives had been accomplished.

---

[18] Nemmers, *United States Army South Staff Ride,* p. 47; Memo, Reed for CG, FORSCOM, and Cdr, CALL, 27 Mar 1990; Intervs, Wright with Ethridge, 10 Jan 1990; with Goss, 13 Jan 1990; author with Reagan, Zebrowski, and Flynn, 21 Jun 1990.

[19] Intervs, author with Lt Col James W. Reed, U.S. Army, 29 Jan 1990, Camp Gator, Panama; with Reagan, Zebrowski, and Flynn, 21 Jun 1990; Memo, Reed for CG, FORSCOM, and Cdr, CALL, 27 Mar 1990.

All the commanders and planners involved agreed on this arrangement. What was not at all clear, however, was when the transfer of operational control of Gator from Snell's Task Force Bayonet to Downing's joint task force would take place. The mechanized battalion's after action report listed noon on 19 December as the time of the transfer, but that entry simply represented an after-the-fact convenience. At the time, no one really knew, least of all Reed. In the inevitable flurry of preinvasion activity on Monday and Tuesday, there was an extended period in which JTF-Panama and Task Force Bayonet, on the one hand, and Downing's headquarters, on the other, simultaneously believed that Task Force Gator was working for them. Reed's attempt to seek a definitive decision on the issue met with conflicting answers or genuine professions of ignorance.[20]

The confusion surrounding the issue would have been merely academic were it not for the fact that, during those two days, the mechanized battalion's commander and staff received contradictory or competing guidance from the two separate chains of command. In one instance, General Cisneros, the JTF-Panama commander, ordered a low-level Sand Flea exercise as another ploy to impart a business-as-usual façade. Downing, however, was concerned that the proposed troop movement would tip off enemy forces as to the coming hostilities. Reed found himself caught in the middle. Unable to obtain a definite answer as to whom he worked for, he went ahead and conducted the exercise, using the Sand Flea to allow his Company D to recalculate the time it would require to reach its H-hour line of departure from its new starting position at Fort Clayton.[21]

The uncertainty of the command arrangements continued into Tuesday, the day of the Sand Flea, with Reed and his staff asking, "Who are we listening to?" "Who's in charge?" As Major Freeze recounted, "Colonel Snell would come down . . . and say, 'Well, you need to do this, this and this.' And in some cases it would conflict with what the JSOTF individuals were saying." At some point on Tuesday, Reed received the information he had been seeking: he was working for Downing. This raised what Freeze referred to as "the other problem," in that Reed now had to convince Colonel Snell "that, no, I'm not working for you as of this point in time, . . . so you need to quit coming down here and telling me what it is that you want me to do, because the task organization is in effect." Overall, the confusion was a source of frustration for all concerned, but at no point did it seriously interfere with preparations for the H-hour mission. Still, as Freeze summarized, "all the time prior [to being told we would execute the plan], we were still being jerked around between whose control are we actually under."[22]

Determining the command and control relationships in effect was but one of many issues that Task Force Gator had to address in the twenty-four hours preceding its attack on the Comandancia. In two early Tuesday morning meetings, Reed reviewed a checklist for the assault with special operations

[20] Intervs, author with Reed, 29 Jan 1990; Wright with Freeze, 6 Jan 1990; Memo, Reed for CG, FORSCOM, and Cdr, CALL, 27 Mar 1990.

[21] Interv, author with Reed, 29 Jan 1990; Nemmers, *United States Army South Staff Ride*, p. 48.

[22] Interv, Wright with Freeze, 6 Jan 1990.

personnel. Afterward, he gathered his own staff and, in the course of the discussion, informed his company commanders that at 0900 they could tell their platoon leaders that BLUE SPOON would be executed and that combat operations were imminent. At 1300, he held another meeting, this one a comprehensive briefing for his staff, commanders, and other key personnel attached to Task Force Gator on their portion of the BLUE SPOON operation order. Around the same time, the troops themselves were locked down and prohibited from making or receiving telephone calls. At 1700, platoon leaders formally briefed the men on the mission and informed them that H-hour would be 0100. In one case, this platoon-level briefing took place prematurely. Company C, 1st Battalion, 508th Infantry, had spent most of the day in a gradual movement—one 2½-ton truck every hour or so—from Fort Kobbe over the Bridge of the Americas to the elementary school in Curundu, from which it would move to its attack position shortly before H-hour. Around 1600, when Captain Flynn arrived at the school after the 1300 meeting with Reed, one of his platoon leaders was already in the process of discussing the contents of the company's battle book with his unit. Not happy at this security violation, Flynn told the platoon that, if the operation was canceled for any reason, they would have to be locked up indefinitely, in that they now knew the details of a top secret plan. He then accepted the fait accompli and had the book distributed to the other platoons as well.[23]

After the soldiers and marines of Task Force Gator had received formal notification late Tuesday afternoon that they were about to go to war, fewer than eight hours remained for them to get ready. During that short time, platoon leaders briefed the plan, and each squad examined its specific role in detail. There was also a plethora of information concerning complex communication arrangements, detailed rules of engagement, vehicle and uniform identification, call signs, near and far recognition signals, challenges, hand signals, and running passwords to memorize. Furthermore, the troops had to be thoroughly hydrated for combat in a tropical climate; glint tape had to be affixed to helmets and vehicles so that friendly forces could be identified; and the M113s and other vehicles had to be checked one last time. There was still equipment to distribute, together with ammunition that the troops had not yet been issued. Against the PDF's AK47s, rocket-propelled grenades [RPGs], sniper rifles, and 90-mm. cannon mounted on V300 Cadillac Gage armored reconnaissance vehicles, the U.S. ground force would amass .50-caliber and M60 machine guns, M16 and sniper rifles, hand grenades and grenade launchers, light antiarmor weapons [LAWs], TOW missiles, and rounds fired from the Sheridans and light armored vehicles. Both sides had riot control gas, although the Americans fervently hoped it would not be employed, since donning chemical masks would reduce further what would already be limited nighttime vision. Given all that had to be accomplished in the few hours before Operation JUST CAUSE, corners were often cut. For example, even though Company D's 2d Platoon would be required to breech the Comandancia wall with explosive charges, constraints on money and time had not allowed the soldiers to train with such devices in Panama. Shortly before midnight with

[23] Memo, Reed for CG, FORSCOM, and Cdr, CALL, 27 Mar 1990; Intervs, author with Reed, 29 Jan 1990; Wright with Freeze, 6 Jan 1990.

H-hour looming, their platoon leader hurriedly gave them a cursory overview of the explosives to be used.[24]

For some units, final preinvasion preparations on the night of the nineteenth also meant moving from one location to another to link up with troops who would accompany them to their H-hour lines of departure. Thus, the Team King armored personnel carriers and the military police platoons attached to Task Force Gator made the journey from Fort Clayton to Corozal, where the mechanized battalion's Company B was situated. In the meantime, two platoons from Company C, 1st Battalion, 508th Infantry, boarded 2½-ton trucks at the Curundu elementary school in order to join Company D, 4th Battalion, 6th Infantry, at the Fort Clayton gymnasium. Special operations personnel made the trip to Clayton as well. The Marine LAV platoon at Rodman also had to link up with the Sheridan platoon at Camp Rousseau before Team Armor would be ready to cross to the east side of the canal. All these and other last-minute troop movements were completed without incident around 2200, but the time they required further depleted precious minutes that could have been used to acquaint the soldiers further with their combat missions, the specifics of which they had learned only a short while before.

That evening, Reed called one last meeting with his staff and troop commanders to give them an intelligence update; to drive home the point that, once they crossed their lines of departure, they needed to keep moving, regardless of the obstacles and opposition they encountered en route to their objectives; and to discuss various courses of action they might adopt should circumstances require them to deviate from the plan.

The plan itself, while intricate in terms of the precise timing it required and the number of small-unit activities it contained, was, in conceptual terms, very straightforward. Basically, the target was divided into two sectors: Area A, which included the Carcel Modelo, and Area B, which contained the Comandancia compound. Reed's Company B and the military police accompanying it would move from Quarry Heights across Fourth of July Avenue "to secure Area A and to isolate the prison area." At the same time, his Company D would take a route through Balboa, then across Fourth of July Avenue to isolate the Comandancia and secure Area B. The troops from Panama-based Company C would move dismounted behind the two mechanized companies to secure an area of operations, labeled AO Spear, in the vicinity of a high-rise apartment building overlooking the Comandancia from the west.

While the three key maneuver companies were thus engaged, the four Sheridans and four LAV–25s of Team Armor would be taking up positions, Bull I and Bull II, on Ancon Hill, looking directly down into the Comandancia compound (*Map 6*). Besides providing supporting fires, the crews were to engage any Cadillac Gage armored vehicles the Panamanian defenders attempted to deploy, either from the Comandancia or as part of a reinforcement column from some other PDF base in the vicinity. A military loudspeaker team would also set up its equipment on Ancon Hill in order to broadcast appeals for the Panamanian defenders in the area to surrender. To monitor these deployments and the subsequent battle, Reed and one of his platoons would establish the

[24] On the lack of training with explosives and the problems that that created the night of the operation, see Telephone Interv, author with Rubin, 5 Apr 1990.

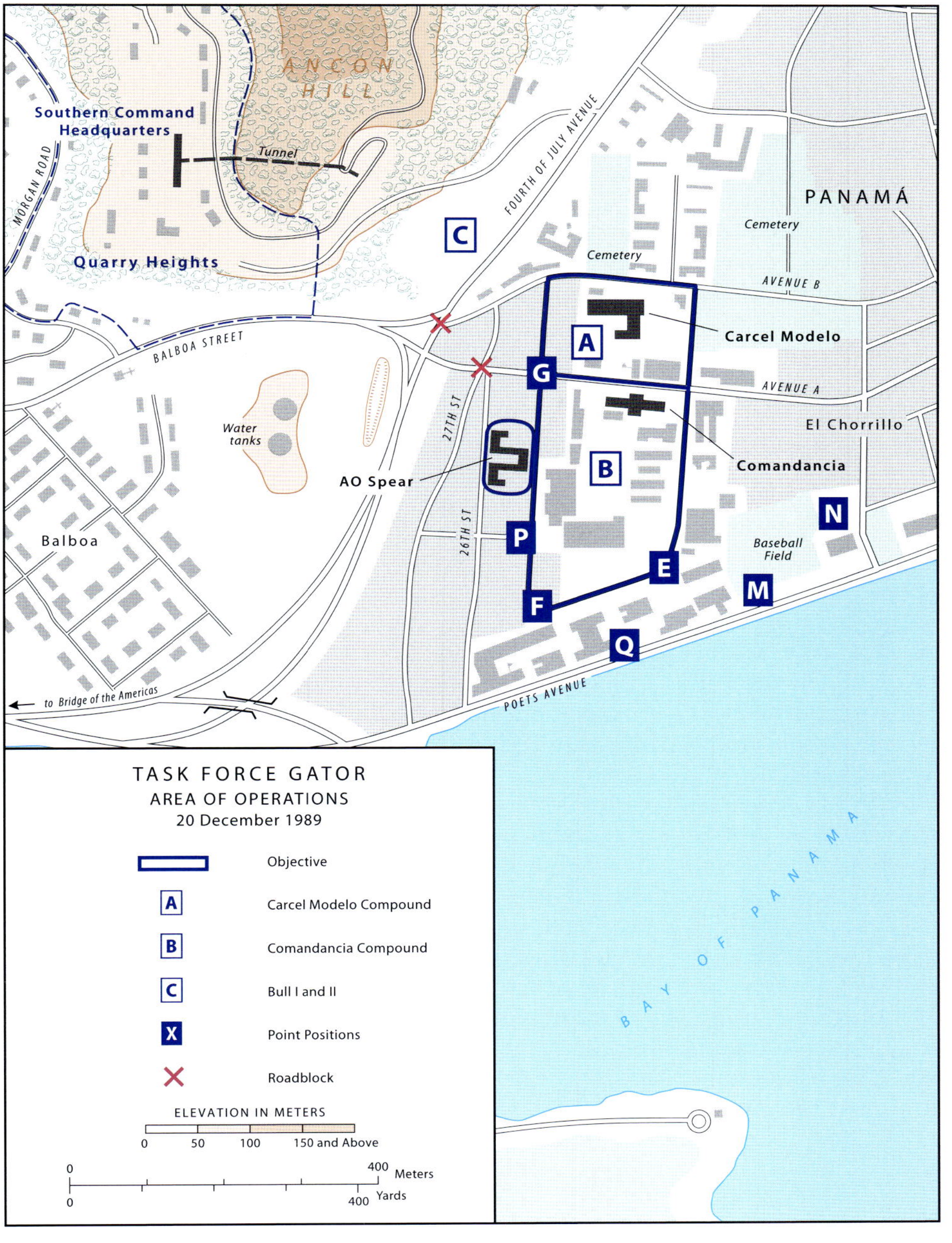
ANCON HILL
Southern Command Headquarters
Tunnel
MORGAN ROAD
Quarry Heights
FOURTH OF JULY AVENUE
C
Cemetery
PANAMÁ
Cemetery
AVENUE B
A
Carcel Modelo
BALBOA STREET
G
AVENUE A
Water tanks
27TH ST
El Chorrillo
AO Spear
B
Comandancia
N
Balboa
P
26TH ST
Baseball Field
E
M
F
Q
to Bridge of the Americas
POETS AVENUE
TASK FORCE GATOR
AREA OF OPERATIONS
20 December 1989
Objective
A Carcel Modelo Compound
B Comandancia Compound
C Bull I and II
X Point Positions
Roadblock
ELEVATION IN METERS
0 50 100 150 and Above
0 400 Meters
0 400 Yards
BAY OF PANAMA

Map 6

Task Force Gator command post at Point G on the operations map, a street intersection that would offer him "the best vantage point" from which to observe the Comandancia directly, and thus "the fires into it." Located with him, elements from an attached engineer platoon would run a collection point for Panamanian prisoners. Nearby, just to the west of Fourth of July Avenue, there would be an intermediate casualty collection and transfer point at which wounded troops could be removed from M113s employed for medical evacuation and put in wheeled ambulances, which would then proceed to the task force's main aid station at Balboa High School. By using the intermediate facility, the M113s would be able to return expeditiously to the battle area to pick up additional casualties.[25]

Following Reed's final meeting with his staff and commanders, the participating units continued their last-minute preparations for the coming battle. Right after midnight on 20 December, the news that had already reached higher headquarters that the Panamanian military knew about the attack and were taking defensive measures began circulating among the troops. Task Force Gator, in other words, should expect a fight en route to its objectives. As the implications of this development were sinking in, word also arrived of the advance in H-hour from 0100 to 0045. Adjusting to the changed time, the M113s of Company B and Team King began moving as a column out of a parking lot at Corozal at 0021 en route to the target area. When they passed the entrance to Albrook Air Station, Panamanians in a police car opened fire on the lead elements with automatic weapons, and a U.S. gunner returned fire with an M60 machine gun. The incident was inconsequential but had its effect on the troops. In the words of Captain Goss, Company B's commander, the PDF "knew we were coming, so we were kind of expecting the worst at that point." When the column reached Quarry Heights, it stopped for five to ten minutes, allowing the Company D column from Fort Clayton to reach Balboa High School. That column, its attached units, and the task force commander had left the fort by the front gate at 0033. When it arrived at the high school, it, too, paused briefly to synchronize times with Company B. Company C (minus), meanwhile, had followed Company D out of Clayton. When it reached the vicinity of the high school, the soldiers climbed out of the deuce-and-a-half trucks, which, unlike the M113s, offered little protection from hostile fire. After some initial confusion, the platoons fell in line for the foot march ahead of them. Team Armor made up the rear of Gator's advance. The Sheridans and LAVs had departed Camp Gator for the swing bridge at 0027, the crews thankful in the knowledge that the bridge, located across from Fort Clayton, had opened in time to obviate taking the backup route over the Bridge of the Americas, which the Panamanians had reportedly blocked.[26]

---

[25] The overview of the plan of attack and the quotes are from Interv, Wright with Reed, 6 Jan 1990.

[26] Nemmers, *United States Army South Staff Ride,* pp. 48, 53, 57–59, 60b; Memo, Reed for CG, FORSCOM, and Cdr, CALL, 27 Mar 1990. Quote from Interv, Wright with Goss, 13 Jan 1990. Intervs, Wright with Ethridge, 10 Jan 1990; author with Reagan, Zebrowski, and Flynn, 21 Jun 1990. Company C, 5th Battalion, 87th Infantry, also fell in behind the Company D column. The unit was part of Task Force Wildcat and had missions near the Task Force Gator line

Minutes after staging and coordinating, the three main maneuver elements moved out toward their respective lines of departure. The Company B column crossed its line at 0045. The Company D column, with 1st Lt. Douglas L. Rubin's 2d Platoon in the lead, met armed resistance before reaching its line when Panamanians in three cars trying to block the intersection of Fourth of July Avenue and Avenue A opened fire with AK47 assault rifles. The platoon returned fire with its .50-caliber machine guns and, after crushing one of the vehicles under an M113's tracks, moved through the intersection at 0044. A minute later, the soldiers of Company C heard a series of explosions as the AC–130s over the Comandancia began their attack. The troops, themselves, took some AK47 fire from a water tower and wood line in Balboa as they walked to the corner of Morgan Avenue and Balboa Street and began to negotiate the hill separating them from AO Spear. As they reached the crest of the hill, they looked in awe upon the fiery sight below them. The battle for the Comandancia was under way.[27]

THE ATTACK

As Panamanian riflemen and snipers on the perimeter of the Comandancia began firing on the advancing U.S. mechanized and dismounted infantry companies, the PDF headquarters itself came under intense fire from the air. Just before H-hour, two AH–6 attack helicopters flew by the sixteen-story high-rise building west of the Comandancia checking for snipers. Because the rules of engagement for Operation JUST CAUSE emphasized the use of minimum force and then only when an unmistakably hostile target presented itself, the helicopters' crews made no attempt to engage Panamanian forces they could see inside the building but did use their miniguns to drive enemy snipers off the rooftops. Both Little Birds then turned to attack the Comandancia with 2.75-inch rockets. In the process, one of the helicopters was hit by hostile fire and began falling to the ground. The two-man crew, Capt. George Kunkel and CW3 Fred Horsley, spied an open spot below them and managed to guide the craft to a hard landing. The AH–6 burst into flames, but both men were able to exit safely. Whatever relief they experienced for having survived the crash dissipated when they realized they had come down within the Comandancia compound. They immediately tried to contact friendly forces via their PRC–90 radios, but, failing that, they moved along a nearby wall, using various buildings in the compound for cover. More than the threat of enemy fire alarmed them. Knowing that the AC–130 attack on the compound was imminent, they began looking for a secure position from which to survive the coming inferno.[28]

---

of departure. For the duration of the movement to its area of operations, however, Company C was placed under Gator's operational control. See Chapter 6.

[27] Telephone Interv, author with Rubin, 5 Apr 1990; Interv, author with Reagan, Zebrowski, and Flynn, 21 Jun 1990; Nemmers, *United States Army South Staff Ride,* pp. 48, 53, 60b; Memo, Reed for CG, FORSCOM, and Cdr, CALL, 27 Mar 1990.

[28] Col. Frank Akers, comp., "A Night in the Comandancia," Operation JUST CAUSE, *The Warriors*; Interv, author with Maj Charles G. McMillan, U.S. Air Force, 31 May 1995, Fort Leavenworth, Kans. McMillan was a pilot in one of the two AC–130s over the Comandancia during the U.S. attack.

The crew of the other Little Bird, not realizing that Kunkel and Horsley had crashed, cleared the two AC–130s circling overhead to begin firing. For the first time in combat, the gunships used a new tactic, called Top Hat, in which one flew at a higher altitude and in a smaller circle than the one below it, a configuration ideally suited to allowing two Spectres to bring all their firepower to bear on a small, highly concentrated target. At thirty seconds after H-hour, the higher plane started firing its 105-mm. howitzer at the two enemy barracks in the compound. Joining in, the lower gunship began to level a devastating fusillade at its targets. After several minutes, the first AC–130 prepared to leave the airspace of the Comandancia to assist Special Operations Forces elsewhere in the search for Noriega. If Panama's dictator was discovered trying to escape in his private jet at Paitilla, the Spectre crew was supposed to interdict him. Noriega would be given a chance to turn back and surrender, but, if he ignored the appeal, the crew had orders to shoot down his plane. As the AC–130 was on the verge of departing the Comandancia area, however, the pilot received word from the assault force at the Carcel Modelo that it was taking fire from across the street. He responded by having his gunners rake the PDF headquarters building with the plane's 40-mm. Bofors gun, after which the gunship left the scene to perform its other missions.

The second AC–130 stayed behind to provide close air support for the ground assault, firing its 105-mm. howitzer into the main headquarters building. A videotape of the assault taken from the plane's camera showed a series of rapid explosions, as volumes of smoke and debris ascended into the night sky. At ground level, the AC–130 onslaught produced the fiery sight witnessed by the U.S. troops en route to their objectives. In reality, though, the attack on the three-story headquarters building wreaked havoc on the top floor only, leaving defenses on the lower two floors largely intact. The howitzer rounds, it turned out, were not designed to penetrate reinforced concrete—they would go through the unreinforced roof but not the floors of the building. Even so, the Spectre preparation fires, while not reducing the target to rubble, caused much damage within the compound and served to suppress the fire of enemy soldiers situated therein.[29]

The main problem facing Task Force Gator in attempting to surround the Carcel Modelo and Comandancia was fire coming not from the two compounds, but from the scores of defenders who had vacated the premises well before H-hour to set up roadblocks, strongpoints, and sniper positions in the streets and within the buildings along the principal avenues the American forces needed to negotiate before they could set up their own series of blocking points. As Reed recollected, the PDF had done a professional job in ensuring that each roadblock or defensive position called for in the operation plan was well covered by machine guns, rocket-propelled grenades, and AK47s. Even with the American advantage in firepower, he anticipated that his task force would suffer around three dozen fatalities. It certainly helped that his troops knew in advance where many of the enemy obstacles were located, thanks largely to the reports of special operations personnel observing the

[29] Intervs, author with McMillan, 31 May 1995; Wright with Reed, 6 Jan 1990; Thomas B. Hunter, "Operation Acid Gambit: The Rescue of Kurt Muse," at http://www.specialoperations.com/Operations/Just_Cause/Acid_Gambit/default2.html.

Comandancia throughout the day from positions on Ancon Hill. But the intelligence, while accurate, was not comprehensive; it could not cover what the observers could not see. Also, knowing what to expect and where to expect it did not preclude a fierce battle. As each U.S. maneuver company crossed its line of departure, the soldiers immediately encountered heavy resistance.[30]

In Area A, the northern sector of the battlefield, the column of armored personnel carriers from Captain Goss' Company B moved down Fourth of July toward Avenue B, a maneuver that meant exposing the vehicles' flanks to the high volume of sniper and small-arms fire from defenders in the buildings along the route. As a countermeasure, the drivers increased their speed, up to thirty or thirty-five miles per hour—"those 113s were just howling," recalled one observer. When the column made a left turn onto Avenue B at those high speeds, though, it literally ran into a roadblock that the special operations observers on Ancon Hill had not been able to see. The obstacle consisted of two dump trucks stretched across the street, and the first M113 in the Company B column could not avoid hitting one of them. While the next three personnel carriers stopped or swerved to miss ramming the first one, the vehicles that followed collided in a chain reaction, putting one M113 out of action and causing one passenger to break his ribs when he was propelled forward from the rear into the crew's compartment. The scene reminded Captain Goss of "a California freeway smashup." As soon as the column stopped, it received concentrated AK47 and automatic weapons fire, together with some RPGs, from the buildings and apartments nearby, especially from a multistoried building at the intersection of Fourth of July Avenue and Avenue B. Goss' men began returning fire and, in the captain's judgment, gained fire superiority within a matter of minutes. The pileup of M113s, in this respect, actually worked to the Americans' advantage, for, although it created the kill zone the enemy wanted, it also, in Goss' words, "helped because we had so much firepower in that small area and we were able to suppress every window that they had in [the large building]." After a couple of minutes, the shooting began to subside, but the column remained stalled at the roadblock. It had been a well-planned but poorly executed ambush because of the poor marksmanship of the enemy, who persisted in shooting high. Still, surviving this initial encounter had been costly for the Americans: Company B received its first fatality, Cpl. Ivan Perez, an M113 commander killed while trying to provide suppressing fire with his .50-caliber machine gun.[31]

The advance of Captain Ethridge's Company D fared only slightly better. After crossing Fourth of July Avenue and moving down Avenue A, the armored column needed to turn right onto 26th Street. As was the case with Company B, the lead element encountered dump trucks blocking the route at the point where 27th and 26th Streets intersected. In this instance, the 2d Platoon's M113s were able to circumvent the obstacles by driving partly on the sidewalk. But as they sped down 26th Street, they experienced what Lieutenant Rubin termed *the Gauntlet*, where machine gun and AK47 fire and a few RPGs fell

---

[30] Interv, Wright with Reed, 6 Jan 1990.

[31] First quote from Interv, Wright with Freeze, 6 Jan 1990. Other quotes from Interv, Wright with Goss, 13 Jan 1990. Intervs, Wright with Reed, 6 Jan 1990; author with Reed, 29 Jan 1990; Donnelly, Roth, and Baker, *Operation Just Cause*, pp. 145–46.

M113 armored personnel carrier in position near the Comandancia compound

on the column from the sixteen-story high-rise in AO Spear and from the two- and three-story buildings on each side of the street. The second M113 in line was struck by an RPG, but the angle was bad and the round did not penetrate the vehicle. Rubin, himself, was hit by a bullet in the shoulder, but, with his adrenaline flowing from the sounds of combat and the "*Star Wars*" images of green and red tracers all around him, he did not immediately realize he was wounded. His main concern was to keep moving and reach his objective, Point E. Behind him, the 2d Platoon from the mechanized battalion's Company C—the platoon attached to Company D as reinforcement—followed in his wake, with Company D's own 3d Platoon after that. In the meantime, Ethridge pulled out of the formation and, with his 1st Platoon and engineers in tow, headed to set up his command post near Point G.

After turning left onto Poets Avenue (*Avenida de los Poetas*), Rubin's platoon reached Point Q, in the process exchanging fire with the crews of two PDF 75-mm. recoilless rifles positioned farther down the avenue. His M113s then crashed through a chain-link fence and moved toward Point E, where, in a brief engagement, the men killed four Panamanian defenders. Meanwhile, the 3d Platoon, after pausing at Point P during the first round of AC–130 preparation fires on the Comandancia, moved with two squads into its positions at Point F. For its part, the 2d Platoon from Company C prepared to maneuver to its objective, Point M, near the baseball field just east of the compound.[32]

As the mechanized platoons under Ethridge were running the Gauntlet on 26th Street, the men of Company C, 1st Battalion, 508th Infantry, were working their way along 27th Street—and down alleyways—toward AO Spear. As they fought past each building on their route, they, like Companies B

[32] Interv, Wright with Ethridge, 10 Jan 1990. Quote from Telephone Interv, author with Rubin, 6 Apr 1990. Memo, Reed for CG, FORSCOM, and Cdr, CALL, 27 Mar 1990.

and D, received heavy fire, especially from the sixteen-story high-rise. Captain Flynn briefly pulled his men back and called for fire support, only to be told to keep moving and return fire. Being on foot, the light infantry found its ability to comply with the directive restricted by the number of civilians crossing through its ranks, mainly refugees trying to flee the fighting that had suddenly engulfed the poor neighborhoods of El Chorrillo, the district in which the Comandancia was situated. As one account related the scene, "The main avenue was a flood of fear, streaming with Panamanians, most of them barefoot and crying." Also a problem was the fact that most of the PDF defenders had shed their uniforms for civilian attire, thus making it difficult for U.S. forces to determine who exactly was shooting at them. Despite these impediments, the company managed to occupy its objective around the high-rise apartment building shortly after 0100, using whatever cover it could, including a ditch running parallel to the main street, to conceal the troops' movements and positions from the snipers above them. Once in place, Flynn's men awaited the engineers who were to provide them the explosives and expertise for blowing a hole in the Comandancia's west wall.[33]

Meanwhile, Company B's efforts to break through or bypass the roadblock on Avenue B finally paid off. For one thing, the Team King M113s and the 3d Platoon simply pulled out of the column and headed for Avenue A, an alternate route to their objectives. Captain Goss thought about sending the rest of the company after them, but when his troops' fire superiority forced a lull in the fighting, he decided instead "to bust through where we were and continue on with the normal plan," rather than risk further confusion and possible friendly fire by rerouting the remaining two platoons. In front of him, an armored personnel carrier rammed one of the dump trucks four times in a futile attempt to move it. Then, in an effort to tow the obstacle out of the column's path, two soldiers dismounted from an M113 and attached a hook and cable to the ten-ton truck. The vehicle barely budged, but what small movement there was created just enough space between the dump truck and the building next it to allow the American vehicles through. A lieutenant in the first M113 proceeded to bypass the obstacle, scraping both truck and building in the process. The second track contained the body of Corporal Perez, so it pulled back and ultimately made its way back to Quarry Heights to deliver his remains there. With some difficulty and several minutes after the first personnel carrier had successfully negotiated the barrier, Goss' 1st Platoon finally slipped past the dump truck. Goss himself followed and then the 2d Platoon. After a brief mix-up concerning the correct location of the 1st Platoon's final positions, all the units in the Company B column arrived at their objectives. The troops unloaded wire and other material and set up roadblocks, while Goss radioed Reed and told him that Area A surrounding the Carcel Modelo compound was secured. The time was 0145.[34]

---

[33] Interv, author with Reagan, Zebrowski, and Flynn, 21 Jun 1990; Nemmers, *United States Army South Staff Ride*, pp. 60b, 60d. Quote from Donnelly, Roth, and Baker, *Operation Just Cause*, p. 147.

[34] Donnelly, Roth, and Baker, *Operation Just Cause*, pp. 145–46; Memo, Reed for CG, FORSCOM, and Cdr, CALL, 27 Mar 1990; Interv, Wright with Freeze, 6 Jan 1990. Quote from Interv, Wright with Goss, 13 Jan 1990.

Once his men were in place, Goss had them dismount and take up positions that would prevent any Panamanian defenders from escaping the ring closing around them. To this end, he also sent several squads into the cemetery across from the Carcel Modelo. The problem, again, was that most of the Panamanian forces were fighting in civilian clothes—Reed could not remember seeing many uniformed enemy soldiers at all—while members of the Dignity Battalions present were, in many cases, civilians to begin with. Thus, whenever a defender determined that the time had come to put down his weapon and steal away, he had little to do but mingle with the throngs of innocent civilians trying to flee both the fighting and the fires that were beginning to spread in the area. Since there was no effective and expeditious way to separate the fighters from the noncombatants, U.S. troops allowed most unarmed persons in civilian clothes attempting to leave the area to do so. The Americans could only hope that they would be able to round up the departing enemy combatants in the days ahead. As for those hostiles who did surrender or who were caught, they were taken to a prisoner-of-war collection area near Point G and turned over to engineers not otherwise involved in the fighting; from Point G, they were escorted to Albrook Air Station for processing.[35]

At the time Company B was moving to secure its positions in Area A, Company D was still encountering problems in Area B. This was particularly true for the 2d Platoon after it reached Point E. Lieutenant Rubin and his men had just exited the platoon's three M113s and were advancing to place explosive charges near the southeast corner of the wall surrounding the Comandancia when they were hit with about twenty rounds of indirect fire. Of the three M113s in the platoon, the first in line, in which Rubin was riding, was disabled, while the second burst into flames. Miraculously, no soldiers were killed, but several were wounded. Rubin got the most serious casualties into the third M113, which was still intact, and sent them to the medical aid station in the rear. Then, he, the remaining wounded, and those who were unscathed pulled back to a seawall near them, leaving behind the damaged and the burnt-out M113s. When Rubin reported his casualties to Captain Ethridge, the company commander contacted the 2d Platoon, Company C, and stopped its advance toward Point M, ordering the men instead to set up a "support by fire position" to cover Rubin's platoon, which was "pretty strung out."[36]

Meanwhile, Rubin and his men headed back toward Fourth of July Avenue, regrouped, and then moved forward again, this time behind the seawall, determined to retrieve the one possibly operable M113 they had abandoned at Point E. What was left of the platoon crossed a road near the personnel carrier shortly before 0200, at which time a second barrage of indirect fire engulfed them. Soon thereafter, a medical M113 arrived, and Rubin was able to evacuate all of the wounded. With the remaining five members of his platoon, he retrieved the damaged personnel carrier and drove it to Point M, where it stayed the rest of the night, providing security. At some point, the 2d Platoon, Company C, arrived, took up its assigned positions, and began clearing the

[35] Intervs, Wright with Goss, 13 Jan 1990; with Reed, 6 Jan 1990; author with Reed, 29 Jan 1990.

[36] Telephone Interv, author with Rubin, 5 Apr 1990. First quoted words from Interv, Wright with Ethridge, 10 Jan 1990. Second quoted words from Interv, Wright with Freeze, 6 Jan 1990.

area around Point N. The platoon leader also detached one of his squads to reinforce Rubin's depleted ranks.[37]

The two instances of indirect fire had devastated 2d Platoon, Company D. During the first episode, Rubin had assumed that the shelling came from enemy mortars or RPGs and reported this up his chain of command, through Ethridge to Reed. When Reed heard the news, he directed his antimortar radar crews to locate the weapons responsible. Some men in the 2d Platoon, however, were not so sure they had been the object of enemy mortars, thinking instead that the lethal fire might have come from the AC–130 that was still circling overhead to provide air cover for the task force. The second attack pretty much removed any doubt about the source: it was the 40-mm. gun of the Spectre gunship. The aircraft used television and infrared sensor systems to locate and identify targets on the ground, and, during the fighting around the Comandancia, the sensor operators switched back and forth between the two systems, depending on the degree to which the smoke and debris emanating from the battlefield below obscured their view. After one switch over, the sensors picked up a target, Rubin's three M113s, that, given reports the Spectre crew was receiving from ground communication elements, were believed to be enemy V300 Cadillac Gage vehicles. The result was friendly fire. After the second attack on the target, the Spectre crew received word to cease firing over Points Q, M, and E, but by then the damage had been done.[38]

With Company B still receiving sporadic fire in Area A, and with Company D still facing heavy fighting in Area B, Reed decided to bolster the firepower available to Goss and Ethridge by providing each of them one LAV and one Sheridan from Team Armor. In part, the decision was driven by his desire to get some use out of a unit that had yet to fire a single round in the battle. There were several reasons for this inaction, although Reed only became aware of some of them in the days to follow. To begin with, the team had been divided into two groups, each with two Sheridans and two LAVs. Both groups were to occupy positions on Ancon Hill overlooking the Comandancia area, with one group going to Bull I and the other, below it, to Bull II. The difficulties started prior to H-hour when Panama Canal Commission regulations forced the team to cross the swing bridge one vehicle at a time, with the subsequent delay causing the column to reach its positions on Ancon Hill only after the battle had begun. By the time the crews were ready to engage the enemy, smoke and flames obscured their view of the target, making friendly fire a distinct possibility given the Sheridan's lack of a nighttime acquisition and fire control system. "We didn't fire because we couldn't see the Comandancia clearly, and we didn't want to cause collateral damage," one gunner later related. Further complicating the situation, the four vehicles at Bull I were positioned behind three mammoth trees. Engineers at the site were supposed to blow down the obstacles but managed to fell only one. With Reed's decision to reinforce Companies B and D, at least half of Team Armor would participate in the street fighting that night. In Reed's words, "It was clear to me that those guys

[37] Telephone Interv, author with Rubin, 5 Apr 1990.

[38] Ibid.; Intervs, author with McMillan, 31 May 1995; Wright with Freeze, 6 Jan 1990; Donnelly, Roth, and Baker, *Operation Just Cause*, pp. 150–52; Nemmers, *United States Army South Staff Ride*, p. 49.

just weren't doing much for us up in that position, so I decided that I'd bring them down."[39]

Reed was not certain that Company B really needed a LAV and Sheridan, but he knew that Company D did. When Ethridge received word that the two vehicles were heading to his position, he directed them to Point E, where he knew Rubin's platoon was in trouble. Moments later, Rubin saw them coming toward him and, not realizing they were for his use, moved to commandeer them. "Now I'm truly combined arms," he thought. In the hour or so that followed, he had the Sheridan fire two 152-mm. rounds at the Comandancia wall and then ram a hole in it. Leaving the LAV and its marines to hold Point E, he moved to the baseball diamond, where he acquired some additional M113s from the Company C platoon and sent them toward another end of the field. There he left the tracks in a defensive position and returned to his own men, whom he had dropped off at Point M. The next task was to build a sandbagged defensive position between Points M and E, which Rubin used as his command center. With that accomplished, all of Area B was deemed secured around 0330.[40]

Either shortly before or right after Rubin's platoon received friendly fire for the second time, there was one of several pauses in the AC–130's bombardment of the Comandancia. Inside the compound, the two downed AH–6 pilots decided the time had come to depart the enemy's enclave and make their way to friendly lines. Winding their way through a maze of buildings, Kunkel and Horsley reached the outer wall and were preparing to scale it when a Panamanian soldier, his hands raised, stepped out from behind some bushes and surrendered to them. His comrades, he explained, were either dead or had fled. Kunkel then threw his flak jacket over the concertina wire on top of the wall and climbed over it, dropping into the street below. Shouting the running password, "Bulldog"—or, as some recollect, words that sounded more like a string of profanity—he made contact with American troops and returned for Horsley and the prisoner. The three were escorted to Ethridge's position, where the Panamanian was taken to the prisoner-of-war site and Kunkel and Horsley found haven in the back of the company commander's M113, their ordeal finally over.[41]

During the remainder of the night, Companies B and D would engage in sporadic fighting, with Goss' men in one instance using their .50-caliber machine guns to repel a V300 Cadillac Gage coming toward them. In the sequence of events since H-hour, Company B had first experienced the heaviest fighting; then Company D had come under an onslaught of enemy and friendly fire. Finally, around the time Ethridge's platoons were establishing their positions and isolating the Comandancia in Area B, it became Company C, 1st Battalion, 508th Infantry's turn to take the brunt of enemy resistance. Spread out around the high-rise in AO Spear, the men waited while the squad of engineers who had arrived in an M113 set a 40-pound cratering charge

[39] First quote from Akers, "Discipline on Ancon Hill." Second quote from Interv, Wright with Reed, 6 Jan 1990. Intervs, author with Reed, 29 Jan 1990; Wright with Freeze, 6 Jan 1990.

[40] Interv, Wright with Ethridge, 10 Jan 1990. Quote from Telephone Interv, author with Rubin, 5 Apr 1990. Hammond and Sherman, "Sheridans in Panama," p. 10.

[41] Akers, "A Night in the Comandancia"; Interv, author with McMillan, 31 May 1995.

against the Comandancia's west wall. Flynn was supposed to give the order to set off the charge, but the engineer's squad leader blew it prematurely. Flynn then radioed Reed that his soldiers were ready to move inside the compound. Reed told the captain to hold his position. Flynn repeatedly tried to get Reed to reverse the decision, but to no avail. The battalion commander's intention was to have troops clear the buildings in the compound only after they had been reduced to rubble. Given the ineffectual performance of Team Armor and the nonpenetrating rounds of the AC–130s, that had not happened. Under the circumstances, clearing operations would have to wait until daylight.[42]

That left most of Company C's men outside the wall with their "butts out in the wind," to use Flynn's phrasing, receiving small-arms and automatic weapons fire from the high-rise and other buildings. The soldiers sought cover, hugged the sides of buildings, and traded shots with the defenders, firing grenades from M203 launchers when an enemy sniper position could be clearly identified. The Panamanians also had grenades, and one thrown from the high-rise knocked an American soldier unconscious and severely wounded another. As the fighting continued, civilians streamed out of the apartment building. To prevent armed PDF and Dignity Battalion members from mingling with the throng and escaping, U.S. troops positioned at the doors ordered adult males to leave the building wearing only pants and shoes. After four hours of this, Flynn told his men shortly after dawn that they would probably remain in the area for a while and to clear nearby buildings that would provide them better cover. During the process, enemy grenades took an additional toll, killing two Americans, Sgt. Michael DeBlois and Pfc. Troy Coats, and wounding several others. It had been a very rough morning for Company C.[43]

Flynn later received another explanation for why he was not allowed to move his men inside the wall after it had been breached. The compound's southern end, he was told, had not been secured because U.S. special operations forces had taken some of Task Force Gator's M113s to go hunting for Noriega in Panama City. This was a dubious assessment. The difficulty in sealing off the compound stemmed mainly from the friendly fire episodes. But it was true that, while Company D was still struggling to isolate the Comandancia, General Downing had called on Reed to provide him with six M113s to be deployed to the U.S. Embassy in Panama City. Reed readily complied, taking Team King's three vehicles and three more, each with a driver and commander, from Company B. Captain Goss further supported the excursion with two squads from his 3d Platoon. Since intelligence reports indicated that the column could meet resistance along the ten-minute trip to the embassy, Reed also provided two LAVs and a Sheridan from Team Armor for additional security. The vehicles moved out under the leadership of the mechanized battalion's operations officer, Major Donivan. Reed expected they would be returned to him within a few hours. To his surprise, the next time he saw them was several days later. Labeled Team Donivan and nicknamed

[42] Intervs, Wright with Reed, 6 Jan 1990; author with Reagan, Zebrowski, and Flynn, 21 Jun 1990; Donnelly, Roth, and Baker, *Operation Just Cause*, pp. 152–53.

[43] Quote from Interv, author with Reagan, Zebrowski, and Flynn, 21 Jun 1990. Donnelly, Roth, and Baker, *Operation Just Cause*, p. 152; Nemmers, *United States Army South Staff Ride*, pp. 60b, 60d.

"Panzergruppen," the column first secured the U.S. Embassy, rescuing some employees in the process, and then joined the search for Noriega around Panama City. Following that, it spent several days engaged in a number of special operations and security missions.[44]

About two hours after parting with the M113s, LAVs, and Sheridan, Reed's task force went back under the operational control of Task Force Bayonet and its commander, Colonel Snell. A couple of hours after that, the units remaining as part of Task Force Gator began preparations for the final assault on the Comandancia. They were not prepared, however, for the torrent of refugees that confronted them. Throughout the night, several hundred Panamanians had fled the area, to get away from both the fighting and the fires spreading through the shantytown that constituted the adjacent neighborhood. After dawn, the hundreds increased rapidly to thousands, as flames engulfed virtually all of El Chorrillo. The source of the conflagration was not immediately known but initially was assumed to have been stray rounds igniting the highly flammable materials used in the construction of many of the neighborhood's houses. Later reports indicated that the most damaging blazes had been deliberately set by the Panama Defense Forces and Dignity Battalion members in retaliation for El Chorrillo's failure to support Noriega's candidates in the May 1989 presidential election. Whatever the cause, the fires left U.S. troops surrounding the Comandancia wondering how to handle the unexpected influx of so many civilians into their lines. As Reed later observed, "I frankly don't think we had anticipated as well as we probably should have the extent of the refugee evacuation problem and the impact that would have on military operations."[45]

It was not that the refugee issue had been ignored during the planning phase. BLIND LOGIC, the Southern Command's operation order for the reconstruction of Panama in the wake of a U.S. invasion, contained an annex with information on the handling of refugees and the proposed locations for several refugee centers. But no one had anticipated the execution of BLUE SPOON during the holiday season in December. At midmonth, BLIND LOGIC was thus bouncing back and forth between Quarry Heights and Fort Clayton to determine whether the Southern Command or U.S. Army, South, would be in charge of implementing the plan should the occasion arise. In the process, officers reviewing the document identified several gaps that needed to be filled with more detailed planning. That requirement became urgent once President Bush made the decision to take military action in Panama.

---

44 Intervs, Wright with Reed, 6 Jan 1990; with Goss, 13 Jan 1990; John Partin with Maj Gen Wayne A. Downing, U.S. Army, 6 Jan 1990, Howard Air Force Base, Panama; Memo, Reed for CG, FORSCOM, and Cdr, CALL, 27 Mar 1990; Hammond and Sherman, "Sheridans in Panama," p. 10.

45 JTF-Panama Input, 4 Feb 1990, to AAR, JTF-South, p. 4; Memo, Reed for CG, FORSCOM, and Cdr, CALL, 27 Mar 1990; Interv, Wright with Reed, 6 Jan 1990. Quote from Interv, author with Reed, 29 Jan 1990. Interv, author with Capt Eric Fatzinger, U.S. Army, 4 Oct 1990, Fort Leavenworth, Kans. For the cause of the fire, the author had a conversation with an officer who had been in the Southern Command intelligence directorate. The officer had photographs taken from Ancon Hill that showed the fires starting and spreading. He also indicated that interviews with residents of the neighborhood revealed that enemy forces had started the fires. Colonel Reed also mentioned in the 29 January interview that fires started by ordnance would not have become so intense so quickly.

Thus, on Tuesday, 19 December, the acting U.S. Army, South, chief of staff directed Col. William J. Connolly, USARSO's deputy chief of staff for resource management, to go to Quarry Heights to help fix the BLIND LOGIC operation order. Once Connolly and others began working on the plan, they decided to focus on the refugee issue, assuming that it would be one of the first problems with which the military would have to cope as the result of combat operations. From a work station in the Tunnel at the Southern Command headquarters, Connolly wrestled with the issue all Tuesday night. Little did he realize that his presence, his work, and his physical proximity to General Thurman would ensure that, when the refugee issue surfaced, he would find himself in the forefront of dealing with it.[46]

His new duties were not long in coming. As several thousand Panamanians made their way through Task Force Gator's area of operations, Colonel Reed gathered some Spanish-speaking U.S. soldiers to intercept them and direct them to Balboa High School nearby. Military police positioned at two roadblocks near Fourth of July Avenue did the same. At the school, Reed had established a battalion administration and logistics center to support him. As part of that role, the facility also served as his medical aid station. When the refugees started to arrive, the medical platoon leader and noncommissioned officers at the site responded by setting up what would become the U.S. Army, South, Refugee Center. The medical staff, while continuing to treat wounded U.S. soldiers, began ministering to injured Panamanians as well. Someone sent word to Fort Clayton of what was transpiring, which led to Maj. Les Knoblock, an Army psychological operations officer, being dispatched to the scene. Knoblock, in turn, contacted Connolly, who also made his way to the school, a short distance from the Tunnel at Quarry Heights.[47]

At some point during the day, the number of refugees at the high school approached ten thousand, a figure far in excess of what BLIND LOGIC planners had estimated. Chaos prevailed as all involved realized that the ad hoc center was not set up to handle the variety of problems, both immediate and looming, caused by this unexpected horde of humanity. There was no organization. There were no ground rules in place. There was either a shortage or a complete lack of urgently needed goods and supplies. The school and its athletic field quickly filled to capacity, with those crowded together finding themselves confronted with backed up toilets and an insufficient food supply. Aggravating the situation, mothers, children, and old people were intermingling with drug dealers, armed criminals, and PDF members who had shed their uniforms. Establishing order and security in the midst of the bedlam would not be easy.

[46] Intervs, author with Col William J. Connolly, 29 Jan 1990, Fort Clayton, Panama; Dolores De Mena with Connolly, 20 Feb 1991, Fort Clayton, Panama. The latter oral history interview is one in a series conducted by U.S. Army, South's command historians before and after Operation JUST CAUSE. Under command historian Dolores De Mena, the interviews were transcribed and made available to the U.S. military.

[47] Intervs, author with Reed, 29 Jan 1990; Wright with Freeze, 6 Jan 1990; author with Lt Col Les Knoblock, U.S. Army, 25 Jun 1990, Fort Clayton, Panama; with Connolly, 29 Jan 1990. In the last cited interview, Connolly mentioned that some refugees thought that the center was being set up at the elementary school in Balboa and went there. Finding no one to help them, they broke into the school, not to loot but to find a place just to get out of harm's way.

The fact that most U.S. troops were committed to combat operations and thus unavailable to assist on the first day did not help matters.[48]

Back at the Comandancia, the disruption created by the refugees did not deflect attention from the most pressing matter at hand: securing all of the battlefield. Snell and Reed met after dawn to discuss the next steps to be taken. Both the prison and the headquarters compounds remained isolated, but there were holes in Company B's northern cordon as a result of the loss of the six M113s and two squads to Downing. Snell decided to bring in the Task Force Bayonet reserve force—a company headquarters and two platoons from Company C, 4th Battalion, 6th Infantry (Mechanized)—to fill the gaps. Furthermore, knowing that General Stiner wanted the operation around the Comandancia wrapped up as soon as possible, Snell and Reed determined that clearing the PDF headquarters, the largest building in the main compound, could best be performed by U.S. Rangers, who had trained for such an operation. Company C, 1st Battalion, 508th Infantry, would clear the remaining buildings. As a result of this decision, Reed received operational control of Company C, 3d Battalion, 75th Ranger Regiment, a reinforced unit commanded by Capt. Al Dochnal. The Rangers received word of their new assignment while en route to Howard Air Force Base after having secured the main terminal at Omar Torrijos International Airport.[49]

While waiting for the ranger company to arrive, U.S. troops around the Comandancia strengthened their positions. Medical teams, meanwhile, picked up wounded soldiers who could not make it to the ambulance exchange site near Point G and collected the remains of the dead Panamanian defenders they could reach. As the fires in El Chorrillo raged out of control, some of the residents began to loot the stores that remained standing, taking food, televisions, liquor, and other items. Elements of Task Force Gator tried to discourage the looters or keep them at bay, but there was little that could be done to stop the lawless behavior while the American troops continued to receive small-arms and sniper fire. As one countermeasure to the continuing resistance, vehicles from what remained of Team Armor reoccupied Ancon Hill and, with daylight restoring their line of sight, began firing rounds into the Comandancia whenever a target could be positively identified. U.S. snipers also inflicted casualties among the remaining Panamanian defenders.[50]

Around midmorning, about forty PDF members emerged from the Carcel Modelo compound and surrendered to the only remaining squad of Company B's 3d Platoon. Once these prisoners of war were cuffed and led off to the collection point, between a hundred and two hundred men who had been incarcerated in the prison emerged. Kept separated from their jail keepers,

---

[48] Interv, author with Connolly, 29 Jan 1990.

[49] JTF-Panama Input, 4 Feb 1990, to AAR, JTF-South; Intervs, Wright with Snell, 1 Jan 1990; author with Reed, 29 Jan 1990; Donnelly, Roth, and Baker, *Operation Just Cause*, pp. 155–56. For the ranger operation at Torrijos International Airport, see Chapter 7.

[50] Intervs, Wright with Freeze, 6 Jan 1990; author with Reagan, Zebrowski, and Flynn, 21 Jun 1990; Donnelly, Roth, and Baker, *Operation Just Cause*, pp. 155–57; Nemmers, *United States Army South Staff Ride*, p. 59.

AH–64 Apache helicopter

they, too, were taken to the prisoner center to be sorted out. Once everyone had left the facility, Captain Goss' men entered the arms room and began removing weapons, which were put onto a truck and taken out of the area. Shortly before noon, the Carcel Modelo compound was considered secured.[51]

The Rangers arrived soon thereafter and began coordinating clearing operations for the Comandancia complex with the task force commander and the Task Force Gator units that would be working inside the compound with them. The final assault was set to begin at 1500, to be preceded by preparation fires from the remnants of Team Armor and two Apache attack helicopters. In order to have more supporting fires for the impending operation, the Rangers also placed one of their squads on top of a gymnasium overlooking the Comandancia. The plan was to wait for the preparation fires and then have the Rangers enter the headquarters, while Company C tended to the other buildings. At the same time, one ranger platoon would recheck the Carcel Modelo.[52]

The Team Armor elements started shelling the compound at 1445, with the LAVs firing 25-mm. rounds through Comandancia windows with remarkable precision, and the Sheridans placing rounds from their 152-mm. gun-launchers near where the building's arms room was reportedly located. One round fell dead on target, resulting in a fire that, in Reed's opinion, probably helped to clear the headquarters of any remaining defenders. The two Apaches were late in arriving, not appearing on the scene until 1545. They quickly began their attack, firing Hellfire missiles and 2.75-inch rockets at the headquarters

[51] Interv, Wright with Goss, 13 Jan 1990; Nemmers, *United States Army South Staff Ride*, p. 54.

[52] JTF-Panama Input, 4 Feb 1990, to AAR, JTF-South; Intervs, Wright with Reed, 6 Jan 1990; author with Reed, 29 Jan 1990; Donnelly, Roth, and Baker, *Operation Just Cause*, p. 158.

The final phase of the Comandancia battle

building, the latter causing some collateral damage, including the wounding of an American soldier outside the compound.[53]

Once the Apaches had finished their work, the ranger company (minus) and Captain Flynn's Company C began clearing operations shortly before 1600. Some of the Rangers had, in fact, rehearsed the mission over the past year, when, prior to the 3 October failed coup, the Comandancia had been considered one of their targets. Thus, although Reed tried to get them to enter the compound from the side, they walked straight through the main gate on Avenue A as had been rehearsed. The choice turned out to make little difference. Using assorted weapons including grenades, LAWs, and their assault rifles and employing rules of engagement that allowed them to shoot first at any hostile target they identified, the Rangers reached the headquarters building without incident and began the laborious process of clearing each floor, room by room. Except for the bodies of several Panamanian defenders, the building was empty.[54]

Meanwhile, Captain Flynn's men, still in AO Spear, had fired two 90-mm. recoilless rifle rounds for "psychological effect" into the hole they had blown in the west wall earlier in the operation, several hours before dawn. His troops also fired or lobbed a few grenades in as well. Then they entered the compound and began clearing the assorted buildings. Company C's operation was still in progress when the Rangers finished their portion of the task, thus placing Reed under some pressure to send them to their next assignment, something

[53] JTF-Panama Input, 4 Feb 1990, to AAR, JTF-South; Intervs, Wright with Reed, 6 Jan 1990; author with Reed, 29 Jan 1990; Wright with Freeze, 6 Jan 1990; Donnelly, Roth, and Baker, *Operation Just Cause*, p. 157.

[54] Intervs, Wright with Reed, 6 Jan 1990, and author with Reed, 29 Jan 1990; Donnelly, Roth, and Baker, *Operation Just Cause*, pp. 158–59.

he would not do until Flynn's men were ready to relieve them. Flynn felt the pressure in the form of a directive to expedite the clearing process, a prospect that made him uneasy, especially since his soldiers would probably spend the night in the compound, potential targets for any undetected defenders. The captain successfully stalled for time, and after his men had wrapped up the operation to his satisfaction, he joined Captain Dochnal to inspect the whole compound. In the process, a remaining defender emerged with a .38-caliber pistol in hand. After a U.S. soldier shot and killed the man, the compound stood ready to be pronounced secured.[55]

For the Task Force Gator units that would move inside the two compounds that night and for those that continued to man roadblocks and other positions outside, sniper fire would persist as a problem into the next day. But the battle of the Comandancia was essentially over by Wednesday evening. For the Americans, casualties included three killed—a figure remarkably lower than Reed had anticipated in the close and deadly confines of urban combat—and thirty-four wounded. The number of Panamanian defenders killed and wounded could not—and still cannot—be accurately stated because most fought in civilian clothes in an area in which a number of innocent bystanders were also killed and wounded. It is clear, however, that the defenders suffered casualties well in excess of those they inflicted on the U.S. troops. In this particular case, the advantage had not gone, as doctrine would have it, to the defender enjoying a one-to-one ratio in troop strength. The American troops, together with the weapons they employed, simply constituted the better force. Reed wondered, though, at what price the inevitable U.S. victory would have been purchased had the Panama Defense Forces been a more professional military organization. "Had we fought a more disciplined, better-trained force," he mused, "the fighting would have been a good deal more intense, it would have gone on longer, and we would have taken a lot more casualties."[56]

Soon after Task Force Gator's success on D-day, Colonel Reed and others took stock of what they had learned from the experience.[57] To begin with, the task force commander praised the weeks, in some cases months, of training and rehearsals the troops had conducted in preparation for going to war in Panama. He similarly touted the suitability of mechanized infantry for fighting in cities, although doctrine at the time for urban combat favored light infantry for street fighting, thus providing little information geared specifically to mechanized units. The mechanized battalion's after action report called for the Army to "develop and refine" such doctrinal guidance, as well as to analyze the way mechanized and light infantry might interact in urban combat. Reed was not altogether pleased with the way his battalion had been task-organized and, at one point, parceled out piecemeal to other task forces. As one platoon leader in the 4th Battalion (Mechanized) observed, the dispersal of Task Force Gator's assets paid little attention to the system in place for meeting the requirements of a mechanized unit's logistics. Breaking up the battalion also deprived Reed

[55] Interv, author with Reagan, Zebrowski, and Flynn, 21 Jun 1990; Donnelly, Roth, and Baker, *Operation Just Cause*, pp. 159–60.

[56] Interv, author with Reed, 29 Jan 1990.

[57] The lessons discussed in the following section pertain exclusively, or nearly so, to Task Force Gator. Broader lessons will be discussed at the end of this book.

of the scout platoon he needed for reconnaissance. In summarizing the "lesson learned," the battalion's after action report stated, "When cross-attaching heavy forces to light forces, task organizing below company level adversely affects the ability of heavy companies to sustain themselves."

Reed did single out for special praise the performance of the M113 armored personnel carrier. Blue Spoon planners had been right in believing that the vehicle would be well suited to the narrow streets of Panama City. But the .50-caliber machine gun mounted on each of the infantry carriers had proved an even greater asset, a godsend even, when confronting sniper and small-arms fire from high-rise buildings. The .50-caliber could fire at much higher angles than the turret cannon of the Bradley fighting vehicle that was replacing the M113. Furthermore, while the .50-caliber provided cover to the M113's front, an M60 machine gunner and riflemen inside could cover the flanks, something that would not have been possible in a Bradley traversing its turret gun. In short, the battalion's after action report recommended that the U.S. Army retain the M113 "in war reserve stocks for use in a low intensity conflict or MOUT environment." In one cautionary note, however, Lieutenant Rubin, alluding to the more protective armor of the Bradley, observed that a direct RPG hit on any one of the M113s in column along the narrow streets near the Comandancia would not only have put the vehicle out of action but would have blocked the M113s behind it as well.[58]

Regarding the other major weapons used to furnish fire support for Task Force Gator, the after action report gave high marks to the accuracy of the AC–130s, the Sheridans, and the LAVs, while noting the inability of the latter two to fire when their nighttime line of sight was obstructed by smoke and debris. The report termed the Spectre gunship "the most effective means of fire support in the MOUT environment," although Reed and certainly Rubin could testify to its inability under certain conditions to determine enemy from friendly targets. One group of weapons that was positioned near Albrook Air Station, ready to be employed as fire support for Task Force Gator, was not used: the battalion's 107-mm. mortars. As the after action report summarized, "Given concerns over collateral damage and troop safety . . . it was understood that they would only be used if all other fire assets were not available." The report recommended retaining the mortar as an urban-combat weapon because of the immediate support it could provide, if needed.[59]

As with many commanders in combat, Reed believed that the intelligence he needed before and during the battle, while generally "adequate," was "not consistently collected and disseminated to lower echelons." One example he cited afterward was the failure to discover the dump trucks blocking Company B's route until Captain Goss' lead M113 had literally run into one of them. That situation, in turn, uncovered another shortcoming: the task force lacked a combat engineer vehicle (CEV) with which to remove formidable roadblocks. Such a

[58] Intervs, Wright with Reed, 6 Jan 1990; author with Reed, 29 Jan 1990; Telephone Interv, author with Rubin, 5 Apr 1990. Quotes from Memo, Reed for CG, FORSCOM, and Cdr, CALL, 27 Mar 1990. JTF-Panama Input, 4 Feb 1990, to AAR, JTF-South.

[59] Quotes from Memo, Reed for CG, FORSCOM, and Cdr, CALL, 27 Mar 1990. JTF-Panama Input, 4 Feb 1990, to AAR JTF-South; Intervs, Wright with Reed, 6 Jan 1990; author with Reed, 29 Jan 1990.

vehicle, the after action report contended, could have used its blade to force the dump trucks aside, thus allowing the column to maintain its momentum. Furthermore, "the 165mm shape charge from the CEV's main gun would have been very effective breaching several walls that required Engineers to breach while exposed to hostile fire."[60]

There were shortages in manpower as well. The two platoons from the 988th Military Police Company, although having arrived in Panama just days before JUST CAUSE, provided enough personnel to man the roadblocks they set up on Fourth of July Avenue but not enough to handle the mass of refugees that began to appear soon after the operation at the Comandancia commenced and that turned into a flood of people after dawn. The mechanized battalion's after action report cited the necessity to "plan for—and allocate forces for—control of civilian refugees." To that recommendation, Colonel Connolly added from firsthand experience the need in the planning phase for some group to rehearse the roles and missions required to run a refugee camp. Besides wishing they had more military police participating in the operation, many officers also said they could have used more snipers, mainly for use against enemy snipers. The report recommended that one sniper be included in each infantry platoon, although some platoon leaders such as Rubin called for more. The report suggested that the rules of engagement for the snipers who were on the scene might have been too restrictive, limiting their effectiveness in protecting friendly units who found their positions exposed to hostile fire.[61]

On the topic of command, control, and communications, Reed reiterated that the time and circumstances determining the transfer of operational control of Task Force Gator from the 193d Brigade to the Joint Special Operations Task Force had never been clearly identified, even after the transfer took place just one day before JUST CAUSE. Moreover, the timing of the transfer of the task force back to Colonel Snell's brigade was determined while the Comandancia battle was still being fought. "This never resulted in a major adverse operational impact," the report conceded, although it did cause Reed a good deal of frustration and "created the potential for adverse operational consequences to occur." Similarly, once Gator had been "chopped" back to Task Force Bayonet, Reed found himself unable to communicate with the six M113s, two LAVs, and the Sheridan of Team Donivan that he had turned over to the special operations commander, nor with an AC–130 that was supposed to come under the battalion's fire support officer. Again, the impact in both instances was mostly one of frustration.[62]

For many in Task Force Gator, the overwhelming lesson was the experience itself. The after action report praised the "judgment, aggressiveness, and initiative of junior leaders" in "a fight that was played out largely at squad and platoon level." While addressing the human element in the battle, this conclusion did not capture, nor did it try, the emotions and actions triggered by the sights and sounds of combat in a built-up, congested, dark, frenetic, and confusing arena.

---

[60] Quotes from Memo, Reed for CG, FORSCOM, and Cdr, CALL, 27 Mar 1990. Telephone Interv, author with Rubin, 5 Apr 1990.

[61] Quotes from Memo, Reed for CG, FORSCOM, and Cdr, CALL, 27 Mar 1990. Telephone Interv, author with Rubin, 5 Apr 1990; Interv, author with Connolly, 29 Jan 1990.

[62] Memo, Reed for CG, FORSCOM, and Cdr, CALL, 27 Mar 1990.

The Comandancia headquarters after the battle

This was a setting in which enemy forces could deliver large volumes of fire from unseen locations overlooking the battleground, in which innocent civilians were intermingled with Panamanian defenders, and in which friendly units were packed together tightly in such a way as to make several incidents of friendly fire virtually inevitable. One officer tried to describe the horror that engulfed him when his M113 ran over an automobile in the confusion and uncertainty of the opening assault, and he realized that innocent civilians were inside the car. And while most soldiers maintained fire discipline called for in the strict rules of engagement, there were those who simply sprayed bullets indiscriminately into suspected enemy positions inside homes and apartment buildings. Lieutenant Rubin, in describing the critical impressions he took away from the fight, talked about how under fire it was the simplest things that worked. Soldiers remembered running passwords but not more complicated means of oral and visual communication, such as hand signals. Moreover, when being shot at, the troops would not do anything that might obstruct their vision, such as donning night-vision goggles, or even sighting their rifles before firing. In his opinion, the principal attribute required of junior leaders was the ability to assess and adapt. He, himself, had demonstrated that ability in doing what was necessary to accomplish his mission at Point E.[63]

In the days after 20 December, the Task Force Gator units taking part in the battle for the Comandancia would pursue a variety of follow-on missions. For a short time, however, they were able to take advantage of the brief yet dangerous respite their efforts had earned them. As they rested in and around the Comandancia, they received news on the progress of other critical battles throughout the country.

[63] Quotes from ibid.; Telephone Interv, author with Rubin, 5 Apr 1990. Other officers discussed some of the more negative aspects of combat around the Comandancia only on the condition of anonymity.

5

Task Force Semper Fi and Task Force Black Devil

☆ ☆ ☆

In the hours before Task Force Gator launched its attack against the Comandancia, its commander considered all the things that could go wrong. Among those uppermost in his mind was the prospect of the Panama Defense Forces sending reinforcements to defend its main headquarters. Given the undesirable one-to-one ratio of attacker to defender that Colonel Reed already faced, his troops could find themselves in a very precarious situation should additional enemy units arrive on the scene. "I had no ability within my own resources to fight the deep battle," Reed readily admitted.[1] Moreover, he had no reserve to commit; that force was controlled by Task Force Bayonet and consisted of just one mechanized infantry company (minus). Whether the reserve would be available if needed or, if available, adequate to defeat a major enemy counterattack was problematic. Gator's superior firepower would give Reed some advantage should enemy reinforcements appear: the Sheridans and LAVs—or at least what was left of them after Team Donivan departed—could dispatch any reasonable number of V150 Cadillac Gage armored cars and V300 armored reconnaissance vehicles. But superior weaponry alone might not be able to repel a large and well-disciplined force such as Battalion 2000, which could reach the Comandancia from Fort Cimarrón in just over an hour. There were several other PDF troop bases and units even closer than that, including the 5th Infantry Company at Fort Amador, within sight of Task Force Gator's area of operations. If these forces were not as formidable as Battalion 2000, they could still create a disruption serious enough to prolong the battle and increase the casualty figures among combatants and civilian bystanders alike. Given these possibilities, Blue Spoon planners repeatedly considered steps that would prevent enemy units in and around Panama City from joining the battle to defend or reclaim their headquarters.

Shortly after the fighting in Just Cause had come to an end, one observer suggested to a group of staff officers from the XVIII Airborne Corps that, on a map, those preventive measures resembled the geometric pattern of a circular shooting target. At the bulls-eye was the Comandancia, the enemy's critical command and control node, surrounded by a series of concentric rings, each containing hostile elements that could influence the Comandancia battle. One of the staff officers

---

[1] Interv, Robert K. Wright Jr. with Lt Col James W. Reed, U.S. Army, 6 Jan 1990, Fort Clayton, Panama.

replied, "I don't know if it was as deliberate as that," although he acknowledged the importance of the Comandancia as a target and "a commensurate need" to "take out or neutralize" PDF installations nearby, especially those in which troops were garrisoned.[2]

The concept of concentric rings, even though an after-the-fact construct, still serves as a useful framework for visualizing U.S. combat operations in Panama City and the surrounding areas. The configuration would have the Comandancia located at the center; the first ring containing Fort Amador, enemy facilities on Ancon Hill, the Bridge of the Americas, and the west bank of the canal; the second ring including PDF garrisons at Las Tinajitas and Panama Viejo; the third ring covering the Torrijos-Tocumen airport complex and Fort Cimarrón; and the outermost ring pulling in the two infantry companies billeted at Rio Hato, nearly forty miles from the capital. Within the first ring, most of the targets listed were the responsibility of Task Force Semper Fi and units in Task Force Bayonet not committed to the Comandancia mission.

Task Force Semper Fi

On the eve of Just Cause, Task Force Semper Fi was the Marine Forces component of JTF-Panama, responsible for overseeing and executing crisis-related U.S. military activities on the west bank of the Panama Canal in the southern half of the country. Headquartered at Rodman Naval Station, the two largest combat elements under its control were a Marine infantry rifle company, the latest one to rotate in as a continuation of the security enhancement troop buildup started in 1988, and a light amphibious company, part of the Nimrod Dancer unit rotations.[3] The latter brought with it light armored vehicles that, ideally suited for operations in the waters of the Panama Canal and the land bordering it, helped to transform Semper Fi "from a reinforced security force into a maneuver force." Bearing 25-mm. Bushmaster chain guns and 7.62-mm. machine guns, the LAVs could also be intimidating, an image the Southern Command and JTF-Panama often sought to project when conducting contingency readiness exercises, Sand Fleas, and Purple Storms.[4]

---

[2] Quotes from Interv, Wright with G–3 Plans Section, XVIII Abn Corps, 30 Mar 1990, Fort Bragg, N.C. While interviewing Lt. Gen. Carl W. Stiner, the JTF-South commander, Robert Wright observed, "It appears to me that it was essentially a concentric circles type of tactical plan. Is that correct?" Stiner replied, "Yes. Well, in a way. The main criterion that we used was—dates back to the 3 October coup. We watched the units that influenced that [counterattack]—what they did. And then we took a look at the other units that could influence the Panama Canal and that could threaten the safety of the US civilians there (we had almost 15,000 US civilians that we had to protect). And then we had to 'get' Noriega, neutralize the Panamanian command and control; we had to control the minds of the people, so we had to take or override or disable certain radio and TV stations. So that dictated what the target list would be." Interv, Wright with Lt Gen Carl W. Stiner, 2, 7, 27 Mar and 11 Jun 1990, Fort Bragg, N.C.

[3] Of 17 light armored vehicles (LAVs), 14 were light assault LAV–25s; 2 were logistics variants, or LAV–Ls; and 1 was a command and control variant, or LAV–C2. Col. Robert P. Mauskapf and Maj. Earl W. Powers, "LAVs in Action," *Marine Corps Gazette* 74 (September 1990): 51.

[4] For the use of LAVs in JTF-Panama exercises and operations and the quote, see the Marine Corps' official history of the Panama crisis, Reynolds, *Just Cause*, p. 15, and also pp.

A Marine light armored vehicle (LAV) on patrol in Panama prior to Just Cause

In command of Task Force Semper Fi was Col. Charles E. Richardson, who had arrived in Panama in September 1989. He came at what was in some ways a propitious time. For well over a year, there had been significant friction between the commander and staff of JTF-Panama, on the one hand, and their Marine Forces component, on the other. The firefight on the Arraiján Tank Farm in April 1988 and subsequent developments had led General Loeffke and his people to question whether combat-oriented marines with their warrior ethos possessed the discipline and savvy to deal with the subtleties and nuances of a politico-military crisis. The marines, for their part, protested having their professional competency questioned and charged JTF-Panama with promulgating restrictive rules of engagement that abridged the Leathernecks' inherent right of self-defense.[5] By the time Richardson arrived, however, the bureaucratic battles caused by these two opposing perspectives had begun to subside. Loeffke and his staff were gone, the tank farm had ceased to be a major concern once Nimrod Dancer began, and Cisneros as the new JTF-Panama commander was inclined to allow the marines more leeway, especially after NSD-17 went into effect. Thus, looking back at the period between his becoming the Task Force Semper Fi commander on 17 September and the eve of Operation Just Cause, Richardson could speak of the rapport that existed between him and his immediate superiors on what had theretofore been divisive issues. Not that the colonel found his new command free from problems. The portion of the canal's west bank for which he bore responsibility contained no major Panamanian combat units. As a result, the Marine task force often stood at a disadvantage vis-à-vis other JTF-Panama components in the recurring competition for manpower and materiel.[6]

16–18, as well as Mauskapf and Powers, "LAVs in Action," pp. 51–54; and Yates, *U.S. Military Intervention in Panama, June 1987–December 1989*, p. 234.

[5] For a detailed account of the incidents on the Arraiján Tank Farm and the friction between the JTF-Panama headquarters and its Marine Force component, see Yates, *U.S. Military Intervention in Panama, June 1987–December 1989*, pp. 66–77, 118–27.

[6] Interv, author with Col Charles E. Richardson, U.S. Marine Corps (USMC), 25 Jan 1990, Rodman Naval Station, Panama.

Richardson's concern over Task Force Semper Fi's assets grew after the failed coup of 3 October resulted in significant changes in the Blue Spoon operation order. As the XVIII Airborne Corps reworked the plan, Semper Fi received more missions that significantly expanded its projected wartime area of operations on the west bank. To acquire the increase in manpower deemed necessary to cover the additional missions and terrain, Richardson, his operations officer Lt. Col. Michael J. Franks, and the Marine liaison officer with JTF-Panama and the XVIII Airborne Corps, Maj. Brod N. Madrigan, appealed to Stiner to deploy more Marine units. The general refused, offering instead to place a battalion of the 82d Airborne Division—later changed to a battalion from the 7th Infantry Division (Light)—under the Marine task force's operational control within two days after the beginning of hostilities. Stiner's planners also approved a similar arrangement for the 534th Military Police Company and the 536th Engineer Battalion, although both of these units were to come under Richardson's operational control at or before H-hour.[7]

As a result of these decisions in the weeks preceding Just Cause, the order of battle for Task Force Semper Fi at H-hour included the following forces: an infantry rifle company, a light amphibious infantry company, a Brigade Service Support Group detachment, a Marine Corps Security Force Company-Panama detachment, a Fleet Anti-Terrorist Security Team (FAST) platoon, a military police company, and an engineer battalion.[8]

During October, the two infantry companies forming the core of Task Force Semper Fi's fighting capabilities underwent unit rotations. On 8 October, Company K, 3d Battalion, 6th Marines, took over as Task Force Semper Fi's infantry rifle company, while Company D, 2d Light Amphibious Infantry Battalion, 2d Marine Division, arrived later in the month with seventeen LAVs. A few weeks later, twelve of the light armored vehicles and crews were employed in Operation Rough Rider, an excursion into several populated areas west of Howard Air Force Base that, before it was over, turned confrontational. Designed to familiarize Company D with the towns of Nuevo Emperador, Nuevo Guararé, Vista Alegre, and Arraiján, the operation also included an Army psychological operations detachment, two U.S. treaty affairs officers, and a helicopter for aerial observation. During the approach to Vista Alegre, the assembled force encountered a roadblock on Thatcher Highway surrounded by a crowd of mostly hostile Panamanians. Following a heated but futile attempt to negotiate a way through the obstacles, Company D's commander, Capt. Gerald H. Gaskins, requested and received headquarters' permission to breach the roadblock with his LAVs. As the vehicles moved forward, one demonstrator was knocked to the ground, after which others in the startled crowd began beating on the vehicles. As Gaskins' men quickly put the crowd behind them and proceeded back to Rodman Naval Station, they had to breach a second roadblock and, following that, persuade a farmer who had single-handedly set up a third to remove his pickup truck from the road. When it was over, Rough Rider had proved an instructive initiation to the

---

[7] Reynolds, *Just Cause*, p. 17.

[8] The list of units is from JTF-South OPLAN 90–2, 3 Nov 1989.

crisis for Company D and, for Task Force Semper Fi as a whole, a source of several tactical lessons in crowd control and the art of removing roadblocks.[9]

In general, Richardson found the Sand Fleas, contingency readiness exercises, and Purple Storms to be of tremendous value. Through these, the recently arrived forces under his command gained a working knowledge of the west bank's terrain, its road networks, its towns and cities, as well as the people who lived in them, to include the location of ranking political and administrative officials in the Noriega regime who would have to be apprehended in the event of hostilities. The exercises and operations also gave the troops an idea of the tactics the Panama Defense Forces and Dignity Battalions would use against them. Most important, perhaps, was the information, including human intelligence, the marines were able to acquire and process as they increased the database that ideally would enable Task Force Semper Fi to achieve its objectives more proficiently and with fewer casualties in the event of war.[10]

Following the 3 October coup attempt, as the likelihood increased of a U.S.-Panamanian military confrontation and as JTF-Panama exercises and operations like ROUGH RIDER began more and more to approximate BLUE SPOON rehearsals, this interaction between intelligence collection and activities in the field became even more crucial, as described in the Marine's official history of the crisis.

> Marine counterintelligence operatives and interrogator-translators continued to collect data on potential threats and targets of all kinds, including specific buildings and individual defense force personalities. . . . With that knowledge in hand, they were able to conduct realistic planning exercises and prepare for operations against a range of potential targets. As part of the process, operations officers drafted sets of orders, written for both offensive and defensive scenarios, which were distributed to all companies.[11]

Other means the marines used to prepare for hostilities included map exercises, sand table exercises, and live-fire demonstrations. In addition, company commanders briefed their troops weekly on developments in the crisis and the effect those might have on the wartime missions the Marine task force might receive. Aside from the antiterrorist platoon, few of the Marine units had extensive training in urban operations, especially room clearing, although Company D tried to gain what experience it could on the limited facilities available to it at Rodman. Ever aware of the manpower shortage he could anticipate on D-day, Richardson, in a measure reminiscent of General Loeffke's approach to JTF-Panama units early in the crisis, prepared his administrative and combat service support personnel for combat. "They practiced basic combat skills and virtually fired their weapons day and night until he was satisfied with their proficiency."[12]

Except for a handful of officers, none of the marines knew precisely what they would be directed to do should war come. Richardson, as a participant in the planning process, was aware that his missions and objectives had pretty much been

[9] Reynolds, *Just Cause*, pp. 18, 43; Mauskapf and Powers, "LAVs in Action," pp. 53–54.

[10] Interv, author with Richardson, 25 Jan 1990.

[11] Reynolds, *Just Cause*, p. 18.

[12] Quote from ibid., pp. 18–19. Interv, author with Richardson, 25 Jan 1990.

finalized by the end of October. In general, Task Force Semper Fi's role under the deliberate BLUE SPOON scenario was to conduct defensive and offensive operations to secure the area west of Panama City and the canal. (*See Map 7.*) Defensive operations included establishing external security for Howard Air Force Base, Fort Kobbe, and the Rodman Naval Station ("organic assets" at those facilities would provide internal security), perimeter security at the Arraiján Tank Farm and the Rodman Ammunition Supply Point just north of it, and roadblocks at the Bridge of the Americas and other transit points on the west bank. Offensive operations involved seizing the Directorate of Traffic and Transportation (the Spanish acronym DNTT) Station 2; the PDF substation and the regime's political headquarters in Arraiján; PDF stations and substations in Veracruz, Rio Potrero, Vista Alegre, and Nuevo Emperador; a Dignity Battalion training facility; a port facility; and a radio tower. A key objective common to several of these operations (and certainly one consonant with the concentric circle model) was to "block the western approach to Panama City to prevent Panama Defense Forces . . . reinforcement."[13]

As the holiday season encroached on the crisis in Panama, the marines welcomed the brief respite afforded by the decreased operational tempo for U.S. forces. Then, on Saturday night, 16 December, word circulated that Panamanian guards had shot and killed an American officer at a checkpoint near the Comandancia. Given the comparatively small number of Marine officers in Panama, few Leathernecks were prepared for the news that it was one of their own who had died. Richardson later described the sadness that descended over his headquarters. At some point, JTF-Panama directed him to place his forces on a higher state of alert, although, unlike several of the units under Task Force Bayonet, Task Force Semper Fi forces remained in place that night.[14]

Over the next three days, Richardson attended JTF-Panama and JTF-South meetings, learning from General Cisneros Monday night that President Bush had decided to execute the BLUE SPOON operation order. Faithful to the guidelines disseminated by General Stiner the next day, Richardson did not reveal this information even to his principal staff officers until 1800 Tuesday evening, the authorized time. Before then, though, he did tell them that Lieutenant Paz had not died in vain and asked them to review Task Force Semper Fi's operation plan to "get the cobwebs out." He also instructed them to go over their ammunition requirements and other needs and to "start prestaging" various assets for any contingencies or exercises that might be conducted. In assigning these tasks, Richardson knew that he was signaling his staff that something significant was likely to happen.[15]

For Marine units at Rodman Naval Station, Tuesday, 19 December, became "a maintenance and logistics day," with platoon leaders being told that their place of duty was the maintenance ramp, where they were to work on essential

---

[13] The list of missions and the quotes are from JTF-Panama Input, 4 Feb 1990, to After Action Report, Operation JUST CAUSE, 20 Dec 1989–12 Jan 1990, JTF-South, 19 Mar 1990 (hereafter cited as JTF-South AAR), an. D. Task Force Semper Fi's missions and targets can also be found in JTF-South OPLAN 90–2, 3 Nov 1989.

[14] Interv, author with Richardson, 25 Jan 1990.

[15] Quotes from ibid. Reynolds, *Just Cause*, p. 20; Maj Brod N. Madrigan, How I Spent the Holidays, unpublished article, n.d.

equipment as though "their lives depended on it." The task was completed by 1800. As for Major Madrigan, his duties as liaison officer with JTF-South took him to a series of meetings at Fort Clayton with the XVIII Airborne Corps staff, General Thurman, and the JTF-South commanders and staff. Madrigan later noted how many loose ends were tied up at these last-minute sessions, even as new problems arose. Illustrative of the latter, Task Force Semper Fi found itself fighting to retain the engineer and military police units Stiner had promised would come under Richardson's operational control by H-hour. Madrigan's efforts to this end were only partly successful. At 2230, just two and a half hours before the start of Operation Just Cause, General Downing's Joint Special Operations Task Force took one of the military police platoons in question. The JTF-South operations officer present during the switch commented that the marines were being hung out to dry. When Madrigan informed Richardson, the colonel simply saw the move as another manpower setback that amplified the uncertainty already surrounding his plans for the impending fight.[16]

Task Force Semper Fi unit commanders briefed their troops on Just Cause at 2100 Tuesday, going over the orders and missions they were about to execute. Richardson briefed the operation as well that night, emphasizing the use of minimal force, the need to avoid unnecessary destruction of historical sites and public works, and the fact that the Panamanian people were not the enemy. Following this, the commanders, staffs, and troops made their final preparations for combat.[17]

Operations began shortly before H-hour when reports reached Rodman around 0040 that a number of enemy V300 armored vehicles were in the area, presumably en route to Panama City. Captain Gaskins, the Company D commander, ordered his troops to establish their assigned blocking positions. Ten minutes later, thirteen LAVs carrying his 1st and 3d Platoons, seventeen marines from the antiterrorist team, and a scout platoon (which brought with it 14 squad automatic weapons [SAWs], 4 shoulder-launched multipurpose assault weapons, and 10 rifle grenade launchers) began their movement along Thatcher Highway from Rodman toward Arraiján. Accompanying them was a U.S. Army psychological operations team in an unarmored HMMWV. As the column approached DNTT Station 2, one of Semper Fi's objectives, the troops received fire, a prelude to the marines' first and deadliest battle of Just Cause.[18]

The mission of seizing the DNTT facility and neutralizing any defenders belonged to 3d Platoon, the lead element in the Company D column.[19] As the remainder of the convoy continued toward Arraiján, three of the platoon's LAVs made a left turn into the transportation directorate's driveway and crashed through the outer gate. As the vehicles approached the building,

---

[16] First quoted words from Reynolds, *Just Cause*, p. 20. Second quoted words from Mauskapf and Powers, "LAVs in Action," p. 54. Madrigan, How I Spent the Holidays; Interv, author with Richardson, 25 Jan 1990.

[17] Reynolds, *Just Cause*, p. 22.

[18] Ibid.; Mauskapf and Powers, "LAVs in Action," p. 57; Donnelly, Roth, and Baker, *Operation Just Cause*, p. 185. As noted in Chapter 4, Gaskins' 2d Platoon with its four LAVs had been "chopped" to Task Force Gator.

[19] This brief account of Marine operations at the DNTT Station 2 is from Reynolds, *Just Cause*, pp. 22–23; Mauskapf and Powers, "LAVs in Action," p. 57; and Maj. Brod N. Madrigan, USMC, "The Attack on the DNTT Station," *Marine Corps Gazette* 74 (April 1990): 9–10.

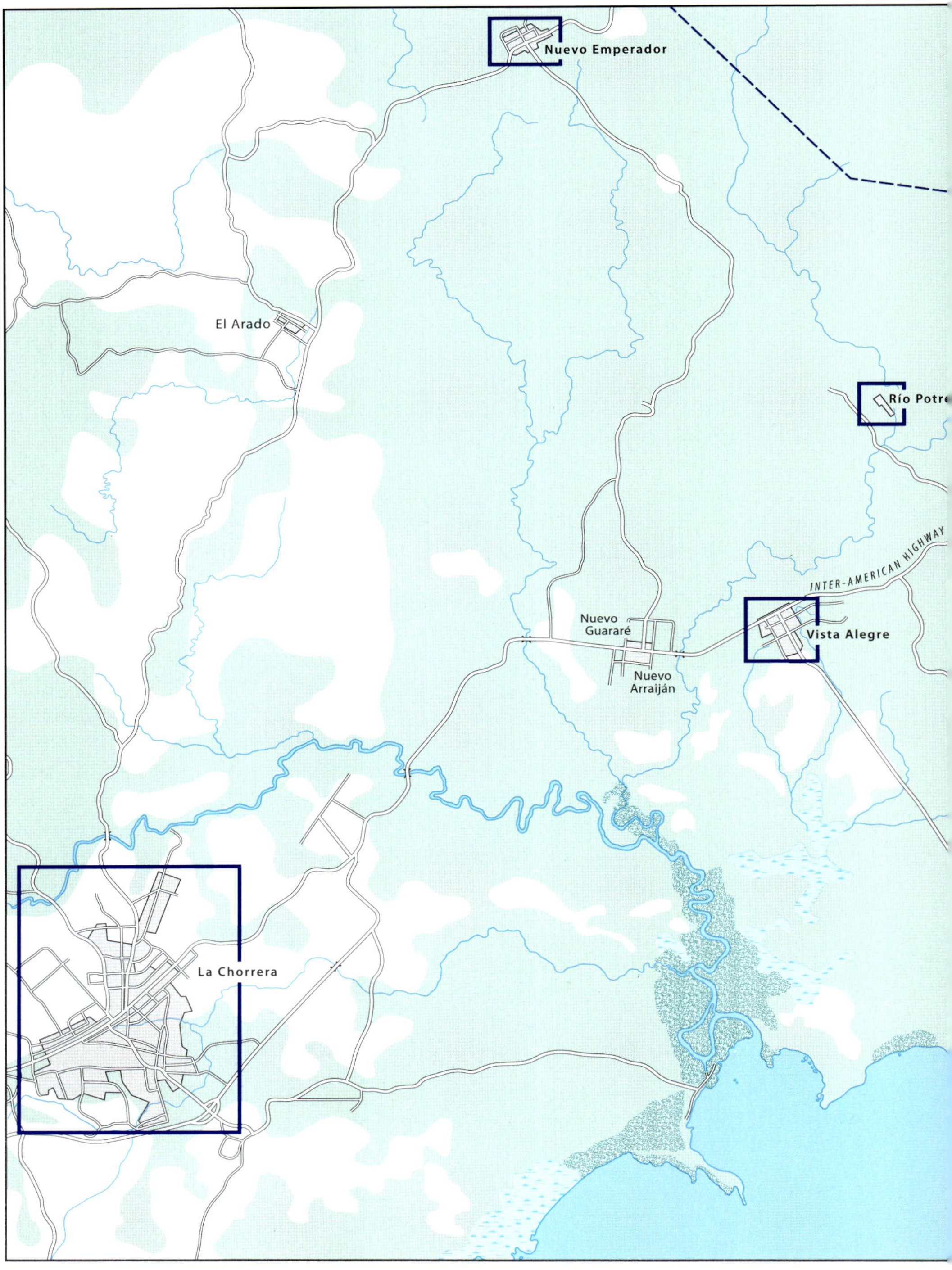
Nuevo Emperador
El Arado
Río Potre
INTER-AMERICAN HIGHWAY
Nuevo Guararé
Vista Alegre
Nuevo Arraiján
La Chorrera

Map 7

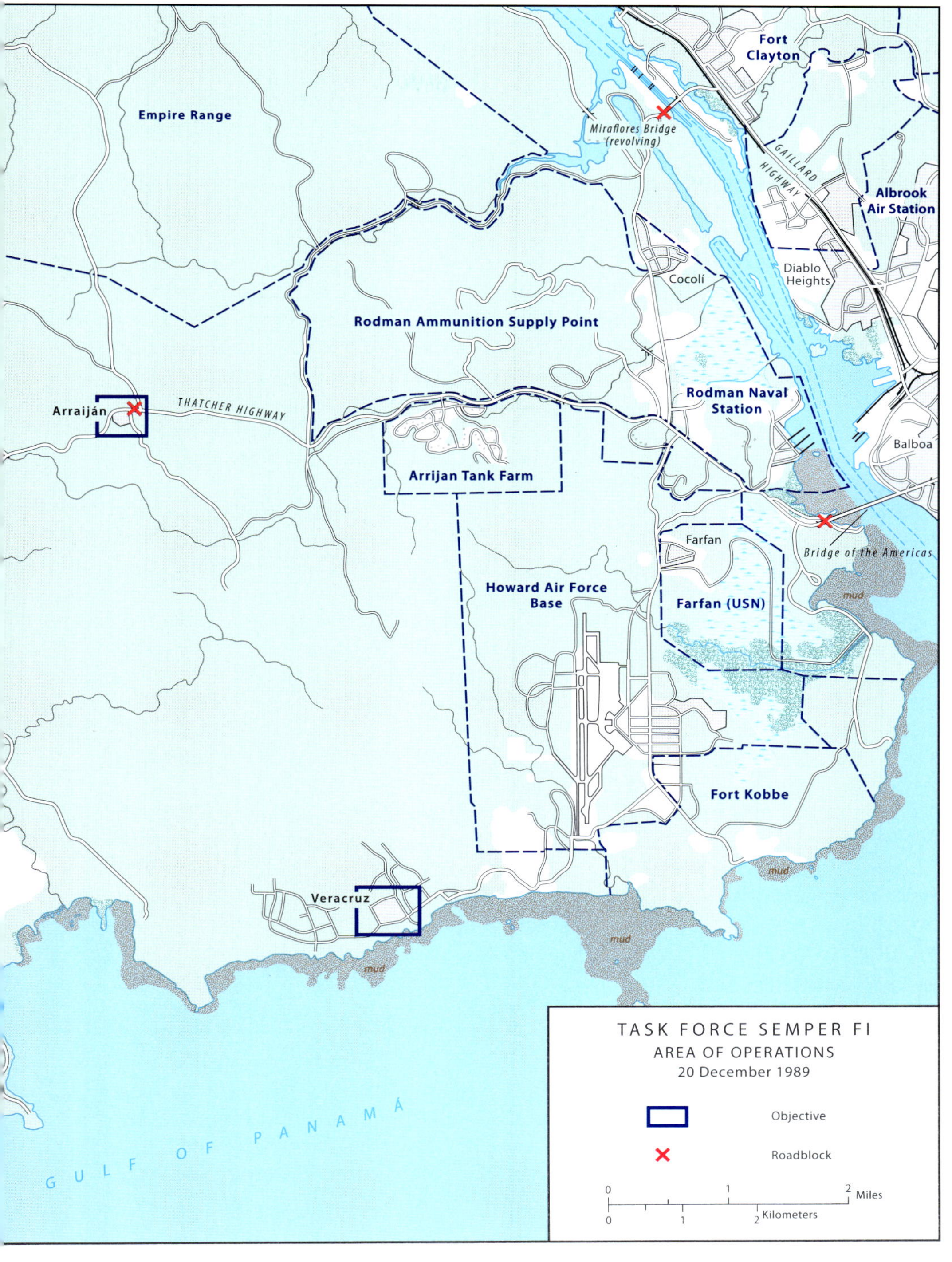
Empire Range
Fort Clayton
Miraflores Bridge (revolving)
GAILLARD HIGHWAY
Albrook Air Station
Cocolí
Diablo Heights
Rodman Ammunition Supply Point
Rodman Naval Station
Arraiján
THATCHER HIGHWAY
Balboa
Arrijan Tank Farm
Farfan
Bridge of the Americas
mud
Howard Air Force Base
Farfan (USN)
Fort Kobbe
Veracruz
GULF OF PANAMÁ
TASK FORCE SEMPER FI
AREA OF OPERATIONS
20 December 1989
Objective
Roadblock
0
1
2
Miles
Kilometers

Marine gunners fired the M60 coaxial machine gun mounted on each LAV to suppress fire from the defenders. (To minimize collateral damage and to avoid destroying PDF radio equipment that the marines could use to monitor enemy transmissions once the station was taken, the LAVs' 25-mm. guns were not employed.) Two of the three vehicles moved to within a short distance of the building so that eight scouts could dismount and seek cover in its entranceway. Once Sgt. Thomas P. Bernius and Cpl. Garreth C. Isaak breached the door with buckshot from their rifle grenade launchers, the scouts entered the DNTT and began exchanging fire with a handful of Panamanians inside. Outgunned, the defenders moved down a central corridor away from the action. The marines followed and, with their rifles and SAWs, began to clear each room along the hallway. Corporal Isaak was the first marine to enter the final room and was hit by several bullets. The reaction of his comrades was immediate. After they sprayed the room with automatic weapons fire, Sergeant Bernius ordered one of the men to toss in a fragmentation grenade, even though Task Force Semper Fi had been told not to use such devices for room-clearing purposes. With the shooting of Corporal Isaak, as one writer later noted, the "Marines learned a costly mistake by interpreting what judicial application of force means. It would not be repeated."[20]

After clearing the last room, the marines took Isaak from the building to have him airlifted to the rear, although it seemed to all that he was dead. Minutes later, one of the makeshift units of administrators, cooks, armorers, and mechanics that Colonel Richardson had cobbled together as an operational force arrived to relieve the 3d Platoon, and the scouts departed to rejoin the column heading for Arraiján. As the small and inexperienced relief force attempted to occupy the DNTT, the men were surprised to find one defender still alive, wounded but ready to resist. Once the defiant soldier had been killed, the station was pronounced secured. The operation had left one marine dead—the only marine killed in Operation Just Cause—and one wounded. In addition to the dead Panamanian, three defenders had been wounded and three taken prisoner.[21]

The initial assault on DNTT Station 2 had taken about ten minutes. Soon afterward, the scouts in the 3d Platoon's light armored vehicles caught up with the main column, only to find that it had been stopped on Thatcher Highway outside of Arraiján by a roadblock of two gasoline tank trucks covered by an undetermined number of enemy gunmen. Captain Gaskins kept his vehicles, especially the unarmored HMMWVs, at a safe distance, outside the range of any rocket-propelled grenades the defenders might have. He also sent scouts forward to reconnoiter the situation. When they returned, they reported there were ten to twenty enemy personnel in the area, at which time Gaskins established an overwatch position with 360-degree security. As the standoff approached fifty minutes, the captain realized he could not remain stationary. He thus had no choice but to fire on the gas trucks, regardless of the inferno that might result. With approval from the Task Force Semper Fi headquarters, he ordered the LAVs to use their 25-mm. guns. To his relief, the trucks turned out to be empty, and once

[20] Madrigan, How I Spent the Holidays.

[21] Ibid.

the marines destroyed them, the defenders fled. The Company D column then resumed its movement into Arraiján.[22]

There were two principal targets in the town: a PDF substation and the local headquarters of one of Noriega's top political supporters, Rigoberto Paredes. At the substation, scouts used only the weapons they carried in attempting to breach the two-story building and the wall that surrounded it. When their efforts failed, they pulled back and let the LAVs' 25-mm. guns create the openings for them. A Marine FAST close-quarters battle team then entered the building and exchanged fire with the defenders, all of whom fled after offering only token resistance. Not wanting to risk another fatality in the process of entering rooms they had to clear, the team again used fragmentation grenades to accomplish the task. This prudent approach had an unintended consequence, for the building caught fire and burned to the ground. At the next target, the political headquarters, the scouts again could not force an entry through iron bars bolted into place, so the antiterrorist team used its shotguns to destroy the locks. They then employed the psychological operations team to broadcast a message in Spanish, urging any defenders inside to surrender. When no one responded, the marines threw a stun grenade inside the one-story edifice, after which a dazed Panamanian emerged. The team moved inside the headquarters, where it cleared the building without incident, taking five prisoners in the process. When more defenders found hiding in a nearby trailer surrendered, the detainees in custody numbered over seventeen, all of whom had to be taken back to Rodman for interrogation. (Most of them, it turned out, were members of Noriega's paramilitary Dignity Battalions.)[23]

Once the targets were secured, elements of Company D blocked off all the ground approaches to and from Arraiján. Twice, cars packed with Panamanian military personnel tried to run the roadblocks but were stopped when the marines opened fire. In one of the cars, the driver was killed and some of the passengers wounded. A marine was also seriously wounded and transferred to the DNTT building, which had become the company aid station. Throughout the night, rumors circulated that the Dignity Battalions were planning an attack on the Marine positions, but, aside from the two unsuccessful attempts to run the roadblocks, the night passed uneventfully. Shortly after dawn, the marines began searching the homes of key Noriega supporters in Arraiján but found the residences vacated.[24]

While Company D, the antiterrorist team, and the psychological operations personnel were conducting operations along the highway from Rodman to Arraiján, other marines were performing security missions throughout the Task Force Semper Fi area of operations.[25] Between the Arraiján Tank Farm and the town of Arraiján, elements from the in-country Marine security force, assisted by U.S. Army engineers, used heavy earth-moving equipment to block Thatcher Highway. At the same time, a task force of about twenty-five marines and Army engineers employed almost identical methods to assert their control over the Bridge

---

[22] Reynolds, *Just Cause*, p. 23; Mauskapf and Powers, "LAVs in Action," p. 57.

[23] Ibid.; Interv, author with Richardson, 25 Jan 1990.

[24] Mauskapf and Powers, "LAVs in Action," p. 57; Reynolds, *Just Cause*, p. 24.

[25] The following two paragraphs are from Reynolds, *Just Cause*, p. 24; 1st Lt. Kenneth M. DeTreux, "Infantry in Low-Intensity Conflict: Observations From Operation Just Cause," *Marine Corps Gazette* 74 (September 1990): 64–65.

of the Americas minutes after H-hour. As the fighting in the capital intensified, hundreds, then thousands, of refugees fled the city westward over the bridge; no Panamanian military units, however, sought to enter the city by moving eastward over the structure.

At the Arraiján Tank Farm, security force marines relieved Company K from its defensive positions, thereby freeing the unit's 1st and 3d Platoons to begin squad-size patrols in the jungle west of Howard Air Force Base, a critical area and one with which the company was very familiar thanks to the exercises and operations mounted by JTF-Panama after 3 October. With thirteen thousand U.S. troops scheduled to fly into Howard during the course of the invasion, providing security around the air base was an essential requirement. While the success of American arms in Just Cause was a virtual certainty, only one or two Panamanian soldiers armed with shoulder-fired surface-to-air missiles in the vicinity of Howard could make it a costly victory. The intensive patrols mounted by the bulk of Company K hoped to prevent such an occurrence. Meanwhile, the company's 2d Platoon moved to secure the town of Veracruz and the PDF station located there. Anticipating a tough fight, the marines instead found evidence—a half-eaten meal, for example—that the building had been hastily abandoned. The platoon consequently pulled back to the outskirts of the town, established blocking positions, and began patrolling the area.

On Wednesday morning, 20 December, Task Force Semper Fi, having accomplished its H-hour missions, received an unexpected order to seize and secure the PDF 10th Military Zone headquarters in La Chorrera, a city of eighty thousand, outside the western edge of the marines' original area of operations. Richardson passed the mission to Company D and the FAST marines attached to it, with Gaskins and his command moving out along the Inter-American Highway around 1530. Fifteen minutes later, the column ran into an enemy roadblock,

Marines wait to attack the PDF headquarters in La Chorrera.

this one composed of buses and small-arms fire from defenders placed among the vehicles and on a hill nearby. Gaskins saw no reason to stop this time and ordered the lead LAVs to advance while firing their main guns. The shock effect was immediate, as the bus drivers drove off, followed on foot by enemy gunmen in the area.[26]

In the meantime, the pilot of an OH–58 hovering over Company D flew off to reconnoiter the block-long compound of four buildings and a six-foot-high wall that constituted the PDF 10th Military Zone headquarters. Under fire from the ground, he observed what appeared to be a formidable defensive position that could inflict significant casualties on Gaskins' convoy. He thus radioed back to Rodman, requesting close air support. Task Force Semper Fi, in turn, passed the request to officers at JTF-South who had the authority to approve or deny the use of close air support in a built-up area. The request was denied. At that point, Major Madrigan, located in the JTF-South operations center at Fort Clayton, personally took the request to General Stiner's operations officer, Col. Thomas H. Needham, who approved the air attack on the condition that it did not violate the rules of engagement for air support. Madrigan returned to the officers who had originally denied the request only to have them do so again, despite Needham's intercession. Undeterred, the major next went to Stiner himself, who quickly approved the mission. Still, Air Force Brig. Gen. Robin Tornow, Madrigan's next interlocutor, questioned the wisdom of the attack because of the collateral damage it might cause. Of his exchange with the general, Madrigan later wrote, "I forcefully told him that the mission was essential and that Marine lives were at stake and that he had better approve it because LtGen Stiner had." Tornow relented, and about an hour later, an OA–37 Dragonfly marked the target, after which two Virginia National Guard A–7 Corsairs flew in for the attack, strafing the target with their 20-mm. cannon. The pinpoint accuracy of the close air support was confirmed later with the discovery that none of the civilian homes near the headquarters had been damaged.[27]

As soon as the A–7s departed, Company D's light armored vehicles entered the compound firing their Bushmaster chain guns. With the exception of a few snipers, the defenders fled, leaving the scouts and FAST marines to search the buildings and to seize documents and weapons. With those tasks accomplished and the headquarters building on fire, the marines pulled out of the compound and headed back to Arraiján. Colonel Richardson's concerns during the planning process about manpower shortages now came into play. As Madrigan observed, "The lack of available forces precluded us from maintaining a presence in La Chorrera." Richardson would acquire that capability only later, after he had assumed operational control of the units the XVIII Airborne Corps planners had promised he would receive a day or two after the invasion began.[28]

In assessing Task Force Semper Fi's performance on the first day of JUST CAUSE, Richardson later praised the manner in which his light armored vehicles, the antiterrorist team, and the psychological operations

[26] Reynolds, *Just Cause*, p. 25.

[27] Ibid., p. 26; Mauskapf and Powers, "LAVs in Action," pp. 57–58. Quote from Madrigan, How I Spent the Holidays.

[28] Reynolds, *Just Cause*, p. 25; Mauskapf and Powers, "LAVs in Action," pp. 57–58. Quote from Madrigan, How I Spent the Holidays.

team had worked together as "a versatile and potent force, particularly for offensive operations." The interaction of these elements had "saved many lives on both sides," and the example the integrated force set deserved further conceptual consideration by the Marine Corps. Less gratifying, in his assessment, was the hour-long delay in getting close air support at La

Marines of Company D, 2d Light Amphibious Infantry Battalion, outside the damaged PDF headquarters in La Chorrera

A marine escorts one of many Panamanian detainees taken in Task Force Semper Fi's area of operations.

Chorrera, despite the fact that the mission had been approved by General Stiner.[29] Still, in looking over what they had accomplished by the early evening of 20 December, members of the task force had reason to feel satisfied: they had effectively executed their principal mission by securing the west bank of the canal on the Pacific Ocean side of the country. The resistance they had encountered in the process had been dispatched, although at the price of one Marine fatality.

TASK FORCE BLACK DEVIL

The marines and Army engineers manning the roadblock at the western end of the Bridge of the Americas had a clear view of the battles on the east bank of the canal, including the one at Fort Amador, where U.S. troops were engaging Panamanian forces sporadically in close-quarters fighting. Of the onlookers, many, if not most, understood how important U.S. control of the fort was for a speedy and decisive outcome to Operation JUST CAUSE (*Map 8*). To begin with, the installation was in the first concentric ring around the Comandancia, with the PDF 5th Infantry Company being billeted there. Not only would the company be just minutes from Task Force Gator's operations at the main PDF headquarters, and thus a threat to intervene in that battle, but, on the fort itself, the troops' barracks were located just across a golf course from a row of houses in which U.S. military personnel and their families lived. If 5th Company infantrymen chose to avoid the firestorm that was engulfing the Comandancia—a conflagration that was easily seen from practically any point on Fort Amador—they could still demonstrate their patriotism and loyalty to Noriega by wreaking havoc on the defenseless residences just yards away or by seizing civilian hostages. Such a possibility had to be taken seriously, given two facts. First, after the 3 October coup attempt, the 5th Company had constructed new defensive positions from which rifle and machine gun fire could be aimed directly at the American housing area. Second, well before JUST CAUSE, U.S. military intelligence had learned of Noriega's plans, code-named *GENESIS* and *EXODUS*, to resort to hostage taking and guerrilla warfare in the event of war with the United States. In addition to these concerns over what the 5th Company (and a small number of Panamanian naval infantry troops located on the same grounds) might do, Fort Amador also claimed the attention of American planners because it lay along the route the PDF's UESAT forces would have to take into Panama City from their training areas on four islands at the end of a causeway extending from the peninsula on which the fort was situated.[30]

Given its importance, Fort Amador had been an ELABORATE MAZE, then BLUE SPOON, objective since the beginning of contingency planning in early 1988. From the outset, securing the installation had been a task relegated to conventional forces under JTF-Panama, although, during the nearly two years of planning that followed, different units had been assigned the mission. On

[29] JTF-Panama Input, 4 Feb 1990, to JTF-South AAR, an. D.

[30] Donnelly, Roth, and Baker, *Operation Just Cause*, pp. 164–65.

Quarry Heights
BALBOA STREET
PANAMÁ
Carcel Modelo
Water tanks
Comandancia
El Chorrillo
Balboa
AO Spear
POETS AVENUE
AMADOR AVENUE
tanks
tanks
Objective Cortizo
BAY OF PANAMA
U.S. Naval Buildings
Main Gate
FORT AMADOR
Building 1
Gymnasium
Quarters 20
ditch
Yacht Club
U.S. Housing
Torrijos Memorial
PDF Headquarters
Objective Piña
ditch
Chapel
Officers' Club
Back Gate
CAUSEWAY
TASK FORCE BLACK DEVIL
AREA OF OPERATIONS
20 December 1989
Objective
Roadblock
ELEVATION IN METERS
0 50 100 150 and Above
0 500 Meters
0 500 Yards

Map 8

the eve of Just Cause, the responsibility resided with a force task-organized around the 1st Battalion, 508th Infantry (Airborne), at Fort Kobbe. Under Task Force Bayonet's operational control, the group would conduct an infiltration and, at H-hour, mount an air assault into Fort Amador.

Since July 1989, Lt. Col. Billy Ray Fitzgerald had commanded the airborne battalion. In his previous assignment, Fitzgerald had served in U.S. Army, South's operations shop. As with most other USARSO staff officers, he was "dual-hatted" in the corresponding JTF-Panama directorate as well, where he had become acquainted with the Blue Spoon operation order for conventional forces. Moreover, by the time he assumed command of the 1st Battalion, JTF-Panama had been conducting freedom of movement and other Nimrod Dancer exercises, training events, and operations for almost two months. Air assault exercises into Fort Amador had become an integral part of these maneuvers, which, besides asserting U.S. treaty rights in Panama, accustomed PDF infantrymen on the scene to the sight of troop-carrying American helicopters landing within sight of 5th Company's barracks. To keep the Panamanians from knowing exactly what would happen in an actual combat assault, the helicopters employed deception, rarely if ever setting down in the landing zones identified in the contingency war plans. After the failed coup of 3 October, the practice air assaults, as with so many other JTF-Panama activities in the field, increasingly became formal rehearsals for the real thing.

In common with most other units slated for combat in Blue Spoon, Fitzgerald's 1st Battalion supplemented its rehearsals with map and sand table exercises—the precision of the Fort Amador table model at battalion headquarters was impressive—intelligence updates, tactical exercises without troops (TEWTs), room-clearing training using rolled up socks as grenades, live-fire exercises, weekly briefings as to how current events might affect the unit's likely wartime missions, and the compilation of detailed battle books. Opportunities for direct observation also abounded. The 29th Military Intelligence Battalion in Panama had three observation and listening posts on the Amador peninsula from which intelligence personnel could acquire information on the number of Panamanian troops based there, as well as on their comings and goings, daily activities, and morale. The 29th also had the capability of monitoring messages the 5th Company commander, Maj. Moisés Cortizo, received from his superiors. U.S. military policemen and their PDF counterparts manned guard stations at the gates to Fort Amador, thus affording the MPs a vantage point from which to observe troop and vehicular movements, engage the Panamanian guards in conversation, and pick up assorted facts, rumors, and gossip. And, of course, there were the U.S. military officers who lived on Amador with their families—this included the USARSO commander, whose house was near the southern end of the fort near the officers' club. And if the 5th Company had a clear view of the American houses across from its base, nothing prevented the inhabitants of those homes from looking back or from using a game of golf or a casual stroll around the installation's grounds to gain an even closer glimpse at 5th Company activities. As the 1st Battalion, 508th Infantry, made its final preparations for Just Cause, it had an abundance of information concerning PDF weapons, troop locations, and

Part of a sand table model used by Task Force Black Devil in preparing for its mission at Fort Amador. The row of red-roofed structures on the right is mostly PDF buildings; the row on the left is American housing. (*North is at the bottom of the photo.***)**

even intentions. In one of the final messages Fitzgerald would receive from the 29th before H-hour, he learned that the task force going into Fort Amador under his command might be facing slightly over five hundred armed enemy troops.[31]

That task force was named Black Devil and, in the JTF-South Operation Plan 90–2, its mission read, "Seize and secure FT AMADOR. Isolate, neutralize, and/or destroy PDF units at FT AMADOR."[32] In contrast to this simple statement, the tasks required to accomplish the mission were numerous and complex. Task Force Black Devil had to defeat the 5th Company before it and other enemy elements at Amador could move to reinforce Panamanian defenders at the Comandancia. This meant, among other things, clearing and

---

[31] Ibid., pp. 161–62, 165, 167, 168; Intervs, author with Lt Col Billy Ray Fitzgerald, U.S. Army, 27 Jan 1990, Fort Kobbe, Panama; with Capt William R. Reagan, Capt Robert Zebrowski, and Capt Timothy John Flynn, all U.S. Army, 21 Jun 1990, Fort Clayton, Panama; Wright with Col Michael G. Snell, U.S. Army, 1 Jan 1990, Fort Clayton, Panama.

[32] Quote from JTF-South OPLAN 90–2, 3 Nov 1989. The 1st Battalion, 508th Infantry, was known as the Red Devils, but since the 82d Airborne Division was slated to field a battalion with an identical nickname, the task force led by Colonel Fitzgerald's battalion became Task Force Black Devil during JUST CAUSE. CD-ROM, *Operation Just Cause: Task Force Bayonet, Dec 1989–Jan 1990* (XVIII Airborne Corps, June 2000). The CD-ROM, which contains interviews and other material on the invasion, was issued by the XVIII Airborne Corps historian.

securing the row of 5th Company buildings and other PDF facilities; closing the front gate to the fort, with U.S. military policemen capturing or killing any Panamanian guards on duty; sealing off the causeway connecting the peninsula with the island-based UESAT forces (themselves a target of U.S. Special Operations Forces); and handling enemy wounded and prisoners. Additionally, Black Devil had to protect the American housing area on post and, if possible, evacuate the inhabitants before H-hour to avoid a hostage situation and civilian casualties.

To accomplish these and other tasks, Fitzgerald had available to him two of his airborne infantry companies, A and B (with C, the third, scheduled to become part of Task Force Gator prior to H-hour). Each company had its rifle platoons, weapons platoon, fire support teams, and medics. The battalion headquarters company contributed snipers, scouts, mortar men, and antitank missile crews, with the headquarters company commander, according to one account, controlling three elements. "One was a makeshift combination . . . consisting of the scout platoon and a platoon equipped with TOW antitank missiles, the result being two scout/antitank platoons," with the arrangement giving the scouts greater mobility. The second was the headquarters mortar platoon, and the third was "a scout detachment of eight M113s" and two improved TOW vehicles taken from Colonel Reed's mechanized battalion and attached to Black Devil. Rounding out the task force, Fitzgerald received a squad from the 193d Infantry Brigade's 59th Engineer Company; a towed 105-mm. howitzer from Battery D, 320th Field Artillery (the crew that accompanied the artillery piece had conducted live-fire exercises with the 1st Battalion, employing the howitzer in a direct-fire mode); and, as a last-minute attachment, a military police platoon. Three AH–1 Cobra attack helicopters and an OH–58 were available to support the air assault, and an AC–130 could be called if necessary.[33]

The Task Force Black Devil plan, briefed up the BLUE SPOON/90–2 chain of command, called for a phased operation. The first phase would have the two scout-antitank platoons close off Fort Amador's main gates—the front entrance in the Balboa district and the rear gate by the causeway at the southeastern tip of the fort. The headquarters mortar platoon would then go house to house on the American side of the fort, alerting the civilian residents and offering to take them to safety. The second phase would be the air assault, with the lead element, Company B, arriving on six UH–60 Black Hawk helicopters and securing the southern end of the fort. Company A would follow behind on eight Black Hawks and secure the northern end, including

[33] Quotes from Donnelly, Roth, and Baker, *Operation Just Cause*, p. 166. See also Intervs, Wright with 7th Howitzer Section, Battery D, 320th Field Arty, 9 Jan 1990, Fort Kobbe, Panama; Sgt Dan Wagner with M Sgt Christopher Hinman, U.S. Army, 18 Oct 1990, Fort Sherman, Panama; Wright with Reed, 6 Jan 1990; author with Fitzgerald, 27 Jan 1990. Concerning the last-minute attachment of the military police platoon to Black Devil, the platoon leader knew next to nothing about the impending operation. Consequently, battalion officers tried to force-feed him the information he needed, so much of it, in fact, that a company commander later quipped, "We tried to get him to take a drink of water from a fire hydrant." Intervs, author with Reagan, Zebrowski, and Flynn, 21 Jun 1990; Capt Joseph M. Nemmers with Maj Robert Pote and Capt Ted W. Mauzey, both U.S. Army, 18 Oct 1990, Fort Sherman, Panama.

the American housing area and some U.S. naval buildings; if required, the company could also seize and clear a PDF housing area in its sector, dubbed Objective Cortizo after the 5th Company commander. Broadcast teams would offer Panamanian troops a chance to surrender. If they refused, their various facilities and defensive positions—labeled Objective Piña—located along a north-south axis on the western side of the fort would be softened up, starting with .50-caliber machine guns and, if resistance continued, escalating to M60 machine guns, 90-mm. recoilless rifle fire, and antitank rounds. The 105-mm. howitzer would be used to blow large holes in the buildings, presumably traumatizing the inhabitants while providing openings through which U.S. forces could enter. The 1st Battalion's Company B would lead the assault, capturing the PDF motor pool and other structures clustered near it at the southern extreme of Piña and then moving northward up a row of numbered 5th Company buildings parallel to the canal. Once the fort was pronounced secured, Task Force Black Devil would begin its follow-on missions.[34]

The event that set this plan in motion, the shooting of Lieutenant Paz, occurred while Fitzgerald, like many other Army officers, was attending USARSO's Christmas formal at the U.S. officers' club on Fort Amador. His immediate superior, Colonel Snell, told him to put his quick-reaction company on heightened alert, a status that the whole battalion assumed as the evening wore on. As part of the precautionary measures directed by JTF-Panama later that night, the mechanized platoon slated to come under Fitzgerald's operational control deployed to Fort Amador. Over the next three days leading up to the invasion, the platoon would remain there, its M113s concealed among the golf carts in the course's caddie shack. The military police platoon that would be attached to Task Force Black Devil was also present at the fort, carrying out its normal duties.[35]

On Monday, 18 December, Snell told Fitzgerald to be sure that his units were prepared "if something were to occur." On Tuesday, around noon, the colonel informed him, "This is it," a message Fitzgerald passed on informally to his company commanders an hour later. (The formal notifications and briefings for Task Force Black Devil took place as directed by General Stiner—at 1800 for the commanders, 2100 for the troops.) On both days, Fitzgerald met with the company commanders who would participate in the Amador mission, going over different scenarios and anticipating the orders they might receive if BLUE SPOON were executed. Once Fitzgerald learned that the operation had been approved, he made arrangements, starting mid-Tuesday afternoon, to insert the headquarters' scout-antitank and mortar platoons onto Amador, where they conducted routine security operations to keep from tipping off the Panamanians about what was to come. Later that night, the soldiers of Companies A and B received their briefings and began their final preparations for the air assault. As they were doing so, Fitzgerald and some of his staff left Fort Kobbe, crossed the Bridge of the Americas, and entered Fort Amador

[34] Donnelly, Roth, and Baker, *Operation Just Cause*, p. 166; Intervs, author with Fitzgerald, 27 Jan 1990; with Reagan, Zebrowski, and Flynn, 21 Jun 1990; Nemmers with Pote and Mauzey, 18 Oct 1990.

[35] Interv, author with Fitzgerald, 27 Jan 1990; Nemmers, *United States Army South Staff Ride*, p. 38; Donnelly, Roth, and Baker, *Operation Just Cause*, p. 169.

around 2200, heading for the task force command post in the basement of Quarters 20. At some point prior to H-hour, he was joined by General Cisneros and his entourage, who would remain at Amador through much of the night.[36]

One problem that arose during Black Devil's preparatory activities concerned ammunition, which most of the task force's troops drew from the Rodman Ammunition Supply Point north of Howard Air Force Base and the Arraiján Tank Farm. To Fitzgerald's consternation, the process of distributing the ammunition accorded no priority to infantry about to go into combat. Equally perplexing, he doubted whether his men really needed the TOW and AT4 rounds being issued to them. "TOW Weapon systems," he maintained, were not needed for his low-intensity mission, while LAWs could cause the same damage as AT4s and were much less cumbersome for airborne and dismounted infantry to carry. "If you talk with our soldiers," he later stated, "they'll tell you, 'Give us the 90mms [recoilless rifles] or the LAWs.'" Once the ammunition was issued, couriers from the headquarters company under Capt. John Hort had to deliver rounds and weapons to the scout-antitank and mortar platoons already in position at Amador. The trucks transporting the grenades, rockets, and machine gun rounds, however, did not begin arriving until 2100. For the next two and a half hours they entered the fort in a clearly visible supply line with some of the ammunition crates wide open. Remarkably, the sight occasioned no reaction from the Panamanian police and soldiers looking on.[37]

If the problems with ammunition bothered Fitzgerald, other developments did little to improve his disposition. Over a half hour before his air assault was scheduled to begin, he and others could hear shots being fired from the vicinity of the Balboa district, just north of Amador. Equally troubling, he learned that at least a couple dozen uniformed and armed Panamanians were boarding a bus behind one of the PDF buildings at the fort. Whether they were planning to head for the Comandancia or just trying to escape the coming battle, Fitzgerald had no way of knowing. What he did know was that he could not let the men depart. After conferring with Snell and Cisneros, with the latter being in direct contact with the JTF-South operations center at Fort Clayton, Fitzgerald at 0025 ordered Captain Hort to have the two cross-attached scout-antitank platoons set up prefabricated barriers to close down Amador's front and back gates.[38]

The platoon responsible for blocking the front gate, Team Recon, was still unloading ammunition trucks when it received Fitzgerald's order. Consequently, the scouts and TOW personnel did not arrive at the gate until after 0032, just in time to help two military policemen at the guard station

[36] Quotes from Interv, author with Fitzgerald, 27 Jan 1990. See also Interv, Nemmers with Pote and Mauzey, 18 Oct 1990; Nemmers, *United States Army South Staff Ride*, pp. 37, 39; Flanagan, *Battle for Panama*, p. 111.

[37] Fitzgerald's frustrations over the ammunition issues are from Interv, author with Fitzgerald, 27 Jan 1990. Quotes from Interv, Capt John Hollins with Fitzgerald, 20 Jun 1990, Fort Clayton, Panama. The story of the trucks carrying the ammunition is from Donnelly, Roth, and Baker, *Operation Just Cause*, p. 169.

[38] Interv, author with Fitzgerald, 27 Jan 1990; Donnelly, Roth, and Baker, *Operation Just Cause*, pp. 169–70.

Three cars (*right*) are lined up at the front gate of Fort Amador.

subdue their Panamanian counterparts. A minute or so later, as the scouts were stringing concertina wire, the bus carrying the two dozen armed men came speeding straight at the Americans, its headlights turned off. Four snipers protecting the scouts opened fire on the bus, after which its passengers, their weapons sticking through the windows, returned the fire. Team Recon joined the fight, which ended only a few seconds later after the driver received a fatal wound. The bus careened through the front gate out of control, hit and "beheaded" a fire hydrant, and then smashed into a tree about three hundred yards farther on. By that time, it was no longer visible to the U.S. soldiers at the gate, even through night-vision goggles and the TOW's thermal sights. Sometime later it became obvious that the survivors had fled, leaving their weapons, equipment, and uniforms behind. There had been no U.S. casualties during the fight, but shots from the bus killed a PDF guard whom the scouts had just placed in custody.[39]

Within minutes of the bus episode, a sedan with six armed Panamanians inside also tried to run the roadblock at the front gate, which had just been strengthened by positioning three HMMWVs there as obstacles. Again, a shootout ensued, with the driver of the car being killed almost immediately. The vehicle then crashed into one of the HMMWVs, nearly severing the leg of

[39] Intervs, author with Fitzgerald, 27 Jan 1990; Wagner with Hinman, 18 Oct 1990; Nemmers, *United States Army South Staff Ride,* pp. 37–38; Donnelly, Roth, and Baker, *Operation Just Cause*, pp. 170–71; Flanagan, *Battle for Panama*, p. 114.

one of the passengers. Two of the remaining occupants continued to fire, and U.S. troops replied with their M16s. When the shooting stopped, three of the Panamanians were dead, including the one with the lacerated leg; the other three were soon in custody.[40]

Word that Team TOW had secured the back gate to the causeway without incident came as good news but did not alter the fact that, with the firefights at the front gate, Task Force Black Devil had lost the element of surprise. Driving home this point, the battle at the Comandancia began minutes thereafter, with the fighting there taking place in full view of the fort across the bay. With hostilities clearly under way, Task Force Black Devil's headquarters and headquarters company and the families in the American housing areas were now at risk and would remain so until the two companies of airborne infantry arrived via helicopters from Fort Kobbe. Knowing this, Fitzgerald had been trying since the shooting started to get higher headquarters to authorize an earlier time for the air assault, but his pleas were in vain. Further changes in the timing of the plan, JTF-South officers at Fort Clayton reasoned, would only increase the possibility of friendly fire incidents or midair collisions in the densely packed airspace over what amounted to a very small portion of southwestern Panama City. The Black Hawks would take off on schedule at 0047, and the airborne assault would begin at 0100, as planned.[41]

One change to the Task Force Black Devil plan that Fitzgerald authorized on his own as a result of the early commencement of combat operations in and around Fort Amador concerned the American families residing on the installation. The plan had originally called for the mortar platoon of the headquarters company to knock on the back door of each home facing the 5th Company and evacuate the inhabitants. With exchanges of gunfire taking place at the fort for over thirty minutes prior to H-hour and with errant rounds and ricochets from the Comandancia fight landing inside the fort, Fitzgerald and his staff, according to the battalion's executive officer, "made the decision on the spot that it was not safe to bring civilians out of their quarters and into that fire." Other considerations affecting the decision included the possibility that, in a last, brutal act of defiance, enemy troops might deliberately target any exposed group of American civilians or that the UESAT units on the nearby islands to the south might, if they fired their mortars at the fort, kill or wound noncombatants huddled in the open. Hence, when soldiers from the mortar platoon began their door-to-door trek in the housing areas, they did offer to evacuate the residents who wanted to leave but discouraged this course of action by pointing out the great risks involved. It would be better, they advised, to move to the rear of the concrete-reinforced homes, away from the fighting that was almost certainly going to get much worse. Only in the U.S.

---

[40] Interv, author with Fitzgerald, 27 Jan 1990; Nemmers, *United States Army South Staff Ride*, pp. 37–38; Donnelly, Roth, and Baker, *Operation Just Cause*, pp. 171–72; Flanagan, *Battle for Panama*, pp. 114–15.

[41] Interv, author with Fitzgerald, 27 Jan 1990; Donnelly, Roth, and Baker, *Operation Just Cause*, pp. 172–73.

Navy housing area, which Fitzgerald knew would be directly in the line of fire, did his soldiers insist on evacuating the Americans living there.[42]

An enemy attack at Fort Amador prior to H-hour never materialized. Meanwhile, at Fort Kobbe, Capt. Robert Zebrowski, the commander of Company B, 1st Battalion, 508th Infantry, and Capt. William R. Reagan, the Company A commander, loaded their men and supplies onto Black Hawks manned by crews from the 7th Infantry Division. The seats had been removed from the helicopters to get more troops aboard—up to seventeen per aircraft, as opposed to the normal eleven or twelve—as well as their heavy rucksacks, several crammed with up to three times a soldier's basic load. Some of the helicopters also carried cumbersome antitank rounds and crates of hand grenades, although Zebrowski refused to load the Dragon rockets offered him on the grounds that the company did not have the sights needed to fire them. As for the four men who would man the task force's 105-mm. howitzer, they were scheduled to fly to the objective in the second wave, which would include the artillery piece slung under one of the Black Hawks. On schedule at 0047, the pilots began lifting off, aware that fierce fighting was already under way at several locations in Panama City. According to plan, the Black Hawks transporting Company B took the lead, followed by those with Company A. The flight route was circuitous, with the blacked-out helicopters flying very low out over the canal and Bay of Panama to avoid detection and antiaircraft fire by UESAT and any other hostile units in the vicinity of Fort Amador.[43]

As the Black Hawks approached their landing zones in the fort, many of the soldiers aboard could see the explosions, fire, and smoke at the Comandancia. More unsettling, they could also see tracer rounds coming straight at them. Some of the rounds were being fired by enemy forces on Fort Amador; others were stray ordnance or ricochets from the Comandancia, where, depending on "the angle of the dangle," as Reagan phrased it, rounds fired by U.S. forces skipped off enemy targets and ended up across that narrow section of the Bay of Panama separating the fort from the southernmost portion of Panama City. To Zebrowski, this had been a major flaw in the plan for Company B. The landing zone for his men was a wide ditch behind the row of American houses, out of sight of the 5th Company barracks but fully exposed to the Comandancia and within range of the weapons being employed there. Now, his worst fears seemed about to be realized as the pilots informed him and Reagan that the landing zones were "hot." Besides the danger to the troops, the situation dictated that the Black Hawks not linger for more than several seconds while the men aboard dismounted. As Reagan described the scene, the Black Hawks landed, rolled for about ten feet, and then took off. If you had not gotten out of the helicopters at that point, it was too late. Some soldiers had to jump eight to ten feet to the ground in making their exit. At least a couple of men were left stranded on board. While several of the Black Hawks

---

[42] Interv, author with Fitzgerald, 27 Jan 1990. Quote from Interv, Nemmers with Pote and Mauzey, 18 Oct 1990.

[43] Intervs, Nemmers with Pote and Mauzey, 18 Oct 1990; author with Reagan, Zebrowski, and Flynn, 21 Jun 1990; Wright with 7th Howitzer Section, Battery D, 320th Field Arty, 9 Jan 1990.

were hit by the ground fire, none suffered serious damage. Nor were there any casualties among the crews and the troops who had been crowded inside.[44]

As soon as the first wave of Black Hawks departed, Zebrowski moved Company B into position at the southern end of Fort Amador. One infantry platoon established itself behind the row of houses that included General Cisneros'; a second platoon took up positions by American housing near the Amador chapel. Both platoons had a clear view of the PDF motor pool and the buildings next to it. Zebrowski's third infantry platoon gathered in an assembly area by the officers' club, while the weapons platoon set up a prisoner collection point in the club's parking lot and prepared the unit's recoilless rifles for firing. With Company B on the ground, the headquarters company's Team TOW, which a half-hour earlier had closed the back gate, was free to move out onto the causeway where it had orders to search three buildings next to the fort, the so-called witches' houses, where Noriega was rumored to lodge his unorthodox consultants on the spiritual and supernatural. The clearing process would take over an hour.[45]

Company A followed Company B into Fort Amador. Captain Reagan quickly established contact with Fitzgerald and then moved his three infantry platoons into position at the northern end of the fort, with one of the platoons linking up with the scouts at the front gate. The maneuver proceeded without incident, although some U.S. military police, unaware that one of Reagan's platoons was heading into their area, came close to opening fire on the unit. Once in place, Company B held its positions until shortly before dawn, providing flank security while closely monitoring the situation.[46]

As Reagan and Zebrowski oversaw the disposition of their troops, the pilot and copilot of the OH–58 Kiowa helicopter, which, together with three AH–1 Cobras, had accompanied the troop carriers in the air assault, observed from above. At one point, they received a radio message directing them to pinpoint the location of an enemy ZPU4 antiaircraft gun that, before the assault, intelligence officers had reported as being hidden somewhere behind the row of 5th Company buildings. As the Kiowa's pilot, Capt. Timothy Jones from the 7th Infantry Division, flew toward the area suspected of concealing the weapon, the helicopter came under heavy ground fire, was hit, and crashed into the canal. Jones freed himself from the wreckage but was unable to find his copilot, CWO2 Andrew P. Porter, who, it was later discovered, had drowned. Over the next half hour, Jones tried to swim to shore, only to hear a group of men speaking Spanish once he managed to reach land. Slipping back into the water, he swam south until he arrived in the vicinity of the U.S. officers' club. Once again coming ashore, he encountered another group of Spanish-speaking

[44] Quote from Interv, author with Reagan, Zebrowski, and Flynn, 21 Jun 1990. Nemmers, *United States Army South Staff Ride,* pp. 39–43; Donnelly, Roth, and Baker, *Operation Just Cause*, pp. 172–74; Flanagan, *Battle for Panama*, pp. 111–12.

[45] Interv, author with Reagan, Zebrowski, and Flynn, 21 Jun 1990; Nemmers, *United States Army South Staff Ride,* pp. 38–39, 43.

[46] Interv, author with Reagan, Zebrowski, and Flynn, 21 Jun 1990. In this interview, Reagan attributed the near miss by the military police to the fact that they had been brought into the operation at the last minute and were not fully aware of all the details of Task Force Black Devil's plan.

men, this time shouting at him. Thinking he had run into more Panamanian soldiers, he scurried back into the water. The troops, in fact, were from the scout-antitank platoon near the causeway, speaking Spanish because they were certain Jones was an enemy soldier. Then Jones heard the men speaking in English, so he called out, "Don't shoot, I'm American." Unfortunately, he did not provide the proper password when challenged, and the ranking U.S. officer on the scene ordered warning shots fired. Frightened and angry, Jones unleashed a torrent of expletives that managed to convince the men he was who he said he was. His ordeal had come to an end.[47]

Around 0200, while Company A held its positions at the northern end of the fort, Company B began to move on its targets at the southern end. Unable to hear the PSYOP appeals being broadcast from the task force command post, Zebrowski's 2d Platoon used its own bullhorns to demand the surrender of anyone inside the PDF motor pool, the unit's first objective. Four mechanics complied. Meanwhile, from his operations center, Fitzgerald extended the twenty minutes he had planned to allow for the battalion's psychological

After the battle, U.S. troops inspect the ZPU4 antiaircraft gun at Fort Amador.

[47] Quote from Donnelly, Roth, and Baker, *Operation Just Cause*, pp. 174–76. There was some speculation that the OH–58C had been shot down by ground fire from the Comandancia, but Zebrowski, for one, believed that the helicopter's position behind the PDF barracks, out of sight and range of the PDF headquarters, made that an unlikely possibility. Probably, Jones and Porter had been hit by ground fire from Panamanian soldiers at Fort Amador. See Interv, author with Reagan, Zebrowski, and Flynn, 21 Jun 1990. One official log contained the following entry: "200218 OH–58 Down in the Canal Vic Ft Amador Pilot Swam Ashore. Co-Pilot Unknown." This would support the pilot's account, which placed the crash within the first thirty minutes after H-hour.

operations to sixty-two minutes, mainly because the people with Cisneros were telling the general that they had Major Cortizo on the telephone and that the 5th Company commander wanted to surrender, a development that could dramatically reduce the prospects for further violence. Fitzgerald also hoped he could coax the cooks in Building 5, the PDF mess hall, to capitulate. Knowing that they had been told American soldiers would kill them if they gave up, he reasoned that "if I could get that first group to come out with their hands up and show that we weren't going to massacre them, there was a very good possibility that there were others that would follow." When none of these initiatives paid off, Company B's 1st and 3d Platoons began their preparatory fires on three PDF buildings, using an antitank rocket followed by .50-caliber machine guns. When Zebrowski received word to start clearing operations, a squad from the 2d Platoon used grenades to enter two structures near the motor pool. When the men encountered no resistance, they began securing the buildings, eschewing the further use of grenades as they went room to room. At some point, operations stopped for about an hour because of reports that a U.S. military policeman was "lost" somewhere on Amador. Once the missing person was found, Company B returned to its task.[48]

Before Fitzgerald ordered the main attack on the 5th Company headquarters and the row of PDF buildings along the fort's western edge, he wanted to be sure that the ZPU4 and two V300s, the largest and most sophisticated weapons in the enemy's inventory at Amador, had been knocked out of action. Accordingly, he directed the two Cobras still flying over the fort to locate the targets, after which he had his Air Force liaison officer request that the on-call AC–130 destroy them. The Spectre arrived around 0430, but its crew initially refused to fire into the congested area below, even when told that Task Force Black Devil could actually see the two V300s. After some heated exchanges, the gunship did engage the vehicles, claiming that one of them had been reduced to a burning hulk of twisted metal. Satisfied that the threat posed by these weapons had been eliminated, Fitzgerald ordered Company B to begin the main attack, starting with Building 9 at the southernmost end of the long row. Not until later did the colonel learn that all three AC–130 targets were still intact, virtually untouched by the Spectre's rounds. As his executive officer, Maj. Robert Pote, summarized the episode, the AC–130 crew had been "uncooperative." "We were under more danger from the Spectre gunship," he went on to say, "than [from] the enemy." Another battalion staff officer was more succinct in describing the AC–130: "It was worthless."[49]

Shortly before dawn, the headquarters company provided more preparation fires while Zebrowski's 1st Platoon, which had been in its assembly area at the officers' club, readied to clear its assigned half of Objective Piña, Buildings 9, 8, and 7 as viewed from the south. The general pattern was for each building to be raked by .50-caliber and M60 machine guns and then hit with two AT4 high-explosive, antitank rounds and one round from the 105-mm. howitzer used in a

[48] First quote from Interv, author with Fitzgerald, 27 Jan 1990. Second quoted word from Interv, author with Reagan, Zebrowski, and Flynn, 21 Jun 1990. Nemmers, *United States Army South Staff Ride,* pp. 41–44.

[49] Nemmers, *United States Army South Staff Ride,* pp. 41–42; Interv, author with Fitzgerald, 27 Jan 1990. Quotes from Interv, Nemmers with Pote and Mauzey, 18 Oct 1990.

Building 9 after the battle at Fort Amador

direct-fire mode. Other weapons and explosives, such as bangalore torpedoes, the 90-mm. recoilless rifles, and C4 charges, were to be used as required. As gunners began directing this formidable array of firepower at each target, they managed to hold collateral damage to a minimum, an achievement that became a point of pride, especially in connection with one symbolic structure, the mausoleum of former Panamanian leader, General Omar Torrijos, located in front of the main PDF barracks. SOUTHCOM and JTF-South officers feared that damaging the monument could have a negative impact on the large number of Panamanians who still considered the general a hero. After Fort Amador was pronounced secured later in the day, the monument was found to be virtually unscathed, with only a slight, almost imperceptible chip on one of its bricks. More important, Company B had avoided any friendly fire incidents. In one close call, though, the .50-caliber machine gunners aiming at Building 9 had to stop their preparatory fires when ricochets from the weapons began landing dangerously close to one Company B platoon in its assault position.[50]

Starting at 0607 and for three hours thereafter, Company B's 1st Platoon moved in sequence into Buildings 9, 8, and 7. The troops methodically searched the rooms inside, took some prisoners, and seized a variety of weapons. Despite the sighting of several armed Panamanian troops in the area, the attackers quickly realized that the number of defenders in no way approached the estimated five hundred plus—admittedly a maximum, worst-case projection—

[50] Nemmers, *United States Army South Staff Ride,* pp. 41–44; Donnelly, Roth, and Baker, *Operation Just Cause*, p. 179; Interv, author with Reagan, Zebrowski, and Flynn, 21 Jun 1990.

Torrijos mausoleum

Fitzgerald had been told to expect. The morning's operations proceeded without incident. Thanks to some timely adjustments made by Captain Reagan, the same could be said of Company A's progress in Objective Cortizo. Reagan had not anticipated having to clear the PDF housing area near his positions, nor was he pleased when he received the mission. "What am I supposed to do with these civilians? Just stroll up to their door and say, 'I'd like to come through your house now?'" Eventually, he formulated a plan that began with a psychological operations team broadcasting instructions for anyone still in the residences to show themselves to the U.S. forces outside. Following these broadcasts, members of his 3d Platoon entered the houses, while the 2d Platoon assumed an overwatch position. During the course of the searches, the troops discovered the makings of a bomb in Major Cortizo's home and a few weapons in other dwellings. But neither Cortizo nor any other enemy personnel were to be found. According to Panamanian civilians in the area, the 5th Company commander had fled an hour or two before the arrival of the American troops. This account was verified later by a PDF medic, the only commissioned Panamanian officer to be found at Fort Amador. Left to defend the fort in the absence of their superiors were a number of enlisted men, some under the threat of death from noncommissioned officers who stayed behind with them, others motivated by a sense of personal, national, and organizational honor.[51]

[51] Quote from Interv, author with Reagan, Zebrowski, and Flynn, 21 Jun 1990. Intervs, author with Fitzgerald, 27 Jan 1990; Wright with Snell, 1 Jan 1990; Nemmers, *United States Army South Staff Ride,* pp. 39–41.

By early afternoon, Task Force Black Devil seemed close to completing its mission with few casualties and little resistance. At Objective Cortizo, Reagan's men continued to inspect the PDF housing area without incident. From his position there, the captain could see the progress being made by Zebrowski's platoons and thought with relief that at least his own men would not have to get involved in "that mess" of going through the 5th Company buildings room by room. At that moment, Company B's 3d Platoon was in the process of securing Building 4, the PDF headquarters. There was some apprehension on the part of the senior officers present that the men might be letting down their guard. If there was a sense of complacency, it ended abruptly when a couple of defenders fired an antitank round at Colonel Snell and Colonel Fitzgerald as they inspected the V300s the AC–130 had failed to destroy. Neither officer was injured. Soon after that, Reagan's 1st Platoon got into a firefight that spilled over into Company B's positions as well, even as Zebrowski's men were also receiving fire from a nearby gymnasium. The enemy troops soon slipped away, presumably into the Amador yacht club, but Fitzgerald had seen enough. After ordering Zebrowski to pull his men back, he directed Reagan to have Company A attack the PDF buildings from the north, starting with the three-story structure, Building 1, to be followed by the gym next door to it. Reagan's men had now become part of the mess the captain had hoped to avoid.[52]

At 1500, Company A moved through Building 1 and then the gym, taking its time to search the numerous rooms, many of them locked. During all this, Fitzgerald found himself under enormous pressure from higher headquarters to speed up the operation, so that his troops would be available to begin follow-on missions. He passed this sentiment along to the two infantry company commanders but refused to order them to hurry. Only when Reagan informed him that Company A wanted to go after any remaining defenders in the yacht club did Fitzgerald, on orders from Cisneros, say no. Reagan, while later admitting to being "pissed" at Fitzgerald's negative response to his request, was told that he would be allowed to enter the club the next day. With the other buildings already cleared and the American housing area well protected, Fort Amador was pronounced secured at 1800 on Wednesday, 20 December. The Americans had suffered 2 fatalities and 6 wounded during the operation; the PDF had 6 confirmed fatalities and 4 wounded, while 141 were taken prisoner. Fitzgerald's men had also captured a substantial number of weapons.[53]

The Task Force Black Devil after action reports contained a number of observations and recommendations, among them the need for better cooperation and coordination with the Air Force, for adequate time before an operation to engage in troop-leading preparations, and for better ways to handle prisoners and detainees. There was also a good deal written on how the 90-mm. recoilless rifle had proved a superb weapon for low-intensity conflict in an urban setting and how, instead of being phased out of the Army's

[52] Quoted words from Interv, author with Reagan, Zebrowski, and Flynn, 21 Jun 1990. Interv, author with Fitzgerald, 27 Jan 1990; Donnelly, Roth, and Baker, *Operation Just Cause*, pp. 180–81.

[53] Quoted word from Interv, author with Reagan, Zebrowski, and Flynn, 21 Jun 1990. The figures are from Nemmers, *United States Army South Staff Ride,* pp. 44–45.

inventory, it should be ranked ahead of the cumbersome Dragons and AT4s as the weapon of choice for certain operations.[54]

One observation did not make it into the official reports. During the Fort Amador operation, firefights and explosions from antitank weapons, howitzer shells, and (however ineffectual) rounds from an AC–130 occurred right in front of houses that the families of American service personnel had not had time to vacate. Once the fighting started, spouses and children found what safety they could in their own dwellings—in closets, under beds, and in bathtubs—and then prayed for the best, while fearing the worst. As the wife of one U.S. Army, South, officer related, "I was certain I was going to die." Her two sources of consolation: foot-thick concrete walls and, at some point, a telephone. Once Amador was secure, fear gave way to anger and, in several cases, marital discord, as some civilians who had heard a cannon being fired in what amounted to their front yards refused to talk to their spouses, whom they blamed for knowingly putting them at risk. Rationalizations claiming operations security did little to assuage the bitterness. The most disturbing aspect of this friction, according to several observers, was that the Southern Command, JTF-Panama, and JTF-South had not anticipated the problem; they therefore failed to provide the needed professional counseling as soon as possible after the fighting had stopped.[55]

On the positive side for U.S. forces, the enemy at Fort Amador, a vital target within the first concentric ring around the Comandancia, had been defeated in a battle far short of the bloodbath anticipated in the worst-case scenarios. The west bank of the canal area was also inside that ring, and it, too, had been secured with low casualty figures for both friendly and enemy forces, thanks to the efforts of Task Force Semper Fi. As for the third target area within the ring, Ancon Hill, that was the responsibility of Task Force Wildcat. The outcome there would follow the pattern set at Amador and on the west bank.

---

[54] A number of Task Force Black Devil's observations can be found in Nemmers, *United States Army South Staff Ride*, pp. 44–45. See also Donnelly, Roth, and Baker, *Operation Just Cause*, p. 163.

[55] Quote from Donnelly, Roth, and Baker, *Operation Just Cause*, p. 163. The problem of marital discord following the fight at Fort Amador (and attacks at other locations near where American families lived) was brought to the author's attention by a former USARSO/JTF-Panama staff officer.

6

TASK FORCE WILDCAT

☆ ☆ ☆

From almost any location in Panama City, one could see Ancon Hill rising up over six hundred feet, bordered on the west by the canal and in the south by the Bay of Panama. Atop the hill was a large but tattered Panamanian flag. It had been new when Panama's government raised it there as a symbol of sovereignty on 1 October 1979, the day much—but not all—of the hill transferred from American to Panamanian control as a consequence of the canal treaties. Looking down on the northeastern slope, an observer could see Gorgas Army Hospital near the base of the hill. On the reverse slope and somewhat higher up were two man-made terraces known as Quarry Heights, home to the U.S. Southern Command.[1] From this location overlooking the capital, U.S. military personnel enjoyed a beautiful panorama of the city and a

Ancon Hill (*right*) as seen from the vicinity of Fort Clayton, the location of JTF-South's operations center

[1] On 1 October 1979, some areas of Quarry Heights were included in the portions of Ancon Hill transferred to the government of Panama. As a result of the change, one of SOUTHCOM's two main gates was moved from the foot of the hill farther up the slope. Under the treaty, the Southern Command, itself, was not required to turn over its facilities until 31 December 1999. In fact, the headquarters relocated from Panama to Miami, Florida, in September 1997.

utilitarian line of sight into PDF facilities such as the Comandancia. The latter had proved valuable during the 3 October failed coup, when SOUTHCOM personnel had gained some insight as to what was transpiring simply by going outside and peering down into Noriega's headquarters.

If being located at Quarry Heights offered the unified command certain advantages during the crisis with Panama's dictatorship, these could possibly be supplanted by certain vulnerabilities in the event of war. There were, for example, no large U.S. combat units stationed on Ancon Hill, while Panamanian infantry companies at the Comandancia and Fort Amador were within quick striking distance of the SOUTHCOM complex. Additionally, there were several PDF facilities ringing Ancon Hill. Although the policemen and military engineers who manned those sites were not trained for combat, that fact alone would not preclude their initiating any number of disruptive activities that could threaten noncombatants, SOUTHCOM installations, and U.S. military operations in the vicinity. In response to these potential threats, the JTF-South Operation Plan 90–2 for deliberate U.S. offensive operations in Panama, as well as the series of High Anxiety fragmentary orders covering a no-notice attack by the PDF, directed Task Force Bayonet to "secure QUARRY HEIGHTS and GORGAS HOSPITAL," "control and demilitarize the area between FT CLAYTON and FT AMADOR," and "secure the ANCON HILL and BALBOA areas of PANAMA CITY."[2]

Colonel Snell, as Task Force Bayonet's commander on the eve of Just Cause, passed almost all of these missions to Task Force Wildcat, a subordinate command he had task-organized around the 5th Battalion, 87th Infantry, 193d Infantry Brigade, headquartered at Fort Clayton and commanded by Lt. Col. William Huff III. Besides the battalion's three infantry companies—A, B, and C—Huff would have operational control over Company A—dubbed Team Track—from Colonel Reed's mechanized battalion. In addition, Wildcat's headquarters and headquarters company included an antitank platoon, a scout platoon, a mortar platoon, and combat service support elements. Also attached to the task force was a platoon of military police.[3]

On battle maps, Huff's principal targets were in a sector labeled Area of Operations Antietam (*Map 9*). Beginning at the southwest base of Ancon Hill and moving clockwise, the battalion's Company B was responsible for the entire Balboa district, including the PDF's Balboa National Investigation Department (DENI) station that was located along Task Force Gator's route to the Comandancia and a portion of the route taken by Task Force Black Devil's ground elements into Fort Amador. North of that, Company C was assigned the main DNTT building and the Ancon DENI station, the latter situated at a critical intersection where two main avenues merged into the eastern edge of Gaillard Highway as

[2] FRAGOs (High Anxiety) to JTF-Panama OPORD 7–88 (Blue Spoon), 20 Jun 1989, 20 Aug 1989, and 11 Oct 1989. Quotes from JTF-South OPLAN 90–2, 3 Nov 1989, but are contained virtually word for word in the High Anxiety fragmentary orders.

[3] The task-organization for Task Force Wildcat is an Encl to Memo, Cdr, 5th Bn, 87th Inf, for Cdr, 193d Inf Bde, 5 Jan 1990, sub: After Action Report, TF Wildcat Operation Pineapple Crush.

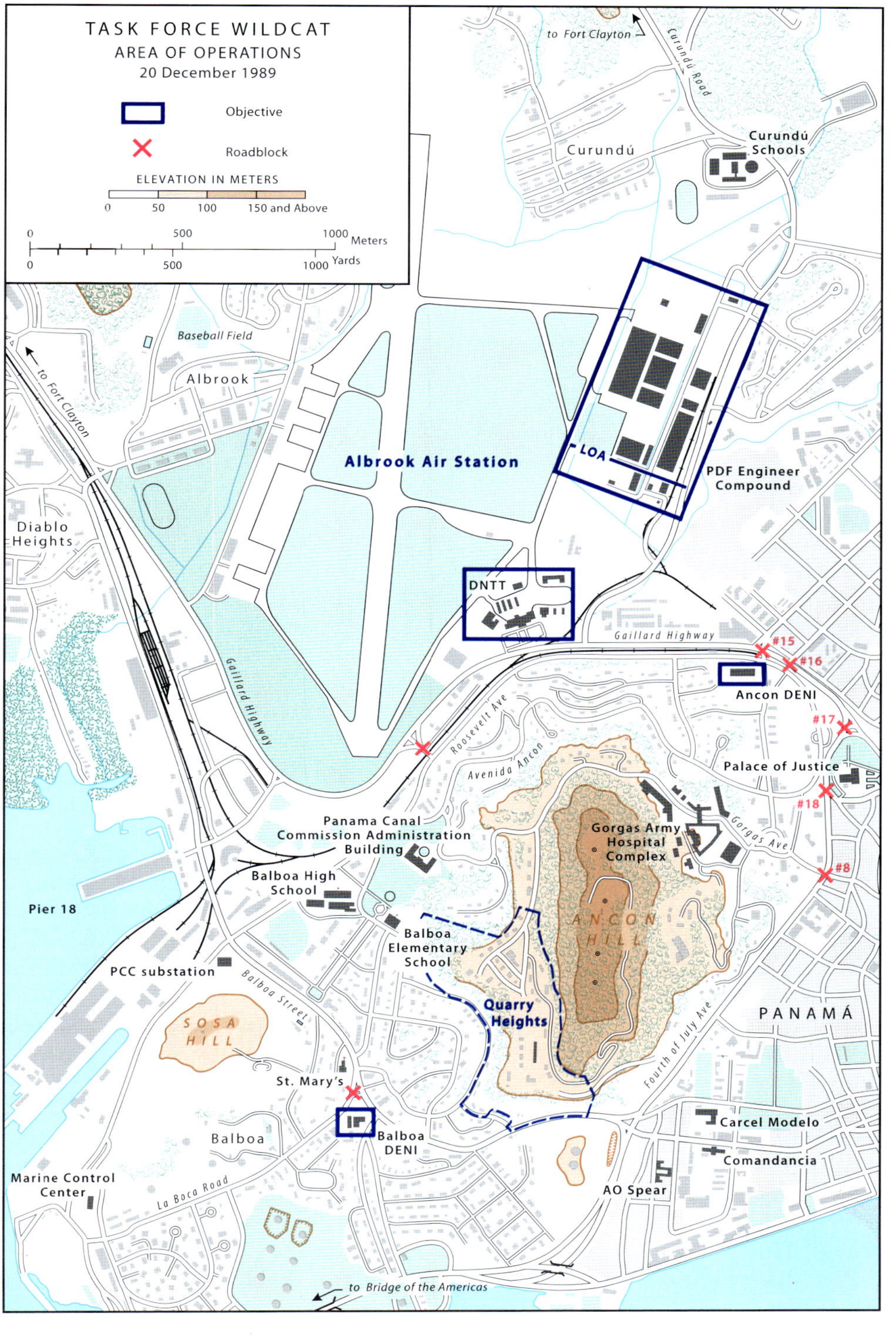
TASK FORCE WILDCAT
AREA OF OPERATIONS
20 December 1989
Objective
Roadblock
ELEVATION IN METERS
0
50
100
150 and Above
0
500
1000 Meters
0
500
1000 Yards
to Fort Clayton
Curundú Road
Curundú
Curundú Schools
Baseball Field
Albrook
to Fort Clayton
Albrook Air Station
LOA
PDF Engineer Compound
Diablo Heights
DNTT
Gaillard Highway
#15
#16
Ancon DENI
#17
Gaillard Highway
Roosevelt Ave
Avenida Ancon
Palace of Justice
#18
Gorgas Ave
Panama Canal Commission Administration Building
Gorgas Army Hospital Complex
#8
Balboa High School
Pier 18
ANCON HILL
Balboa Elementary School
PCC substation
Balboa Street
Quarry Heights
PANAMÁ
SOSA HILL
Fourth of July Ave
St. Mary's
Carcel Modelo
Balboa
Balboa DENI
Comandancia
Marine Control Center
La Boca Road
AO Spear
to Bridge of the Americas

Map 9

it ran alongside the northeast base of the hill. Company A was given responsibility for the PDF engineer compound farther off to the northeast, mainly because the company commander, Capt. William C. Flynt III, lived near the target and knew the terrain. For its part, the mechanized company attached to Wildcat was to set up five roadblocks at key intersections on the eastern edge of Antietam to isolate the battlefield and block any PDF reinforcements—either from the Ancon Hill area or from Las Tinajitas, Panama Viejo, and Fort Cimarrón—that might attempt to join in the fighting at the Comandancia. Finally, the military police were assigned to secure Gorgas Army Hospital, while an infantry platoon would do the same for the Panama Canal Commission (PCC) administration building and several power and communications facilities on the hill itself.

Preparations and Initial Movements

Prior to assuming command of the 5th Battalion, 87th Infantry, in June 1988, Huff had been in the SOUTHCOM Policy, Strategy, and Programs (J–5) Directorate. As a result of both assignments, he had been well positioned to witness the evolution of the crisis, for a year at the strategic and operational levels, then a year and a half at the tactical level. Because part of this experience involved the Prayer Book, he had participated in a simulation of the initial Blue Spoon operation orders while at Quarry Heights, briefing General Woerner on projections for battlefield and collateral damage, and he had helped analyze plans for the follow-on civil-military operations. After he took command of the 5th Battalion at U.S. Army, South, he continued to be involved in the planning process once Blue Spoon was passed down for battalion-level scrutiny in August 1988. Fleshing out the contingency plan for combat accelerated during Operation Nimrod Dancer following the arrival in May 1989 of the mechanized battalion from Fort Polk and the decision to place one of that battalion's companies under Huff's operational control in the event of war. By the time planners were making their final revisions to Blue Spoon in the weeks after the 3 October coup attempt, Task Force Wildcat's targets had been clearly identified and the tactics for dealing with them formulated and, as much as practicable, rehearsed.

The "trickiest part" of the plan, according to Colonel Snell, was if defensive operations detailed in the High Anxiety fragmentary orders and in Phases I and II of the JTF-South operation order preceded offensive operations. In that case, the mechanized battalion would initially occupy defensive positions to secure Quarry Heights and Balboa. Once combat operations began, the mechanized companies assigned to Task Force Gator and the Comandancia mission would have to vacate these positions, which soon thereafter would be reoccupied by Task Force Wildcat as it set up to attack its targets around Ancon Hill. Snell was concerned that complications during the changeover might impede the launching of Gator's attack on the Comandancia or leave American citizens and U.S. facilities on Ancon unprotected for an unspecified interval. This apprehension subsided only after he and Huff learned that they would not be executing the defensive phases of Blue Spoon. On a lighter note, at some point during the planning process, the operations assigned to Huff's task force acquired the code name PINEAPPLE CRUSH, a play on the derisive nickname—*La Piña*—Noriega had acquired because of his pockmarked face.[4]

To increase the readiness of the troops to execute their assigned missions, Huff's battalion received training similar to that of other Panama-based and NIMROD DANCER units operating in the country. Since May, the 5th Battalion had been conducting its share of Purple Storm, Sand Flea, contingency readiness exercises, and other operations, several of which had taken Huff's troops into the Ancon area. For the most part, the maneuvers had proved uneventful. After the 3 October failed coup, the training and other preparations in which Huff's soldiers participated intensified. There were more security missions and show-of-force operations to run, many of them tantamount to BLUE SPOON rehearsals, as was the case with other JTF-Panama forces. As part of the accelerated training program, the 5th Battalion used the Empire Range on the west side of the canal to improve its ability to conduct military operations on urbanized terrain. As one staff officer recalled, "We actually constructed wooden buildings, and we conducted live-fire raids, clearing room by room, until every soldier knew exactly what he needed to do for MOUT operations." Another officer, one of Huff's company commanders, praised the "extensive and realistic MOUT live fires including the bounding of fire teams, squads, and platoons." The simple "SOPs developed during these live fires were followed by the men and leaders," he added in the wake of JUST CAUSE. While the limited training area did not permit calling in close air support and indirect fire and while not all the ground maneuver elements had the opportunity to hone their MOUT skills at night, the program did allow "the live fire employment of almost every weapon system." The result was that each of the units acquired the status of *trained* at the end of the course.[5]

Since Task Force Wildcat units executing BLUE SPOON would be scattered throughout the Balboa and Ancon districts on terrain featuring a large hill and several built-up urban areas, Huff, his staff, and commanders had to pay special attention to communications. Beginning in October, the battalion's

---

[4] For Huff's involvement with BLUE SPOON planning and his and Snell's concerns about the plan, see Intervs, author with Col William H. Huff III, U.S. Army, 29 Jan 1990, Fort Clayton, Panama; Robert K. Wright Jr. with Huff and Senior Officers of the 5th Bn, 87th Inf, 193d Inf Bde, 5 Jan 1990, Fort Clayton, Panama. Quoted words from Interv, Wright with Col Michael G. Snell, U.S. Army, 1 Jan 1990, Fort Clayton, Panama. See also Task Force Wildcat's version of the BLUE SPOON operation order, OPLAN 90–1 (PINEAPPLE CRUSH), 17 Oct 1989, for a description of Phases I through IV operations. In Wright's lengthy and invaluable interview with Task Force Wildcat personnel, the officers on the battalion staff joining Colonel Huff were the executive officer, Maj. Carter Thomas; the operations officer, Maj. James Woods; the intelligence officer, 1st Lt. Kevin Huggens; the S–4, Capt. Robert Colon; the adjutant, 1st Lt. Robert E. Vikander; Chaplain (Capt.) William E. Knight; and Cmd. Sgt. Maj. Eric L. Haney. Also present were Capt. Robert W. Jones Jr., the commander, Headquarters and Headquarters Company; Capt. Mark Conley, commander, Company B; and Capt. Donald S. Currie, commander, Company C.

[5] First quote from Interv, Capt Joseph Nemmers with Capt Robert E. Vikander, 5th Bn, 87th Inf, 193d Inf Bde, 16 Oct 1990, Fort Clayton, Panama. Last quote from AAR, Task Force Wildcat, and second and third quotes from AAR, Co A, 5th Bn, 87th Inf, both Encls to Memo, Cdr, 5th Bn, 87th Inf, for Cdr, 193d Inf Bde, 5 Jan 1990, sub: After Action Report, TF Wildcat Operation PINEAPPLE CRUSH.

communications officer started reconnoitering the task force's projected target areas to determine what locations would best allow Huff and his subordinates to stay in contact with each other and with higher headquarters, meaning Snell's Task Force Bayonet. In a related measure, the officer and others also established several brigade and battalion retransmission stations. To test the results, Wildcat units employed the officer's recommendations and arrangements during Sand Flea exercises, in the process confirming or modifying where Blue Spoon company and task force command posts could best be located.[6]

Besides urban combat training, communications planning, and ongoing exercises, the lengthy list of activities constituting the 5th Battalion's Blue Spoon preparations between early October and mid-December also included the creation of detailed unit battle books; a variety of live-fire exercises geared to each echelon; briefings up the chain from company commanders to their battalion and brigade superiors; a battalion-level war game that approximated the plan; a jeep exercise and a "van mounted TEWT with commanders"; "several alerts, marshalling, and deploying exercises"; map exercises; and a number of "walk-thrus" and "talk-thrus." Not on the list of formal activities was the simple function of just being conscious of one's daily environment. As one sergeant in the battalion recalled, "You know, you drive by this place every morning that you know is going to be a target, on the way to work. You just pass the same place every day. . . . Although [an] actual rehearsal wasn't done, we almost did it every day. No problem taking a look at this stuff."[7]

In the opinion of Huff and his subordinates, the battalion was ready to perform its Blue Spoon/90–2 combat missions. The soldiers had been well trained and morale was high, the latter in part due to the prospect at long last of putting a stop to the intimidation and coercion leveled at them by the Noriega regime. Since 3 October, U.S. military personnel had experienced increased levels of police harassment, verbal abuse and obscene gestures, and, in general, a more belligerent environment. The troops were "tired of all this," according to one staff officer, and wanted a chance to respond. As a noncommissioned officer put it, "So when this thing finally started to kick off, everybody was pretty happy just to get it going. Everybody was up for it. Let's just do it and get it over with, so we can get back to some kind of normal life around here." Helping to keep morale high were intelligence reports indicating that most Panamanian units would not fight and that a U.S. victory was foreordained.[8]

---

[6] Interv, Nemmers with Vikander, 16 Oct 1990.

[7] The list of preparatory activities and the first through fourth quotes are from AAR, Task Force Wildcat, Encl to Memo, Cdr, 5th Bn, 87th Inf, for Cdr, 193d Inf Bde, 5 Jan 1990, sub: After Action Report, TF Wildcat Operation Pineapple Crush. Interv, author with Huff, 29 Jan 1990. Last quote from Interv, Sgt Daniel C. Wagner and Dolores De Mena with Sfc Gerald Stewart, U.S. Army, 16 Oct 1990, U.S. Army, South, Panama.

[8] For the morale issue and the first quote, see Interv, Nemmers with Vikander, 16 Oct 1990. Second quote from Interv, Wagner and De Mena with Stewart, 16 Oct 1990. Interv, Wagner with 2d Lt Paul H. Fredenburgh and 1st Lt Robert M. Mundell, both U.S. Army, 18 Oct 1990, Fort Davis, Panama; AAR, Task Force Wildcat, Encl to Memo, Cdr, 5th Bn, 87th Inf, for Cdr, 193d Inf Bde, 5 Jan 1990, sub: After Action Report, TF Wildcat Operation Pineapple Crush.

As with so many other Army officers in Panama, Huff was enjoying himself at the USARSO Christmas formal when he received word about the Paz shooting. Right after Cisneros canceled the remainder of the celebration, the colonel went to his headquarters on Fort Clayton. As an alert went out to all JTF-Panama units, Huff activated his battalion's tactical operations center and awaited further developments. For the next three days, the battalion was confined to the fort. On Monday, once Colonel Snell learned of the impending operation, he called in his battalion commanders and, mindful that they were not to be officially informed until the next day, wrote on a piece of paper, "We're going to war." According to Huff, Snell then shredded the note.[9]

On Tuesday at the authorized time, Huff disclosed the secret to his company commanders, telling them to inform their platoon leaders at 1800 that evening. As was the case in other units, some officers later complained that company commanders and platoon leaders would have been better able to conduct a final round of reconnaissance and surveillance and to prepare their troops for battle had they been given a 24-hour heads-up rather than only a few hours' notice before JUST CAUSE was scheduled to begin. Supporting this point was the plight of several company-grade officers, who in trying to use what precious minutes they had before 1800 to write the operation orders they would distribute to their soldiers, were often interrupted to attend critical meetings and tend to other tasks, thus delaying completion of the orders. This, in turn, meant that the troops would have even less time to receive the necessary briefings and complete their preparation checklists. Serving to complicate matters further was the realization that the joint communications-electronics operating instructions provided by the XVIII Airborne Corps contained several duplicate frequencies that, once identified, had to be altered at the last minute.[10]

Although morale remained high as H-hour approached, a "real serious atmosphere" pervaded the barracks. The troops had to attend briefings, acquire ammunition and essential equipment, and review and discuss procedures and numerous other details. Having trained in urban operations, the men knew how complicated and potentially deadly clearing buildings could be. Yet they also knew that the professionalism and firepower they would bring to the coming battle far outweighed anything their adversary possessed. With between 650 and 700 troops in the 5th Battalion, riflemen would begin the operation with their M16s and about 240 rounds per soldier. Each squad would have two or more LAWs, and grenades were handed out to squad and team leaders. Each

---

[9] Interv, author with Huff, 29 Jan 1990.

[10] Intervs, Nemmers with Vikander, 16 Oct 1990; Wagner with Fredenburgh and Mundell, 18 Oct 1990; Wright with Huff and Senior Officers of the 5th Bn, 87th Inf, 193d Inf Bde, 5 Jan 1990; Nemmers with Capt Stephen A. Smith, U.S. Army, 18 Oct 1990, Fort Davis, Panama; AAR, Task Force Wildcat, Encl to Memo, Cdr, 5th Bn, 87th Inf, for Cdr, 193d Inf Bde, 5 Jan 1990, sub: After Action Report, TF Wildcat Operation PINEAPPLE CRUSH. Commenting after JUST CAUSE in the interview with Wright on the communications-electronics operating instructions (CEOI) problem, Huff gave an illustrative example: "The CEOI that we thought was all going to be set up for this thing then got dicked up as we got new frequencies. . . . We found that the battalion command [frequency] was the same as some MP platoon [frequency]—unbelievable screw-ups at the last minute."

combat company in the task force would have a sniper team using the M24 sniper system. Also in the task force's inventory were .50-caliber machine guns mounted on M113s and other vehicles, M60 machine guns, 90-mm. recoilless rifles, AT4 portable antitank weapons, M203 grenade launchers, and TOW II antitank missiles. Each company from the 5th Battalion had two 60-mm. mortars, while the battalion's four 81-mm. mortars would be infiltrated onto the athletic field at Albrook Air Station to provide on-call fire support. Finally, Huff could request artillery and close air support if he believed a situation required such assistance.[11]

During the morning and afternoon of the nineteenth, the colonel and his principal staff officers joined other task force commanders and their staffs at the various meetings and briefings held throughout the day. Then, at 1800, Huff helped brief platoon leaders on their roles in the upcoming conflict. After he had finished speaking to one group, Cisneros arrived, himself making the rounds from unit to unit and talking to officers on the brotherhood of war. A few hours later, Huff visited each company in Task Force Wildcat, including Company A from the mechanized battalion, seeing as many soldiers as possible, reviewing their missions with them, and, in general, helping to keep morale high. During the course of what he later called his Patton speech, he also reminded the troops of the long-term consequences of the action in which they were soon to be engaged, a point that Snell had emphasized to him and Task Force Bayonet's other commanders. Once the enemy was defeated, Huff stated emphatically, there would be the nation-building phase, the Panama Defense Forces would be reconstructed in one form or another, and the U.S. military would continue to be stationed in Panama well into the 1990s. "We're going to have to pat these guys on the back and say, 'All is forgiven,' after we do this thing," he told the troops, instructing them on the importance of treating prisoners and the enemy dead with respect and leaving Panamanian flags flying.[12]

Shortly after the talks and briefings, elements of Task Force Wildcat began their initial movements. At 2200, the mortar platoon boarded HMMWVs and, with its 81-mm. mortars and ammunition aboard, left Fort Clayton by the back gate and made its way into Albrook Air Station next door, stopping briefly while the platoon leader and a reconnaissance party checked out the primary mortar firing position on the facility's baseball field. Given that the platoon was on a fenced-in U.S. installation with Air Force policemen to guard and, if necessary, defend the athletic field, the site selected for the mortars seemed safe enough. The only problem the platoon encountered once it arrived related to operations security: the mortar crews could not properly adjust the base plates

---

[11] Quote from Interv, Wagner with Fredenburgh and Mundell, 18 Oct 1990. Intervs, Nemmers with Vikander, 16 Oct 1990; with Smith, 18 Oct 1990; AAR, Task Force Wildcat, Encl to Memo, Cdr, 5th Bn, 87th Inf, for Cdr, 193d Inf Bde, 5 Jan 1990, sub: After Action Report, TF Wildcat Operation PINEAPPLE CRUSH.

[12] Quote from Interv, author with Huff, 29 Jan 1990. Intervs, Nemmers with Vikander, 16 Oct 1990; Capt John Hollins with Huff, 20 Jun 1990, Fort Clayton, Panama.

of their weapons without firing registration rounds, something that was out of the question.[13]

About thirty minutes later, two platoons from the 5th Battalion's Company A—nicknamed the "Jaguars"—also departed Fort Clayton to rendezvous with the company's third infantry platoon, which was conducting roving patrols in the Curundu housing area. Having been responsible for security in Curundu for a couple of months, Company A was familiar with the area, which allowed its officers to pick a guardhouse located there as an ideal assembly point if needed. By moving in staggered intervals on single trucks, the troops en route to the point late on 19 December sought to convince any observer who had become familiar with the unit's routine movements in the area that they were engaged in changing the guard or bringing in supplies. Once all the trucks arrived and the company assembled, the guardhouse provided superb cover for the soldiers and vehicles. Captain Flynt, the company commander, "thought the troops were entirely too noisy off loading the trucks" but found the issue to be "academic" in that "the trucks were much louder than the soldiers." It would have been better, he realized, had the transport stopped "about 500 meters away down the road," allowing the troops to walk to the guardhouse. In any case, he concluded that the commotion had not jeopardized tactical surprise. As for those troops who now had up to a two-hour wait before moving to their attack positions, the time passed uneventfully, with plenty in the way of additional preparations to keep them busy.[14]

As H-hour approached, Huff wanted up-to-the-minute intelligence on the targets in Task Force Wildcat's area of operations, as well as the status of the roadways leading into it. The scout platoon of his headquarters company had surveillance assets, while the company's antitank platoon had transportation. Sometime between midnight and 0030, the two units left through the back gate at Fort Clayton to reconnoiter the Ancon-Balboa area. So as not to attract unwanted attention, the soldiers wore soft caps and placed their weapons on the floor boards of their vehicles. At first, the reconnaissance was routine, with the scouts reporting that the route to the PDF engineering compound was clear, although the compound itself was blacked out, the entrances were blocked by vehicles, and armed men in civilian clothes could be seen in firing positions. Almost identical reports described conditions at the transportation building and the Ancon DENI. The scouts proceeded to Gorgas Army Hospital, reported on the situation there, then retraced their route back to the blacked-out Ancon DENI station, where a group of ten men dressed in civilian clothes and armed with AK47s "zeroed in on the convoy" just as the vehicles approached a stoplight that had turned red. The platoon leader considered

[13] Intervs, Wright with Huff and Senior Officers of the 5th Bn, 87th Inf, 193d Inf Bde, 5 Jan 1990; Nemmers with Smith, 18 Oct 1990; Dolores De Mena with Spec William Anthony Martin, U.S. Army, 27 Sep 1991, Fort Clayton, Panama; AAR, Task Force Wildcat, Encl to Memo, Cdr, 5th Bn, 87th Inf, for Cdr, 193d Inf Bde, 5 Jan 1990, sub: After Action Report, TF Wildcat Operation PINEAPPLE CRUSH.

[14] Quotes from AAR, Co A, 5th Bn, 87th Inf, Encl to Memo, Cdr, 5th Bn, 87th Inf, for Cdr, 193d Inf Bde, 5 Jan 1990, sub: After Action Report, TF Wildcat Operation PINEAPPLE CRUSH. Interv, Wright with Huff and Senior Officers of the 5th Bn, 87th Inf, 193d Inf Bde, 5 Jan 1990.

running the light but then decided to "play it cool," as though his presence was "normal business." To his relief, the armed men held their fire.[15]

Within a few minutes of passing the Ancon DENI, a sense of déjà vu swept the reconnaissance group as armed men approached the convoy while it waited at a second stoplight. "Light changes thank God!" the platoon leader later recorded in his unit's chronology. The feeling of relief, once again, was short-lived. Right after the light changed, the scout and the antitank platoons became distressingly aware of U.S. military police "rousting" the Pier 18 area of Balboa Harbor, arresting PDF administration personnel at the location. Although the policemen secured the area, their actions were premature, part of a misunderstanding rooted in the attempts by JTF-South to change H-hour. The ensuing ruckus caught the attention of Panamanian forces around the harbor and piers, the area Wildcat's reconnaissance group was just entering. PDF and Dignity Battalion members in the vicinity began locking and loading their weapons and pointing them at the Americans, who tried to appear calm. Then, believing that their "cover was blown," they prepared to "run the gauntlet" but were spared that ordeal when the armed Panamanians chose not to open fire. Several minutes later, right after the battle for the Comandancia commenced, the scouts concluded their patrol and proceeded immediately to their assembly areas, where they sought cover from the heavy indirect fire passing over their heads. Later, they would move onto Ancon Hill to provide security for Huff's jump tactical operations center and to serve as the task force's only reserve. From Huff's perspective, the reconnaissance mission had been a huge success. The reports relayed by the platoons contained what he considered to be the most valuable intelligence he received prior to hostilities. But he also realized, as one officer later observed, that the whole effort "could have gone very bad very quickly."[16]

As Huff scrutinized the information he was receiving from the scout and antitank platoons, Task Force Wildcat units still located at Fort Clayton began moving to their assembly areas. At 0040, Company A from the mechanized battalion left Clayton by the back gate and headed to the Curundu Elementary

[15] Intervs, Wright with Huff and Senior Officers of the 5th Bn, 87th Inf, 193d Inf Bde, 5 Jan 1990; author with Huff, 29 Jan 1990. Quoted words from AAR, Task Force Wildcat, Encl to Memo, Cdr, 5th Bn, 87th Inf, for Cdr, 193d Inf Bde, 5 Jan 1990, sub: After Action Report, TF Wildcat Operation Pineapple Crush. Videocassette tape, 5th of the 87th After Action Just Cause, 1990. Colonel Huff and his subordinates video-recorded the VHS tape shortly after Just Cause. The recording features statements from Huff, his company commanders, and their platoon leaders and noncommissioned officers, as well as video footage of the targets and a brief narrative on how each was neutralized.

[16] First through fourth quoted words from AAR, Task Force Wildcat, Encl to Memo, Cdr, 5th Bn, 87th Inf, for Cdr, 193d Inf Bde, 5 Jan 1990, sub: After Action Report, TF Wildcat Operation Pineapple Crush. While the Task Force Wildcat tactical operations center remained at Huff's battalion headquarters on Fort Clayton, his mobile or "jump" tactical operations center allowed him to go to Ancon Hill where he could direct the task force's actions from a better vantage point. For brevity in the remainder of this chapter, the term *operations center* will refer to the jump tactical operations center. Last quote from Lawrence A. Yates, "Operation JUST CAUSE in Panama City, December 1989," in *Block by Block: The Challenges of Urban Operations*, ed. William G. Robertson and Lawrence A. Yates (Fort Leavenworth, Kans.: U.S. Army Command and General Staff College Press, 2003), p. 348.

School, a location that would put the troops in position, once the shooting started, to move quickly down Curundu Road and past the PDF engineer compound before the battle there got under way. They would then maneuver to set up the five roadblocks assigned to them on the stretch of Gaillard Highway beginning north of the Ancon DENI and veering south to Fourth of July Avenue. As Company A departed Clayton, Task Force Wildcat's Company C and a platoon from Company B left by the front gate, turning onto the segment of Gaillard Highway running past the fort. For movement purposes only, these units headed south in the wake of Task Force Gator, briefly coming under the operational control of Colonel Reed. Minutes later, the situation, according to Huff, "turned to s——t." At 0045, Wildcat units began to hear mortar fire and other explosions and to see tracer rounds in the vicinity of the harbor and the Comandancia. Similar to the reconnaissance group, the troops leaving Clayton and those still preparing for departure realized that within minutes they would encounter PDF and Dignity Battalion elements in the Ancon-Balboa area that would have had time to prepare for the arrival of U.S. forces.[17]

With Operation JUST CAUSE now officially under way, Snell quickly radioed Huff, telling him to "go ahead and jump." The remaining troop trucks of Company B immediately moved out Clayton's front gate onto Gaillard, with four military police vehicles, Huff's operations center composed of two HMMWVs, and a support platoon from Company C behind them. At this point, all the main elements of Task Force Wildcat were en route to their lines of departure, with Company A and the Team Track mechanized company traveling toward the area of operations via Curundu Road, Company C and the Company B platoon following in the wake of Task Force Gator, and the column led by Huff's Company B (minus) bringing up the rear. As the troops of Company C drove past Albrook Air Station, they noticed three PDF police cars. If there were *tránsitos* inside, they chose not to open fire, and the company convoy proceeded to Ancon Hill.[18]

When Company B attempted to move through the same area, an intense firefight erupted. Armed Panamanians inside a bus also traveling south down Gaillard Highway, but in the northbound lane, opened fire on the column parallel to them and then tried to maneuver the vehicle into the southbound lane in such a way as to block Company B's progress. The occupants of one or two of the PDF police cars in the vicinity also joined in the shootout. Unable to avoid an engagement, U.S. troops at the head of the column returned fire. Soon thereafter, Air Force security police at Albrook, believing that they were under attack, began shooting as well. Behind Company B, the military police

---

[17] Quoted words from Interv, Wright with Huff and Senior Officers of the 5th Bn, 87th Inf, 193d Inf Bde, 5 Jan 1990. AARs, Task Force Wildcat, and Co C, 5th Bn, 87th Inf, both Encls to Memo, Cdr, 5th Bn, 87th Inf, for Cdr, 193d Inf Bde, 5 Jan 1990, sub: After Action Report, TF Wildcat Operation PINEAPPLE CRUSH; Nemmers, *United States Army South Staff Ride,* pp. 29–32. For information on Task Force Gator, see Chapter 4.

[18] Quote from Interv, Wright with Huff and Senior Officers of the 5th Bn, 87th Inf, 193d Inf Bde, 5 Jan 1990. AARs, Task Force Wildcat, and Co C, 5th Bn, 87th Inf, both Encls to Memo, Cdr, 5th Bn, 87th Inf, for Cdr, 193d Inf Bde, 5 Jan 1990, sub: After Action Report, TF Wildcat Operation PINEAPPLE CRUSH.

and the men in Huff's HMMWVs found themselves coming under intense fire, with no idea if it was the enemy or the Air Force police shooting at them. According to Huff, the firefight could have been disastrous save for the fact that the bullets flying into his column's "kill zone" seemed to be hitting everything—rifle butts, stocks, and sites; swivels; squad automatic weapons; equipment; and the vehicles—except the soldiers themselves, three of whom were wounded. He used terms such as "unbelievable" and "miraculous" to describe the phenomenon in which "people just escape[d] with their lives by flukes." The ambush came to an abrupt end when Company B pushed past the bus, which went off the highway and crashed. Three of the men inside were killed, either during the fighting or in the crash. Although there had been no U.S. fatalities in the convoy, a Department of Defense Dependent Schools schoolteacher, Gertrude Kandi Helin, was caught in the sudden crossfire and killed outside the gate at Albrook, her death being the first reported American civilian casualty of Just Cause.[19]

Toward the end of Huff's column, the battalion support platoon heard the shooting and pulled off to the side of the highway to assess the situation. With tracer rounds flying about, the platoon leader, 1st Lt. Robert M. Mundell, expressed some concerns about entering Albrook, where the plan called for him to establish a first aid station and facilities for resupplying Task Force Wildcat with water, fuel, and ammunition. He was "apprehensive" because "we did not know that the Air Force understood what our friendly signs were and we didn't understand what their friendly signs were." His anxiety might have increased had he known that, in the shootout just yards ahead of him, the U.S. convoy suspected friendly fire from the air station as the source of some of the rounds impacting around it. The support platoon finally began moving again, with Mundell's worst fears being allayed after his men entered Albrook without incident.[20]

The remainder of the column continued to its objectives, with Huff's operations center making its way up Ancon Hill to a clearing just below Quarry Heights and above the PCC administration building. As Huff approached the position, he could see that it was being shelled by mortars, although by the time he arrived, the firing had subsided. With him were the 5th Battalion's operations officer, its fire support officer, an air liaison officer, a communications officer, two radio telephone operators, and the battalion sergeant major. From this select location, Huff could keep in contact with higher headquarters and with the four scattered companies constituting the bulk of his task force. He also could easily visit each company's area of operations if he required more information or believed his presence to be necessary.[21]

---

[19] Quoted words and quote from Interv, Wright with Huff and Senior Officers of the 5th Bn, 87th Inf, 193d Inf Bde, 5 Jan 1990. Nemmers, *United States Army South Staff Ride,* pp. 26–28; AARs, Task Force Wildcat, and Co B, 5th Bn, 87th Inf, both Encls to Memo, Cdr, 5th Bn, 87th Inf, for Cdr, 193d Inf Bde, 5 Jan 1990, sub: After Action Report, TF Wildcat Operation Pineapple Crush.

[20] Quote from Interv, Wagner with Fredenburgh and Mundell, 18 Oct 1990. AAR, Task Force Wildcat, Encl to Memo, Cdr, 5th Bn, 87th Inf, for Cdr, 193d Inf Bde, 5 Jan 1990, sub: After Action Report, TF Wildcat Operation Pineapple Crush.

[21] Intervs, author with Huff, 29 Jan 1990; Nemmers with Smith, 18 Oct 1990.

Around and below him, these units were preparing to carry out their assignments—the mechanized company, to install roadblocks; the three rifle companies, to isolate four hard targets and two large neighborhoods. At each of the targets, the tactics to be used mirrored those employed at most JUST CAUSE objectives. A heavy broadcast team would issue appeals for Panamanian defenders to surrender, accompanied by warnings of dire consequences should the entreaties be ignored. Against any enemy forces choosing to resist, U.S. units would conduct what was termed a carefully orchestrated firepower demonstration designed to convey the devastating effects of their weapons. Following the demonstration, there would be a new round of surrender broadcasts. Only if alternating the carrot and the stick became an exercise in futility would American troops actually launch an attack to seize an objective. The proximity of civilian residences to targets in the Ancon-Balboa districts dictated such a patient approach, as did the long-term considerations, including nation building, that Huff had earlier voiced to his troops. Also, as Snell later remarked, "We did not want to kill unnecessarily a lot of soldiers because we knew the PDF were there to put bread on the table, and many, we suspected, were not strong supporters of Noriega."[22]

TEAM TRACK AND COMPANY C

Around the northern and eastern slopes of Ancon Hill were three of Task Force Wildcat's hard targets: the PDF engineer compound, the DNTT building housing the PDF's main *tránsito* and transportation headquarters, and the Ancon DENI station. Company C, 5th Battalion, bore responsibility for fixing and, on order, neutralizing the latter two targets. If necessary, the company commander, Capt. Donald S. Currie, could request assistance from Company A, 4th Battalion, 6th Infantry—Team Track—whose principal mission was to isolate the battlefield, both at the Comandancia and around Ancon Hill, by establishing five roadblocks on Currie's right flank.

Having moved from Fort Clayton into the Curundu area twenty minutes earlier, Team Track crossed its line of departure at 0100 and started heading down Curundu Road on a route that would take the M113s past the PDF engineer compound and the DNTT on their right flank to the intersection with that segment of Gaillard Highway running around the northeastern base of Ancon Hill. The foray would not go unopposed. Just seconds into the column's movement, three or four Panamanians lying in ambush opened fire with machine guns. The team's 2d and 3d Platoons were best positioned to return fire, while the entire convoy accelerated to twenty-five miles per hour. A second ambush occurred as Company A passed the front gate of the engineer compound. Again, only two or three enemy personnel were involved, and, as a few minutes before, U.S. troops returned fire without stopping. When the vehicles reached Gaillard Highway, they made a left turn and proceeded east

[22] Ibid. *Firepower demonstration* was the term employed by U.S. commanders in JUST CAUSE to describe the employment of various weapons to show their destructive effects, not so much to inflict casualties, but to intimidate enemy defenders at several locations into surrendering. Quote from Donnelly, Roth, and Baker, *Operation Just Cause*, p. 153.

past the Ancon DENI station and then south to positions where the 1st Platoon would erect the first two roadblocks—numbers 8 and 18—at the intersection of Gorgas Avenue and Fourth of July Avenue. The 2d Platoon at some point would turn back north and establish roadblock number 17 at the intersection of Gaillard and Calle 9 de Enero. The 3d Platoon would set up the last two roadblocks (numbers 16 and 15) at two intersections along Gaillard right next to the Ancon DENI, but only after the 1st Platoon from the 5th Battalion's Company C had seized the station. Similarly, the mechanized company commander, Capt. Isadore Bowers Jr., could not set up his command post at the Palace of Justice while the DENI station next door was still in enemy hands. Huff later observed without elaboration that making the completion of Team Track's mission contingent on seizing the Ancon DENI was "probably a screwed-up tactical plan." Yet, given the layout of the objectives, the road network around them, and the Comandancia's location to the southeast, an alternative plan had not readily presented itself.[23]

As soon as the lead elements of Bowers' mechanized company reached the southernmost point on their route, two Panamanians with AK47s started firing at the troops from a three-story building. According to the unit's after action report, elements from the headquarters and from the 2d and 3d Platoons "pounded the area with M16s, M60s, and 50 cal fire." The two men stopped shooting, and the mechanized 1st Platoon, with the help of engineers from Huff's battalion, prepared to set up roadblocks 8 and 18. Before they could get started, though, someone detected movement in a window over the nearby Ancon Inn. Bowers responded to the potential threat by directing his troops to remain in their M113s until the location of any additional enemy snipers had been identified. Although a scan of the floors and rooftops of adjacent buildings revealed nothing, the captain decided to keep the company moving along Gorgas Avenue to a point closer to the U.S. Army hospital complex. He then sent the 1st Platoon back to set up roadblocks 8 and 18 and to occupy the area around them. Meanwhile, a few blocks to the north, at Calle 9 de Enero, the 2d Platoon set up roadblock 17. At 0137, Captain Bowers reported that the three roadblocks were in place. Meanwhile, the 3d Platoon awaited word of developments at the Ancon DENI. Once the units responsible for seizing the station reported it was secured, the platoon would retrace its movements along the march route, cross Phase Line Red that separated Task Force Wildcat's infantry and mechanized companies, and establish the final two roadblocks behind the facility. If needed, the soldiers of the 3d Platoon could also serve as reinforcements in the DENI operation.[24]

Across Phase Line Red, while Team Track was defending its position and setting up checkpoints, the 1st Platoon of Captain Currie's Company C had been making its way from the northwest to the northeast side of Ancon Hill.

[23] AAR, Co A, 4th Bn, 6th Inf (Mech), Encl to Memo, Cdr, 5th Bn, 87th Inf, for Cdr, 193d Inf Bde, 5 Jan 1990, sub: After Action Report, TF Wildcat Operation Pineapple Crush. Quote from Interv, Wright with Huff and Senior Officers of the 5th Bn, 87th Inf, 193d Inf Bde, 5 Jan 1990. Videocassette tape, 5th of the 87th After Action Just Cause, 1990.

[24] Quote from AAR, Co A, 4th Bn, 6th Inf (Mech), Encl to Memo, Cdr, 5th Bn, 87th Inf, for Cdr, 193d Inf Bde, 5 Jan 1990, sub: After Action Report, TF Wildcat Operation Pineapple Crush. Nemmers, *United States Army South Staff Ride,* pp. 31–33.

Moving across Avenida Ancon from the vicinity of the canal commission administration building, the infantrymen arrived at a position from which they could get out of their trucks and approach the DENI station from the south. As with the company's other platoons, the 1st had an antitank team and a medic attached to it. In addition, it had a forward observer, a sniper team from the headquarters company, and an eighteen-wheel tractor-trailer. Given its attack positions on the northeast slope of the hill, the 1st Platoon was the farthest removed from the Company C command post near the PCC building and was thus well beyond Captain Currie's line of sight. Currie would have to rely on radio transmissions to keep him abreast of developments at the Ancon DENI. Yet, when he initially tried to contact the platoon, he discovered to his dismay that he had lost communications with the unit. For almost an hour, he tried unsuccessfully to radio the platoon leader.[25]

When Currie finally reestablished contact with his 1st Platoon, he learned that, as soon as the troops had arrived in the vicinity of the Ancon DENI, they had come under what he later described as "a heavy volume of fire, probably the heaviest of the first day." The news came as something of a surprise because intelligence reports available to Huff and his subordinates had indicated most PDF units, including the special police at the DENI, would not fight. Whether that assumption was incorrect or whether some other, more motivated Panamanian force was in control of the station seemed a moot point at the moment. Of greater consequence to the 1st Platoon leader was the fact that just two of his three squads were in their assigned positions when the shooting commenced. Consequently, instead of surrounding the station, he could only cover it from three directions, with one squad located to the south of the DENI, and the other having maneuvered to higher ground on the east side. The west side, in effect, was blocked by the whole of Task Force Wildcat, just by being located on Ancon Hill, but positions on the north side were still unoccupied. Seven or eight defenders, realizing that they would be safer elsewhere and seeing no threat north of them, took advantage of the situation by exiting the station on that side and crossing Gaillard Highway into a business area, from which they began firing at the Americans. The company snipers Currie had placed with the 1st Platoon responded, killing up to seven men during the night. Before any other defenders could leave the station, the third squad finally arrived and moved into position, using a bus and tractor-trailer it brought with it to block the northern escape route.[26]

Since the platoon was still under heavy fire from defenders in the station, the 1st Platoon leader turned to Captain Currie for support, prompting Currie to radio Bowers at his Team Track command post and request execution of a contingency plan that the two had worked out in advance. The arrangement called for the mechanized unit to send one or two of its armored personnel

---

[25] AAR, Co C, 5th Bn, 87th Inf, Encl to Memo, Cdr, 5th Bn, 87th Inf, for Cdr, 193d Inf Bde, 5 Jan 1990, sub: After Action Report, TF Wildcat Operation PINEAPPLE CRUSH; Interv, Wright with Huff and Senior Officers of the 5th Bn, 87th Inf, 193d Inf Bde, 5 Jan 1990.

[26] AARs, Task Force Wildcat, and Co C, 5th Bn, 87th Inf, Encl to Memo, Cdr, 5th Bn, 87th Inf, for Cdr, 193d Inf Bde, 5 Jan 1990, sub: After Action Report, TF Wildcat Operation PINEAPPLE CRUSH. Quote from Interv, Wright with Huff and Senior Officers of the 5th Bn, 87th Inf, 193d Inf Bde, 5 Jan 1990.

Ancon DENI station

carriers to join the fight at the Ancon DENI. In an area where mortars could not be employed because of the target's proximity to civilian residences, the M113's mounted .50-caliber machine gun could be used to devastating effect in any firepower demonstration and would augment the 1st Platoon's LAWs and formidable 90-mm. recoilless rifles. Bowers readily agreed to Currie's request and dispatched one armored personnel carrier to the scene. It took over thirty minutes for the linkup to occur, during which time Currie asked the Task Force Wildcat operations center to telephone the Ancon DENI with a demand to surrender. Several minutes later, he was informed that no one inside the station had picked up the phone. When the M113 eventually arrived, its crew members learned that they could not immediately engage the enemy. In front of their position, troops from the 1st Platoon's 3d Squad who had been located north of the station were moving to the higher ground to the east. Given the potential for friendly fire, the M113 remained stationary and its gunner refrained from shooting until the squad in motion had completed its maneuver. Meanwhile, Currie, thinking that the delays at the Ancon DENI might portend a "loss of momentum" on the part of the 1st Platoon, sent his executive officer to "energize" the unit.[27]

During this period, when the platoon was making its last-minute adjustments prior to employing more than small arms against the DENI station, Task Force Wildcat suffered its only fatality, the result of an exchange of gunfire in the vicinity of the mechanized company's 2d Platoon, which was manning the number 17 roadblock position. Following three single shots from

[27] AAR, Co A, 4th Bn, 6th Inf (Mech), and quoted words from AAR, Co C, 5th Bn, 87th Inf, both Encls to Memo, Cdr, 5th Bn, 87th Inf, for Cdr, 193d Inf Bde, 5 Jan 1990, sub: After Action Report, TF Wildcat Operation Pineapple Crush. Nemmers, *United States Army South Staff Ride,* pp. 29–33; Interv, Wright with Huff and Senior Officers of the 5th Bn, 87th Inf, 193d Inf Bde, 5 Jan 1990.

The DNTT compound

a high-powered rifle fired from the fifth floor of a nearby building, bursts from two AK47s raked the platoon's M113 directly below. The bullets wounded two U.S. soldiers and killed another, Pfc. Kenneth D. Scott. Muzzle flashes from the weapons gave away the snipers' positions, and the platoon pounded them with a variety of munitions. In short order, the three U.S. casualties were evacuated to Gorgas Army Hospital just a very short distance away, while the 2d Platoon moved across Gaillard Highway to a more secure position from which to guard the roadblock.[28]

Back at the Ancon DENI station, Team Track's M113 was finally able to begin its firepower demonstration, shooting 1,200 .50-caliber rounds into the target in two bursts of 600 rounds each. Twenty minutes later, Company C's 1st Platoon joined in, using a 90-mm. recoilless rifle against the building's entrance. By 0330, the platoon was ready to assault the objective and clear the station room by room, with elements of the 3d Platoon from Captain Bowers' mechanized company establishing a "support by fire" position outside, then moving onto Gaillard Highway to set up the company's last two roadblocks, numbers 15 and 16. By 0445, the operation was over and the Ancon DENI cleared without incident, thus neutralizing one of Company C's two hard targets.[29]

Efforts to subdue the other, the DNTT, were still in progress, and, as had been the case with the Ancon DENI, the rifle units involved—Company C's 2d

[28] AAR, Co A, 4th Bn, 6th Inf (Mech), Encl to Memo, Cdr, 5th Bn, 87th Inf, for Cdr, 193d Inf Bde, 5 Jan 1990, sub: After Action Report, TF Wildcat Operation PINEAPPLE CRUSH.

[29] AAR, Co A, 4th Bn, 6th Inf (Mech), and quoted words from AAR, Co C, 5th Bn, 87th Inf, both Encls to Memo, Cdr, 5th Bn, 87th Inf, for Cdr, 193d Inf Bde, 5 Jan 1990, sub: After Action Report, TF Wildcat Operation PINEAPPLE CRUSH; Nemmers, *United States Army South Staff Ride,* pp. 29–33; Interv, Wright with Huff and Senior Officers of the 5th Bn, 87th Inf, 193d Inf Bde, 5 Jan 1990.

and 3d Platoons—had come under enemy fire from the outset of the operation. The first to arrive at the target was the 2d Platoon, having turned off Gaillard Highway onto the road leading up to the PCC administration building, and then proceeding along Roosevelt Avenue, which ran parallel to Gaillard but at a higher elevation, to a release point just southwest of the target. As soon as the first vehicle stopped, it and those behind it encountered automatic weapons fire from the nearby Explansa building, causing the troops to vacate their trucks in disarray. "I've never seen people de-truck so quickly in my life," Captain Currie later exclaimed. "I mean there was stuff everywhere, and people were just jumping over one another; going over the cab." The platoon leader, 2d Lt. Paul H. Fredenburgh, called the scene "a chaotic mess in the road on the ground." "Thank God the bullets were high and no one got hit," he added. Arriving just minutes later, soldiers in the 3d Platoon positioned their vehicles right next to the Explansa building, exited, and returned fire. The threat from that location subsided but, in quick succession, both platoons received fire from a bunker and guard shack on the PDF side of Albrook Air Station. Again, the 3d Platoon, this time employing its 90-mm. recoilless rifle, silenced the shooters. In the meantime, the remainder of the lengthy Company C convoy moved into an area west of the administration building. The task force's antitank team and two roadblock personnel provided security for the company positions by covering the Gaillard Highway approaches and by setting up a roadblock on Roosevelt Avenue to prevent any hostile force from advancing from the rear.[30]

Once the initial shooting stopped, Fredenburgh prepared to deploy his men to two fixed positions across from the DNTT and just south of Gaillard. He would lead part of his platoon into a lengthy ditch or gutter about a hundred fifty to two hundred yards from the building near railroad tracks running parallel to the highway, while the remainder of the unit would occupy a position farther down the ditch on his right flank. As the troops began to move, however, they once again came under fire, this time from inside the transportation headquarters. According to Fredenburgh, "That's when the soldiers got scared. When the real bullets were fired at them and cracking over their heads, that's when I saw fear in the faces of my soldiers." To keep the troops moving into their positions required, in his words, "a little kicking in the ass," but, after about five minutes, "the initial shock was over," as the troops calmed down and began to carry out their platoon's portion of the plan.[31]

With the soldiers in the 2d Platoon facing the front of the DNTT, the 3d Platoon moved into position next to a group of buildings from which it could make a flanking attack from the west to clear the target. Before that, though, the platoon would wait for the cycle of surrender appeals and firepower demonstrations intended to persuade the defenders to give up and avoid bloodshed. The process, which would take over four hours, began as

[30] Currie quotes from Interv, Wright with Huff and Senior Officers of the 5th Bn, 87th Inf, 193d Inf Bde, 5 Jan 1990. Fredenburgh quotes from Interv, Wagner with Fredenburgh and Mundell, 18 Oct 1990. Videocassette tape, 5th of the 87th After Action JUST CAUSE, 1990; AAR, Co C, 5th Bn, 87th Inf, Encl to Memo, Cdr, 5th Bn, 87th Inf, for Cdr, 193d Inf Bde, 5 Jan 1990, sub: After Action Report, TF Wildcat Operation PINEAPPLE CRUSH.

[31] Interv, Wagner with Fredenburgh and Mundell, 18 Oct 1990.

soon as both platoons were in position. In issuing the first "verbal warning," the attached heavy broadcast team told whoever was inside the transportation building that they had five minutes to come out. If they did so, they would not be harmed; if they disregarded the ultimatum, they would be fired on. In the absence of a public address system, the team used bullhorns to broadcast the message, the result being that a number of defenders, as U.S. officers learned later, could not hear it. Once the five minutes passed without compliance, Captain Currie ordered the 2d Platoon to begin its first firepower demonstration, which was limited to small arms (M60s, M16s, and squad automatic weapons). Soon thereafter, a second surrender appeal was made, again without positive results, followed by a second firepower demonstration, this time with 90-mm. recoilless rifles, LAWs, and small arms. One hour into the standoff, Currie requested that the 5th Battalion operations center at Fort Clayton contact the defenders directly by telephoning the DNTT to deliver the warnings. The call was placed, but whoever answered the phone hung up.[32]

When PSYOP broadcasts failed to achieve the desired response, Currie concentrated on intensifying the firepower demonstrations. At one point, an AC–130 became available, and he gave it directions for firing down into the DNTT. The first marking round hit near the target, and Currie provided information for adjusting the fire. But the second round went even farther astray, exploding in one of the buildings around which his 3d Platoon had assembled. "We're very lucky that nobody was killed in that one," he later commented. He also noted that "despite glint tape and light signals which I used, Spectre seemed to have a difficult time locating the target and the friendlies." After-the-fact explanations for the near disaster tended to focus on the communications setup in effect at the time of the AC–130's arrival over the DNTT. One problem, according to a 5th Battalion staff officer, was that there was no ground liaison officer aboard the gunship, someone familiar enough with the array of structures in the area of operations to help the Spectre's gunners identify the proper targets. There was, to be sure, an Air Force ground liaison officer at Huff's operations center, but the communications equipment connecting him with the aircraft was not working properly. Furthermore, in the opinion of the 5th Battalion fire support officer, Capt. Stephen A. Smith, proximity—or the lack thereof—was a problem, in that Currie's command post and Huff's operations center where Smith was positioned were not collocated. Also, because the noncommissioned officer serving as Currie's own Air Force liaison for fire support was at that moment with the antitank platoon, Smith, who in his own words "should have been controlling aircraft for [Currie's] fire support NCO," was not even in touch with that individual. That left Currie providing directions to the gunship, said Smith, "in a manner that was not exactly correct. It was confusing. . . . You cannot give a left-right, add-drop correction as you do [with] artillery and mortar fire." The noncommissioned Air Force liaison officer, Smith

[32] Quoted words from AAR, Co C, 5th Bn, 87th Inf, Encl to Memo, Cdr, 5th Bn, 87th Inf, for Cdr, 193d Inf Bde, 5 Jan 1990, sub: After Action Report, TF Wildcat Operation PINEAPPLE CRUSH. Intervs, Wagner with Fredenburgh and Mundell, 18 Oct 1990; Wright with Huff and Senior Officers of the 5th Bn, 87th Inf, 193d Inf Bde, 5 Jan 1990; Videocassette tape, 5th of the 87th After Action JUST CAUSE, 1990.

argued, should have been located with the company commander. By the time Currie was able to contact the man and get him to the company command post for better coordination, the Spectre, which had been flying overhead for about five minutes, had been pulled off to perform another mission—a development that Currie, having just witnessed a near tragedy, found little cause to regret.[33]

Having had no success with the AC–130, Currie around 0245, turned for support to the battalion's 81-mm. mortar platoon located in the Albrook Air Station baseball diamond. Members of the unit were busy digging and fortifying their mortar pits, filling sandbags that would protect them from random PDF mortar rounds that occasionally landed in the area surrounding the athletic field. Currie had to wait for higher headquarters to authorize employment of the mortars, but, once that occurred, the platoon at Albrook fired a couple of spotter rounds. After Currie communicated the adjustments needed to hit the target, twelve more were fired for effect. These rounds, in the captain's opinion, "did a fairly decent job on the eastern section of the objective," so he asked the platoon to shift its fire in order to hit the western side. But when the second barrage came in, the rounds landed farther to the east, not west, in an area between the company's 1st and 2d Platoons. After checking to be sure that the explosions had not caused any friendly casualties, Currie decided to dispense with further support from the 81-mm. mortars. For the next firepower demonstration against the DNTT, he turned instead to the 60-mm. mortar section attached to Company C. A spotter round from one of the two mortars landed far from the target mainly because the weapons' base plates had not been settled. Adjustments were made, after which three rounds fired for effect scored direct hits inside the transportation building.[34]

Currie next contacted Captain Bowers and, impressed by the M113's performance at the Ancon DENI, requested an armored personnel carrier to execute the last firepower demonstration prior to an infantry assault on the DNTT. Currie had to wait over an hour for the vehicle to arrive. In the meantime, he ordered his 3d Platoon into its attack position, a move that took it from the vicinity of the Explansa building to a large furniture warehouse across the street. By this time, the target had gone cold; no hostile fire was coming from the transportation headquarters, although U.S. troops could see movement inside the building. At one point during

[33] First quote from Interv, Wright with Huff and Senior Officers of the 5th Bn, 87th Inf, 193d Inf Bde, 5 Jan 1990. Second quote from AAR, Co C, 5th Bn, 87th Inf, Encl to Memo, Cdr, 5th Bn, 87th Inf, for Cdr, 193d Inf Bde, 5 Jan 1990, sub: After Action Report, TF Wildcat Operation Pineapple Crush. Interv, Nemmers with Vikander, 16 Oct 1990. Remaining quotes from Interv, Nemmers with Smith, 18 Oct 1990.

[34] Intervs, De Mena with Martin, 27 Sep 1991; author with Huff, 29 Jan 1990. Currie quote from Interv, Wright with Huff and Senior Officers of the 5th Bn, 87th Inf, 193d Inf Bde, 5 Jan 1990. AAR, Co C, 5th Bn, 87th Inf, Encl to Memo, Cdr, 5th Bn, 87th Inf, for Cdr, 193d Inf Bde, 5 Jan 1990, sub: After Action Report, TF Wildcat Operation Pineapple Crush. In the mortar platoon's contribution to the Task Force Wildcat after action report, the platoon leader listed as a lesson learned the fact that "throughout the operation, my FDC [Fire Direction Center] did not receive one correct call for fire." See AAR, Task Force Wildcat, Encl to Memo, Cdr, 5th Bn, 87th Inf, for Cdr, 193d Inf Bde, 5 Jan 1990.

the night, an automobile with two people inside had tried to enter the parking lot in front of the target, but the defenders' own barricades slowed its movement, allowing U.S. forces to fire on it, killing the driver and wounding his passenger. Sometime after that, U.S. troops severely wounded a motorcycle rider trying to enter the parking lot. Currie later noted with irony how the defensive barricades at the entrances to the DNTT actually worked to Company C's benefit by keeping the defenders from escaping and reinforcements from entering.[35]

The M113 Currie requested arrived around 0430. Once it reached its firing position in front of the DNTT, Currie climbed on top, and, with his hand on the .50-caliber machine gunner, "we basically raked the objective from right to left or east to west about four or five times," in the process expending hundreds of rounds of ammunition. When the preparatory fire stopped, the 3d Platoon began its attack around 0512. Thinking that the defenders expected to be hit from the south, the platoon instead breached a fence on the building's west side and cleared a small building en route to the main headquarters. In the process, the soldiers suffered two casualties, one hit in the knee from a friendly fire ricochet, the other suffering a severe hand cut from a piece of glass. With the outer building secured, the platoon moved on to the DNTT.[36]

Currie likened the front facade of the main building to a triangle. "It's got single floors on the sides, then it goes up to two floors and then three floors at the top." Doctrine directed that the edifice should be cleared from the top down, but the 3d Platoon had no means to get to the roof. The troops therefore advanced from the bottom floors upward. Initially, they encountered no resistance and managed to clear about a quarter of the building without incident. Then, three or more snipers who were concealed inside opened fire, pinning the men down and halting their progress. For twenty minutes, the platoon leader tried to determine the source of the shots, but, in the dark and complex interior of the large building, that proved impossible. Outside, Currie guessed that the snipers had to be on the east side of the structure and, knowing the 3d Platoon had not advanced that far, ordered a recoilless rifle round fired into that area. "We didn't hear any more snipers," he recounted. Soon thereafter, three men in sniper gear surrendered in the parking lot.[37]

The 3d Platoon secured the DNTT around 0800 after methodically searching what Currie likened to a "bee's nest" with "so many little catacombs." Colonel Huff made reference to the movie *Aliens*, a science-fiction thriller, when discussing the sensory experience the troops shared during the clearing

---

[35] AAR, Co C, 5th Bn, 87th Inf, Encl to Memo, Cdr, 5th Bn, 87th Inf, for Cdr, 193d Inf Bde, 5 Jan 1990, sub: After Action Report, TF Wildcat Operation Pineapple Crush; Interv, Wright with Huff and Senior Officers of the 5th Bn, 87th Inf, 193d Inf Bde, 5 Jan 1990.

[36] AAR, Co C, 5th Bn, 87th Inf, Encl to Memo, Cdr, 5th Bn, 87th Inf, for Cdr, 193d Inf Bde, 5 Jan 1990, sub: After Action Report, TF Wildcat Operation Pineapple Crush. Quote from Interv, Wright with Huff and Senior Officers of the 5th Bn, 87th Inf, 193d Inf Bde, 5 Jan 1990.

[37] Interv, author with Huff, 29 Jan 1990. Quotes from Interv, Wright with Huff and Senior Officers of the 5th Bn, 87th Inf, 193d Inf Bde, 5 Jan 1990. AAR, Co C, 5th Bn, 87th Inf, Encl to Memo, Cdr, 5th Bn, 87th Inf, for Cdr, 193d Inf Bde, 5 Jan 1990, sub: After Action Report, TF Wildcat Operation Pineapple Crush.

process. Electrical power to the building had been cut, so the lights were out. Using flashlights, the soldiers could see steam escaping from pipes, and they could hear the hissing, an eerie sound in the near dark, compounded by the noise of various heavy objects crashing to the floor in the battered building. As the platoon moved from hallway to hallway, the flashlights cast bizarre shadows. The cumulative effect of all this was a dark, hot, and surreal ambience that elevated one's sense of fear, which, in turn, made the troops even more cautious, causing them to slow down. The men were determined to be safe but thorough, as they shone their flashlights into corners, down hallways, and around doorways to see if the light would draw fire. When it did, they answered by lobbing grenades toward the source, waiting, and then resuming their movement once the sniping stopped. After three hours of this activity, Currie observed that the men were, understandably, "completely exhausted."[38]

According to plan, once the DNTT had been secured, the 2d Platoon began clearing two small buildings nearby, finishing the task around 0900. The remainder of the day was spent improving defenses around the complex, seizing and securing an air traffic control building and a radio station in the area, and examining official papers and other material discovered while clearing the target. The troops took turns being search teams and security teams, with the former turning up large sums of money (nearly a half-million dollars, according to one account), weapons caches, and, in Currie's words, "a veritable motherload of intelligence information from video tapes, to safes, to confidential and secret documents." U.S. troops at the Ancon DENI had made

The PDF engineering compound

[38] Parts of this paragraph are taken almost verbatim from Yates, "Operation JUST CAUSE in Panama City," pp. 350–51. See also Interv, author with Huff, 29 Jan 1990, for the reference to *Aliens*. Quotes from Interv, Wright with Huff and Senior Officers of the 5th Bn, 87th Inf, 193d Inf Bde, 5 Jan 1990. AAR, Co C, 5th Bn, 87th Inf, Encl to Memo, Cdr, 5th Bn, 87th Inf, for Cdr, 193d Inf Bde, 5 Jan 1990, sub: After Action Report, TF Wildcat Operation Pineapple Crush.

similar discoveries, with the addition of drugs, "sex-oriented objects," and the accoutrements of an illegal passport and immigration operation. Huff visited each site to evaluate the findings and determine how to handle the material. At the same time, he was concerned about the linkup that would occur between Company C and Company A once the latter finished its mission at the PDF engineer compound, just north of the DNTT and northwest of the Ancon DENI. To eliminate the chances of friendly fire occurring as the two units and the mechanized company converged, Huff instituted certain fire control measures that "worked very well."[39] By that time, two of Task Force Wildcat's hard targets had been secured, and 5th Battalion's Company A was on the verge of securing a third.

Company A, 5th Battalion, 87th Infantry

The headquarters for a PDF engineer battalion was in a compound containing fourteen main buildings of various sizes, including two large aircraft hangars. Consequently, as Huff was quick to point out concerning Task Force Wildcat's mission, "A Company was probably the most detailed clearing operation of the three companies, because [it] had the longest—the engineer compound was very, very large, a couple of football fields." Captain Flynt, the Company A commander, offered a more precise measurement: 1,100 yards long and 550 yards wide. Because, in Huff's words, "we owned Curundu," a residential area overlooking the complex from the north, Flynt possessed highly detailed maps and photographs of the objective, as well as intelligence obtained from around-the-clock observation (and from coordination with the 470th Military Intelligence Brigade). Curundu also offered Company A a relatively secure assembly area and, farther south past the junior high school, positions from which two of the company's platoons could launch their attacks to clear most of the compound's buildings should the broadcast appeals to surrender fail.[40]

Flynt's concept of the operation was straightforward. His 1st Platoon, reinforced by M60 machine guns and two 90-mm. recoilless rifle teams, would move into an overwatch position east of the compound. Meanwhile, the 2d and 3d Platoons would take up positions to the north. The 2d Platoon, at the northwest corner of the complex, would also be able to bring under direct fire an airfield west of the buildings, a facility also covered by the task force's mortars at Albrook. Flynt had no units at the southern end of the compound; there he would rely on Company C's DNTT and DENI operations across the company phase line to seal off potential escape routes

---

[39] First quote from AAR, Co C, 5th Bn, 87th Inf, Encl to Memo, Cdr, 5th Bn, 87th Inf, for Cdr, 193d Inf Bde, 5 Jan 1990, sub: After Action Report, TF Wildcat Operation Pineapple Crush. Second quote from Interv, Nemmers with Smith, 18 Oct 1990. Interv, Wagner with Fredenburgh and Mundell, 18 Oct 1990. Last quote from Interv, Wright with Huff and Senior Officers of the 5th Bn, 87th Inf, 193d Inf Bde, 5 Jan 1990.

[40] Quotes from Interv, Wright with Huff and Senior Officers of the 5th Bn, 87th Inf, 193d Inf Bde, 5 Jan 1990. Videocassette tape, 5th of the 87th After Action Just Cause, 1990.

for any defenders. As his concept statement succinctly directed, "Nobody leaves or enters the engineer complex."[41]

Around 0040, Flynt ordered the three rifle platoons to begin moving out of the Curundu assembly area. "I modified the time as I saw fit," he later reported, "due in part to my eagerness to 'get on with it,' and also [due to] new calculations of estimated travel time to the attack positions." The initial movement did not go smoothly, with personnel from the different platoons intermingling, but, as Flynt observed, "the soldiers knew at night who their squad leader was and there were no lost soldiers." Almost as soon as the troops left the assembly area, they came under small-arms and sniper fire, but not with sufficient volume or accuracy to disrupt their advance. To the east of the compound, in the overwatch position, the 1st Platoon leader placed two squads up front, each with two M60 machine guns, and his third squad in the rear for security. At the northwest corner of the compound, in front of a stretch of cuna grass, the 2d Platoon leader sent one squad at a time to its assigned position, placing an M60 machine gun between his 1st and 2d Squads. The 3d Platoon also moved without incident into its attack position.[42]

At this point in the operation and for the next four hours, the mission was only to fix the engineer compound. For some soldiers, most of this time was spent in a prone position near the objective. For others, the period was more eventful. On three separate occasions, automobiles trying to enter the compound were stopped by the 1st Platoon's accurate and deadly machine gun fire. (As was the case at the DNTT, barriers around the buildings meant to keep attackers out also obstructed entry to individuals trying to assist the defenders.) As soon as Flynt's men were in position, the process of alternating surrender appeals with firepower demonstrations began. In Company A's sector, the loudspeaker broadcasts informed the defenders that they were surrounded and urged them to give up. Once an appeal had been made, a five-minute countdown began, after which, if there had been no response, the 1st Platoon opened fire on selected buildings. The first broadcast, indeed, failed to produce any prisoners, and the platoon's first firepower demonstration included a hundred rounds from each of the M60s and a 90-mm. recoilless rifle round fired into the engineer battalion's headquarters building. A second broadcast also proved unsuccessful, and another firepower demonstration ensued.[43]

Soon afterward, Flynt, like Currie, sought to intensify the demonstrations by calling in additional firepower. But this proved difficult. He was aware of

---

[41] AAR, Co A, 5th Bn, 87th Inf, Encl to Memo, Cdr, 5th Bn, 87th Inf, for Cdr, 193d Inf Bde, 5 Jan 1990, sub: After Action Report, TF Wildcat Operation Pineapple Crush. Quote from Nemmers, *United States Army South Staff Ride,* p. 24, and see also p. 25.

[42] Nemmers, *United States Army South Staff Ride,* p. 22. Quotes from AAR, Co A, 5th Bn, 87th Inf, Encl to Memo, Cdr, 5th Bn, 87th Inf, for Cdr, 193d Inf Bde, 5 Jan 1990, sub: After Action Report, TF Wildcat Operation Pineapple Crush. Videocassette tape, 5th of the 87th After Action Just Cause, 1990.

[43] AAR, Co A, 5th Bn, 87th Inf, Encl to Memo, Cdr, 5th Bn, 87th Inf, for Cdr, 193d Inf Bde, 5 Jan 1990, sub: After Action Report, TF Wildcat Operation Pineapple Crush; Videocassette tape, 5th of the 87th After Action Just Cause, 1990.

Company C's misfortunes in communicating with the AC–130 gunship, so he ruled out that option. He knew that demands were "not heavy" for fire support from the 81-mm. mortars at Albrook, but Huff told him that those tubes had not been entirely accurate in supporting Company C's actions earlier at the DNTT. "Due to the proximity of my soldiers to the desired impact area," Flynt explained, "I didn't feel confident in calling in the 81mm mortars on a non-registered TRP [target reference point]." He also expressed disappointment that there were no attack helicopters he could summon, especially since in his opinion the engineer battalion was "particularly suited to accommodating the splash effects" of an attack helicopter's guns. "In the final analysis," he later observed, "the only fire support asset that worked for Co A was the organic 60mm mortar section." One of the section's members monitored both the company command communications network and the company mortar network, which was essential to imparting "the tactical situation as it unfolded, increasing [the section's] understanding of why fire missions were called for." Given this setup, Flynt chose to employ fires from the mortars just before he issued orders to his three rifle platoons to begin seizing and clearing the compound's fourteen buildings.[44]

The first of those orders went to the 2d Platoon between 0530 and 0600, directing it to seize what had been designated on Company A's photographs and maps as Building 1, in the northwest corner of the complex. With the 1st Platoon laying down suppressing fire, the 3d Squad, 2d Platoon, moved into a long ditch or gutter, headed south in it to the target, then entered and cleared the building. The platoon's other two rifle squads followed behind, joining up with the 3d Squad to lay down suppressive fire for the 3d Platoon as it moved out to seize Buildings 2 and 3 on the northeast corner of the compound. For the first of these, the 3d Platoon leader had one squad make the assault, while keeping two squads in support and a fourth squad composed of nine engineers in reserve. Preparatory fires employing an AT4 projectile, a LAW rocket (to knock the door to the building off its hinges), and M16 and M203 rounds against guard posts in the vicinity lasted a little over a minute prior to the attack, which began when the designated squad "lined up on the door" and a team leader threw several fragmentary grenades inside. The troops then cleared the building, after which they moved without significant opposition into Building 3, the PDF engineer battalion headquarters, finding seventy to eighty weapons inside, along with a large amount of ammunition. With the headquarters building secured, the 2d Platoon received orders to seize the first of the large hangars, labeled on the maps as Building 4.[45]

This pattern of attack continued into the early daylight hours. Flynt described the tactic as "leapfrogging" his platoons, "always keeping the platoon that just finished clearing a building in reserve, resting and redistributing ammunition" and other supplies received throughout the operation from the company's supply trains at Curundu and the task force

---

[44] AAR, Co A, 5th Bn, 87th Inf, Encl to Memo, Cdr, 5th Bn, 87th Inf, for Cdr, 193d Inf Bde, 5 Jan 1990, sub: After Action Report, TF Wildcat Operation PINEAPPLE CRUSH.

[45] Videocassette tape, 5th of the 87th After Action JUST CAUSE, 1990. In this videotape, each rifle platoon leader provides a narrative overview of the actions his platoon took during the night of 20 December.

support platoon at Albrook. "This way," he explained, "there was always a rested, prepared platoon for the next mission."[46]

During the 2d Platoon's sweep of the Building 4 hangar, the troops uncovered a major arms cache consisting of grenades, several hundred AK47s, LAWs, Uzi submachine guns, thousands of rounds of ammunition, and other ordnance. They also took three prisoners, followed by twelve more who arrived from Building 5, the adjacent hangar, with their hands raised. In accordance with standard procedures in effect for Task Force Wildcat, the fifteen men were processed and then moved to the prisoner-of-war site at Albrook. Before being sent back, they were questioned as to any useful information they might impart. Most interrogations revealed little of value, but, in one exception, that of a wounded and cooperative lieutenant, Flynt learned that there were around seventy engineers left in the compound, along with an unspecified number of infantry, perhaps special operations troops, from one or more PDF combat units.[47]

Up to the time when the 2d Platoon secured Building 4 and 3d Platoon cleared Building 5, 1st Platoon's role had been to lay down suppressive fire. Then, once the second hangar had been taken, the platoon leader received orders to seize Buildings 6 and 7, with suppressive fire being provided by the 2d Platoon. In keeping with the leapfrog tactic, once those two building were secured, the 2d Platoon moved through them to advance on Buildings 9 and 10. The last building to be cleared was number 14, again the work of the 2d Platoon. The entire attack had taken several hours mainly because of the size of the objective but also because, at some locations, resistance was significant, if not particularly heavy. As Huff observed,

> It seemed like people would not give up. They would literally wait till [U.S. troops] came into the building, and then spring up, either with their hands up or, in one case, with a rifle in his hands, not particularly to shoot at anybody, but I don't know what they were thinking of. One guy kept hopping around like a rabbit; hid underneath a wood pile like a rabbit, and they had to take the plywood out to drag this guy out. Just scared; thought he was going to die; did not want to particularly fight, but wanted to hide. And that's the way a lot of these guys were—hard to talk them out.[48]

Despite the resistance, the number of defenders, and the dangers inherent in seizing and securing fourteen buildings of various sizes and configurations, Company A suffered no casualties on 20 December. After Building 14 was cleared, some troops continued to move south to the company's limit of advance, while others shifted north and took up defensive positions. At 1240,

---

[46] AAR, Co A, 5th Bn, 87th Inf, Encl to Memo, Cdr, 5th Bn, 87th Inf, for Cdr, 193d Inf Bde, 5 Jan 1990, sub: After Action Report, TF Wildcat Operation Pineapple Crush.

[47] AAR, Co A, 5th Bn, 87th Inf, Encl to Memo, Cdr, 5th Bn, 87th Inf, for Cdr, 193d Inf Bde, 5 Jan 1990, sub: After Action Report, TF Wildcat Operation Pineapple Crush; Videocassette tape, 5th of the 87th After Action Just Cause, 1990; Interv, Wright with Huff and Senior Officers of the 5th Bn, 87th Inf, 193d Inf Bde, 5 Jan 1990.

[48] Videocassette tape, 5th of the 87th After Action Just Cause, 1990. Quote from Interv, Wright with Huff and Senior Officers of the 5th Bn, 87th Inf, 193d Inf Bde, 5 Jan 1990.

Balboa DENI station

Captain Flynt reported up his chain of command that the PDF engineer compound, the largest of Task Force Wildcat's four hard targets, was secured.[49]

COMPANY B, 5TH BATTALION, 87TH INFANTRY

If Company A, 5th Battalion, 87th Infantry, drew the largest of the targets, Company B—the "Jungle Cats"—confronted the busiest. The Balboa district in the southern portion of Task Force Wildcat's Antietam area of operations had in previous decades been the unofficial capital of the Canal Zone and a thriving commercial and residential area. In the north, the district ran from the Panama Canal Commission administration building and housing in the east to Cerro Sosa in the west. In the southwest, Balboa came up against the western edge of El Chorrillo, which contained the Comandancia and Carcel Modelo. Of special concern to U.S. planners looking at their maps prior to the invasion was the Balboa DENI complex, home of the only significant PDF element in the district, a police agency. Although a small organization, it was located directly along the route Task Force Gator would take to the PDF main headquarters. The station thus became another of the hard targets Huff's troops had to seize. Numerous other critical sites in the area had to be secured as well, to include the Balboa Heights PCC housing area, a PCC electrical

[49] Videocassette tape, 5th of the 87th After Action JUST CAUSE, 1990; Interv, Wright with Huff and Senior Officers of the 5th Bn, 87th Inf, 193d Inf Bde, 5 Jan 1990; AAR, Co A, 5th Bn, 87th Inf, Encl to Memo, Cdr, 5th Bn, 87th Inf, for Cdr, 193d Inf Bde, 5 Jan 1990, sub: After Action Report, TF Wildcat Operation PINEAPPLE CRUSH.

substation, a water pump station, the canal commission's telephone exchange, a Marine Control Center, and certain facilities on Cerro Sosa.[50]

The plan that Company B's commander, Capt. Mark Conley, carried with him into the district on 20 December called for his executive officer and 1st Platoon to occupy positions at the Balboa Elementary School, below the canal commission's administration building on its southwest side, while the 3d Platoon would move to the commission's electrical substation at the intersection of Balboa Street and Roosevelt Avenue. An antitank section attached to the platoon would initially close off a potential enemy escape route by securing a roadblock position to the west on La Boca Road, an avenue running right by the DENI station on one side and by St. Mary's Church, the parking lot of which contained Conley's command post and a support element, on the other side. The command post overlooked the DENI station from the northwest, while the three squads in the 2d Platoon would fix the complex from positions to its northeast, southeast, and southwest. Because two of these locations would be close to Gaillard Highway, the 2d Platoon would move from Fort Clayton into Balboa, with Company C under the operational control of Task Force Gator, a means of orchestrating the two units' deployment in such a way as not to interfere with the mechanized convoy heading for the Comandancia. After the platoon entered Balboa, however, it would link up with the rest of Company B and operational control would revert back to Task Force Wildcat.[51]

As the situation developed on 20 December, the advance of Company B (minus) down Gaillard was disrupted only slightly, the cause being the brief firefight with the occupants of the bus and police cars in front of Albrook Air Station and, soon thereafter, some fire the company received, easily suppressed, from the vicinity of the PDF engineer compound. Rushing through the ambushes, most squads and other elements from the company's three rifle platoons arrived at their positions pretty much on schedule. The same was not true of the maneuver group that included Conley, his headquarters, the 1st Squad from the 2d Platoon, and a support element consisting of the 2d Platoon sergeant armed with a 90-mm. recoilless rifle. As they approached the parking lot at St. Mary's, they again came under enemy fire, this time from an estimated three or four snipers who were shooting into the 2d Platoon's positions. The shots were not accurate, but, with rounds ricocheting about, Conley and the other troops got out of their trucks and decided to low-crawl to the designated command post area, a movement that, in Conley's words, "took us a long time." Once he arrived, the captain tried to radio Huff's command post but experienced "a little bit of a commo problem." By this time, about 0140, the 2d

---

[50] Brief descriptions of the Balboa district were at http://www.moon.com/planner/panama/mustsees/balboa.html and http://www.greatestcities.com/Central_America/Panama/Balboa_city.html. The list of sites in Balboa that planners specifically identified for U.S. protection during Operation Just Cause is in AAR, Co B, 5th Bn, 87th Inf, Encl to Memo, Cdr, 5th Bn, 87th Inf, for Cdr, 193d Inf Bde, 5 Jan 1990, sub: After Action Report, TF Wildcat Operation Pineapple Crush. See also Interv, Wright with Huff and Senior Officers of the 5th Bn, 87th Inf, 193d Inf Bde, 5 Jan 1990.

[51] AAR, Co B, 5th Bn, 87th Inf, Encl to Memo, Cdr, 5th Bn, 87th Inf, for Cdr, 193d Inf Bde, 5 Jan 1990, sub: After Action Report, TF Wildcat Operation Pineapple Crush; Interv, Wright with Huff and Senior Officers of the 5th Bn, 87th Inf, 193d Inf Bde, 5 Jan 1990.

Platoon's other squads were already in place. Its 3d Squad was located at the Christian Science reading room near the front of the larger of the two main buildings in the DENI complex. Upon arriving at the position, the troops took fire from the building, prompting the squad leader to order "a couple rounds" fired in return. As this was going on, the 2d Squad with a 90-mm. recoilless rifle assumed its positions at the Young Men's Christian Association (YMCA) building on the DENI's right flank. Finally, as the platoon sergeant's support element took its position near Captain Conley on the hill overlooking the DENI, the 1st Squad completed the platoon's encirclement of the complex by moving to the Marco Polo restaurant southwest and to the rear of the station. By this time, around 0200, all elements of Company B were either engaged in their assigned missions or making last-minute preparations to do so.[52]

For the 1st Platoon, the mission translated into having soldiers block the entrance to the Panama Canal Commission housing area with a tractor-trailer, using one squad armed with antiarmor weapons and a 90-mm. recoilless rifle to guard the roadblock and another squad to conduct roving foot patrols through the rows of houses in search of any enemy groups or individuals who might have taken refuge there. Unknown to the men on patrol, twenty to thirty PDF personnel, upon hearing the premature firing from Pier 18 around 1230, had already discarded their weapons, shed their uniforms, and fled into the housing area. Although there would be no contact with them that night, several in the group would be captured later, during daylight hours. As for the remainder of the 1st Platoon, one squad guarded the platoon command post at the elementary school and secured the commission's telephone exchange and water pump station, while another group of soldiers helped set up the Company B prisoner-of-war collection point.[53]

The mission for the 3d Platoon involved employing one squad to secure the Marine Control Center containing computers that monitored traffic in the canal; another squad to protect the electrical substation that provided power to the Balboa, Diablo, Los Rios, and Corozal districts; and a security element to guard a microwave relay repeater station atop Cerro Sosa. In none of these assignments did the troops encounter any serious problems. The same could not be said of the attached antitank section covering the La Boca roadblock. Shortly after the checkpoint was set up, a car containing two people approached and failed to slow down. The troops opened fire, killing the driver, Richard A. Paul, who turned out to be a college student on his way home from the block party held for American college students returning to Panama

[52] AAR, Co B, 5th Bn, 87th Inf, Encl to Memo, Cdr, 5th Bn, 87th Inf, for Cdr, 193d Inf Bde, 5 Jan 1990, sub: After Action Report, TF Wildcat Operation PINEAPPLE CRUSH; Nemmers, *United States Army South Staff Ride,* p. 26; Videocassette tape, 5th of the 87th After Action JUST CAUSE, 1990. Quotes from Interv, Wright with Huff and Senior Officers of the 5th Bn, 87th Inf, 193d Inf Bde, 5 Jan 1990.

[53] AAR, Co B, 5th Bn, 87th Inf, Encl to Memo, Cdr, 5th Bn, 87th Inf, for Cdr, 193d Inf Bde, 5 Jan 1990, sub: After Action Report, TF Wildcat Operation PINEAPPLE CRUSH; Videocassette tape, 5th of the 87th After Action JUST CAUSE, 1990; Interv, Wright with Huff and Senior Officers of the 5th Bn, 87th Inf, 193d Inf Bde, 5 Jan 1990.

over the Christmas holidays to visit their parents. The passenger, while injured, survived the tragic incident.[54]

As for the 2d Platoon, the troops, having fixed the Balboa DENI, prepared to neutralize the complex, following the same procedure that elements of Companies A and C were employing at their respective hard targets: the combination of surrender appeals and firepower demonstrations. After an attempt to contact any defenders inside the station by telephone failed to elicit a response, the scripted broadcasts, initially made with a bullhorn, began at 0208 from the 1st Squad's position at the Marco Polo restaurant. Defenders were told they had five minutes to surrender or face the consequences of U.S. firepower. At the end of the five minutes, another appeal was made, giving the defenders fifteen seconds to come out. When no one appeared, the 2d Squad at the YMCA fired its M60 machine guns and three 90-mm. rounds at the main building. In this and subsequent firepower demonstrations, the plan was to reduce the risk of friendly fire and collateral damage by aiming the small arms at the windows and doors and the larger weapons—the recoilless rifles and AT4s—at the building's superstructure. When the first firepower demonstration produced the same results as the surrender appeals, the process was repeated, again with the broadcasts going unheeded (or, possibly, unheard given the use of only a bullhorn). The second firepower demonstration was more intense, with all positions firing their M249 squad automatic weapons, M60s, AT4s, and recoilless rifles at the objective. (Two to three LAWs were also fired at two automobiles parked in front of the main building.) When no one surrendered, Conley requested support from the heavy broadcast team then located with Company A. The task force acceded to the request, although Maj. James Woods, the operations officer, believed it to be "a moot point by that time, because we realized nobody was really coming out because of the HB team." While waiting for the team to arrive, the 2d Platoon conducted a third firepower demonstration. After that, shortly before 0300, Conley received word to have the platoon seize and clear the two DENI buildings. Just before the assault began, the newly arrived heavy broadcast team, in Conley's words, "gave the last loud blast of 'surrender or die.'"[55]

To prevent one squad from firing into another during the assault, one of Conley's lieutenants decided that it would be best to link up two of the squads to "lay down a base of fire and go in." The linkup took time and perhaps gave the defenders an opportunity to escape. Conley could not be sure if they fled

[54] AAR, Co B, 5th Bn, 87th Inf, Encl to Memo, Cdr, 5th Bn, 87th Inf, for Cdr, 193d Inf Bde, 5 Jan 1990, sub: After Action Report, TF Wildcat Operation Pineapple Crush; Videocassette tape, 5th of the 87th After Action Just Cause, 1990; Intervs, Wright with Huff and Senior Officers of the 5th Bn, 87th Inf, 193d Inf Bde, 5 Jan 1990; author with Maria Len-Rios, 29 Nov 1991, Fort Leavenworth, Kans.; Maria Len-Rios, "Operation Just Cause: An Eyewitness Account," *Focal Point* (Winter 1990): 10–11, 25–27. In the interview with Huff and his commanders and staff, the colonel said that an autopsy of the student revealed alcohol in the blood, suggesting that he may not have been fully aware of the situation that cost him his life. For more information related to the block party incident, see Chapter 2.

[55] AAR, Co B, 5th Bn, 87th Inf, Encl to Memo, Cdr, 5th Bn, 87th Inf, for Cdr, 193d Inf Bde, 5 Jan 1990, sub: After Action Report, TF Wildcat Operation Pineapple Crush. Quotes from Wright with Huff and Senior Officers of the 5th Bn, 87th Inf, 193d Inf Bde, 5 Jan 1990.

just before the attack or during the firepower demonstrations. "I guess they just jumped over the fence or whatever else," he mused afterward. With the two main buildings empty, the clearing process, thorough as it was, took almost two hours. By 0500, the 2d Platoon had set up a security perimeter around the complex, and the fight for the DENI appeared to be over. Colonel Huff decided to take a look, but, when he arrived and entered the main building, he quickly became aware that there were still snipers in the area. As one of his staff recalled, "So, we had to reclear [the building] during the night." In the process, what registered most in that officer's mind was the large number of rooms—"I don't think we realized just how many there were"—in which the enemy could easily hide without detection. "For one company to clear all that is just unbelievable," he recounted. Still, by dawn, another of Task Force Wildcat's hard targets, the Balboa DENI, appeared to be under U.S. control. Daylight, however, would bring additional challenges for the soldiers of Company B.[56]

Two episodes occurred almost simultaneously. At 0730, troops in the 2d Platoon security perimeter began to receive sniper fire. In response, two soldiers, a specialist and a first sergeant, moved toward the source of the shooting and killed the sniper. Meanwhile, reports from the 1st Platoon sector described hundreds of refugees arriving at the elementary school seeking shelter, food, and assistance. These were the same people who had flooded Task Force Gator's area of operations earlier in response to the uncontrolled fire in El Chorrillo. Company B soldiers at the school set up a treatment center in an effort to help in any way they could, but the numbers soon overwhelmed them. Whereas Conley, in notifying Huff of the problem, first reported about 400 refugees in the area, he was soon reporting 1,000 and then 3,000. By 1000, the estimate was 5,000 and still climbing. The figures, according to Huff, "just went up exponentially." The lieutenant colonel sent the scout platoon that was providing security at his command post to the school to look into the situation, which he soon learned was rapidly getting out of control. Without food to distribute and without the facilities to determine who needed medical treatment and who did not, or which people were friendly and which were PDF or Dignity Battalion members, troops from the 1st Platoon did what they could but soon moved the ad hoc aid center over to the high school next to them, a larger facility. After breaking into the gymnasium and depositing the refugees there and on the adjacent athletic field, the troops turned responsibility for their well being over to USARSO's Maj. Les Knoblock. "We washed our hands of that business as quickly as we could and got out of there," Huff observed, while lamenting that USARSO's plan to handle refugees was not "formulated to the degree that we had hoped."[57] In the days to come, he would

---

[56] AAR, Co B, 5th Bn, 87th Inf, Encl to Memo, Cdr, 5th Bn, 87th Inf, for Cdr, 193d Inf Bde, 5 Jan 1990, sub: After Action Report, TF Wildcat Operation Pineapple Crush. First three quotes from Interv, Wright with Huff and Senior Officers of the 5th Bn, 87th Inf, 193d Inf Bde, 5 Jan 1990. Last two quotes from Interv, Nemmers with Vikander, 16 Oct 1990.

[57] Interv, Nemmers with Vikander, 16 Oct 1990. Quotes from Interv, Wright with Huff and Senior Officers of the 5th Bn, 87th Inf, 193d Inf Bde, 5 Jan 1990. On the fire in El Chorrillo, the initial phase of the refugee problem, and Major Knoblock's role in running the Balboa High School refugee center, see Chapter 4.

discover that extricating his men from the refugee business was not as easy as it first seemed.

Compounding the confusion created by the east-to-west movement of an unexpectedly large number of refugees was the countermovement after dawn of several hundred people released from the prisoner-of-war center at Albrook. A number of enemy prisoners had been taken during the night but so, too, had a number of suspicious-looking civilians or just innocent bystanders. In general, U.S. troops had adhered to a cautious approach: if Panamanian males of military age were apprehended and could not prove they were not PDF or Dignity Battalion members, they were taken to the prisoner center. With daylight, a hasty determination was made as to who really needed to be kept in custody. Many of the rest were released, most moving west-to-east into Task Force Wildcat's area of operations. Captain Conley noted that the numbers involved were "pennies" compared to the traffic created by the refugees, but the intrusion nonetheless disrupted Company B's operations.[58]

Before things began to quiet down for Company B, the troops of the 2d Platoon manning the perimeter around the cleared DENI complex had to contend with a fire in the main building. Around 0900, a pungent odor gave away the fire's existence, but no one sent to determine its location could find it despite gagging on the smell. The men engaged in the search prudently called off their effort after someone suggested that, if there were a gas leak in the building, an explosion might be imminent. The inability to find the fire did not prevent speculation on what caused it. Of the two main explanations put forward, one held that, during the firepower demonstration, a 90-mm. round might have gone under the building's roof and lodged in the wooden superstructure; the second contended that U.S. rounds might have hit an electrical box inside the building. On the scene that morning, Conley concluded the latter when he radioed for help, indicating that "we can't fight electrical fires." "It's not real bad," he continued. "If we can get the fire department in here right now, we'll be able to save this room." A lieutenant in the battalion operations center at Fort Clayton called the Corozal fire station and, surprisingly, found a crew on duty. When he asked the firemen to go to Balboa, however, the answer was not so surprising: "Hell, no. We aren't going anywhere. . . . You all fight the fire!" By the time word reached 1st Lt. Kevin Huggens from the 5th Battalion intelligence shop that the Corozal fire trucks were not coming, flames could be seen engulfing the roof of the DENI. Huggens and others recognized the value of the building for intelligence purposes, so he joined Conley in urging the troops to find a manned fire department in Balboa and escort whoever was on duty to the scene of what was turning into a conflagration. After some time passed, a Panamanian fire truck actually arrived. The six-man crew belonged to the civilian wing of the Panama Defense Forces. Two of the men tried in vain to put out the fire; the other four set about looting whatever they could lay their

[58] Interv, Wright with Huff and Senior Officers of the 5th Bn, 87th Inf, 193d Inf Bde, 5 Jan 1990; Yates, "Operation JUST CAUSE in Panama City," p. 351.

hands on. In the end, the 2d Platoon simply watched the building burn top to bottom, with only its scorched outer walls left standing.[59]

There was a footnote to the operation at the Balboa DENI. In front of the main building was a large, makeshift, thatch-roofed structure that sheltered a life-size Nativity scene. While the DENI itself had been subjected to three firepower demonstrations and the main building had burned down, the display remained unscathed. As Huff said later, "The last communication to occur was not to hurt the Nativity scene, . . . and the goat with it." In discussing the outcome with an interviewer, General Stiner agreed that the untouched status of the scene stood as a tribute to the "measured application of force" by U.S. troops.[60]

By 1300, a partial calm had descended over the Balboa district. As with all of Task Force Wildcat, Company B would have snipers, fleeing PDF and Dignity Battalion members, prisoners, looters, arms caches, refugees, security, and other issues with which to contend, but, for the remainder of the day, many of the troops could rest, replenish their supplies, and await follow-on missions.

On 20 December, each element in Task Force Wildcat performed tasks unique to its mission. Yet, in taking stock of the overall experience, Huff, his commanders, his staff, and his troops found that certain experiences were common throughout the Antietam area of operations. To begin with, most units at the company level and below had problems with their tactical communications. Leaders often found it difficult to call in fire support, while units, once separated by built-up structures or after entering buildings, found it nearly impossible to talk to one another or their superiors via handheld radios. In contrast, at higher headquarters, communications between brigade and battalion, and battalion and company were generally good, with the retransmission systems set up in advance of hostilities being properly located and functioning well.[61]

As was the case with Task Force Black Devil, almost all the commanders in Task Force Wildcat from Huff to squad leaders praised the 90-mm. recoilless rifle as a weapon ideally suited to urban combat, both because of its accuracy and because of the effects of its impact. Huff added that the weapon was loud and thus preferable to AT4s and LAWs in subjecting the enemy to psychological stress in the confined spaces of the rooms, hallways, attics, and closets so prevalent in urban combat. Like his Task Force Gator counterpart, however, he was skeptical of the AC–130's utility over a congested urban area in which friendly forces were often just yards away from the enemy, especially

[59] Quotes from Interv, Wagner and De Mena with Stewart, 16 Oct 1990. Intervs, Wright with Huff and Senior Officers of the 5th Bn, 87th Inf, 193d Inf Bde, 5 Jan 1990; author with Huff, 29 Jan 1990. In the Company B after action report, Captain Conley states that he was the one who ordered one of his platoons to go find a fire truck and crew. This does not rule out the possibility that Huggens did so as well.

[60] First quote from Interv, Wright with Huff and Senior Officers of the 5th Bn, 87th Inf, 193d Inf Bde, 5 Jan 1990. Second quote from Wright with Lt Gen Carl W. Stiner, 2, 7, 27 Mar and 11 Jun 1990, Fort Bragg, N.C. In his interview with Wright, Huff noted that at the Balboa DENI there were also "literally, houses feet away that didn't take a single round."

[61] Intervs, Nemmers with Vikander, 16 Oct 1990; author with Huff, 29 Jan 1990.

if the gunship did not have a liaison officer aboard intimately familiar with the targets on the ground. As an alternative to the Spectre, at least two of Huff's company commanders found they could enhance the intensity and accuracy of their firepower by calling on the 60-mm. mortars available to them.[62]

At the hard targets Task Force Wildcat was directed to neutralize, a common pattern emerged where the Panamanians chose to resist. For starters, the defenders were not always the enemy units Huff and his commanders expected to encounter. For example, the colonel later noted that he had seen no policemen at the DENI stations in Ancon and Balboa. Rather, infantrymen, usually with UESAT officers giving them orders, manned the sites. As Wildcat units began to neutralize these targets and take prisoners, they would ask the men why they had not surrendered after the PSYOP broadcasts and firepower demonstrations. The answers across the board were similar. First, the defenders in some cases either had not heard the surrender appeals or believed—often after having been told by their officers—that U.S. troops would kill them if they tried to give themselves up. Second, several officers themselves threatened to kill anyone who tried to surrender. After the firepower demonstrations, however, most of the officers fled, making it easier for those who remained behind to give up the fight. Finally, Task Force Wildcat encountered many Panamanian military personnel who fought simply from a sense of national honor.[63]

Perhaps the observation most common to the officers and troops of Task Force Wildcat was their universal praise for the training they had received in urban operations two or three months prior to Just Cause. City fighting was something few if any of them had experienced before, and, as they discovered in the early hours of 20 December, the phenomenon was complex and contained its own variety of risks and dangers. It could also be unnerving, especially at night. Given all that, the training the troops had received beginning in October had been invaluable and had helped ensure a favorable outcome at a very low cost in lives and destruction; by midday on the twentieth, the task force reported only one soldier killed and a handful wounded. But those who praised the training also recognized that the low casualty rate was also the result of the U.S. preponderance of firepower and, more important, the fact that the Panama Defense Forces and Dignity Battalions simply did not constitute a first-rate military opponent. Had they been better disciplined, well armed, and determined to fight for every square foot of the positions they held during the first hours of Just Cause, the overall outcome—their defeat—would have been the same, but, undoubtedly, the U.S. victory would have come at a considerably higher price.[64]

By midday on 20 December, enemy forces and installations in the first ring of targets surrounding the Comandancia bulls-eye had been neutralized by

---

[62] Yates, "Operation JUST CAUSE in Panama City," p. 352.

[63] Interv, Wright with Huff and Senior Officers of the 5th Bn, 87th Inf, 193d Inf Bde, 5 Jan 1990.

[64] AAR, Task Force Wildcat, Encl to Memo, Cdr, 5th Bn, 87th Inf, for Cdr, 193d Inf Bde, 5 Jan 1990, sub: After Action Report, TF Wildcat Operation Pineapple Crush; Nemmers, *United States Army South Staff Ride,* pp. 22–25; Interv, Nemmers with Smith, 18 Oct 1990.

U.S. forces already in Panama at H-hour. Targets in the second ring, to include PDF units and facilities at Panama Viejo, Las Tinajitas, and Fort Cimarrón, were the responsibility of the 82d Airborne Division, due to arrive from Fort Bragg. But before the paratroopers could execute their portion of the plan, the airport complex at Torrijos-Tocumen, itself a second-ring target, needed to be seized. That was one of the missions of Task Force Red-T.

7

Task Force Black and Task Force Red-T

☆ ☆ ☆

While several task forces—Gator, Semper Fi, Wildcat, Black Devil, and White—neutralized enemy targets in the immediate area of the Panama Canal, other units directed their attention to the third concentric ring of objectives surrounding the Comandancia. These targets included PDF elements at Panama Viejo, Fort Cimarrón, and Las Tinajitas; the television tower on Cerro Azul and the bridge over the Pacora River; and the Torrijos-Tocumen airfield complex. As called for in JTF South's Operation Plan 90–2, the U.S. Special Forces constituting Task Force Black would seize the television tower and perform reconnaissance and surveillance operations at Fort Cimarrón, Las Tinajitas, and the Pacora River bridge. (The 470th Military Intelligence Brigade was responsible for observing Panama Viejo.) As for Torrijos-Tocumen, U.S. Rangers of Task Force Red-T would launch an assault at H-hour and, after securing the airfields, make immediate preparations for the arrival of a combat brigade from the 82d Airborne Division (*Map 10*). The paratroopers, for their part, would drop onto the captured facility, board waiting helicopters, and launch air assaults against the three PDF garrisons under surveillance. If all went well, the attacks at Panama Viejo, Fort Cimarrón, and Las Tinajitas would take place at night and require only a few hours to achieve a successful outcome.[1]

Task Force Black

Task Force Black would be led by the commander, Special Operations Command, South, who at the time of Just Cause was Col. Jake Jacobelly. His operations center would be in a hangar at Albrook Air Station, which would also house the headquarters element of the 3d Battalion, 7th Special Forces Group (Airborne). The battalion, which was stationed in Panama, made up the operational core of the task force, with reinforcements from the 7th's Company A, 1st Battalion, based (as was the group) at Fort Bragg. Aviation support for Task Force Black would consist of MH–60 Black Hawks from the 617th Special Operations Aviation Detachment; UH–60s from U.S. Army,

[1] Briefing Slides, Blue Spoon: The Evolution of the Plan for Just Cause, n.d.

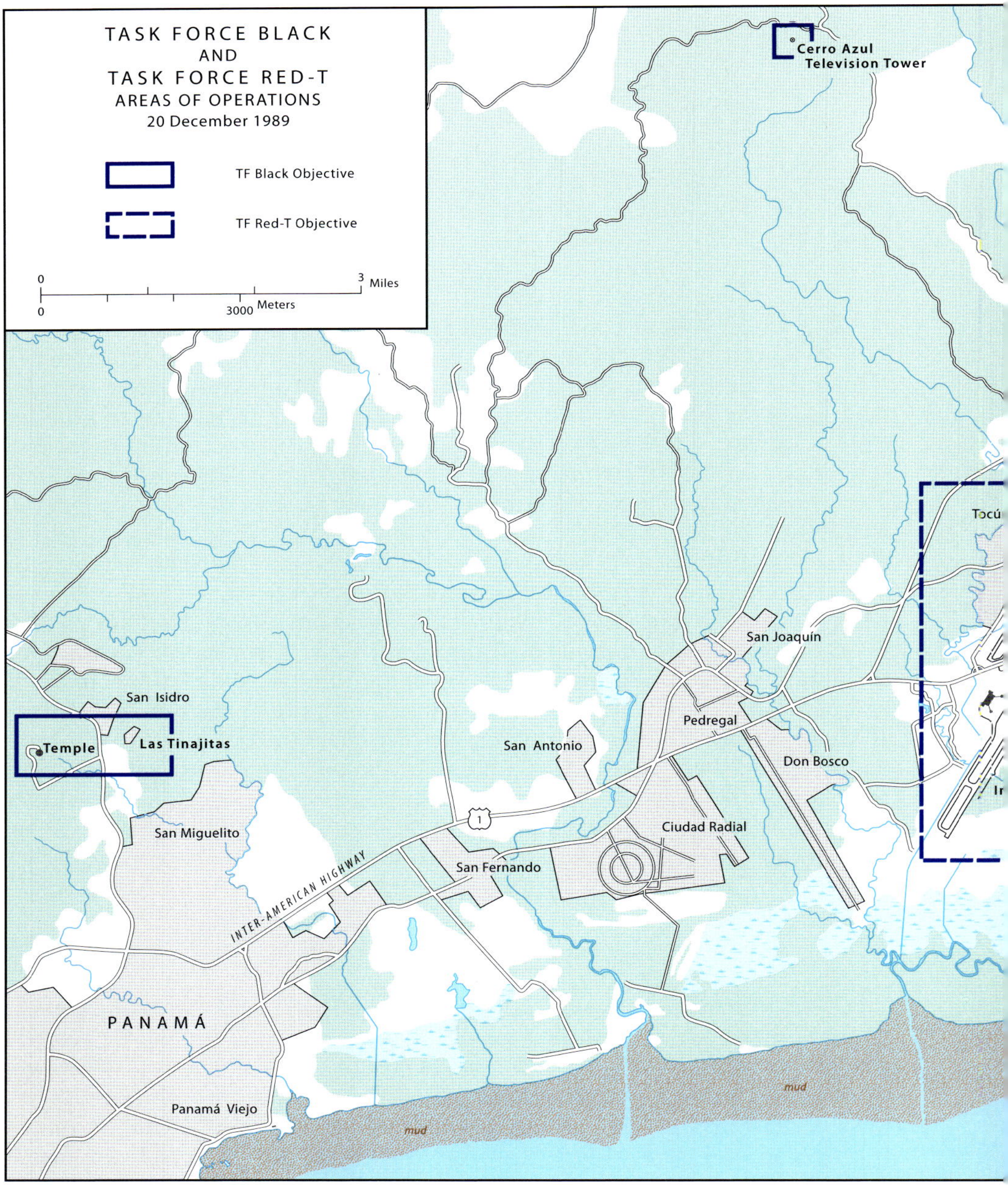
TASK FORCE BLACK
AND
TASK FORCE RED-T
AREAS OF OPERATIONS
20 December 1989
TF Black Objective
TF Red-T Objective
0
3 Miles
0
3000 Meters
Cerro Azul
Television Tower
San Joaquín
San Isidro
Temple
Las Tinajitas
Pedregal
San Antonio
Don Bosco
San Miguelito
Ciudad Radial
1
INTER-AMERICAN HIGHWAY
San Fernando
PANAMÁ
Panamá Viejo
mud
mud

Map 10

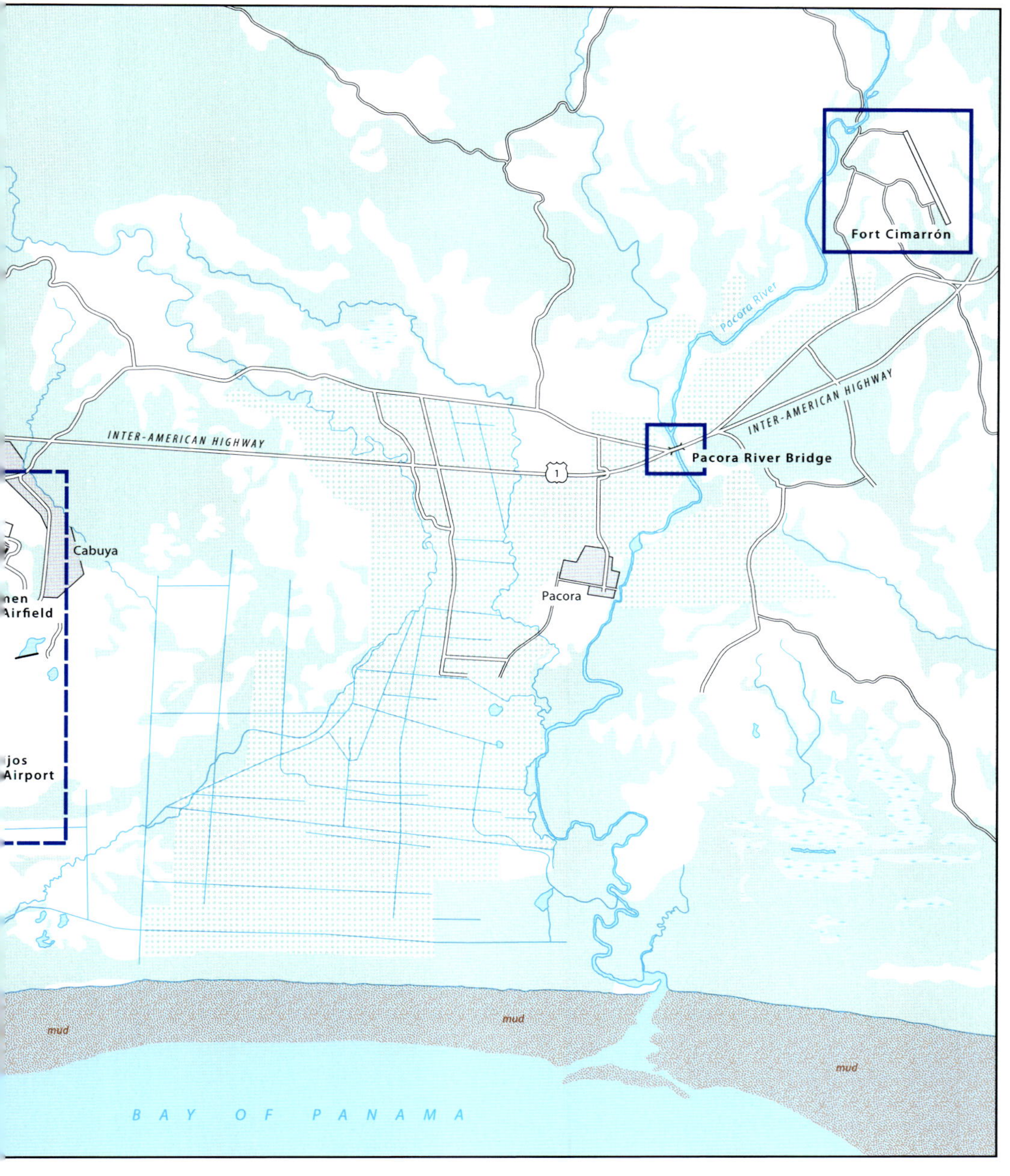

Fort Cimarrón
Pacora River
INTER-AMERICAN HIGHWAY
INTER-AMERICAN HIGHWAY
1
Pacora River Bridge
Cabuya
nen
Airfield
Pacora
jos
Airport
mud
mud
mud
BAY OF PANAMA

Albrook Air Station

South's 1st Battalion, 228th Aviation, 193d Infantry Brigade; and Air Force AC–130s from the 1st Special Operations Wing.[2]

Lt. Col. Roy R. Trumbull commanded the 3d Battalion, the combat power of which centered on eighteen twelve-man A-Teams. The small size of the team dictated the kinds of missions it could perform: reconnaissance and surveillance; ambushes, raids, sabotage, and other kinds of surgical strikes and small-scale offensive actions; and foreign internal defense. Anything more, especially direct combat against a large, disciplined conventional force, entailed unacceptable risks or was just plain suicidal. Conventional-oriented planners, however, generally unfamiliar with an A-Team's capabilities but aware that there were a dozen available in a Special Forces battalion, occasionally assigned them too many missions or ones that were extremely inappropriate. To avoid overcommitment and misuse of the Special Forces, Jacobelly and his staff put in long hours to screen, analyze, and discuss each tasking. As Trumbull noted in December, "We have not had a day off since the 3rd of October," the failed coup against Noriega. Despite these efforts, Jacobelly knew to his dismay that during the weeks leading up to Just Cause planners had parceled out too many targets, over a dozen, to Task Force Black.[3]

After SOCSOUTH remonstrations about its limited resources and the dangers of this overload, some of these missions would be dropped. Even so, on the day President Bush ordered the invasion of Panama, seven still remained for the Green Berets to carry out: finding and furnishing security for Guillermo Endara, Ricardo Arias Calderón, and Billy Ford, the three men who would head the new Panamanian government; seizing Panama's main television

[2] U.S. Special Operations Command, *History, 1987–2007*, 20th anniversary ed. (MacDill Air Force Base, Fla.: U.S. Special Operations Command, n.d.), pp. 41–42; Donnelly, Roth, and Baker, *Operation Just Cause*, p. 76.

[3] Donnelly, Roth, and Baker, *Operation Just Cause*, p. 122. Quote from Interv, John Partin with Lt Col Roy R. Trumbull, Maj Kevin M. Higgins, Capt Charles T. Cleveland, Cmd Sgt Maj Robedeaux, and Capt Bailey, all U.S. Army, 28 Feb 1990, Panama; Interv, Partin with Col Robert C. Jacobelly, U.S. Army, 26 Feb 1990, Panama.

tower at Cerro Azul north of the capital; and providing reconnaissance and surveillance at Fort Cimarrón, Las Tinajitas, and the Pacora River bridge. (Each of the three politicians and the three reconnaissance locations were regarded as separate tasks.) Even with this pared down list, Jacobelly felt uneasy, mainly about sending his men into Panama City to round up the opposition leaders, dubbed The Three Musketeers, in the midst of what could be heavy fighting. When the plan changed at the last minute—just as the Green Berets were preparing to perform "habeas grabbis," as one U.S. officer quipped—the SOCSOUTH commander was greatly relieved. The switch by higher authorities to the ruse of inviting the three Panamanian leaders to dinner at Howard Air Force Base not only obviated the highly risky security mission but allowed Jacobelly to reassign the troops freed from it to the Task Force Black reserve.[4]

The day before The-Three-Musketeers assignment was reallocated, there had already been a significant change to Task Force Black's orders, this one affecting what the Special Forces would do at the Pacora River bridge, about twenty miles northeast of Albrook Air Station. Originally, four Green Berets were to be transported to the objective in a helicopter, where, once on the ground, they would conduct reconnaissance and surveillance. Specifically, they were to look for any Battalion 2000 units from Fort Cimarrón, about 7½ miles farther east, heading toward the bridge en route to the Torrijos-Tocumen complex or downtown Panama City. With approximately two hundred soldiers, at least twelve mortars, and nine V300 Cadillac Gage light armored vehicles mounting 90-mm. cannon, Battalion 2000 constituted the PDF's largest and most formidable fighting force. U.S. military intelligence knew that Noriega had relocated some of its units to the Comandancia and Panama Viejo, but, even so, whatever remained of the battalion at Fort Cimarrón could disrupt American combat operations in the capital if allowed to leave the base and cross the Pacora River. To prevent this possibility, the Green Berets watching the bridge were to radio Task Force Black's operation center if they spotted Battalion 2000 forces on the move. Jacobelly's staff would pass the information to JTF-South, which presumably would dispatch an AC–130 or U.S. Rangers to deal with the threat.[5]

Concerned about the ramifications of a Battalion 2000 breakout, especially if it was attempted before the Rangers were finished with their H-hour missions at Torrijos-Tocumen, BLUE SPOON planners belatedly decided on an alternative to the four-man reconnaissance team. In its place, a larger group of Special Forces would fly from Albrook to the bridge in helicopters, land, and seize the structure. On Monday afternoon, 18 December, Jacobelly received confirmation of the new mission, which he entrusted to Maj. Kevin M. Higgins, commander of Company A, 3d Battalion, 7th Special Forces Group. That evening Higgins

[4] Donnelly, Roth, and Baker, *Operation Just Cause*, p. 122. Quote from Interv, Partin with Trumbull et al., 28 Feb 1990. For details concerning the dinner invitation to the three Panamanian leaders, see also Chapter 3 of this book.

[5] Donnelly, Roth, and Baker, *Operation Just Cause*, p. 124; Memo for Cdr, 3d Bn, 7th Special Forces Group (Airborne) (SFG [Abn]), 21 Dec 1989, sub: Mission Direct Action #03A, The Seizure of the Pacora River Bridge, 20 December 1989; Interv, Partin with Trumbull et al., 28 Feb 1990.

worked with Trumbull and the battalion operations officer to devise a tactical plan. There were only two Black Hawks available for the insertion, which limited the participants to sixteen, eight on each specially rigged helicopter. The number was deemed adequate, however, because an AC–130 would be overhead from the outset to help engage any PDF column approaching the bridge. Furthermore, the revised mission did not require much additional military intelligence: the photography, maps, and information gathered for the four-man team could be used by the larger force. Shortly before midnight, the company's operation order for an air assault on the Pacora River bridge was ready for dissemination. The next morning, Higgins briefed it up his chain of command for his superiors' approval.[6]

As Jacobelly, his staff, and commanders coped with last-minute adjustments, the soldiers of Task Force Black, like their conventional colleagues under JTF-South, spent most of Tuesday receiving a variety of briefings, going over their orders and execution checklists, marshaling their equipment, and preparing themselves for a fight. One of the first elements to move to its H-hour position was the four-man team assigned to carry out reconnaissance and surveillance at Las Tinajitas, an area in northeast Panama City that was home to the PDF's 1st Infantry Company. Within its weapons inventory, the 1st possessed 120-mm. mortars that could hit Torrijos-Tocumen, Fort Clayton, Quarry Heights, and even Howard Air Force Base. For that reason alone, Las Tinajitas demanded careful observation from H-hour until the paratroopers who were charged with seizing it arrived. The mission sounded straightforward enough but, in fact, posed a significant problem: the difficulty of inserting the surveillance team into a position that would afford the four men a good view of the installation. In the words of General Stiner, the 1st Company barracks at Las Tinajitas was within "an area of suburbs, surrounded by suburbs, that were reported to be [the] most loyal of all throughout Panama to Noriega." Any helicopter carrying the Green Berets into the area would likely be spotted and reported. Not only would the observation team be placed in jeopardy, but the element of surprise for the whole of Operation Just Cause could be compromised. With so much at stake in even this very small undertaking, the decision was made to have the four men walk most of the ten miles from Albrook Air Station to Las Tinajitas.[7]

Traversing the approach route by foot required an early 1800 departure from Albrook, a time when most of the other Task Force Black members were still attending briefings and checking their gear. The first leg of the trek was eased by the use of four rental vehicles—three cars and a truck—that delivered each team member to a rendezvous point in the vicinity of Albrook's back gate. From there, the group moved out on foot in what turned out to be a difficult hike through unexpectedly rugged terrain and jungle. Soon the four men realized that they would not arrive at Las Tinajitas by H-hour. In fact, they arrived an hour and a half late. The delay, at first, seemed a moot

[6] Memo for Cdr, 3d Bn, 7th SFG (Abn), 21 Dec 1989, sub: Mission Direct Action #03A, The Seizure of the Pacora River Bridge, 20 December 1989.

[7] Quote from Interv, Robert K. Wright Jr. with Lt Gen Carl W. Stiner, U.S. Army, 2, 7, 27 Mar and 11 Jun 1990, Fort Bragg, N.C. Flanagan, *Battle for Panama*, p. 76; Donnelly, Roth, and Baker, *Operation Just Cause*, p. 123; Interv, Partin with Trumbull et al., 28 Feb 1990.

point. Once they began their surveillance, they witnessed no enemy activity. The quiet was broken only later, shortly before dawn, when the enemy fired two mortar rounds from a concealed position near the barracks (the rounds exploded near Stiner's operations center in the USARSO headquarters at Fort Clayton). Attempts to locate the mortar through the use of two AN/TPQ–36 counterbattery radars set up around Howard Air Force Base initially proved unsuccessful because, after firing, the mortar crew had placed the weapon on a flatbed truck and moved. No more rounds would be fired from Las Tinajitas that night. Once the Special Forces team ascertained this, it had little to do but wait for the arrival—hours behind schedule, as it would later learn—of the 82d Airborne Division assault force.[8]

While the Las Tinajitas team was pushing its way through the jungle prior to H-hour, another four-man Special Forces team from the 3d Battalion's Company B departed Albrook on a Black Hawk around 2100 en route to a reconnaissance and surveillance position situated some three miles from Fort Cimarrón. Unlike the team whose assignment to watch the Pacora River bridge had been canceled with the decision to seize the structure, this group's task remained essential. Located about midway between the fort and the bridge, the four men would occupy an ideal vantage point for the early detection of Battalion 2000 troop movements, information which they would relay to Task Force Black headquarters. They could have done so, that is, had their communications equipment been working. A malfunctioning radio meant that Jacobelly would not receive up-to-the-minute information from these Special Forces troops on the scene; he would only be able to provide Higgins sporadic reports about Battalion 2000 from less well-situated sources.[9]

Higgins' Company A was the principal attack force for taking the bridge. Earlier in the day, after receiving approval from his superiors for the operation order he had helped write for the new "direct action" mission, he informed several of the participants, including the helicopter pilots, one of whom recommended that "the team pre-load equipment during daylight to save time and cut down on confusion." Higgins agreed. Other meetings followed, including an intelligence update that incorporated an infrared videotape of Fort Cimarrón and the Pacora River bridge. Higgins found the visuals "useful in developing a mental picture of our actions at the objective." At 2000, he briefed his entire team on the mission and an hour later went over the execution checklist. At 2315, the group assembled for its final equipment inspection.[10]

Just minutes before midnight, Higgins received word of two significant changes to the plan. The first came as welcome news: an additional helicopter would be available to transport the troops. In total, there would now be a

[8] Donnelly, Roth, and Baker, *Operation Just Cause*, pp. 123–24; Interv, Partin with Trumbull et al., 28 Feb 1990; U.S. Special Operations Command, *History, 1987–2007*, p. 41. Mortar rounds, usually fired two at a time, continued to hit Fort Clayton intermittently in the early morning hours of 20 December. According to one U.S. officer in Stiner's operations center, the counterbattery radars finally located the weapons, and an AC–130 destroyed them. Interv, author with Lt Col Robert Pantier, U.S. Army, 21 May 1991, Fort Leavenworth, Kans.

[9] Interv, Partin with Jacobelly, 26 Feb 1990.

[10] Memo for Cdr, 3d Bn, 7th SFG (Abn), 21 Dec 1989, sub: Mission Direct Action #03A, The Seizure of the Pacora River Bridge, 20 December 1989.

UH–60 Black Hawk from 1st Battalion, 228th Aviation, and two MH–60 Black Hawks from the 617th Special Operations Aviation Detachment. (Both aviation units were based in Panama, the former as part of USARSO, the latter as part of SOCSOUTH.) The change necessitated some hasty alterations in what would go aboard each helicopter, but the mission executive officer "quickly reorganized the load plan in a logical manner before launch time." Most important, Higgins' assault force had just grown from sixteen to twenty-four men.[11]

The second piece of news was disturbing. At 0013, just as the two MH–60s landed alongside the UH–60 already at the Albrook helipad, Lt. Col. David Wilderman, the mission's operations officer, told Higgins that H-hour for Just Cause had been advanced fifteen minutes. The reason, Wilderman said, was that intelligence reports indicated "a column of vehicles (unknown type and quantity) had departed the Cimarrón Cuartel heading toward Panama City." "It was imperative," Higgins was told, that "the team cut the column off at the Pacora River Bridge." After the major informed his men of the new H-hour, the twenty-four Green Berets raced to board the waiting transports whose deafening noise, in Higgins' words, "created an unsettling atmosphere." As the three helicopters lifted off, they became targets for a few enemy soldiers in the vicinity who had been sporadically peppering areas of the air station for five minutes with small-arms fire. The helicopters took some hits, none serious. The troops on board sustained no casualties.[12]

Once airborne, a hurried conversation between Higgins and the pilot of his helicopter, CWO John Estep, a special operations Night Stalker, led to a further modification to the plan. As approved earlier that day, the blueprint for seizing the bridge had the Special Forces conducting an air insertion at the northeast end of the structure, on the Fort Cimarrón side of the riverbank. From there, the men would split into five smaller groups. The largest of these, Team X, consisted of ten men who would set up an ambush position near the landing zone (*Map 11*). Their weapons included 6 AT4 antitank rockets, 3 light antitank weapons, and 2 squad automatic weapons. Three men constituting Team Y would provide security to the east. Because this team would be the farthest out on the perimeter around the bridge, an Air Force technical sergeant would go with it to keep the AC–130 overhead informed as to where the "limit of friendly troops" was located. Four men composing Team Z would move westward across the bridge and, with concertina wire and chemlites, set up a barrier across the highway to prevent civilian traffic from crossing. Another four men, Team U, would cover Team Z's movement and serve as a quick reaction force (QRF). Finally, Team W would provide command, control, and communications as a three-man headquarters element.[13]

When Higgins quickly described the proposed plan to Estep, the chief warrant officer indicated that he considered an air insertion at the planned landing zone unsafe. The designated area on the northeast side of the bridge was too small, the terrain too uneven, and power lines too numerous for three

[11] Ibid.

[12] Quotes from ibid. Donnelly, Roth, and Baker, *Operation Just Cause*, pp. 124–25.

[13] Memo for Cdr, 3d Bn, 7th SFG (Abn), 21 Dec 1989, sub: Mission Direct Action #03A, The Seizure of the Pacora River Bridge, 20 December 1989.

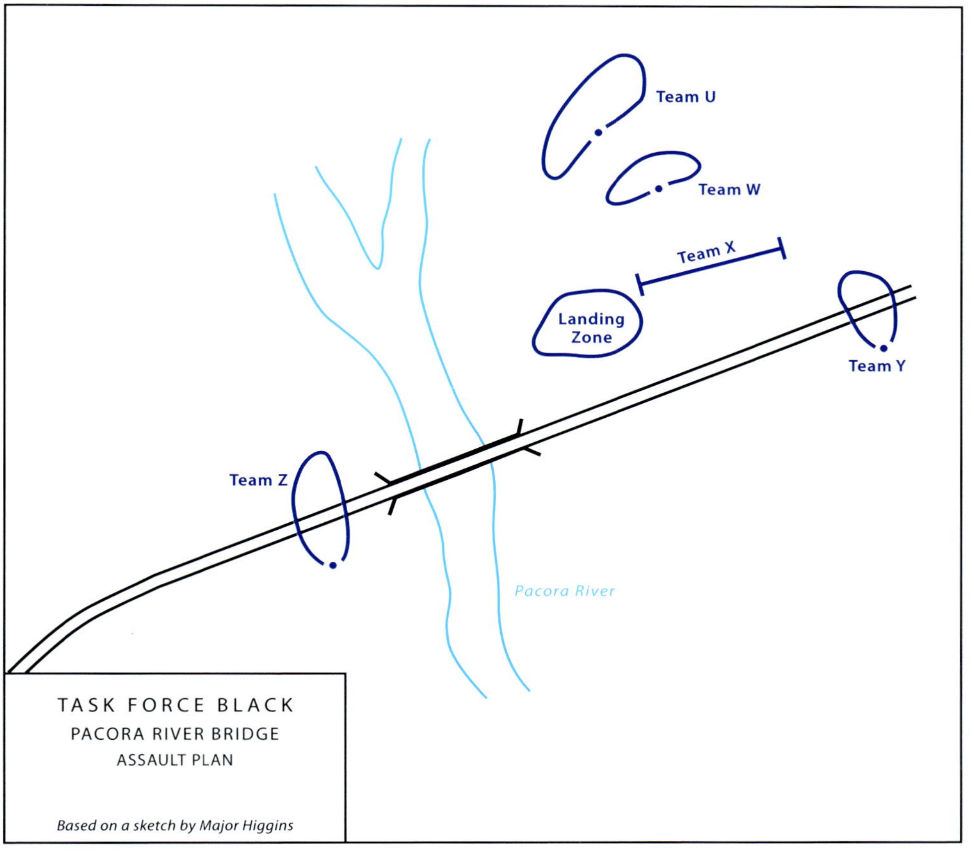

Map 11

helicopters to set down safely. Better, he said, to relocate the landing zone to the southwest side of the bridge. Higgins, while reluctant to authorize yet another last-minute change, recognized the logic of Estep's advice. As the major explained the next day, "I went with his recommendation realizing that it would be easier for us to adapt to this change than to try to recover from a slow, unsafe [infiltration] with a PDF column bearing down on us." In some ways, the new landing zone made good tactical sense as well. It would allow Higgins' men more time to establish their positions while the vehicles from Fort Cimarrón approached; it also placed the riverbed between the Americans and the enemy force, making the bridge a chokepoint and the PDF column an inviting target for the AC–130.[14]

Estep passed word of the new landing zone to the other helicopter crews, while Higgins informed the men with him. Time was now the critical factor. The assault force had to reach the bridge as soon as possible in case the intelligence

[14] Quote from ibid. Donnelly, Roth, and Baker, *Operation Just Cause*, pp. 125–26; Comments, John Partin, on draft of this chapter, 13 Apr 2012.

concerning Battalion 2000 elements on the move was accurate. Even if the reports were false, the beginning of combat operations in Panama City would alert the enemy battalion to a U.S. invasion, conceivably prompting units at Fort Cimarrón to take offensive countermeasures. Estep knew of Higgins' concern, yet the pilot could not guarantee a timely arrival on target. To the contrary, given the change in H-hour and the estimated 25-minute flying time to the Pacora River landing zone, he quickly calculated that the helicopters had little chance of getting there at or before 0045. To shave a few minutes off the flying time, Estep decided to modify the route, flying straight to the objective rather than in a zigzag pattern. This initiative, according to Higgins, "was a calculated risk . . . because [Estep] knew that all flight plans were closely pre-planned to avoid mid-air collisions that could potentially occur during the scheduled H-Hour massive airborne operations and helo insertions." Moreover, there was "also the potential for confusion" since the pilots from the 228th in one of the other helicopters "did not habitually work Special Operations with the 617 AVN unit." Fortunately, as Higgins noted, the 228th pilots "responded admirably."[15]

As a result of Estep's initiative, the three helicopters carrying Higgins' men approached the Pacora River bridge right at 0045, the revised H-hour. Limited visibility made positive identification of the objective difficult, but

The Pacora River bridge

[15] Quotes from Memo for Cdr, 3d Bn, 7th SFG (Abn), 21 Dec 1989, sub: Mission Direct Action #03A, The Seizure of the Pacora River Bridge, 20 December 1989. Although the special operations pilots and the conventional pilots had not worked together, pilots in the 617th Special Operations Aviation Detachment had been "paired" together for six to eight weeks prior to the invasion. "They planned, briefed, and flew together continually," the Task Force Black after action report stated. "Crews were able to build on each pilot's strengths and eliminate weaknesses." Memo, Col Robert C. Jacobelly, Cdr, Task Force Black, for Cdr, XVIII Abn Corps, n.d., sub: Joint After-Action Report for TF Black's Execution of OPORD 1–90, Operation JUST CAUSE.

PDF vehicles already heading toward the structure with their headlights on lit up the area enough for Estep to orient himself as the helicopters flew over the enemy column. Higgins estimated that the enemy procession was about a hundred yards from the bridge and just under four hundred fifty yards from the new landing zone. The site was "clearly a hot LZ," he observed. The question was whether the approaching convoy possessed the kind of weapons "capable of blowing the [helicopter] formation out of the sky." Estep had the authority to abort the mission, and Higgins would have supported the decision, but the Night Stalker instead yelled, "We're going in, we're going in!" Under enormous pressure caused by the poor visibility and the PDF presence and despite a barbed-wire fence that suddenly appeared right in the middle of the landing zone, Estep and the other pilots put their Black Hawks down exactly at the infiltration site, unloaded the troops, and took off in what to Higgins "seemed like a millisecond." From that point on, the major knew, "The ball was in our court."[16]

Higgins' principal concerns at the moment were that the Battalion 2000 column would reach the bridge before his men had positioned themselves and the AC–130 had arrived overhead. He also worried that the enemy vehicles, which he could hear but not plainly see, might be armored personnel carriers armed with .50-caliber machine guns, a weapon capable of decimating his small force.[17] The first of these fears was partially dissipated once the equipment-laden Green Berets made their way up a steep 35-foot embankment, and Team X set up its ambush position. The PDF column turned out to be trucks, not armored personnel carriers, and, as it approached from the east, one member of Team X moved to the center of the road at the western end of the bridge and, "looking head-on into the headlights of the lead vehicle," fired the first round of the battle, a LAW, at the oncoming target. The rocket missed but forced the column to halt for a moment. When it resumed its movement, Higgins worried that the vehicles, which he still could not see clearly, might have enough armor to deflect the AT4 and LAW rounds. A Special Forces captain fired another LAW at the lead vehicle, causing the PDF column to halt again. Quickly, a third Green Beret positioned himself in front of the enemy vehicles and fired an AT4. Other members of the team then opened up with a SAW and 40-mm. grenades.

At that point, Higgins decided the time had come to bring the AC–130 into the fight. The gunship had just arrived on station and was ready to open fire. The question was whether it could do so without hitting the Special Forces. Higgins conferred with the Air Force technical sergeant attached to Team Y and

---

[16] Memo for Cdr, 3d Bn, 7th SFG (Abn), 21 Dec 1989, sub: Mission Direct Action #03A, The Seizure of the Pacora River Bridge, 20 December 1989.

[17] Jacobelly later estimated that Higgins' group had only about a minute on the ground before the PDF column reached the bridge. Interv, Partin with Jacobelly, 26 Feb 1990. The account of the Pacora River bridge battle that follows is primarily from Memo for Cdr, 3d Bn, 7th SFG (Abn), 21 Dec 1989, sub: Mission Direct Action #03A, The Seizure of the Pacora River Bridge, 20 December 1989. Unless otherwise noted, the quotes are from the memo. Several other accounts are also based on this document, but include additional information as well. These include Donnelly, Roth, and Baker, *Operation Just Cause*, pp. 128–30; U.S. Special Operations Command, *History, 1987–2007*, pp. 41–42; Flanagan, *Battle for Panama*, pp. 77–80.

in communication with the gunship, only to be told that the margin of safety was very narrow. Both men agreed, however, that employment of the Spectre was "the only method open to us to effectively engage and neutralize the column." As the Air Force sergeant requested fire support from above, Higgins made sure that his men were on the south side of the road at the western end of the bridge, as called for in the plan they had amended after boarding their helicopters (*Map 12*). Seconds later, the AC–130 fired its first burst with what Higgins described as "devastating accuracy." Immediately thereafter, a Special Forces sergeant launched another AT4 at the column, prompting one of the captains with Higgins to suggest that the ground force conserve its antitank weapons. Higgins concurred. "It was prudent," he recalled, "to let the AC–130 take over." But the decision led to what the major described later as the tensest moment of the evening. To begin with, there was the realization that a stray round from the Spectre could have tragic consequences. A half-hour barrage from the gunship, however, remained as accurate as the first burst. Still, notwithstanding the damage inflicted by the AC–130, Higgins had to assume that most Panamanian soldiers in the column had already dismounted their vehicles and were probably trying to flank his very

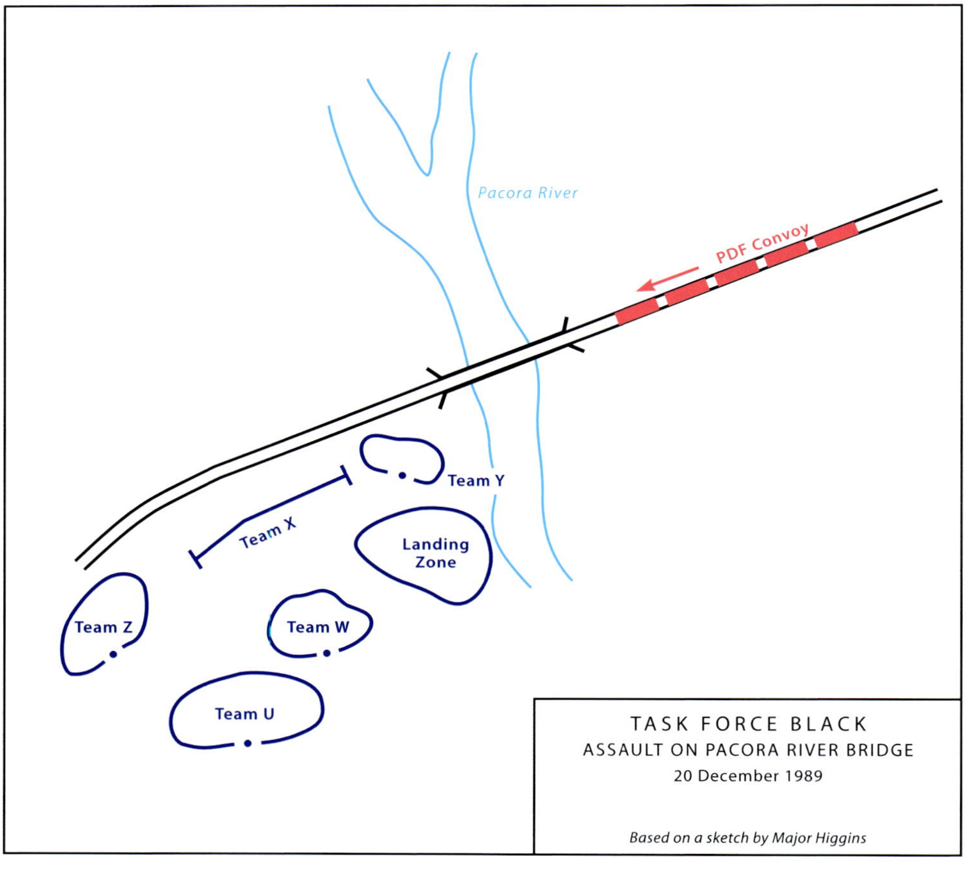

Map 12

vulnerable position, especially from the rear, which was entirely exposed. Fueling this concern, the AC–130 crew reported enemy troop movement in the riverbed near the bridge, clearly a flanking maneuver. As a countermeasure, Team U, the group's quick reaction force, began lobbing grenades from M203 launchers into the riverbed. The Spectre also pounded the area.

As the skirmish continued, a second column of three vehicles approached the Special Forces' positions from the west. Higgins assumed that it was the enemy, but he could not be sure. To "give them the benefit of the doubt," he had members of Team Z fire several warning shots. He then received reports from the AC–130 that the advancing troops were dismounting their vehicles and moving into elephant grass near the road. The news convinced the Green Berets that the force was hostile, and they opened up with their SAWs and grenade launchers, firing several hundred rounds. Confronted with PDF elements moving on him from two directions, Higgins readily granted the combat controller's request to call in another Spectre. The aircraft, diverted from Torrijos-Tocumen, arrived overhead about ten minutes later. The combined firepower of the two planes proved overwhelming, causing the Panamanians on the western approach to the bridge to climb back aboard their transportation and flee the area.[18]

Through all of this, there had yet to be any direct head-on enemy assault from the eastern approach. To thwart such a move, two Special Forces sergeants crawled toward the southeast corner of the bridge to set up a claymore mine. Before they could complete their maneuver, they saw three armed Panamanian soldiers walking toward them. One of the sergeants fired at the three with his M16, killing one and wounding two. One of the wounded men jumped off the bridge into the water fifty feet below. The other climbed down the bridge and lodged himself underneath, where he briefly remained until noise from his movements gave away his position. The two sergeants thinking a possible PDF assault force was responsible for the sounds threw a few hand grenades under the structure. Their concussion caused the wounded Panamanian to fall thirty-five feet onto the riverbed, the impact of the landing breaking both of his legs.

Following this incident, the situation "stabilized somewhat," according to Higgins, so he contacted Jacobelly's operations center to inquire about sending the Task Force Black reserve to the scene. The task force's hangar at Albrook, it turned out, had also been the target of enemy fire, although only briefly. The time of the shooting was never firmly established, although most reports had it occurring at 0030 or soon thereafter. By that time, Higgins' group had already departed Albrook, after itself coming under fire. Most likely, the Panamanians engaged in the running firefight with soldiers from Task Forces Gator and Wildcat in the vicinity of the air station were the source of the shots into Jacobelly's operations center. Whatever the time and source, the people working with the colonel in the hangar found themselves diving for cover. In the process, Jacobelly's briefing board fell over on him, possibly saving his life when a .50-caliber bullet whistled through his hair. A U.S. squad providing

[18] Jacobelly's recollection regarding the second AC–130 differs from that of Higgins. According to the Task Force Black commander, the second Spectre was dispatched because the first was running low on fuel. For several minutes the two gunships provided fire support for the ground force, and then the first departed for refueling. Interv, Partin with Jacobelly, 26 Feb 1990.

perimeter security at the hangar returned fire, and the Panamanians quickly departed. There were no serious U.S. casualties.[19]

Once Higgins contacted Jacobelly during the lull in the fighting at the bridge, the two agreed to defer a decision about sending a relief force until the situation there became clearer. The arrival of a friendly unit in an area still not secured posed too many risks, including friendly fire, and Higgins, unsure of enemy troop dispositions on the east side of the river, did not want "to expose the QRF to a hostile LZ." Only around 0530 did Jacobelly think the time right to dispatch reinforcements, at which point his headquarters asked Higgins if he still wanted them. Higgins said he did, but not until after some of his men had had a chance to reconnoiter the east side of the river. Before he could initiate that action, though, he received word that the relief force, a 44-man element commanded by Maj. Gilberto Perez, was on its way.

The reinforcements arrived shortly after 0600. As Major Perez positioned his men along the southeastern riverbank, Team U from Higgins' force forded the Pacora just over three hundred yards north of the bridge and set up a security position. The relief force then maneuvered "to sweep and secure the far side." In the process, one of Higgins' men heard cries for help from the wounded PDF soldier who had fallen from underneath the bridge. The man had lost a considerable amount of blood, had a bullet in his femur, and was in excruciating pain from his two broken legs. Despite the severity of his injuries, the Green Berets were able to stabilize him. When able to speak, he told his captors that the bulk of Battalion 2000 had left Fort Cimarrón and crossed the Pacora River at 2200, three hours before Higgins had arrived. The column the Special Forces had confronted at the bridge was the "trail party." Where the earlier and larger PDF force had ended up was anybody's guess at that point.

Later that morning, one of Higgins' teams established a control point on the west side of the river and captured three enemy soldiers fleeing Panama City, "perhaps in hopes of reconstituting at Cimarrón." In the sweep Major Perez's men conducted on the other side of the objective, they found several uniformed Panamanians hiding in farmhouses. By noon, when elements of the 82d Airborne Division began their belated assault on the fort, the Special Forces troops at the river had compiled statistics that reflected their accomplishment. Confirmed enemy casualties numbered four killed and one wounded in the battle. Seventeen more Panamanians had been taken prisoner. A large number of weapons had been seized, including machine guns, recoilless rifles, mortars, grenade launchers, gas grenades (confirming Higgins' suspicions that the Panamanians had the capability to use chemical weapons against his force), and various small arms. Four 2½-ton trucks and a pickup truck that were in the column heading toward the bridge from the fort were also in American hands.

---

[19] The shooting incident at Jacobelly's operations center at Albrook Air Station is also recounted in Donnelly, Roth, and Baker, *Operation Just Cause*, p. 125. The authors place the time of the shooting nearly one hour before Jacobelly's estimate, 0110, when Higgins' men would have still been boarding their Black Hawks. Interv, Partin with Jacobelly, 26 Feb 1990. For official accounts by several of the officers involved in providing security for the Task Force Black hangar, see Memo, Maj David E. McCracken, through Cdr, 3d Bn, 7th SFG (Abn), for Cdr, Task Force Black, 25 Dec 1989, sub: Hangar 450 Firefight—20 Dec 89. Times for the shooting range from 0030 to 0115.

After coordinating with Jacobelly's operations center, Higgins' men and the PDF prisoners boarded two CH–53 Sea Stallion helicopters and two Black Hawks for the flight back to Albrook Air Station. They arrived there at 1800, and, following a team debriefing, their mission formally ended. Of the three Task Force Black missions aimed at PDF units and fortifications, this had easily been the most difficult and the most dangerous. It had also been highly successful.

The fourth Task Force Black mission, the seizure of the television tower at Cerro Azul, called for a Special Forces element to disable the tower's transmitting capability temporarily, thus rendering it useless as a propaganda vehicle for Noriega and his followers. On order, the Green Berets would restore transmissions for "exploitation" by U.S. psychological operations forces. Operational Detachment Alpha 785, commanded by Capt. John M. Custer, would carry out the mission. Once augmented by technical experts, Custer's team numbered eighteen men. At 0050, five minutes after the revised H-hour, the team arrived at the objective northeast of the capital, fast-roped to the ground, and, after the PDF guards fled, took control of Panama's Channel 2 television complex. The technicians with the team then disabled the station. Up to that point, the mission had gone exactly as planned. Any celebration, however, was muted by two concerns. The first was the realization that radio stations controlled by the regime were still broadcasting Noriega's recorded messages to the population at large. The second, "poor PSYOP coordination and implementation," delayed any U.S. use of the television transmitter for four days and caused Jacobelly to claim in his after action report that the "PSYOPS campaign for Operation Just Cause was poorly executed/implemented."[20]

Jacobelly's report contained other observations on D-day operations. There was, for example, praise for the support provided Task Force Black by the Defense Mapping Agency element located at Albrook and for the various intelligence products received from JTF-Panama, particularly the imagery that the Special Forces required in large quantities. In contrast, the report criticized the dearth of "push intelligence"—intelligence that is provided to the user not as the result of a formal request but on the grounds that it may be useful—once the invasion began. With respect to the task force's reconnaissance and surveillance teams, Jacobelly indicated that, during the planning phase, "no specific information requirements, except the location of Noriega for some, were specified," thus limiting a team's "effectiveness as an intelligence collection asset." On the subject of communications, the report called for Special Operations Command, South, to have "an organic in-theater communications support element" that would supersede having to rely on quick reaction teams from the Fort Bragg–based 112th Signal Battalion—teams that required nine hours to deploy, thus negating "the 112th's utility to SOCSOUTH in a time sensitive environment which requires the establishment of immediate communication." Jacobelly also submitted another communications lesson: the need for standardized Special Operations Forces joint communications-electronics operating instructions.

[20] U.S. Special Operations Command, *History, 1987–2007*, p. 42. Quotes from Memo, Jacobelly for Cdr, XVIII Abn Corps, n.d., sub: Joint After-Action Report for TF Black's Execution of OPORD 1–90, Operation Just Cause.

The communications-electronics instructions execute order issued less than a day before H-hour did not support specific missions, he charged. With respect to the operation orders for those missions, Jacobelly noted that the Joint Special Operations Task Force and Task Force Black had written some draft orders—possibly referring to the new Pacora River bridge order—before coordinating with the SOUTHCOM commander and JTF-South. As a result, "mission conflictions occurred and some support taskings/requirements were not understood." "Only a few hours prior to H-hour was TF Black able to deconflict missions and arrange/lock-in supporting assets." The lesson learned was that "publishing subordinate unit orders without orders from higher causes conflictions in mission planning and could result in mission delays or failure."[21]

TASK FORCE RED-T

Three of Task Force Black's H-hour missions involved keeping track of PDF units in the vicinity of Panama City, primarily to ensure that they did not attempt to reinforce their colleagues fighting at the Comandancia. In the case of Battalion 2000, planners had an additional concern: that a well-armed force from Fort Cimarrón would descend on the Torrijos-Tocumen airfield complex where U.S. Rangers would be attempting to seize and secure two key runways and the buildings next to them. One of the two facilities, the Tocumen airfield, largely served military aircraft. The Torrijos International Airport adjacent to it was Panama's main hub for commercial flights. With Panama City ten miles to the southwest of the complex and Fort Cimarrón sixteen miles to the northeast, Torrijos-Tocumen was closer to Battalion 2000 than the PDF headquarters.

For several reasons, U.S. staff officers and commanders had targeted both airports from the outset of ELABORATE MAZE and BLUE SPOON contingency planning, and, unlike some of the original objectives identified in 1988, the complex had only increased in importance as the crisis evolved. To begin, there was the need to prevent Noriega from fleeing the country in the event of hostilities. Although the dictator's private jet was at Paitilla, there were numerous aircraft at Torrijos-Tocumen capable of whisking him to a friendly country, such as Cuba, from which he might orchestrate opposition to any Panamanian government coming to power after the collapse of his regime. A second reason for seizing the complex was to prevent the enemy from using it as a staging area for intervening in the battles downtown. PDF units flying from Rio Hato to Panama City during the October coup attempt had done just that, the result being Noriega's rescue and continuation in power. Battalion 2000 and the infantry companies at Rio Hato were the most likely units to use Torrijos-Tocumen as a staging area during a U.S. invasion, but other PDF garrisons could not be discounted, especially the Panamanian air force squadron and the 2d Infantry Company stationed on the premises.

[21] Memo, Jacobelly for Cdr, XVIII Abn Corps, n.d., sub: Joint After-Action Report for TF Black's Execution of OPORD 1–90, Operation JUST CAUSE.

Tocumen airfield

The air force element at Tocumen consisted of an estimated one hundred fifty personnel, mainly pilots, security officers, and maintenance crews, and up to thirteen fixed-wing aircraft and twenty-one Huey helicopters. The infantry company of around two hundred soldiers was barracked within six hundred yards of the Tocumen airstrip and constituted the greater threat, partly because the company included in its arsenal at least one, possibly two (later confirmed), Soviet-built ZPU4 air defense guns.[22]

Another compelling reason for a U.S. H-hour assault on Torrijos-Tocumen stemmed from the need for a secure airfield besides Howard from which planned follow-on operations could be mounted and on which additional troops and supplies could be landed. Operation Plan 90–2 called for a brigade of the 82d Airborne Division to begin air assaults against Panama Viejo, Fort Cimarrón, and Las Tinajitas within an hour or so after the commencement of Just Cause. Several hours later, at least one brigade from the 7th Infantry Division from Fort Ord would be arriving in Panama. The transports carrying the light fighters could always land at Howard, but during the planning process the question arose concerning what would happen if that runway became inoperable, either from enemy mortar and sniper fire (to which the facility was extremely vulnerable) or from an overcrowding of friendly aircraft using the base. Maj. Gen. William A. Roosma, the XVIII Airborne Corps' deputy commanding general, was worried that the initial plans relied too heavily on Howard—putting "all their eggs in one basket"—for bringing in supplies and reinforcements. What about Torrijos-Tocumen? he asked. "Have you thought

[22] U.S. Special Operations Command, *History, 1987–2007*, pp. 33–34; Donnelly, Roth, and Baker, *Operation Just Cause*, pp. 188–89; Gordon L. Rottman and Ron Volstad, *Panama 1989–90* (London: Osprey Publishing, 1991), pp. 11–12.

of using that?" A few months after Just Cause, General Stiner reflected, "The reason for taking Torrijos-Tocumen was [that] we needed another base upon which we could build up combat power very rapidly. The only other base we had was Howard, and there was the possibility that it could be interdicted." Revising the original plans to accommodate troop carriers and supply aircraft required only a minimal effort, given that Torrijos-Tocumen was already on the list of H-hour targets.[23]

From the beginning of the planning process, seizing and securing the complex had been tasked as a mission for U.S. Rangers, specialists in executing such operations. The rangers' headquarters was the 75th Ranger Regiment, located at Fort Benning, Georgia. At the time of Just Cause, the regiment consisted of three combat battalions and was commanded by Col. William F. "Buck" Kernan. From mid-1987 to late 1988, Kernan had been commander of the 1st Ranger Battalion, located at Hunter Army Airfield, Georgia, and he had worked on the early contingency plans for Torrijos-Tocumen. As regimental commander, he saw that the target remained assigned to his old battalion, now commanded by Lt. Col. Robert W. Wagner. Reinforcing Wagner in the assault would be Capt. Alfred E. Dochnal's Company C from the ranger regiment's 3d Battalion; loudspeaker teams and other assets already in Panama from the 4th Psychological Operations Group, headquartered at Fort Bragg; a civil affairs team from the 96th Civil Affairs Battalion, also in Panama from Bragg; two AH–6 attack helicopters from the 160th Aviation Group; and Air Force combat controllers, pararescuemen, and an AC–130 Spectre from the 1st Special Operations Wing.[24]

In his short time as the JSOTF commander, General Downing had made a number of decisions affecting the Torrijos-Tocumen mission, including the type of transports the Rangers would use in flying from Hunter Airfield to the objective. According to Stiner, the U.S. Military Airlift Command had C–141 Starlifters and C–130 Hercules available for the Rangers, but not enough of either aircraft to use only one type for both ranger missions, Torrijos-Tocumen and Rio Hato. Downing, therefore, decided to request C–141s for the 1st Battalion at the airfield complex. C–130s would follow carrying the 3d Battalion's Company C. The Hercules would also be used at Rio Hato. His logic was simple enough. The Starlifter was a much larger transport than the Hercules, and more Rangers would be dropping onto Torrijos-Tocumen than at Rio Hato. Moreover, the drop zone at Torrijos-Tocumen was longer, mainly because of the end-to-end layout of the military and commercial runways. Thus, even though a C–141 would need more time than a C–130 to airdrop its passengers, the descending Rangers still stood a good chance of landing on target. The C–130s, in turn, were well suited to the shorter drop zone at Rio Hato.

---

[23] U.S. Special Operations Command, *History, 1987–2007*, p. 34; Donnelly, Roth, and Baker, *Operation Just Cause*, pp. 188–90. First and second quotes from Interv, Wright with Maj Gen William A. Roosma, U.S. Army, 15 Mar 1990, Fort Bragg, N.C. Third quote from Interv, Wright with Stiner, 2, 7, 27 Mar and 11 Jun 1990. Interv, author with Pantier, 21 May 1991.

[24] Briefing Slides, Blue Spoon: The Evolution of the Plan for Just Cause; U.S. Special Operations Command, *History, 1987–2007*, pp. 33–34; Donnelly, Roth, and Baker, *Operation Just Cause*, p. 190; Flanagan, *Battle for Panama*, p. 157.

For the Rangers assigned the Torrijos-Tocumen mission, they, like other U.S.-based troops named on the BLUE SPOON force list, rehearsed aspects of the operation on several occasions. Of two major dress rehearsals they conducted, the "second round" occurred in mid-December at Duke Army Airfield at Hurlburt Field, Florida, just days before President Bush ordered the U.S. invasion of Panama. Detailed mock-ups added to the realism of the rehearsal, as did the employment of fire support from an AC–130 and two AH–6s. The Rangers performed well, although among the regimental after action comments was an expression of regret that some security elements "earmarked to participate in the operation" did not take part in the practice runs, to the detriment of "team building" efforts.[25]

Once notified of the president's decision, Colonel Wagner and his battalion in Georgia went through pretty much the same procedures as the U.S. forces stationed in Panama, with one exception. He had to be prepared to launch his battalion and augmentation units by late afternoon Tuesday, given the seven-hour flight to Panama. Captain Dochnal's Company C from the 3d Battalion flew into Hunter from Fort Benning on Monday morning, and a command team from regimental headquarters led by the 75th's deputy commander, Lt. Col. Henry L. T. Koren Jr., arrived the next day by bus. Much of Monday and Tuesday was filled with coordination meetings, which went "flawlessly" in Dochnal's opinion, although one complaint recorded later was that the late arrival of execution checklists for the mission made some of the required coordination more difficult than it should have been. Those attending the meetings also went over the operation orders. According to one source, Wagner and his staff studied maps and photographs of their target, poured over the intelligence data available to them, and discussed the plan of attack, which they briefed to Colonel Kernan and General Downing. Next to be informed were company commanders, who, "in turn, briefed their platoons in painstaking detail, including troop-leading procedures." Squad leaders also received their guidance. In Dochnal's company, for example, squad leaders were handed maps of their individual objectives, with orders to come up with workable small-unit assault plans for each target.[26]

By midafternoon Tuesday, most of the Rangers had received word that the enterprise on which they were about to embark was no exercise. Any

[25] Intervs, Wright with Stiner, 2, 7, 27 Mar and 11 Jun 1990; author with Brig Gen William Hartzog, U.S. Army, and Col Thomas A. Braaten, U.S. Marine Corps, 29 Jun 1990, Quarry Heights, Panama; author with Pantier, 13 Jun 1990; Telephone Interv, author with Maj David Huntoon Jr., 27 Jun 1990; Durant and Hartov with Johnson, *The Night Stalkers*, p. 136; Donnelly, Roth, and Baker, *Operation Just Cause*, p. 194. Quotes from After Action Review Comments, Command and Control Teams, 75th Ranger Rgt, n.d.

[26] Handwritten notes, Purpose: To provide information on C 3/75 [Co C, 3d Bn, 75th Ranger Rgt] actions at Torrijos International, n.d. (but apparently written in late December 1989 or January 1990 and attached to HQ, 75th Ranger Rgt, 26 Jan 1990, sub: Historical Information for Commander 44th Historical Detachment). Quotes from Flanagan, *Battle for Panama*, p. 158. Donnelly, Roth, and Baker, *Operation Just Cause*, pp. 194, 196; Handwritten notes, 1st Ranger Bn S–3 section, n.d., which state, "Execution check lists were late getting to the chain of command. This made our check list late." The notes went on to say, "To get the most out of the X-list we must have the time to take our list and the list from higher and get together with the chain of command and TOC elements and conduct a talk through rehearsal and war game session."

lingering doubts were quashed with the distribution of live ammunition—they were going into combat. "I was nervous," one Ranger simply said, expressing the feelings of most of the others. Remaining preparations and the weather, however, did not allow time for brooding. The rangers' preboarding activities included practicing jump procedures in "the skeleton of a C–130 that sits near the airfield." Then, they had to wait in an ice storm that pelted the southeast United States throughout the day and evening of 19 December. Although the same storm would have a decidedly negative impact on the deployment of the 82d Airborne Division at Pope Air Force Base, North Carolina, there were no delays at Hunter Army Airfield, only soaked, cold, and miserable Rangers. A draft of the 1st Battalion's after action report addressed the issue with the observation that "during deployment preparation 1/75 Rangers experienced cold weather in conjunction with a harsh and long lasting rain." "Deploying into a hostile environment is difficult enough," the report continued, "without further hindering our force by subjecting them to cold wet weather with no chance to escape except by finally deploying. . . . Why should pre-jump etc have to be conducted outdoors in inclement weather?" For future operations, the battalion's operations directorate recommended that "a large shelter be built where the task force can chute up and conduct rehearsals and briefbacks."[27]

The same draft report went on to commend the Rangers for taking with them only "the equipment necessary to complete the mission." Even so, some of the planes were overcrowded in that they had to accommodate more passengers than was normally the rule—for example, seventy, not sixty, Rangers were crammed into each of the four C–130s transporting Captain Dochnal's company. Colonel Wagner's 1st Battalion required seven C–141s for the troops and five more for equipment and supplies. The ranger air armada began taking off from Hunter on schedule at 1900 on the nineteenth, with the C–141 troop transports in the lead, followed by the equipment transports, and Company C's four C–130s. Two of the latter aircraft included a nine-man team from Fort Campbell, Kentucky, that, once on the ground in Panama, would have the responsibility of setting up a forward arming and refueling point near the Torrijos-Tocumen runways for use by the two AH–6 Little Bird light attack helicopters assigned to the mission. As the Rangers departing Hunter began their flight, most tried to sleep. Few combat plans, they knew, are executed flawlessly. The extent of the deviations they would experience played on their imaginations.[28]

In determining how the airborne assault would unfold, the tactical planners considered the three main threats the Rangers would face at Torrijos-Tocumen (*Map 13*). At the top of the list was the 2d Infantry Company, the PDF unit on the scene most likely to offer determined resistance. The Panamanian air force squadron came next but was not expected to fight with any enthusiasm, if at all. Finally, there would probably be three or four dozen security personnel

[27] First and second quotes from Donnelly, Roth, and Baker, *Operation Just Cause*, pp. 196–97. Remaining quotes from Handwritten notes, 1st Ranger Bn S–3 section, n.d. On the 82d Airborne Division's deployment difficulties, see Chapter 8.

[28] Quote from Handwritten notes, 1st Ranger Bn S–3 section, n.d. Donnelly, Roth, and Baker, *Operation Just Cause*, pp. 194, 197; Durant and Hartov with Johnson, *The Night Stalkers*, pp. 135–37.

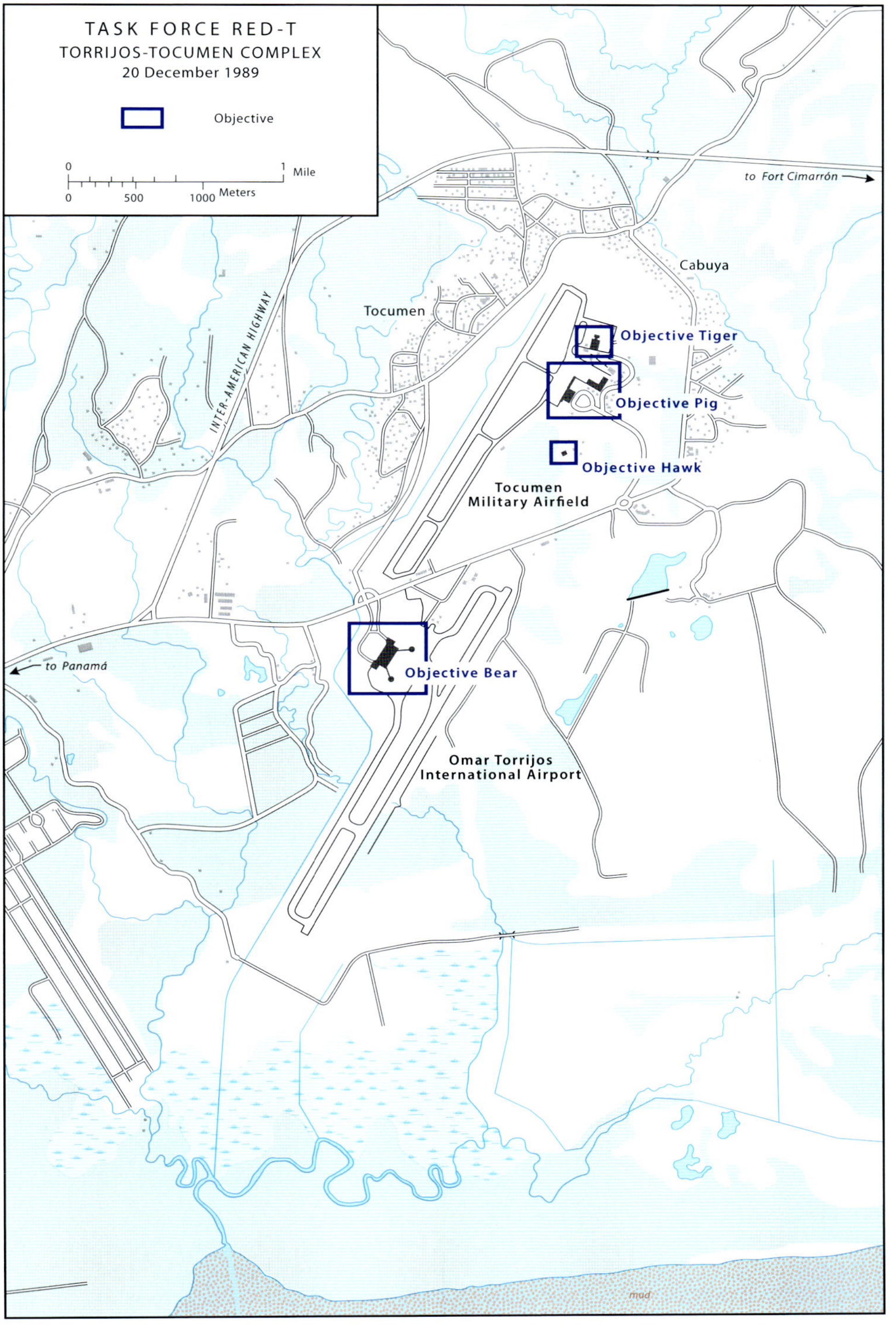
TASK FORCE RED-T
TORRIJOS-TOCUMEN COMPLEX
20 December 1989
Objective
0
1 Mile
0
500
1000 Meters
to Fort Cimarrón
Cabuya
Tocumen
Objective Tiger
INTER-AMERICAN HIGHWAY
Objective Pig
Objective Hawk
Tocumen
Military Airfield
to Panamá
Objective Bear
Omar Torrijos
International Airport
mud

Map 13

in the main terminal at Torrijos International Airport who would have to be subdued.[29] The U.S. attack would begin precisely at H-hour with an AC–130 Spectre and two AH–6 Little Birds bringing under heavy fire the 2d Infantry Company barracks, labeled Objective Pig, and adjacent sites. Three minutes later, the Rangers would parachute into the complex. The 1st Battalion would focus on the Tocumen military airfield, with Company A seizing the air force headquarters, Objective Tiger, and Company C moving against what was left of the infantry company barracks, securing the center of the airfield, and seizing the field's control tower. As for Company B, it would secure the airfield's perimeter, set up roadblocks around the facility, and begin preparing the Tocumen runway for aircraft carrying the light fighters, scheduled to arrive at 0800. Also located on the premises of the complex was the Ceremi Recreation Center, Objective Hawk, thought to be an armory for the paramilitary Dignity Battalions. Unknown to the assault force (or any other U.S. personnel), this was the building Noriega was in when the attack began.

While the 1st Battalion was eliminating resistance and securing Tocumen, the four platoons of Company C from the 3d Ranger Battalion would concentrate on the commercial facility at the southern end of the complex. The goal was to cordon the main terminal, Objective Bear, keeping inside the security personnel and any late-night patrons of the 24-hour duty-free shops. The last commercial flight into Torrijos International on 19 December was scheduled to arrive before 2300, plenty of time for the passengers to get their luggage and clear customs. One of the reasons for picking 0100 as H-hour had been to ensure a nearly empty terminal.

That expectation evaporated while the Rangers were in midflight. Word radioed from Panama informed Colonel Koren that a Brazilian airliner was making a late landing at Torrijos International. At 0100, the 376 passengers would still be inside the terminal building. This news followed on the heels of very alarming messages to the effect that operations security had broken down, and the Panama Defense Forces knew an invasion was imminent. To the Rangers, the loss of surprise likely meant greater enemy resistance, not to mention ample time for the defenders at Torrijos-Tocumen to prepare the shoulder-fired, surface-to-air missiles (SAMs) they were rumored to have. The low-flying C–141s and C–130s would be vulnerable to a SAMs crew possessing even marginal training.[30]

Despite Stiner's decision to advance H-hour by fifteen minutes for the U.S. attacks at the Comandancia and some other targets, the air assault at Torrijos-Tocumen stayed on schedule. Precisely at 0100, the AC–130 opened fire with its 105-mm. howitzer and 40-mm. Bofors cannon on the infantry company barracks. Once that building collapsed, the Spectre destroyed one of the airfield's two ZPU4 air defense guns, a Dignity Battalion car, and three of four machine gun positions. Meanwhile, as an estimated thirty Panamanians

[29] This description of the plan for the air assault on Torrijos-Tocumen is from Donnelly, Roth, and Baker, *Operation Just Cause*, pp. 190, 192–94. After Just Cause, a Panamanian officer conducted a "hot wash" at Fort Clayton in which he told the assembled U.S. personnel that, had the defenders possessed more sophisticated weapons, they could have "played havoc" with Rangers parachuting onto the facility. Hot Wash, 19 Jun 1990, Fort Clayton, Panama.

[30] Donnelly, Roth, and Baker, *Operation Just Cause*, pp. 197–98.

tried to shoot them down, the AH–6 Little Birds strafed the control tower, the Tocumen terminal building, and two guard posts. The intense fire from the gunship and the helicopters killed several of the enemy and caused others to flee, including a group of around forty air force pilots and quite a few members of the infantry company who had remained in the area. (Many of their colleagues had already gone to the surrounding hills once word of U.S. troop movements had first reached Tocumen thirty minutes earlier.) Despite the devastation, the helicopters had failed to disable the facility's radar, but, that notwithstanding, the three-minute preparatory fire had served its purpose.[31]

At 0103, over seven hundred Rangers began to parachute onto the target. Just hours before, they had been freezing in a cold, blustery rain and ice storm; now, as they jumped out the doors of their transport planes, a blast of hot, humid tropical air welcomed them to Panama. To minimize the time in the air where they would be helplessly exposed to hostile fire, they jumped from five hundred feet, rather than from the usual seven hundred fifty feet or higher. The risk here was that a parachute might not open in time to slow a soldier's descent adequately, or, if it did not open at all, there would be no time to open the reserve. (The draft after action report on the operation, in fact, recommended dispensing completely with the reserve parachute in drops planned for five hundred feet.) The low-level drop resulted in over a dozen impact injuries, as Rangers suffered broken legs, fractures, and torn ligaments.[32]

The troop transport planes almost without exception dropped the Rangers right on target. (The last C–130 with Captain Dochnal and a portion of Company C aboard was blown slightly off course, so that most of the group

A Ranger manning a roadblock at Torrijos-Tocumen complex

[31] Ibid., p. 189; Flanagan, *Battle for Panama*, pp. 159–60; Handwritten notes, Purpose: To provide information on C 3/75 actions at Torrijos International, n.d.

[32] Donnelly, Roth, and Baker, *Operation Just Cause*, pp. 198–99; Handwritten notes, 1st Ranger Bn S–3 section, n.d.; Flanagan, *Battle for Panama*, p. 158.

dropped into tall elephant grass about a hundred yards from the Torrijos runway.) Once the Rangers shed their parachutes and equipment harnesses, donned their rucksacks, and readied their weapons, they moved toward their designated assembly areas. The 1st Battalion command element gathered on the Tocumen runway, somewhere near the center of the complex, as the battalion's three infantry companies headed for their objectives. At the northern end, Company A advanced on Objective Tiger, where the Rangers quickly subdued twenty airmen who had chosen to stay behind in one of the hangars. Also falling into the company's hands was "virtually the entire Panamanian Air Force on the ground." Simultaneously, from the middle of the runway, Company C headed toward Objective Pig. The structure lay in ruins, and there was no one, living or dead, inside. From concealed positions near the wreckage, though, several Panamanian riflemen fired into the group, only to have the rangers' return fire claim one fatality. Several of the others surrendered. Company C also received heavy sniper fire as it tried to surround the Tocumen control tower. In the exchange of shots that ensued, Pfc. James William Markwell, a medic, was killed, the ranger battalion's only fatality that night.[33]

While the Rangers of Companies A and C, 1st Battalion, were engaging PDF defenders, Company B was providing perimeter security through means that included the hasty establishment of a number of blocking positions. Minutes after the barricades went into place, several Panamanian vehicles approached heading toward the airfield's exits. Warning shots persuaded most drivers to turn back, although the Rangers had to shoot the tires of one car to get it to stop. The driver of another car ignored the warning shots and escaped. Ironically, a passenger in the only vehicle to get away was Noriega, fleeing the Ceremi Recreation Center.[34]

After suffering one fatality and a small number of wounded and injured, the Rangers at Tocumen had secured the military airfield shortly after 0200. The operation, while not uncontested, had been fairly straightforward and, in its essentials, mimicked the rehearsals. The same could not be said for the ranger mission at Torrijos International Airport, where Captain Dochnal's Company C, 3d Battalion, encountered a number of problems. The most serious of these arose from factors inherent in urban warfare—the presence of multistoried, complex buildings and large numbers of civilians. Others derived from the friction of war, such as the errant landing of the captain and many with him in the five- to twelve-feet-high elephant grass west of the commercial airport's runway. "This slowed assembly," a report later noted with unintentional understatement, as thirty minutes passed before Dochnal could reach the rendezvous point. His situation could have been worse. A few of his men had landed north of the grass in a large ditch that contained a tidal pool and were not extricated until later that morning by U.S. soldiers searching for some lost equipment.[35]

---

[33] Donnelly, Roth, and Baker, *Operation Just Cause*, pp. 199–200. Quote from U.S. Special Operations Command, *History, 1987–2007*, p. 34.

[34] Donnelly, Roth, and Baker, *Operation Just Cause,* p. 200; U.S. Special Operations Command, *History, 1987–2007*, p. 34.

[35] Donnelly, Roth, and Baker, *Operation Just Cause*, p. 203. Quote from Handwritten notes, Purpose: To provide information on C 3/75 actions at Torrijos International, n.d.

As Dochnal finally exited the tall grass, the bulk of Company C was already on the move and encountering enemy resistance at some objectives. North of Torrijos International, the 3d Platoon moved toward the terminal building but stopped first to converge on a fire station along the route. Before the Rangers could reach the building, a fire truck burst forth from one of the station's doors. As the vehicle gained speed, a ranger squad leader told one of his men to fire several warning shots in front of it. The truck immediately returned to the fire station. The driver fled to the rear of the building, where he joined a group of his colleagues "screaming obscenities at the American troops." The U.S. squad leader rejected one suggestion to use a hand grenade to coax the firemen out. Instead, he employed an interpreter who persuaded fifteen Panamanians inside to surrender.[36]

While the 3d Platoon was thus occupied, the 1st and 2d Platoons moved toward the airport terminal from the south. The main part of the building was three stories high. Extending from it were two long concourses, one on the north side, the other on the south, each two stories high with a circular waiting area at the end, where the arrival and departure gates were located. The 1st Platoon was to take and hold the main entrance to the terminal, but first it had to secure a nearby restaurant in which a number of people in civilian clothes could be seen. Were they employees of the establishment or Panamanian defenders in mufti? In the absence of hostile fire, the platoon's 1st Squad scaled a chain-link fence, approached the eatery, and had a Spanish-speaking Ranger try to reassure those inside. After he convinced the occupants they would not be harmed, they unlocked the front door and turned themselves over to the troops. The eighteen were flexcuffed until their exact status could be ascertained. The 1st Squad then cleared the building, while the platoon's 3d Squad took up security positions around it.[37]

In the meantime, 1st Platoon's 2d Squad cut through two chain-link fences and continued its advance toward the Torrijos terminal. The squad reached the objective just as a gun jeep sent by the ranger battalion at Tocumen arrived in support. With the vehicle situated to provide covering fire, the 2d Squad cleared a guardhouse near the terminal entrance, finding two enemy soldiers who had been killed in the AH–6s' preparatory fires. While the clearing operation was under way, the 3d Squad moved from the restaurant to the west side of Torrijos International, where it set up an "observation/battle position to overwatch the terminal and parking lot."[38]

The 3d Platoon, after it secured the fire station, advanced to the terminal to establish its segment of the cordon around the building. As the platoon approached the objective, two Panamanian soldiers opened fire from inside the end of the north concourse, shooting through and shattering the glass that enclosed the waiting area. Most of the Rangers dove for cover, but four members of one squad—Sgt. David Reeves, Spec. Michael Eubanks, and Pfc.

[36] Quote from Donnelly, Roth, and Baker, *Operation Just Cause*, pp. 205–06. Handwritten notes, Purpose: To provide information on C 3/75 actions at Torrijos International, n.d.

[37] Handwritten notes, Actions of 1st Pltn, C Co., at Omar Torrijos Intern'l Airport, n.d. These notes were with the other handwritten notes attached to HQ, 75th Ranger Rgt, 26 Jan 1990, sub: Historical Information for Commander 44th Historical Detachment.

[38] Ibid.

William Kelly, with Pfc. Farber in trail—were able to scale the maintenance stairway into the terminal. Others in the squad followed. Inside, they saw the two shooters heading into a women's restroom. Reeves approached the facility, pulled the pin on a grenade, and opened the bathroom door. To his chagrin, he found himself staring at another door. With no time to open it as well, he flung the grenade down the empty concourse, where the explosion ripped a hole in the floor and shattered more windows. When Reeves and Farber charged through the second door, one of the enemy soldiers shot Reeves three times with an AK47 at point-blank range, the bullets hitting his shoulder and collarbone. Farber tried to return the fire, but his weapon malfunctioned.

With Reeves lying on the floor seriously wounded, his assailants moved to the rear of the large restroom, and Farber used the lull to slip back outside the door and report the situation. Next, Eubanks, Kelly, and Sgt. Thorland entered the darkened restroom unchallenged and began to pull Reeves out. As they were dragging him, one of the Panamanians opened fire again, hitting Kelly twice in his Kevlar helmet. With his adrenaline pumping and focused on the task at hand, Kelly did not even feel the impact, and he and Eubanks finished extracting Reeves, who then received first aid. Eubanks, with Kelly behind him, went back into the restroom, preparing the way with a hand grenade. The two defenders saw the projectile and dodged the blast by seeking shelter in two stalls. Once Eubanks located them, he tried to engage them with his squad automatic weapon, but it jammed. The noise gave away his position, and one of the Panamanians fired three shots at him with a pistol. All three bullets missed, and Eubanks hastily exited the room.

The two Rangers began the next assault with another grenade, followed by small-arms fire as they fought their way back into the bathroom. At one point, the shooting stopped, and Eubanks tried to talk the two into surrendering. The Panamanians responded with curses, but, when one of them exposed too much of his upper body in an attempt to hurl another obscenity, Eubanks shot him in the neck. The man was not dead, and the minutes that followed witnessed hand-to-hand combat between the two pairs of adversaries. When it was over, one of the Panamanians was dead from a gunshot to the head. The other had been thrown out of the second-story window onto the tarmac below, where he was killed by a Ranger with an M60 machine gun. The restroom fight, according to one account, lasted five minutes.[39]

The task of clearing and securing the whole terminal fell to the 2d Platoon, which moved up from its positions to the south of the building and waited for the go ahead to enter. Before the Rangers received the order, they came under mortar fire, prompting them to move into the terminal without official permission simply as a matter of self-preservation. Once inside, the platoon's 1st Squad headed upstairs to the third floor, where it immediately came under hostile fire. The enemy soldiers then ran into the floor's security room and, using furniture and other flammable material, started a fire in order to burn

[39] This account of the battle in the concourse bathroom is based on Donnelly, Roth, and Baker, *Operation Just Cause*, pp. 206–09; Handwritten notes, Purpose: To provide information on C 3/75 actions at Torrijos International, n.d., and 3rd Plt. Outline. For slightly different accounts of the fight, see Flanagan, *Battle for Panama*, p. 163; U.S. Special Operations Command, *History, 1987–2007*, p. 35.

documents. The Rangers moved to the room and exchanged shots with the defenders inside. A metal door limited the effectiveness of small arms from both sides, so the Rangers used a grenade to blow an opening at the bottom of the door, through which they then tossed another grenade and killed five Panamanians. It also contributed to the intensity of the fire, which the Rangers tried to put out once they gained entry to the room. Although their frantic efforts met with no success, the blaze itself set off the room's sprinkler system, which contained the fire. The Rangers then cleared the remainder of the third floor without incident.[40]

Below them, the 2d Platoon's 3d Squad arrived on the first floor of the terminal after having searched an Eastern Airlines commercial jet and the bottom floor of the southern concourse. The squad had found no enemy or civilians in either place. Once in the terminal, one team went to the right, another to the left. "All the doors we checked were locked," one of the team leaders reported later, "and we were told not to bust or blow them down." The reason for the caution soon became apparent. In the building's baggage area, behind a set of unlocked double doors, the team discovered Panamanian soldiers, who opened fire in the dark room. The team leader was about to return fire when "a woman started screaming in English not to shoot." He quickly went back through the double doors and informed the squad leader of an apparent hostage situation. From there, word went up to the platoon leader, company commander (Dochnal), and the ranking officer on the scene (Koren). Dochnal made his way to the terminal, where his 2d Platoon leader gave him a hurried account of the situation, including the willingness of the airport's security manager, who declared himself to be pro-American and anti-Noriega, to help. Dochnal accepted the offer but also called in "our Spanish speakers as a backup to ensure this guy wasn't double crossing us."[41]

The squad leader then told the team leader, another soldier, the security manager, and the Spanish-speaking Rangers to remain in the baggage area and negotiate with the soldiers holding the civilians. The remainder of the squad resumed clearing and securing the first floor of the terminal. At the hostage scene, the security manager tried to persuade the Panamanian soldiers to surrender, but they refused, threatening to kill the civilians if the Rangers attempted any rescue by force. At some point in the talks, the team leader concluded that the hostages included at least two American women and a

---

[40] Handwritten notes, Purpose: To provide information on C 3/75 actions at Torrijos International, n.d.; Flanagan, *Battle for Panama*, p. 161; Donnelly, Roth, and Baker, *Operation Just Cause*, pp. 209–10.

[41] This account of the hostage situation at the Torrijos International Airport terminal building is primarily based on the following: last quote from Handwritten notes, Purpose: To provide information on C 3/75 actions at Torrijos International, n.d. (this section of the notes, at least, was written by Dochnal, as he refers to himself in the first person); first and second quotes from Handwritten notes, no title (written by the team leader who first encountered and reported the hostage situation), n.d. (also included with HQ, 75th Ranger Rgt, 26 Jan 1990, sub: Historical Information for Commander 44th Historical Detachment). Filling out the reconstruction of events are versions of the episode found in Flanagan, *Battle for Panama*, pp. 161–62; Donnelly, Roth, and Baker, *Operation Just Cause*, pp. 210–12; U.S. Special Operations Command, *History, 1987–2007*, p. 35.

baby. After several more verbal exchanges, the security manager drew some conclusions of his own, one of which he shared with the team leader: there were only two Panamanian soldiers behind the doors. He asked whether the team leader could slip inside and "take out the PDF guys."[42] The team leader answered that he probably could and maneuvered into a darkened area between two glass doors. From this vantage point, he could make out three of the enemy—thereby discounting the security manager's hunch—in a small room beyond the door in front of him but could not see any of the hostages. He thus refrained from shooting. Then, as he was preparing to crawl forward into the room, lights above him came on, forcing him to retreat back outside the first set of double doors.

As talks with the enemy troops dragged on for over two hours, Dochnal provided on-the-scene guidance, reinforcements were brought in, uniformed U.S. civil affairs and psychological operations personnel added their skills to the negotiations, and Koren's command post at Tocumen began considering a "surgical strike" to resolve the situation. Meanwhile, a group of Rangers managed to break into a room next to where the hostages were being held. In the process, one of the Rangers jumped over a conveyor belt, landing right on top of a PDF soldier. Alerted by the commotion to the presence of American troops next door, the other Panamanians opened fire. With that, Dochnal vented his frustration. "If you don't come out," he shouted once the shooting stopped, "I'm just going to kill you! You've got five minutes to come out or I'm just going to kill you!" The hostage-takers required only three minutes to reach a decision. They emerged through the door, surrendered their weapons, and dropped to the floor in a prone position. As it turned out, there were eight Panamanians and a Cuban with a diplomatic card. The hostages also came out, were examined, and taken to safety. The affair ended at 0500.[43]

In the nearly four hours the Rangers took to clear and secure the terminal building at Torrijos International, they, like their counterparts in the Gator and Wildcat task forces, had to cope with the unexpected presence of far more civilians than the Blue Spoon planners had anticipated. The situations at the Comandancia and in Balboa were worse, of course, with thousands of refugees passing through American lines during combat operations, but that fact, had they known it, would have been of little solace to the Rangers at Torrijos who, in looking for enemy soldiers, were initially finding dozens, then several hundred civilians in their way. Captain Dochnal, according to one account, was "dismayed." "I ran downstairs, and I'm getting more reports, 'Hey sir, we just found another 50 [civilians] in this area' and all

[42] Handwritten notes, no title (written by the team leader who first encountered and reported the hostage situation), n.d.

[43] Most of this paragraph, including the quotes, is from Donnelly, Roth, and Baker, *Operation Just Cause*, pp. 211–12. U.S. Special Operations Command, *History, 1987–2007*, p. 35, also notes the rangers' threat "to come in shooting." As for the number of hostages being held, accounts differ. The team leader later wrote that there were two American women, a Panamanian woman, and a baby. The Special Operations Command history says there were two American girls. The Donnelly, Roth, and Baker book indicates there were one American woman and a baby.

of a sudden it's 150, 200, then it's 300. And I'm going upstairs, saying holy s——t. We're taking people out from behind the baggage counter and all over the airport. There were no reports of any civilians getting killed, but they were scattered to the four winds." When discovered, civilians were escorted to an assembly area set up in a parking lot and loading area near the airport entrance. There they could receive medical treatment, food, and reassurances.[44]

U.S. troops at Torrijos-Tocumen after dawn on 20 December. In the background is the secured Torrijos International Airport terminal.

During the night, while the Rangers were carrying out their missions, paratroopers from the 82d Airborne Division had dropped onto Torrijos-Tocumen. At 0700, the division's commander, Maj. Gen. James H. Johnson Jr., assumed responsibility for the Torrijos terminal area. This allowed Dochnal's 1st Platoon to turn over the 398 civilians in its care to the 82d's 820th Military Police Company, together with all prisoners and weapons that had been captured. Later that morning, General Johnson took operational control of the 1st Ranger Battalion, while Dochnal's Company C from the 3d Ranger Battalion was sent into Panama City to help clear the Comandancia compound.

In the fighting at Torrijos-Tocumen, the Rangers had accomplished their missions at a cost to themselves of one fatality and eight wounded. While thirteen of the enemy had been killed in the attack and fifty-four taken prisoner, the number of PDF wounded remained unknown. By daylight on

[44] Dochnal quote from Donnelly, Roth, and Baker, *Operation Just Cause*, p. 211. Handwritten notes, Purpose: To provide information on C 3/75 actions at Torrijos International, n.d.

20 December, the airfield complex had been seized and secured and thus could be used as a staging area for the follow-on airborne brigade to launch its planned air assaults against the three major enemy objectives remaining in the Panama City area: Panama Viejo, Las Tinajitas, and Fort Cimarrón.[45]

---

[45] The casualty figures are from Donnelly, Roth, and Baker, *Operation Just Cause*, pp. 212–13. On clearing the Comandancia compound, see Chapter 4.

8

THE 82D AIRBORNE DIVISION ON D-DAY

☆ ☆ ☆

By the time soldiers from the 82d Airborne Division began their parachute drop onto the Torrijos-Tocumen airport complex, all but three of the principal D-day targets in proximity to the fighting in downtown Panama City had been engaged. The exceptions—Panama Viejo, Fort Cimarrón, and Las Tinajitas—had been assigned to the 82d's 1st Brigade, which was supposed to launch its nighttime attacks at those locations soon after hitting the ground. The drop was scheduled to take place at 0145, forty-five minutes after the Rangers were to have started securing the airfields' facilities. When abysmal weather in North Carolina delayed the armada of troop transports, the ripple effect spread through subsequent phases of the operation. Most significantly, the brigade's three air assaults would not begin until well after sunrise, thus magnifying the risks involved, a reality not lost on the commander and staff of JTF-South or on the paratroopers who had the mission.

PLANNING AND PREPARATIONS

The parachute drop on Torrijos-Tocumen was the first combat jump the 82d Airborne Division had made since World War II, a milestone XVIII Airborne Corps planners attributed to plain common sense. The mission of attacking Panama Viejo, Las Tinajitas, and Fort Cimarrón was going to be "one of the most challenging . . . military operations" because the troops conducting it would have to be flown from the United States, to assemble once on the ground in what promised to be a "hot" combat zone, to board waiting helicopters at the scene, and to take off to assault three separate targets—all of this presumably with "no breathing space." Since paratroopers jumping from planes would reach the ground and be able to assemble faster than soldiers landing in transports, an airdrop seemed an obvious course of action for the units assigned the mission. That, however, meant significant revisions to a long-standing version of the BLUE SPOON operation order.[1]

[1] The milestones noted were taken from the 82d Airborne Division's official Web site at http://www.bragg.army.mil/82DV/History.htm. Quotes from Interv, author, Robert K. Wright Jr., and Joe D. Huddleston with Lt Gen Carmen Cavezza, U.S. Army, 30 Apr 1992, Fort Lewis, Wash. As for Panama being the first combat drop for the 82d since World War II, paratroopers

The earliest versions of the plan drafted in 1988 did not include Panama Viejo, Las Tinajitas, and Fort Cimarrón among the high-priority H-hour objectives. Rather, the three were listed as on-order targets, to be attacked if necessary by a force taken from the 7th Infantry Division (Light), a unit whose previous experience with operations in Central America and frequent interaction with the Southern Command made it a logical choice for the assignment. A series of developments—the designation of the XVIII Airborne Corps as the executive planning agency for the Panama crisis, Thurman's assumption of command at Quarry Heights on 30 September 1989, the failed coup three days later, and approval of a new Blue Spoon concept of operations—changed all that. The three PDF bases became high-priority targets, the neutralization of which would fall to paratroopers from the 82d. The light fighters from the 7th, with the exception of the brigade headquarters and one battalion already in Panama as a result of Operation Nimrod Dancer, relinquished their H-hour missions and, exchanging places with the 82d, became a follow-on force.[2]

Both divisions could trace their lineages back to World War I, both had experience in Latin America, and, in the 1980s, both came under the operational command and control of the XVIII Airborne Corps, the Army's contingency corps. Yet, despite this comparable background, after the switch in Blue Spoon roles was formally announced at one of Thurman's planning sessions in Panama, a number of officers privy to the decision expressed displeasure, complaining of an "airborne mafia" at Fort Bragg, with its "godfather" now in residence at Quarry Heights. As General Kelly in the Pentagon observed, "A lot of people went around saying, 'Well, you know Max is an old airborne guy and Carl Stiner can't spell anything but airborne.'" Kelly dismissed the complaints, going on to explain, "The fact is, we could get an airborne division on the ground in ten minutes or we could get an airlanded brigade in a day and a half." At Fort Ord, Maj. Gen. Carmen Cavezza, the commanding general of the 7th Infantry Division, later stated that he had accepted the decision, even though there "was no reason given except that the plan called for quick, decisive involvement." "Once General Thurman came in and made General Stiner his war fighter," Cavezza continued, "there was no doubt in my mind that that was going to be that way. I didn't have any problems with it. If they left the plan as originally drafted I think it would have worked as well. I'm not going to say it would have worked better; I don't know that. But I think it would have worked as well."[3]

from the division did participate in the U.S. intervention in the Dominican Republic in 1965 and in the U.S. invasion of Grenada in 1983 but in both cases airlanded in the area of operations.

[2] On the early and revised plans concerning the 82d Airborne Division, see Briefing Slides, Blue Spoon: The Evolution of the Plan for Just Cause, n.d.; Interv, Wright with G–3 Plans Section, XVIII Abn Corps, 30 Mar 1990, Fort Bragg, N.C.; Telephone Interv, author with Maj David Huntoon Jr., 27 Jun 1990.

[3] Kelly's quotes from Donnelly, Roth, and Baker, *Operation Just Cause*, p. 89. Cavezza's quotes from Interv, author, Wright, and Huddleston with Cavezza, 30 Apr 1992. The ties between two key planners—Maj. David Huntoon of the XVIII Airborne Corps and Maj. William Caldwell of the 82d—epitomized the close working relationship between the corps and the division in Blue Spoon planning. The two officers had been classmates in the School of Advanced Military Studies at Fort Leavenworth. At Fort Bragg, according to Caldwell, each time Huntoon

General Johnson (*center*) at a briefing at Tocumen airfield on 20 December

Maj. Gen. James H. Johnson Jr., the 82d's commander since October 1988, was briefed on the new missions Stiner's staff envisaged for his paratroopers just prior to the first of Thurman's postcoup planning meetings in Panama. As a consequence, according to Brig. Gen. Joseph W. Kinzer, the assistant division commander for operations, "We started planning in earnest," an undertaking that involved staff officers from the 82d's intelligence, operations, and logistics shops. Before leaving for the initial session at Quarry Heights, Johnson and his staff looked closely at what was then Operation Plan 90–1 and "developed implementing plans and began to look at concepts for employment, the troop list, the mission analysis of the targets, . . . the battle tasks that would be associated with our involvement." They employed the "backward planning sequence," first determining how they wanted to launch the air assaults from Torrijos-Tocumen and then working back from that to ascertain the best way to arrange the airdrop there so that a brigade of paratroopers and their equipment "could be assembled and quickly moved to pickup zones (PZs) on the taxiway to the west of the main runway at Torrijos."[4]

In Panama, Johnson and his staff learned more details of the plan, which, once back at Bragg and in subsequent trips to Thurman's headquarters, they continued to refine. One critical issue revolved around the number of aircraft needed to carry an airborne brigade and its equipment to the drop site at Torrijos-Tocumen in a single formation, or serial, rather than piecemeal over the period of a day or so that a brigade from the 7th Infantry Division would

"went somewhere with the corps commander he'd come back and back brief me in detail." The friendship "really made the planning effort smooth," Caldwell concluded. Interv, Col Kevin Benson (Ret.) with Lt Gen William Caldwell, U.S. Army, 27 Apr 2009, Fort Leavenworth, Kans.

[4] First quote from Interv, Dolores De Mena with Brig Gen Joseph W. Kinzer, U.S. Army, 20 Apr 1992, Fort Clayton, Panama. Remaining quotes from Interv, Wright with Maj Gen James H. Johnson Jr., U.S. Army, 5 Mar 1990, Fort Bragg, N.C.

Joseph Kinzer at his later promotion from brigadier general to major general

have required. The deploying unit from the 82d would be the Division Ready Brigade, a unit held in the highest state of preparedness at Fort Bragg, relieved of routine demands on its schedule and capable of lifting off from adjacent Pope Air Force Base within eighteen hours of receiving an alert. The ready brigade assignment rotated among the division's three combat brigades on a six-week cycle, so without knowing when Blue Spoon might be executed—although the best guess was that would happen sometime in January—no one could say which brigade would carry out the mission. Planners, however, estimated the minimum number of transports needed to deploy the bulk of any one of the three units at forty-eight C–141s. Because the Air Force had only 110 aircraft designated for use in conventional operations, there was some concern that the Military Airlift Command might not be able to meet the minimal requirement. To resolve the matter, Stiner at some point conferred with the military airlift commander, who pledged to have the forty-eight planes available within sixty hours of receiving notification. This was welcome news but did not address another problem: without at least fifty-two C–141s, a brigade from the 82d would have to leave behind all or part of certain organic assets, such as field artillery. Also, in a related issue, the summary of one planning session flatly stated, "Aerial refuel will be tough, if not impossible."[5]

Johnson and his staff also used the trips to Panama to reconnoiter the assigned targets and the Torrijos-Tocumen complex. As the 82d's chief of plans, Maj. William Caldwell, recalled, "At night, a Blackhawk would come and pick us up and we'd go to various points around the country." One of the group's priorities was to identify landing zones for troop-carrying helicopters at each of the three objectives. The concept was straightforward enough: "We were trying to select landing zones that were out of direct fire range of the

[5] Intervs, Wright with Johnson, 5 Mar 1990; author with Lt Col Robert Pantier, U.S. Army, 13 Jun 1990, Fort Clayton, Panama, and 21 May 1991, Fort Leavenworth, Kans. Quote from Memo for J–3, 20 Oct 1989, sub: Major Points from XVIII Planning Meeting.

targets, to get the troops on the ground. And then the notion was that we would seal off, isolate the objective, and commence to try to neutralize, if necessary destroy—whatever it took to get them to surrender." On the ground in Panama, the actual selection process proved difficult. The PDF base at Las Tinajitas was "up on a high ground," which made it "a tough objective to get to." Panama Viejo, on the other hand, was near downtown Panama City and had "very constrained areas to get into." Only the Fort Cimarrón landing zones met the criteria without complications. As for the proposed drop zone at Torrijos-Tocumen, Johnson received "pretty good imagery" from satellite photographs. He also flew near the airfield complex in a helicopter but "never got a good look" at it. Operations security dictated that the "fly-bys" take place "at a distance of 2,000, 3,000 meters or more, flying away from it" so as not to "hover over anything" in such a manner that would "give away what we were looking at." From the photographs and air reconnaissance, the general developed concerns about the "tall elephant grass and swamp" to the east of the Torrijos-Tocumen runways. If the paratroopers landed in that terrain, the results would be unfortunate, but hardly catastrophic. The men, he was confident, would extricate themselves in a timely manner—"we can deal with anything," he later asserted. But if the equipment landed in the swamp, it could get stuck. Worse, as he would discover, it could also disappear.[6]

Besides examining and selecting landing zones, Johnson and his people worked with the air mission commander in planning more details of the 82d's air assaults on the three objectives. Receiving special emphasis were the procedures to be followed at the pickup zones—the points at which the paratroopers taking part in the attacks would board the helicopters waiting at Torrijos-Tocumen—and the flight routes the aircraft would take once airborne. After several coordination sessions in Panama, the air mission commander, together with the commanders of various other aviation elements, visited Fort Bragg to fine-tune the arrangements. Ranger officers cleared to see the plan made that pilgrimage as well. Their participation was crucial because Johnson's men would have to link up with ranger units securing an airhead at Torrijos-Tocumen, possibly while combat operations were still in progress. Furthermore, some units from the 82d would be moving to Las Tinajitas, Panama Viejo, and Fort Cimarrón not by helicopter but via ground transport and might require a ranger escort. According to Johnson, "We did the necessary coordination and liaison with Rangers to help facilitate the rapid movement of ground convoys."[7]

Under Operation Plan 90–2, the commander of the 82d Airborne Division was listed as the leader of Task Force Pacific, another component of JTF-South. The task force was responsible for over twenty assignments, key among them being a forced entry at Torrijos-Tocumen at H plus 45 minutes to reinforce the Rangers if required; assumption of control over the airport complex; air assaults to "isolate, neutralize, and/or destroy" Panamanian Defense Forces at Fort Cimarrón, Panama Viejo, and Las Tinajitas; and a linkup with, and relief of, Special Operations Forces at Cerro Azul, Paitilla airfield, and other

[6] First quote from Interv, Benson with Caldwell, 27 Apr 2009. Remaining quotes from Interv, Wright with Johnson, 5 Mar 1990.

[7] Interv, Wright with Johnson, 5 Mar 1990.

locations as needed. Other critical duties included the placement of one infantry company in reserve to conduct on-order air assaults, engagement of PDF reinforcements trying to enter the division's area of operations or to reach Panama City, and preparations to secure and hold other targets and relieve other forces as required.[8]

In the task organization appendix of the operation plan, the order of battle for Task Force Pacific in the deployment phase consisted of one airborne brigade, with nine loudspeaker teams from the 4th Psychological Operations Group attached to it. At some point after the commanding general of the 82d was on the ground, the JSOTF units at Torrijos-Tocumen, together with Task Forces Bayonet and Semper Fi, would come under his operational control. Not itemized in the plan was exactly what the airborne brigade would bring with it; Johnson would have to determine that, based in part on the actual number of C–141s the Air Force could provide and in part on what was required to execute the assigned missions. After considering several alternatives, he requested that the brigade deploy as a task force with "its normal slice of artillery, engineer, air defense, military police, signal and so on." That he would receive approval for this configuration was by no means a foregone conclusion. Stiner's headquarters saw no need for the 82d to take field artillery pieces into Panama. Given that the plan called for most of the fighting to take place in or around built-up, populated areas, the rules of engagement and various restrictions designed to minimize collateral damage all but excluded the use of indirect fire weapons other than mortars. Johnson understood and accepted these constraints, yet did not like to deploy for combat without his artillery—"You never know what you're going to go into"—so he set aside Stiner's "initial guidance" and authorized the deployment of four M102 105-mm. howitzer tubes. "We took the capability of split battery," he said later in explaining his decision, "and intended to use it out to the east at Fort Cimarron if necessary" where it would be "less likely to induce collateral damage." There was never any intention, he emphasized, to use it in or near Panama City.[9]

Johnson also decided that his principal combat missions in Panama required the deploying brigade to take more than its normal complement of M551 Sheridan light armored reconnaissance vehicles. Besides the platoon of four M551s that had deployed secretly in November and would be used at the Comandancia, Johnson wanted eight more of the tanks to protect the troops and equipment that would be moving in ground convoys to support the 82d's missions. He also authorized seventy-eight HMMWVs for employment as troop carriers, TOW transports, convoy escorts, and command-and-control vehicles. Of the HMMWVs designed to mount the TOWs, several would be outfitted instead with .50-caliber machine guns, a configuration with which the division had experimented successfully before JUST CAUSE. The 82d's 180 TOWs were "great in the Middle East" and for other contingencies, Johnson knew, "but not necessarily of any value in built-up areas or in the Central or South American or Caribbean Basin type of scenario." Better to have a more practical weapon in those situations, and, while the Army's M60 machine gun

[8] JTF-South OPLAN 90–2, 3 Nov 1989.

[9] Ibid. Quotes from Interv, Wright with Johnson, 5 Mar 1990.

could be mounted on an HMMWV, the .50-caliber had "a lot more breakdown capability, blow-down of buildings, more accurate, more destructive." By December 1989, the number of .50-caliber machine guns in a battalion had increased from five to around thirteen (the goal being twenty).[10]

The day the Joint Chiefs approved Operation Plan 90–2, they also authorized rehearsals for the units likely to be involved in its execution. The highly elaborate practice runs arranged for the Special Operations Forces based in the United States occurred mainly in the southeastern part of the country at a cost of what Stiner estimated as $3 million. In Panama, U.S. forces conducted rehearsals as part of their routine contingency readiness exercises, joint Purple Storm exercises, and Sand Flea maneuvers. For the 82d Airborne Division, Fort Bragg provided a readily available rehearsal site. There, the Division Ready Brigade was already scheduled to conduct one of its emergency deployment readiness exercises (EDRE) in late November, code-named Operation BLACK KNIGHT. With some changes, the exercise became a brigade battle task rehearsal using "a 100% replica of the operation we would execute in Panama" and closely resembling what the revised BLUE SPOON plan contemplated for the paratroopers. Only personnel with top secret clearances and a need to know—which included the ranger liaison officers in attendance—understood the full implications of what was taking place.[11]

During BLACK KNIGHT, 2,100 paratroopers from the Division Ready Brigade jumped onto the fort's Sicily drop zone, where "a set-up that somewhat replicated the Torrijos-Tocumen" complex had been created, including the roads the ground convoys would use. The Air Force had made twenty C–141 troop transports available for the jump, the exact number that Johnson would have during Operation JUST CAUSE, but not the additional craft needed to drop the brigade's heavy equipment during the rehearsal. To compensate for the shortage, much of the equipment was pre-positioned on the ground. Once the paratroopers had landed, those participating in the air assaults assembled, moved to their pick-up zones, and boarded waiting Black Hawk helicopters from the division's aviation brigade. Those that would travel overland in HMMWVs with the Sheridan escorts de-rigged the vehicles and equipment and prepared to move out.[12]

---

[10] Quotes from Interv, Wright with Johnson, 5 Mar 1990. Corps Historian's Personal Notes Recorded During the Operation [JUST CAUSE], n.d. The several pages of typed notes were written by Robert K. Wright Jr. and can be found on the U.S. Army Center of Military History Web site at http://www.history.army.mil/documents/panama/notes.htm.

[11] Quote from Interv, Benson with Caldwell, 27 Apr 2009. Intervs, Wright with Johnson, 5 Mar 1990, and with Lt Gen Carl W. Stiner, U.S. Army, 2, 7, 27 Mar and 11 Jun 1990, Fort Bragg, N.C.

[12] Intervs, Wright with Johnson, 5 Mar 1990; with Stiner, 2, 7, 27 Mar and 11 Jun 1990; De Mena with Kinzer, 20 Apr 1992. Quote from Interv, Wright with Col Thomas H. Needham, U.S. Army, 6 Mar 1990, Fort Bragg, N.C. In discussing the rehearsal in his 5 March 1990 interview, Johnson noted that, as with any "normal" training event, the seats in the Black Hawks were not removed. "The only seats-out training we did rehearse, though, [was] static," he continued. "[We rehearsed] load training with seats out of the Blackhawk a number of times with those units that were in mission status. We had to do static load training with seats out so that our soldiers were accustomed to that. We are not allowed in peacetime to train, however, with seats out. Therefore, we didn't in rehearsal. Could I have requested to do that rehearsal? Maybe. But

Although rehearsal sites for Las Tinajitas, Panama Viejo, and Fort Cimarrón were more difficult to replicate than the Torrijos-Tocumen landing zones, division planners could at least acquaint the troops with the kind of terrain and weaponry they would face if deployed for combat operations in Panama. Thus, the site representing Fort Cimarrón featured an "enemy" equipped with armored vehicles, while the Las Tinajitas simulation conducted in Bragg's Northern Training Area emphasized the enemy's heavy mortar capability. Since the division's third objective, Panama Viejo, was located near downtown Panama City, the fort's urban operations training site provided a realistic environment, including the presence of civilian bystanders. "We wanted to expose our soldiers to what they'd be running into there, a lot of noncombatants," Johnson explained.[13]

The 82d's rehearsal went "pretty darn well," according to reports received by the operations officer at the XVIII Airborne Corps, so well, in fact, that Stiner's staff made "no adjustments" to the plan. Stiner voiced one concern, however, but it stemmed from the practice runs taking place in Panama for the air assaults, not at Fort Bragg. U.S. Army, South's aviation battalion—the 1st Battalion, 228th Aviation—was not only "down in strength" but, in the general's estimate, was "not proficient for doing night operations" with night-vision goggles. Because that battalion would participate in five of the anticipated nighttime air assaults, including those involving the 82d Airborne Division, efforts were made to raise the unit's "level of proficiency." Among other arrangements, this meant sending Black Hawk, Huey, and Chinook crews from Fort Bragg to Panama, together with door gunners from the 82d, to participate in "a very comprehensive training program" with the battalion.[14]

Unlike the corps, Johnson's people did make some modifications to the plan based on issues they had identified. The first problem they had to tackle, one detected before the rehearsal, concerned the Division Ready Brigade. The 3d Brigade held that status in late November but would relinquish it to the 1st Brigade soon thereafter. In all probability, the unit that had actually practiced the mission at Bragg was not likely to be the one to execute it. Yet, given the expense of a rehearsal and the demand such an event generated for resources—some, such as aircraft, beyond the corps' or division's control—there was no practicable way for the 1st Brigade (or the 2d Brigade, for that matter, which would follow the 1st as the ready brigade) to stage a comparable exercise. Thus, as the 3d Brigade prepared for Operation Black Knight, Stiner and Johnson conferred on how to handle the problem, with the corps commander instructing his subordinate "to include his leadership, down to battalion level from the next brigade to assume the Division Ready Brigade mission, which was the 1st Brigade. And to put them in as controllers or something so they could see every aspect of it." Johnson needed no convincing, and the outcome proved highly satisfactory. "We used the 1st Brigade to evaluate the 3d Brigade during the conduct of their EDRE," he observed, after which officers from both

I did not want to, did not want to make it look like we were rehearsing an OPLAN [operations plan] because I didn't want to 'blow' the operation." The brackets are in the original transcript.

[13] Interv, Wright with Johnson, 5 Mar 1990.

[14] First and second quoted words from Interv, Wright with Needham, 6 Mar 1990. Remaining quotes from Interv, Wright with Stiner, 2, 7, 27 Mar and 11 Jun 1990.

units had to "brainstorm" issues raised during the rehearsal, incorporating the agreed-upon solutions into the division's tactical plan.[15]

In mid-December, the changeover in the Division Ready Brigade from the 3d to the 1st Brigade took place. By that time, the 82d's participation in the planning sessions both in Panama and in the United States, together with the dress rehearsal at the Sicily drop zone, had reassured Johnson and those of his officers familiar with the division's potential missions in Panama that the paratroopers were well prepared to accomplish what would be required of them in the event of hostilities. Late on 16 December, that prospect seemed all the more likely, as news of the death of Lieutenant Paz reached Fort Bragg.

DEPLOYMENT

Realizing what the tragic news could portend, Johnson had his chief of staff, Col. George Crocker, gather the division's senior staff officers for a midnight meeting. Johnson used the session to bring the group up-to-date on what had happened in Panama earlier that night. So far as the officers knew, the National Command Authority had not yet had time to assess and determine its response to the incident, but, according to General Kinzer, the officers present began "considering the options. . . . We just kind of reviewed the bidding in terms of our mission and role in a JUST CAUSE–type operation." The meeting lasted under an hour, with Johnson saying he would keep them abreast of further developments.[16]

Sunday evening, after receiving the telephone call from General Kelly in Washington informing him that the president had approved the invasion, Stiner contacted Johnson, who was attending the Winter Formal of the 82d's aviation brigade, and told him to report to a meeting in the airborne corps' planning area at 2200. When Johnson and his operations officer, Lt. Col. Dan McNeill, arrived, most of the other participants had already gathered. This was the meeting at which Johnson suggested using an emergency deployment readiness exercise as the cover story for the troop and aircraft activity that would be taking place at Bragg and Pope within a matter of hours. Among the several advantages offered by this subterfuge, one concerned the news media. Stiner knew that, in the wake of the shooting in Panama, there was no

[15] First quote from Interv, Wright with Stiner, 2, 7, 27 Mar and 11 Jun 1990. Remaining quotes from Interv, Wright with Johnson, 5 Mar 1990. As Johnson made clear in his interview with Robert K. Wright, the continuous changeover in the Division Ready Brigade affected the 82d's planning process as well as the rehearsal. Specifically, once Stiner's planners tapped the division for D-day combat operations, the 2d Brigade commander, Col. Jack Hamilton, and his staff wrote the initial operation plan. When the 3d Brigade became the Division Ready Brigade, its commander, Col. Glynn Walter Hale, and his people "developed a more comprehensive plan for the employment of the brigade and its assets and resources, based on the airlift that we had requested, what we would be able to take with us." Hale also accompanied Johnson to at least two of the planning sessions in Panama, something that Hamilton had not had the opportunity to do. On Hale's second trip, he was accompanied by Col. Jack Nix, the 1st Brigade commander, who, as events turned out, would execute the mission.

[16] Interv, De Mena with Kinzer, 20 Apr 1992.

way of stopping speculation about U.S. military intervention, but Johnson's suggestion at least held out some hope for minimizing rumors until the last minute.[17]

The next morning, at 0600, Johnson met with General Kinzer and Colonel Crocker, telling them, "The president made the decision to execute. We will put the plan in motion for planning and rehearsal." He also informed them about the cover story. As was the case during Stiner's briefing, operations security dominated the discussion, with Johnson concerned "about creating a signature and compromising the exercise."[18] Yet, as they all knew, the twin nemeses of fog and friction would weigh in at some point. What they perhaps did not realize at that early stage of their preparations, those factors were already making their presence felt and would continue to do so with a vengeance. An indication of one immediate disruption to the timetable came with the premature arrival that morning of the C–141s from the Military Airlift Command and the adverse effect this had on the 82d's N-hour sequence. There were also forecasts predicting extremely bad weather to take into account.[19]

At 0900 Monday, the XVIII Airborne Corps operations officer, Col. Thomas H. Needham, formally transmitted to Colonel McNeill, his counterpart in the 82d Airborne Division, an order to execute Operation Plan 90–2. At that time, Johnson called his principal staff and brigade commanders together "to tell them that this was a real-world operation, not to be discussed with anyone, that we would be running an EDRE to cover the operation, that we would brief our troops at battalion level and below once we got into the marshaling area." Until then, such as at an 1100 briefing required by the N-hour alert sequence, Johnson "used the same scenario as the one that we ran in the rehearsal to be executed here at Fort Bragg, just your normal EDRE, and just treated it as a normal training and readiness exercise using the troop list; that which was to be rigged, and that way we got the rigging started, and that way we got the troops down in the marshaling area isolated, so that we could then brief them and begin the troop-leading procedures and rehearsals, the issue of ammunition and MCI [mission contingency items], pre-jump training, and all the rest that goes into a real-world deployment." Among the officers attending the briefings was Col. Jack P. Nix Jr., who, as commander of the 1st Brigade, had just days before seen his unit assume the role of Division Ready Brigade.[20]

At some point during this avalanche of information Monday morning, General Kinzer learned that he would lead the division advance party to Panama that afternoon. His entourage of about thirteen officers (including the division's provost marshal), noncommissioned officers, and enlisted

---

[17] See Chapter 2 in this volume for more details on the Sunday evening meeting at Fort Bragg. See also Intervs, Wright with Stiner, 2, 7, 27 Mar and 11 Jun 1990; with Johnson, 5 Mar 1990; Operation Just Cause: Significant Activities, n.d. [XVIII Abn Corps chronology of events]; Corps Historian's Personal Notes Recorded During the Operation [Just Cause], n.d.

[18] First quote from Interv, De Mena with Kinzer, 20 Apr 1992. Second quote from Interv, Wright with Johnson, 5 Mar 1990.

[19] For the C–141 issue and the weather forecasts, see Chapter 2 of this volume.

[20] Quotes from Interv, Wright with Johnson, 5 Mar 1990. Interv, De Mena with Kinzer, 20 Apr 1992; Corps Historian's Personal Notes Recorded During the Operation [Just Cause], n.d.; Flanagan, *Battle for Panama*, pp. 58–59.

men would not constitute the full "package" for the assault command post that Johnson would need. The group did include enough people, once they arrived at Howard Air Force Base, moved to Fort Clayton, and linked up with a few of the division's staff members who had preceded them, to begin setting up a tactical operations center in Building 200, the headquarters of U.S. Army, South's 193d Infantry Brigade. The arrangement made sense given the plan, once the division's ready brigade and command, control, and communications elements arrived on the scene, to have Task Force Bayonet come under Johnson's operational control in his role as commander of Task Force Pacific.[21]

As the advance party prepared for its departure, Johnson was focused on the airlift operations. Despite last-minute adjustments to the 82d's readiness posture to accommodate the revised airflow schedule, the potential for operational friction remained. This became apparent when the twenty-eight C–141s slated to pick up the ready brigade's heavy equipment and three container delivery planes arrived at Pope during the course of Monday afternoon. In a departure from normal procedure, all of the heavy-drop aircraft had been configured in advance to accept any cargo. The only problem was that the equipment was still not ready for immediate loading, much to the chagrin of the planes' crews. In

Pope Air Force Base

[21] According to Kinzer, he arrived at Howard with General Stiner the evening of the eighteenth. At the air base, all the general officers received a briefing from General Hartzog, the SOUTHCOM operations officer, after which Stiner and General Thurman "issued final planning guidance, and some additional options were discussed, details of the plan updated, latest intelligence information was disseminated." After this 45-minute session, another briefing took place at Fort Clayton, which did not end until 0200 the next morning (see Chapter 2). Interv, De Mena with Kinzer, 20 Apr 1992; Corps Historian's Personal Notes Recorded During the Operation [JUST CAUSE], n.d. Quote from Interv, Wright with Johnson, 5 Mar 1990.

planning sessions with the airborne headquarters at Fort Bragg, the Air Force, it seems, had received the impression that the division would have at least twenty-five loads—enough to fill five Starlifters—pre-rigged and ready to put aboard the aircraft soon after the parties involved had been notified that they would execute the Panama mission. The anticipated twenty-five loads were not ready, however, in part because of the early arrival of the C–141s and in part because the recent handoff of Division Ready Brigade responsibilities from the 3d to the 1st Brigade had not allowed time for the pre-rigging. Once begun, rigging the artillery, Sheridans, various vehicles, ammunition, and supplies proceeded apace, although the steady progress was not without some hitches. As Johnson later lamented, "We did not cover ourselves in glory in some of the rigging areas, particularly in the upload of the Sheridan tanks with live ammunition." In one piece of good news, the forecasted bad weather held off throughout the day, allowing the thirty-one C–141s to depart Pope with the heavy-equipment loads between 2030 on 18 December and 0845 the next morning.[22]

In the midst of these departures, the twenty C–141s that would transport the paratroopers to Panama began to arrive at Pope, with the first touching down at 2115 Monday and the last at 0300 Tuesday. By that time, Colonel Nix was well into the process of assembling the assault force. He had begun soon after being notified of the readiness exercise by having two of his 1st Brigade's battalions, the 1st Battalion, 504th Infantry, and the 2d Battalion, 504th, make the two-mile journey by cargo trucks and trailers from their barracks on Fort Bragg to the personnel holding area at Pope Air Force Base. Because the six Quonset huts and three canvas-covered buildings in the holding area could only accommodate the two battalions, the brigade's third infantry battalion had to stage at the corps marshaling area nearby. Normally, this would have been the 3d Battalion of the 504th, but that unit was already deployed in Panama at the Jungle Operations Training Center on the Atlantic side of the canal area, where it would come under the operational control of Task Force Atlantic. Johnson and Nix thus tapped the 4th Battalion, 325th Infantry, to bring the 1st Brigade up to strength. (That battalion, as it turned out, had a company conducting exercises in the Joint Readiness Training Center at Fort Polk, so Company A, 3d Battalion, 505th Infantry, was attached to round out the 325th.) These three airborne infantry battalions would provide the nucleus of the 82d's combat force, with each being assigned one of the principal objectives targeted for air assaults. Augmenting the force would be two field artillery batteries and a conglomeration of armor, engineer, maintenance, medical, supply, intelligence, signal, and military police units (*Table 2*).[23]

---

[22] Corps Historian's Personal Notes Recorded During the Operation [Just Cause], n.d. Quote from Interv, Wright with Johnson, 5 Mar 1990. Monday night, while some of the heavy-drop C–141s were still being loaded at Pope Air Force Base, other Starlifters arrived from Charleston Air Force Base, picked up a number of HMMWVs, and flew them to Howard Air Force Base in Panama, returning to Charleston early Tuesday morning. Houston Myers, A First Lieutenant Navigator's Experience in Panama (Student paper, U.S. Army Command and General Staff College, Fort Leavenworth, Kans.) Lieutenant Myers was a navigator on one of the C–141s that flew to Howard.

[23] An overview of activities at the personnel holding area at Pope can be found in Flanagan, *Battle for Panama*, pp. 59–60. The order of battle for the 82d Airborne Division (minus) is taken

TABLE 2—82D AIRBORNE DIVISION (MINUS) UNITS FOR TASK FORCE PACIFIC

HEADQUARTERS AND HEADQUARTERS COMPANY, 82D AIRBORNE DIVISION (MINUS)
1ST BRIGADE, 82D AIRBORNE DIVISION (PLUS)
Headquarters and Headquarters Company, 1st Brigade, 82d Airborne Division
1st Battalion, 504th Infantry
2d Battalion, 504th Infantry
4th Battalion, 325th Infantry (minus) (plus)
Company A, 3d Battalion, 505th Infantry
Battery A, 3d Battalion, 319th Field Artillery (minus)
Battery A, 3d Battalion, 4th Air Defense Artillery (minus)
Company C, 3d Battalion, 73d Armor (minus)
Company A, 307th Engineer Battalion
Company A, 782d Maintenance Battalion
Company B, 307th Medical Battalion
Company A, 407th Supply and Service Battalion
Company A, 313th Military Intelligence Battalion
COMPANY B, 82D SIGNAL BATTALION (MINUS)
82D MILITARY POLICE COMPANY (MINUS)

Source: History Office, XVIII Airborne Corps and JTF-South, Operation JUST CAUSE: Task Organization of Joint Task Force South, at http://www.history.army.mil/documents/panama/taskorg.htm.

The paratroopers assembled in the holding area as the rigging of the heavy equipment was still in progress. Throughout the remainder of Monday and well into the next day, they went through the process of drawing ammunition,

from Interv, Wright with Johnson, 5 Mar 1990; and History Office, XVIII Airborne Corps and JTF-South, Operation JUST CAUSE: Task Organization of Joint Task Force South, at http://www.history.army.mil/documents/panama/taskorg.htm.

together with the gear and equipment they had not brought with them from their barracks. Because they would be going from a cold U.S. winter into the tropical heat, Johnson and his staff decided the troops would deploy "light"; they would not be overloaded. Flak jackets, water wings, grenades, claymore mines, and light antitank weapons were not issued. In addition, each paratrooper received less than a basic load for his individual weapon, while mortar and crew-served ammunition was also cut back. None of these items were likely to be needed at Torrijos-Tocumen, a ranger objective, so, in Johnson's words, they could be "dropped [in] by bulk to be issued later." The troops also had some time to put personal affairs in order, just in case they were actually going into combat, something they would not learn until just prior to boarding the transports. Most of the men would have liked to make a telephone call home to talk with their parents or families, but, as they knew, operations security prohibited any contact with the outside world, even during an exercise, once the troops were "locked down" in the holding area. They could, however, receive briefings on the "readiness exercise"; they could also zero in their personal weapons in facilities designated for that purpose. Then, at 1700 Tuesday afternoon, the time came to begin boarding the C–141s, which were waiting nearby at the Green Ramp. In a departure from normal procedure, the paratroopers received their parachutes at their aircraft when they went aboard via the rear ramp to begin their descent into the belly of the plane. Once situated, they waited, crowded together elbow to elbow, for the sound of the plane's engines, which would signal an imminent takeoff. The wait, as it turned out, was a lengthy one.[24]

The "potpourri" of thoughts that raced erratically through one officer's mind as he huddled aboard the transport probably reflected those of his companions. Initially, there was the fact, now known, that this would be a combat mission, and the officer tried to recall details from the briefing charts and acetated maps that he had viewed before boarding. Those images kept being replaced, though, by the faces of his family, friends, and coworkers. Doubts surfaced in his mind as to whether he had taken care of all the personal, family, and financial matters that are part of the preboarding routine. He mentally gave his one-year-old son a hug but then returned to the issue at hand, thinking of how this would be his first combat jump. In the swirl of contradictory emotions surrounding that essential fact, at one moment he considered himself lucky; at another, he acknowledged, "This is not fun." There was one recurring image he found impossible to suppress: "the vision of a burning C–141 with paratroopers scrambling to get out." The Panamanian military, he knew, had effective air defense weapons. At some point, he nodded off and then awakened. His "mental gyrations continued on and on."[25]

Of his concerns, there was one shared by all U.S. officers around the country and in Panama awaiting the deployment of the 82d Airborne Division units: the weather. As Generals Stiner and Kinzer began setting up their operations centers at Fort Clayton; as General Powell and Kelly prepared to monitor

[24] Flanagan, *Battle for Panama,* pp. 59–60. Quotes from Interv, Wright with Johnson, 5 Mar 1990. Corps Historian's Personal Notes Recorded During the Operation [JUST CAUSE], n.d.

[25] Victor M. Rosello (then a major in the 82d Airborne Division), An MI Paratrooper's Memory of Operation JUST CAUSE, n.d.

events from the Pentagon; as General Johnson and Colonel Nix stepped aboard the first and second C–141s in line; as the pilots and crews at Charleston and Pope went through their flight checklists; and as the paratroopers huddled aboard the Starlifters waiting for takeoff—all were grimly aware, directly or indirectly, that the predicted ice storm, the severest one in that part of the country in twenty years, had hit the southeastern coast of the United States. What impact the storm would have on JUST CAUSE deployments was an open question. In Stiner's operations center at Clayton, the senior Air Force officer on the scene, Brig. Gen. Robin Tornow, promised that his service would not let the command group down. He exuded confidence, in contrast to several of the other officers present who appeared less sanguine. At the behest of Colonel Needham, a piece a butcher paper on which the stateside departure of each C–141 could be recorded was taped to a door in the room.[26]

At Charleston Air Force Base, the crews who would fly the heavy-load and container delivery planes received their principal briefing at 1700. When the time came for questions, most centered on the enemy threat, which was reported to include surface-to-air missiles and medium antiaircraft defense weapons. In the estimate of the briefing officers, the airmen could expect to lose one or two planes to enemy action. While absorbing that disconcerting news, the crews were driven through the winter storm to their waiting aircraft, which Charleston maintenance crews began de-icing. According to one crew member, the process proceeded quickly and smoothly, in part because of the professionalism of the maintenance crews and in part because the Air Force base shared runways with a commercial airport that contributed its de-icing trucks and equipment to the effort. As a result, the Charleston contingent of C–141s experienced no major delays in getting airborne.[27]

That, however, did not mean that they were on schedule. Around 2000, the Starlifters began departing Charleston. Because of the ice storm and the heavy equipment on the planes, the takeoff interval between each aircraft was increased from twenty to thirty seconds. While this deviation from the norm seemed virtually imperceptible to the participating crews, the cumulative effect on all thirty-one aircraft was to prolong the time required to join in formation and to reach their high-level cruise altitude. Then, when word reached the crews that H-hour had been advanced by fifteen minutes, they realized just how far behind schedule they were. "In fact," one navigator wrote later, "we knew we wouldn't be able to make it all up, so we flew balls to the wall to do the best we could." The result was "a very high fuel flow burn rate" that made midair refueling seem imperative. According to the same airman, "We were briefed there would be tankers all over the place after we conducted the airdrop," but, the young lieutenant admitted, "my faith in tanker reliability had already been tainted at this short point in my career." Added to these concerns were problems with the station-keeping equipment (SKE)—that is, the use of radio channels to keep the transports in formation during inclement weather. Because of an SKE malfunction in the storm, the last C–141 missed a number of commands, including "the command to descend from a mid-level

[26] General Tornow's comment was overheard by the author, who was admitted as an observer to the corps' operations center at Fort Clayton several hours before Operation JUST CAUSE began.

[27] Myers, A First Lieutenant Navigator's Experience in Panama.

altitude to the low-level structure." Despite these troubles, the formation remained on course, if not on time.[28]

The impact of the ice storm was worse at Pope Air Force Base, where twenty-degree temperatures ensured that any precipitation would quickly become drizzle, ice, and sleet. On the Green Ramp, the troops were covered with ice as they waited in rain jackets to board the C–141 personnel transports. Given the adverse conditions, "somebody made the decision that we would rig aboard the aircraft," a lieutenant colonel in one of the lead planes recalled. "That was a wise decision. . . . It got the troops out of the weather and it protected the parachutes. . . . I remember when we got on . . . somebody had heated up the aircraft. . . . And I thought 'bless you.'" As the soldiers boarded with their heavy rucksacks and other gear, they could see the icicles forming on the planes' tails and were aware of the hurried efforts by airmen from the 317th Tactical Airlift Wing and Combat Support Group to de-ice each Starlifter. At first, only four trucks or so were available for the procedure, but the number doubled as the situation continued to worsen. When some of the aircraft had to be de-iced a second and even a third time, General Johnson and others realized that the planes would not take off on schedule. Johnson conferred with the air mission commander, who was with him on the lead aircraft, and told him, "Let's get going with what we've got." The C–141s would lift off as they were able, with those delayed the longest having to "catch up as best they can." Following this decision, eight Starlifters, not quite half of the troop-carrying serial, began taking off from Pope at 2130, almost an hour behind schedule. As for the remainder, some would not leave the ground for another three hours. In hindsight, Johnson conceded that "legally, I don't know if they should have taken off. It was a safety thing."[29]

In the lead C–141, Johnson had the communications he needed, including a tactical satellite link that enabled him to talk with Colonel Nix's C–141; with General Roosma, Stiner's deputy, aboard a C–130 Airborne Command and Control Center; and with General Stiner in Panama. Once Johnson realized that what was supposed to be a single-column formation of C–141s carrying his paratroopers to Panama was actually going to be up to five separate clusters, he contacted Stiner and recommended that the 82d's airdrop over Torrijos-Tocumen, already behind schedule, be delayed further until the Starlifters in the subsequent serials could catch up with the first, thus allowing the troops to jump simultaneously as planned. Stiner considered the request but, wanting to keep "as closely as possible to the original timetable," decided against it. "Proceed with what you've got," were his instructions to Johnson. As the JTF-South operations officer, Colonel Needham, later explained, "We had a very

[28] Quotes from ibid. Myers contends that the official time listed for the beginning of the takeoffs from Charleston was incorrect. He may have been referring to the time of 2030 listed in two official chronologies that appeared after Just Cause, thirty minutes later than the time he recorded. Operation Just Cause: Significant Activities, n.d. [XVIII Abn Corps chronology of events]; Corps Historian's Personal Notes Recorded During the Operation [Just Cause], n.d.

[29] Donnelly, Roth, and Baker, *Operation Just Cause,* pp. 201–02. Third, fourth, and fifth quotes from Interv, Wright with Johnson, 5 Mar 1990. First and second quotes from Interv, Wright and Maj Dennis Levin with 3d Bn, 319th Field Arty, 8 Feb 1990, Fort Bragg, N.C.; Operation Just Cause: Significant Activities, n.d. [XVIII Abn Corps chronology of events]; Corps Historian's Personal Notes Recorded During the Operation [Just Cause], n.d.

detailed execution checklist. Once you start adjusting times, your execution checklist falls out, and so, I think, that it was pretty much the staff's position and the boss's position—let's stick with the plan."[30]

The heavy-load and container Starlifters arrived in Panama's sky fifteen minutes ahead of the first eight troop carriers. Flying at a low level, the planes made their approach to Torrijos-Tocumen. The leading edge of the runway served as the "point of impact" for the drop; any equipment hitting the ground before that would end up in the mud. The first green light came on around 0158, with the last plane discharging its load seven minutes later. The formation then broke into smaller groups and gained altitude, with the crews looking for the promised refueling tankers only to find that there were none in sight. Cockpit gauges registering low fuel levels made for a nervous return trip, with many of the planes being forced to land at Air Force bases in Florida, instead of flying directly back to Charleston. Only later did the crews learn that the heavy-equipment drop had not gone well.[31]

The plan called for the equipment to be dropped into an area that ran from a line just east of, and parallel to, the runways at Torrijos-Tocumen out to a perimeter fence about fifty yards away. By confining the loads and their parachutes to that drop zone, the facilities' runways would be kept unobstructed for follow-on aircraft scheduled to land at Torrijos-Tocumen after daylight and for planes that might have to divert there if Howard Air Force Base became disabled or overcrowded. Also, as General Johnson observed, "we had helicopters coming in, twenty of them, blades turning, and we didn't want the chutes to get caught in that." The ground

Jump into Night, Torrijos Airport, **by Al Sprague, 1990**

[30] First and second quotes from Interv, Wright with Johnson, 5 Mar 1990. Last quote from Interv, Wright with Needham, 6 Mar 1990. Donnelly, Roth, and Baker, *Operation Just Cause,* p. 202; Cole, *Operation Just Cause*, p. 33.

[31] Operation JUST CAUSE: Significant Activities, n.d.; Corps Historian's Personal Notes Recorded During the Operation [JUST CAUSE], n.d.; Myers, A First Lieutenant Navigator's Experience in Panama.

in the drop zone was hard and mostly flat, ideally suited to receive the impact of rigged heavy equipment. Beyond the perimeter fence, however, elephant grass and swampy areas dominated the terrain. When the C–141s out of Charleston reached the objective, they were flying low over an area that was still very dark. It was "like flying into a black hole," one of the pilots remarked. What Johnson may not have realized was that the Air Force standard for a successful drop was being able to place the cargo within 325 yards of the drop zone. That meant that some of the equipment, perhaps up to 40 percent of it, landed "long," meaning to the north of the airfields or in the grassy and swampy areas outside the fence to the east. This included two of the eight Sheridans and numerous HMMWVs and trucks. Compounding the problem, mechanical difficulties aboard one of the transports caused one HMMWV to be released well beyond the drop zone. Worse, two containers were not dropped at all. The dispersion of the equipment made it highly likely that, once the late-arriving paratroopers were on the ground, there would be even further delays in launching the air assaults against Panama Viejo, Las Tinajitas, and Fort Cimarrón.[32]

This likelihood became a certainty when the paratroopers experienced the same dispersion as the equipment. The troop jump began once the first serial of C–141s arrived over Torrijos-Tocumen at 0211, just minutes after the formation from Charleston had finished the heavy drop. As the first man at the door of the first plane, Johnson "noted that we were already . . . way off to the east and already significantly beyond the southern end of the runway, so I went right away." Others followed, experiencing as had the Rangers who had jumped over an hour before the blast of tropical air, like "a sauna," that greeted them as soon as they exited the planes into the moonlit night. Once clear of their transports, they could see the outlines of the airfield buildings below. The tracers and fires they saw and the shooting they heard indicated that the Rangers had not yet overcome enemy resistance. Torrijos-Tocumen was still "hot." To the paratroopers' immediate relief, they realized that the airfields' defenders were not employing air defense weapons. The nightmare conjured up by that prospect abated, in some cases to be replaced by another,

[32] Operation Just Cause: Significant Activities, n.d. [XVIII Abn Corps chronology of events]. First and second quotes from Interv, Wright with Johnson, 5 Mar 1990. Last quoted word from Corps Historian's Personal Notes Recorded During the Operation [Just Cause], n.d.; Myers, A First Lieutenant Navigator's Experience in Panama. According to one account, some of the Rangers and paratroopers present at Torrijos-Tocumen believed that the dispersion of the equipment could be attributed to the heavy-drop C–141s flying too low and in a diagonal route over the airfields. Donnelly, Roth, and Baker, *Operation Just Cause,* p. 202. According to Johnson in his 5 March 1990 interview, in the days leading up to Operation Just Cause, he had requested that, if the Air Force crews "were going to err," that they "err on the side of flying closer to the runway, not away from the runway." Afterward, he expressed his belief that "they erred flying away from the runway because of a concern they had about leaving equipment and having equipment laying on the runway, which would have rendered the runway unusable." In an interview after Just Cause, Col. James M. Galyen, the air mission commander, indicated that all the heavy loads except for one HMMWV landed in the drop zone, an assessment possibly made on the basis of the Air Force's standard for a successful drop. Portions of the Galyen interview are included in MFR, Lt Col Dunbar, Twenty-First Air Force, to Maj David Huntoon, XVIII Abn Corps, 24 Jul 1990, sub: Request for Information, 998 HMMWV Malfunction During Operation Just Cause.

depending to a large extent on where one landed—on the drop zone or in the swamp and elephant grass. The results of the first serial were mixed.[33]

The same was true of the jumps that followed at 0351, 0400, and 0451. The final group from the 82d did not arrive until 0517, just before dawn, mainly because the seven Starlifters they were in had to make three passes over Torrijos-Tocumen—which required an aerial refueling between the second and third passes—before the troops received the green light. In all of the jumps, a significant portion of the paratroopers—again, up to an estimated 40 percent—landed north or well to the east of their landing zones. The Air Force air mission commander attributed much of the inaccuracy to the fact that the men had been told there would be no "red light" to stop the jump process; they were to continue exiting the aircraft, even if the planes had already flown past the landing zones. Hours would pass before many of the last soldiers to jump would work their way through the swamp and grass to their assembly areas. En route, they generally encountered fellow paratroopers, as well as an occasional enemy soldier or two fleeing the battle. (In cases involving encounters of the latter kind, the paratroopers generally held their fire unless the Panamanians posed a direct threat.) The troops who reached the perimeter fence used it to maneuver to a place where, lacking bolt cutters but using their rucksacks as ladders, they could scale the barrier somewhere near their assembly points. Despite these miscues and difficulties, 2,179 soldiers from the 82d had arrived in the combat zone by 0530 and were preparing to execute their assigned missions.[34]

ON THE GROUND

While descending over Torrijos-Tocumen during the first wave's jump, General Johnson could see heavy equipment "considerably to the right of the runway." He himself landed in the elephant grass but inside the perimeter fence, a fact that allowed him to reach the runway soon after he had shed his parachute and harness. Not every paratrooper was so fortunate. Some landed outside the fence. Some landed on the designated drop zone but suffered foot and leg injuries from the concrete surface. (In all, there were an estimated thirty-seven minor injuries, including broken toes and legs, sprained ankles, and pulled back muscles.) One military intelligence officer during the jump could not lower his 85-pound rucksack, the weight of which caused his

[33] Operation JUST CAUSE: Significant Activities, n.d. [XVIII Abn Corps chronology of events]. First quote from Interv, Wright with Johnson, 5 Mar 1990. Corps Historian's Personal Notes Recorded During the Operation [JUST CAUSE], n.d. Second quote from Rosello, An MI Paratrooper's Memory of Operation JUST CAUSE. Flanagan, *Battle for Panama*, p. 173. According to the 82d's provost marshal's report, "During the parachute assault, the PDF did not use any air defense weapons. The ZPU–4s reportedly at the airfield were not utilized." Rpt, Lt Col David L. Patton, Provost Marshal, 82d Abn Div, Operation JUST CAUSE, 15 Mar 1990.

[34] Corps Historian's Personal Notes Recorded During the Operation [JUST CAUSE], n.d.; Interv, Wright and Levin with 3d Bn, 319th Field Arty, 8 Feb 1990; Donnelly, Roth, and Baker, *Operation Just Cause,* pp. 214–15; Flanagan, *Battle for Panama*, pp. 164–65.

parachute to "oscillate," which in turn caused his head, fortunately covered by a Kevlar helmet, to hit hard on the runway.[35]

Regardless of where a paratrooper landed at Torrijos-Tocumen, the first order of business was to reach the assembly point assigned to his unit. Moving from south to north, the southwest end of the runway was the assembly point for the 2d Battalion, 504th Infantry; the area just north of that toward the center of the runway was for the 1st Battalion, 504th; then came the division assault command post; and finally at the northeast end, the 4th Battalion, 325th Infantry. General Johnson, after making his way out of the grass with a Signal Corps captain he had "picked up . . . along the way," was the first soldier from the 82d to reach the runway. As other paratroopers began to arrive—some after moving through ranger units still engaged in combat operations at Tocumen—the general turned his attention to pulling his assault command post together. Within an hour, he estimated, it was operational. Behind a security perimeter established by a small detachment of military police, he had his operations and intelligence personnel and the radio operators that enabled him to talk directly to General Stiner and to begin sending reports. Other men arrived to round out the group, after which Johnson moved the command post into the Eastern Airlines maintenance building near the main terminal at Torrijos International Airport, "our new Division Headquarters home away from home."[36]

While Johnson and his staff got situated, the three platoons of the military police company that had jumped in the first wave with the general headed to their assigned areas. For the 1st Platoon, that meant splitting into three squads, with each one going to a different infantry battalion assembly area to participate in the planned air assaults. Elements of the 2d Platoon continued to provide security for the assault command post, while other military police set out to secure the airport and to find the transportation they would need to perform their "area security and battlefield circulation control missions." No vehicles had been dropped for the military police by the C–141s, and, because of competing airlift priorities for follow-on troops and equipment, none were expected for several days. Knowing this in advance, the 82d's provost marshal had devised a "fix"; he trained certain personnel to "procure" the necessary vehicles, which essentially meant hot-wiring any civilian cars or vans in the area. While some of the military police in the 2d Platoon were applying these newly acquired skills, the 4th Platoon moved to the main terminal at Torrijos to provide security, to guard the terminal's shopping mall whose duty-free shops were still open but unsupervised, and, most important, to take control of the prisoners and civilians the Rangers had assembled. This last assignment

[35] First quote from Interv, Wright with Johnson, 5 Mar 1990. Second quote from Rosello, An MI Paratrooper's Memory of Operation JUST CAUSE. The injury estimate can be found in MFR, Dunbar for Huntoon, 24 Jul 1990, sub: Request for Information, 998 HMMWV Malfunction During Operation JUST CAUSE.

[36] First quote from Interv, Wright with Johnson, 5 Mar 1990. Second quote from Rosello, An MI Paratrooper's Memory of Operation JUST CAUSE. Rpt, Patton, Operation JUST CAUSE, 15 Mar 1990.

became what in retrospect was described as "clearly the most challenging task" the military police company faced at the airfields.[37]

That challenge emanated from the unexpectedly large numbers of airline passengers and airport employees gathered in the terminal and from the fact that several enemy soldiers had shed their uniforms and were mingling with the civilians. Separating the former from the latter meant interrogating everyone, up to four hundred people. A seven-man counterintelligence and prisoner-of-war interrogation team, which had been task-organized to the military police company's 4th Platoon, did not have nearly enough personnel to carry out the assignment, so civil affairs teams attached to ranger units and two squads from the military police company's 2d Platoon provided needed assistance. Even then, the process lasted all day and into the following night, with the strain visible on the faces of the questioners and their subjects. To mitigate the stress, the intelligence personnel and military police permitted "liberal use of the airport bathroom facilities" and handed out "plenty of MREs [meals, ready to eat]." Enemy soldiers identified during the interrogations were removed from the larger assemblage and taken to the prisoner collection point. As for the civilians, some did not appreciate having been placed in such danger, not to mention having their holiday travel plans disrupted. Others accepted the inevitable with good grace. One U.S. officer, for example, was touched when a Nicaraguan woman told him that, when she returned to her country, "I will tell my story to the Sandinista government. I will tell them that their descriptions of the US Army as a bunch of blood-thirsty killers is false. We have been treated with utmost courtesy the entire time here. Thank you." This gratitude and the display of respect that accompanied it reportedly mollified many other resentful and disgruntled civilians caught up in the ordeal.[38]

Downstairs from the interrogation area in the terminal, medics from the 82d set up what was supposed to be a temporary dispensary but which, over the next few days, became a "fully functioning field hospital." During the dispensary's operation, two Panamanian babies would be born, and military personnel and civilians could be treated for a variety of ailments. One officer recalled the incongruous sight of one of his men grimacing from receiving a gamma globulin shot in his backside, while just ten yards away a pregnant woman was receiving a pelvic examination.[39]

As paratroopers began replacing the Rangers in the main terminal, in the Eastern Airlines building next door, Johnson was acquainting himself with the military situation on the ground in and around Panama City. As soon as he had operable communications, he talked with the JTF-South operations center and with his division's tactical command post in Building 200 at Fort Clayton. At the latter location, General Kinzer had been closely monitoring the activities of Task Force Bayonet in Panama City and Task Force Semper Fi on the west side of the canal. Although both groups were slated to come under the operational control of Johnson's Task Force Pacific once the division commander had established his headquarters at Torrijos-Tocumen,

---

[37] Rpt, Patton, Operation JUST CAUSE, 15 Mar 1990.

[38] Ibid. Quotes from Rosello, An MI Paratrooper's Memory of Operation JUST CAUSE.

[39] Rosello, An MI Paratrooper's Memory of Operation JUST CAUSE.

Kinzer was having second thoughts about the arrangement. As the general later explained, "It just appeared to me not to be prudent to change command relationships as the battle was ongoing." Both task forces had "real fights" on their hands and both were faring very well. "If we were to change [the existing] relationship and plug it into the 82nd Division, and had everybody change their lines of commo," he continued, "I felt it would be counterproductive and confusing." Even before Johnson's arrival in-country, Kinzer conveyed his assessment to Colonel Needham in the JTF-South operations center and convinced him that the command arrangements for Task Forces Bayonet and Semper Fi should not change. Needham passed the request to Stiner, who agreed, thus settling the issue. When Johnson arrived and was informed of Kinzer's recommendation, he readily concurred. For the next few days, the two task forces in question would continue to work directly for the JTF-South commander. Such was not the case with the Rangers at Torrijos-Tocumen, however. According to plan, they came under Johnson's operational control once the objective was pronounced secured. The formal time for the change in the command relationship was recorded as 1100 on the twentieth.[40]

Throughout the predawn hours, while the remaining four C–141 serials dropped the rest of the 1st Brigade and supporting units, several paratroopers on the ground, especially those in the armor and artillery elements that would move to the division's three objectives in overland convoys, received orders to find the heavy equipment that had fallen beyond the designated drop zones. Each of the packages had been rigged for a night drop, which meant that they had chemical lights and other devices that would enable soldiers to locate them. "The problem," according to one paratrooper, "was the grass was so high, and unless we happened to run—physically run—into the platform, we would not know it was there." Some loads proved readily retrievable. Others took hours to find. Of the eight Sheridans dropped around 0200, the first one was not found until 0600. Another had been disabled in the drop, and a third was discovered stuck in swampy mud so thick and deep that the vehicle could not be extracted and had to be destroyed. As for the four M102 105-mm. howitzers dropped that morning, not one had been retrieved at the time the artillerymen from Battery A, 3d Battalion, 319th Field Artillery, began to assemble on the runway. In fact, the first piece, once recovered, was not ready to fire until 0800; a second piece by 0915. (These were the only two tubes that would deploy with the Fort Cimarrón convoy later that morning.) Finally, of the nearly eighty HMMWVs dropped, most were found to be in usable condition. One that had landed upside down could not be employed on D-day, while another was simply lost, with one report claiming that it had been dropped in the ocean. (Seven months later, Kuna Indians would discover the vehicle in the swamps and report it to USARSO's 193d Brigade; the vehicle was extracted with all its weapons—except four AT4 rockets—and equipment inside.)[41]

---

[40] Quotes from Interv, De Mena with Kinzer, 20 Apr 1992. Interv, author with Brig Gen Joseph W. Kinzer, U.S. Army, 1990, Fort Clayton, Panama.

[41] Quotes from Interv, Wright and Levin with 3d Bn, 319th Field Arty, 8 Feb 1990. Donnelly, Roth, and Baker, *Operation Just Cause,* p. 203; Corps Historian's Personal Notes Recorded During the Operation [Just Cause], n.d. Information on the discovery of the missing

While the search for the heavy equipment was under way but before all the paratroopers had dropped into Torrijos-Tocumen, General Johnson concluded that the 2d Battalion, 504th, had assembled enough manpower on the runway around 0400 to proceed with its air assault mission at Panama Viejo. But when he called Stiner and requested the necessary airlift, including a command-and-control helicopter, he learned that the aircraft were not available. Also conveyed to him was a concern about bringing in the required Black Hawk helicopters while two to three serials of troop transports were still en route from Pope. So Johnson waited while more and more of his soldiers and their equipment reached their assembly points. By dawn, the last of the paratroopers had jumped, and he could survey the status of each unit "just by looking up and down the runway." Both the 1st and 2d Battalions had assembled about 80 percent of their personnel and equipment and were ready to move out. More time passed, though, before the Black Hawks, the air mission commander, and Colonel Nix, the brigade commander, were ready to begin. As feared, the unfortunate combination of the ice storm and the errant equipment and personnel drops meant that all three air assaults would occur during daylight.[42]

THE AIR ASSAULTS

The 82d's air assaults did not launch simultaneously, but in a sequence determined by the priority given to each of the targets (*Map 14*). During most of the planning process, Fort Cimarrón occupied first place on the list thanks to the presence there of the mechanized, motorized, and armored Battalion 2000. After the attempted coup of 3 October and the relocation of part of the PDF's better-armed and -equipped UESAT to the military garrison at Panama Viejo, the home of the PDF 1st Cavalry Squadron, that location on the Bay of Panama replaced Fort Cimarrón at the top of the list. In fact, with the concurrent movement of several Battalion 2000 units to the Comandancia after the failed coup, the fort fell to last place. On the morning of 20 December, therefore, the 82d's targets, from first to last in priority, were Panama Viejo, Las Tinajitas, and Fort Cimarrón. Around 0630, Colonel Nix's 2d Battalion was ready to take off for the first of the three.[43]

The plan was elementary in concept. Two of the battalion's companies, A and C, would be airlifted onto two landing zones, Bobcat and Lion. The first was a strip of grass north of the PDF barracks and training center; the second on the shoreline to the target's south. Intelligence estimates indicated that, as the paratroopers arrived, they would be confronted by up to 250 Panamanians (180 from the cavalry squadron, 70 from UESAT), armed with small arms, sniper weapons, rockets, explosives, two V300s, and a ZPU4 antiaircraft platform. Once the U.S. soldiers were on the ground, according to Lt. Col. Harry B. Axson, the battalion commander, they would surround the garrison and employ a psychological operations loudspeaker team to persuade the enemy to surrender. In the meantime, Axson's Company D would be heading

HMMWV can be found in one of the attachments to MFR, Dunbar for Huntoon, 24 Jul 1990, sub: Request for Information, 998 HMMWV Malfunction During Operation JUST CAUSE.

[42] Interv, Wright with Johnson, 5 Mar 1990.

[43] Ibid.

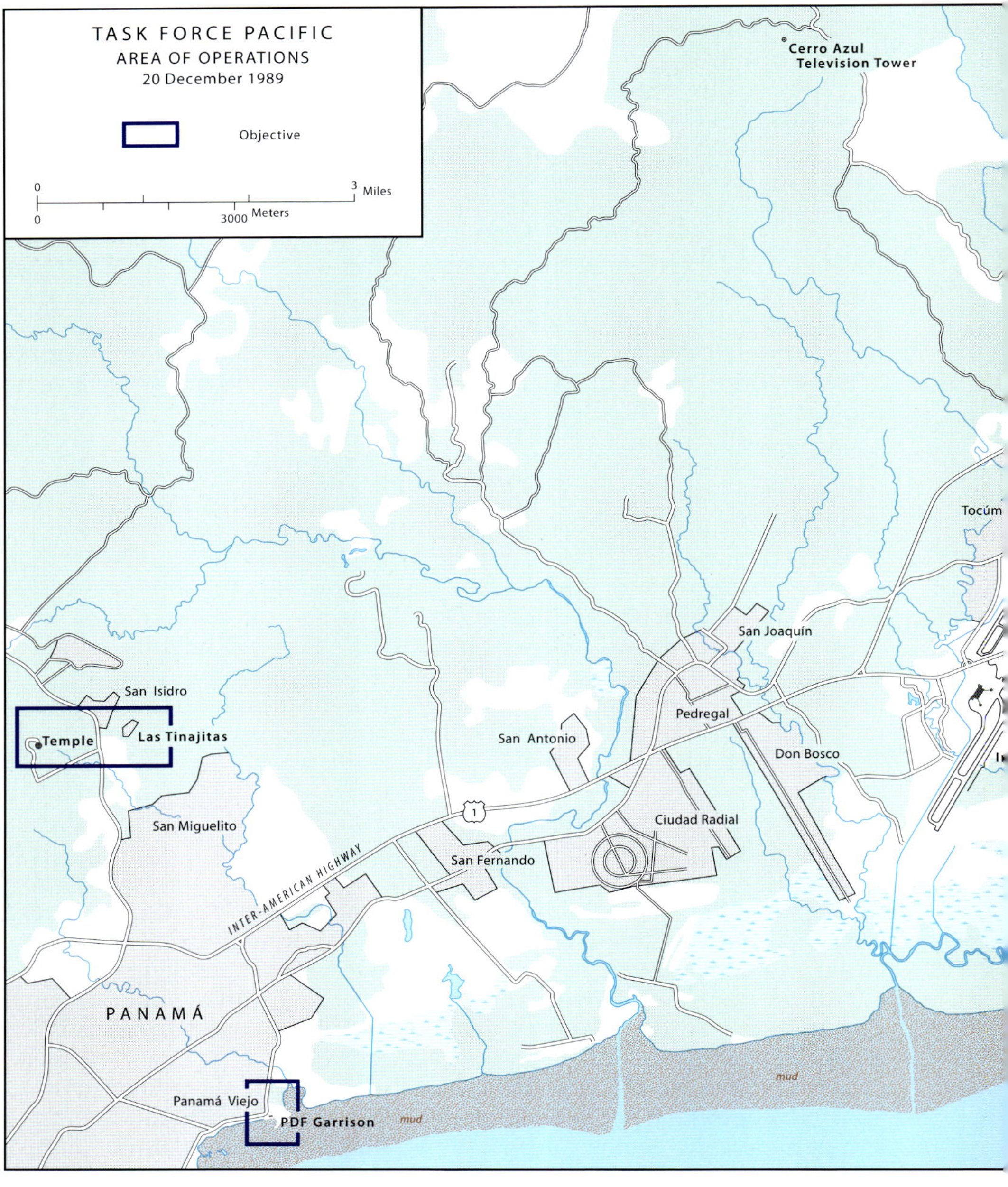
TASK FORCE PACIFIC
AREA OF OPERATIONS
20 December 1989
Objective
0
3 Miles
0
3000 Meters
Cerro Azul
Television Tower
Tocúm
San Joaquín
San Isidro
Temple
Las Tinajitas
San Antonio
Pedregal
Don Bosco
San Miguelito
Ciudad Radial
1
San Fernando
INTER-AMERICAN HIGHWAY
PANAMÁ
mud
Panamá Viejo
PDF Garrison
mud

Map 14

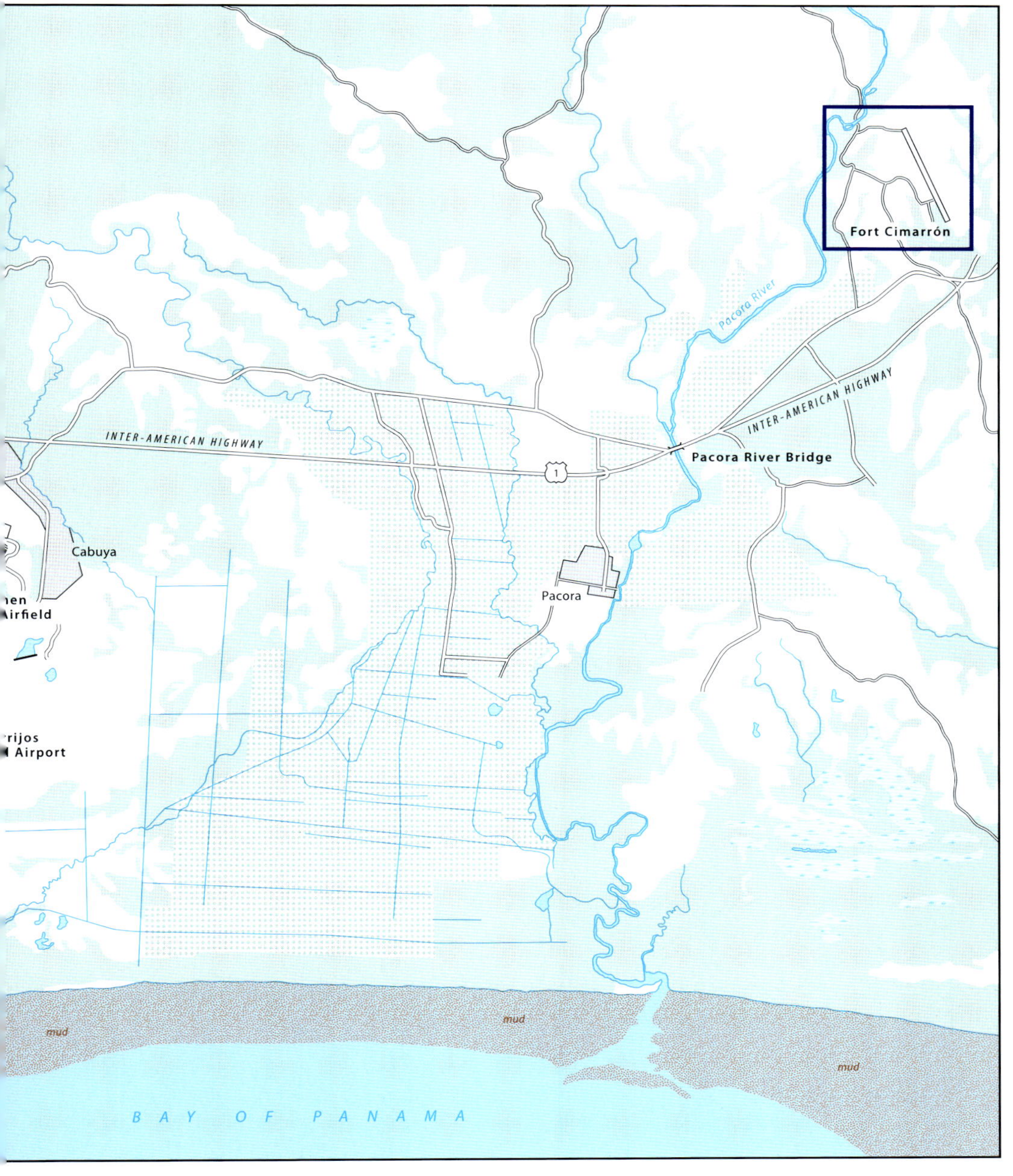
Fort Cimarrón
Pacora River
INTER-AMERICAN HIGHWAY
INTER-AMERICAN HIGHWAY
1
Pacora River Bridge
Cabuya
Pacora
ıen
\irfield
rijos
Airport
mud
mud
mud
B A Y O F P A N A M A

The PDF 1st Cavalry Squadron compound at Panama Viejo. After seizing the compound, the paratroopers used the ruins next to it as their tactical operations center.

to the objective in a ground convoy consisting of four .50-caliber mounted HMMWVs, the remainder of the battalion's mortar platoon, two Sheridans, a command-and-control HMMWV, and a cargo HMMWV. Once the convoy linked up with the air assault element, the attacking force would have ample firepower to defeat the enemy if the surrender appeals had gone unheeded.[44]

Across the canal, on the Empire Range, the Black Hawks that would fly the battalion's two companies to their landing zones were awaiting word to pick up the troops. Several of the helicopters belonging to the Pacific-side contingent of Task Force Aviation had participated in the air assault at Fort Amador and had refueled at the range, the crews thankful they had made it thus far without being hit. Other crews at the range would be flying their first mission in JUST CAUSE. All were apprehensive that, in the fast-approaching operation, enemy antiaircraft fire would become more intense and more accurate than had been experienced to that point. These fears escalated when word arrived that the 82d had been delayed and that air assaults scheduled for nighttime would now take place after dawn. As one flight platoon leader put it, "finding out that we were going to be doing this at daylight right away increased the 'pucker factor' so to speak because I knew we'd be sitting ducks."[45]

---

[44] "The Day Of Bronze Stars," in *Soldiers in Panama: Stories of Operation Just Cause* (Washington, D.C.: Office of the Chief of Public Affairs, Department of the Army, 1990), p. 22; Rpt, Patton, Operation JUST CAUSE, 15 Mar 1990; Briefing Slides, Committed PDF Forces, n.d.; Donnelly, Roth, and Baker, *Operation Just Cause,* p. 217; Flanagan, *Battle for Panama*, pp. 176–77.

[45] Quote from Interv, Wright with 1st Lt Lisa M. Kutschera, U.S. Army, 8 Jan 1990, Fort Kobbe, Panama. Interv, Wright with Lt Col Douglas I. Smith, U.S. Army, 12 Jan 1990, Fort Kobbe, Panama.

UH–60 Black Hawk helicopter in Panama

When the first group of Black Hawks finally took off and arrived at Torrijos-Tocumen to pick up as many troops as possible from the 2d Battalion and attached elements, several portents added to this uneasiness. For starters, as the lead helicopters flew in, their rotor wash caused discarded parachutes lying by the runway to start floating in the air, compelling the crews in the following craft to engage in some split-second maneuvering to dodge the debris and keep it from getting tangled in their blades. Once on the ground, loading soldiers and equipment took longer than expected, largely because only nine of the required twenty-one Black Hawks had arrived and because the troops were lined up to take their seats on the helicopters, as they had practiced during exercises at Fort Bragg, only to discover that, for going into combat, the seats had been removed. "It took them a while to get rearranged and get themselves and all the gear on to the aircraft," one pilot recalled. In the process, anywhere from twenty to thirty paratroopers, instead of the usual eleven, found themselves crammed together on a single aircraft.[46]

The ten-minute flight from the pickup zone to the two landing zones at Panama Viejo was uneventful save for the shared awareness aboard the helicopters that the defenders would be able to see the Black Hawks coming. Just how many enemy soldiers would actually be at their posts and prepared to fight was anyone's guess and another source of anxiety. Unknown to the 2d Battalion, about half the garrison at Panama Viejo had donned civilian clothes and, around 2300, had headed north and out of the area with their weapons. About fifteen to twenty Panamanians remained in the barracks, while many more occupied civilian homes in the vicinity. Nix's paratroopers

[46] Quotes from Interv, Wright with Kutschera, 8 Jan 1990. Flanagan, *Battle for Panama*, p. 176.

would not be facing the large force they were anticipating. Those enemy combatants who had not fled, however, were well armed. They were also well positioned in that the proximity of the 1st Cavalry Squadron compound to Panama City had greatly narrowed the 82d's selection of landing zones near the target—in General Johnson's words, the choice was less than "a piece of cake." Though the paratroopers approaching Panama Viejo did not fully realize just how precarious the location of the two landing zones was, most of them still experienced a strong sense of foreboding that things were about to get "hairy."[47]

The fears proved warranted, but only partly because of the resistance mounted by the Panamanians. Around 0700, as the paratroopers reached their landing zones, all the helicopters came under small-arms fire, but this turned out not to be the paratroopers' principal problem. At Bobcat, the landing zone consisted of grass growing well above the head of the soldiers, a condition not detected during pre–Just Cause overhead reconnaissance flights, usually conducted at night with the aid of night-vision goggles. Once in the grass, the paratroopers lost sight of one another; moreover, they found it impossible to determine the positions of enemy troops firing at them. The situation was even worse at Landing Zone Lion. By the time the assault got under way, the tide from the Bay of Panama had ebbed. From the air, the mud flats revealed by the receding water appeared hard and flat. When the men of Company A dropped with their gear from the Black Hawks, many found themselves up to their chests and armpits in a dark ooze that seemed to suck them in. As the men struggled, one Black Hawk remained behind, hovering low over the mud flats while crew members tried to pull the soldiers onto solid ground.[48]

During the time U.S. forces at both landing zones were under fire, the defenders did not open up with a .50-caliber machine gun they had on the barracks roof, or with the ZPU4 antiaircraft weapon near the beach. (Some paratroopers at Bobcat later claimed that they had heard the ZPU fire three rounds before jamming.) If either weapon had been employed, the result would have been a "nightmare" or "bloodbath." The small-arms fire was intense, yet even when a PDF position was detected, the rules of engagement often precluded bringing it under fire. There were simply too many civilians in the area, many of whom had gathered outside to view the American attack. The constraints on returning enemy fire if innocent bystanders might be harmed extended as well to the Cobra and two Apache attack helicopters flying overwatch positions above Panama Viejo. While cursing the civilian presence that initially seemed to aid the defenders, however unintentionally, the troops and helicopter crews soon had reason to be grateful, as many of the onlookers formed human chains, using their hands and ropes to extract the immobile paratroopers at Landing Zone Lion from the mire.[49]

---

[47] Rpt, Patton, Operation Just Cause, 15 Mar 1990. Quotes from Interv, Wright with Johnson, 5 Mar 1990.

[48] "The Day Of Bronze Stars," in *Soldiers in Panama,* p. 23; Flanagan, *Battle for Panama,* pp. 178–79; Donnelly, Roth, and Baker, *Operation Just Cause,* pp. 217, 219.

[49] First quoted word from Flanagan, *Battle for Panama,* p. 178. Second quoted word from Donnelly, Roth, and Baker, *Operation Just Cause,* p. 219. Lieutenant Kutschera elaborated on the rules of engagement for the aviators. "We weren't supposed to return the fire unless we could

While the drama at Bobcat and Lion was unfolding, the Black Hawks had returned to Torrijos-Tocumen to pick up another load of paratroopers—soldiers, the crews believed, who were waiting to launch against Las Tinajitas. Instead, the pilots received word that Colonel Nix wanted the remainder of the 2d Battalion air assault force, meaning the rest of Companies A and C, the mortar platoon, and now some elements from Company B as well, taken to Panama Viejo first. The news was received with little enthusiasm. For the returning Black Hawk crews, it meant retracing their route to two hot landing zones. "It would give everybody that missed us the first time a second chance," one pilot observed sardonically. Indeed, after depositing the troops during the second lift, three of the Black Hawks were hit. When the helicopters returned to Tocumen, the air mission commander ordered them shut down and checked out for damage. The brief respite before those that were still able to fly took off again would be used by the paratroopers, who one pilot described as "rather disorganized at that point," to improve their "PZ posture."[50]

After the second and third troop lifts to Panama Viejo, there were five hundred U.S. soldiers at the objective, enough to defeat any Panamanian force in the vicinity. Moreover, the paratroopers at Bobcat had managed to advance from their landing zone after one of the sergeants in Company C finally located an isolated enemy position and, through the use of a grenade launcher and directed machine gun fire, neutralized what had been a source of heavy and continuous fire. That notwithstanding, there was still a good deal of confusion, especially when elements of Company B landed at Lion instead of their intended landing zone, Bobcat, thus splitting the unit and commingling it with Companies A and C. In time, Colonel Axson sorted out the command-and-control problems, and the three companies began maneuvering against the PDF compound, pausing at times to exchange fire with groups of determined defenders. Once Company A gained a foothold in the PDF headquarters building and Company B had caused enemy troops within the barracks area to flee into civilian neighborhoods nearby, Company C could begin clearing the remaining buildings.[51]

While 2d Battalion's paratroopers were engaging the enemy at Panama Viejo, the unit's ground convoy of HMMWVs and Sheridans that had started out from Torrijos-Tocumen at 0910 ran into a roadblock near a bridge. Two burned-out cars stopped the column's movement, while an estimated forty Panamanians armed with automatic weapons and rocket-propelled grenades opened fire from a building close by, forcing the U.S. troops to dismount their

definitely identify who it was, if it was any hostile fire [coming] our way, and shoot only as much as it took to suppress that fire. Because they did tell us that after this was all over with, we're going to try and convince the good, honest members of P.D.F. to back the new government, the elected [Endara] government, and that by us just continuing to hammer them was going to make that a lot more difficult. So it was kind of a difficult situation for us in not wanting to hurt any innocent civilians, not wanting to hurt anybody that wasn't hurting us, but still having to protect our own lives and aircraft." Interv, Wright with Kutschera, 8 Jan 1990.

[50] Flanagan, *Battle for Panama*, p. 179. Quotes from Interv, Wright with Kutschera, 8 Jan 1990.

[51] "The Day Of Bronze Stars," *Soldiers in Panama,* pp. 22–23; Flanagan, *Battle for Panama*, p. 179.

vehicles. A platoon from Company D removed the roadblock, but, during the firefight, an enemy machine gun in the building claimed the life of Spec. Alejandro J. Manrique-Lozano. The platoon leader responded by destroying the machine gun with a LAW, after which his men entered the building. In the room-to-room fighting required to clear the premises, eleven Panamanians were killed. Two others were killed behind the structure. Shooting near the bridge continued, however, with the remaining defenders being routed only after reinforcements consisting of forty men and an additional Sheridan arrived from Torrijos-Tocumen. Once the firing subsided, the Company D commander ordered his men back in the HMMWVs, but, instead of proceeding to Panama Viejo, the convoy received orders to return to the airport complex, where it linked up with the convoy from the 1st Battalion getting ready to set out for Las Tinajitas.[52]

At 1140, Colonel Axson reported the Panama Viejo garrison under the control of American troops, who had captured eight of the enemy. Real and potential threats to U.S. forces, however, were still in evidence. There was, for starters, the group of armed PDF soldiers from the cavalry squadron who arrived at their headquarters Wednesday morning apparently oblivious to the fact that the United States had invaded their country. Outnumbered and incredulous, most offered no resistance before surrendering. More serious were the intermittent sniper and mortar rounds, as well as the drive-by shootings directed at American positions throughout the day. In most cases, the drive-bys proved little more than suicide missions, as U.S. forces successfully repulsed nine of them, usually with well-aimed and deadly small-arms volleys. In a slightly more imposing show of PDF resistance, an armored V300 headed directly for an American position but was disabled by a soldier standing in its path with an antitank rocket. Later, the two Apaches flying above Panama Viejo in overwatch positions made their own contribution to disarming the enemy by destroying a "miniature motor pool" composed of two trucks and two V300s, all camouflaged. By Wednesday evening, the battalion's perimeter at Panama Viejo was well established, and the area was generally quiet.[53]

As Axson's men were clearing Panama Viejo during the afternoon and evening of the twentieth, word spread among them about the deadly situation the brigade's 1st Battalion had encountered in its efforts to subdue the PDF 1st Infantry Company at Las Tinajitas, the second target on the 82d's priority list. That air assault mission had begun sometime after 0800 with the first of three troop lifts that employed some of the same Black Hawks that had carried

---

[52] Flanagan, *Battle for Panama*, pp. 181–82; Corps Historian's Personal Notes Recorded During the Operation [Just Cause], n.d.

[53] "The Day Of Bronze Stars," *Soldiers in Panama*, p. 23; Flanagan, *Battle for Panama*, pp. 181–82; Corps Historian's Personal Notes Recorded During the Operation [Just Cause], n.d.; Operation Just Cause: Significant Activities, n.d. [XVIII Abn Corps chronology of events]. Quote from Maj Michael L. Hedegaard, Task Force Wolf in Operation Just Cause: First AH–64 Apache Helicopters in Combat (Student paper, U.S. Army Command and General Staff College, 2 Mar 2001). Major Hedegaard's paper draws on his own experiences together with interviews he conducted with eight participants in aviation operations during Operation Just Cause.

The PDF 1st Infantry Company compound at Las Tinajitas, with a residential area below to the left.

the 2d Battalion to Panama Viejo.[54] Before the morning was over, the fighting at Las Tinajitas would take the lives of two more paratroopers. A much larger number of U.S. airmen and soldiers would be wounded in the battle.

What made the 184-man PDF 1st Infantry Company a high-priority target in Operation JUST CAUSE was the unit's mortar inventory, especially the 81-mm. medium mortars and 120-mm. heavy mortars capable of disrupting U.S. activities at numerous locations, including Howard Air Force Base, Torrijos-Tocumen, Fort Clayton, and Quarry Heights. The mission of the 82d's 1st Battalion, 504th Infantry, therefore, was to "isolate, neutralize, and/or destroy" the enemy company.[55] Obstacles to achieving this objective included the inaccessible location of the company's concrete garrison near the top of a high hill surrounded at the base by ring upon ring of civilian shanties, particularly in the poor, pro-Noriega San Miguelito neighborhood. As at Panama Viejo, the presence of the housing areas negated the use of indirect preparatory fires against the target, restricted the landing zones that the Black Hawks could

[54] Sources differ as to the exact time the Las Tinajitas airlift began, although all have it starting between 0800 and 0830.

[55] JTF-South OPLAN 90–2, 3 Nov 1989.

use, and limited the ability of door gunners and soldiers to return fire against suspected enemy positions.

With these complications in mind, the 1st Battalion commander, Lt. Col. Renard Marable, sought to implement a plan whereby his troops would be taken by Black Hawks to two landing zones. The bulk of the force would land at Landing Zone Leopard, located in a ravine about 750 yards south of the garrisoned hill. One company would then fix the PDF compound from the front, as the battalion's other infantry companies maneuvered to outflank it. Meanwhile, a single helicopter would proceed farther west to Landing Zone Jaguar, dropping a machine gun element on a ridge near another hill on which the PDF was suspected of using a religious temple as a mortar site. The task of the machine gunners was to eliminate the mortar emplacement and provide covering fire for the main attack. Overhead, Kiowa, Cobra, and Apache helicopters would provide assistance as needed.[56]

The actual operation, in the understatement of one participant, "turned out to be just a tad different" from the basic plan. The first of three troop lifts out of Torrijos-Tocumen—with each lift employing six Black Hawks—made its way to Panama City, then turned northward toward Las Tinajitas. Throughout the ten-minute flight, two Cobras provided an escort. Meanwhile, a scout-gun team of two Apaches and a Kiowa flew over San Miguelito and Las Tinajitas from the northwest. Neither the troop carriers nor the attack helicopters encountered enemy fire. Indeed, a paratrooper aboard one of the Black Hawks looked down on the early morning skyline of Panama City and the water sparkling in the sun and thought, "Wow, that's a pretty sight." Given the beautiful panorama, the smoke rising from certain battle sites in and near the city seemed "weird" to him. Other passengers with a view of the city recalled seeing Panamanians waving at them. The deceptive calm then gave way to chaos as the Apaches and the Kiowa flew back over the houses of San Miguelito and came under fire from a machine gun position concealed on a hilltop—apparently at the temple suspected of harboring the mortar position—just northwest of the residential neighborhood. The Kiowa took the first hits as it flew lower than a hundred feet over the gun emplacement. Both Apaches in the team suffered a similar fate. None of the scout-gun team was shot down but damage to the Kiowa and one Apache forced them to return to Howard Air Force Base. As they departed the area, an attack team of additional Apaches arrived on the scene and began looking for the weapon that had caused so much havoc. Requiring a specific military target in their sights before they could open fire, the crew of the lead AH–64 finally spotted the uniformed PDF machine gun crew and, with permission from the task force commander at Howard, opened up with bursts from their 30-mm. cannon, followed by a 2.75-inch rocket. The machine gun position was destroyed, with ten of the eleven Panamanians manning the position killed.[57]

[56] Interv, Wright with Capt Dave Taylor Jr., U.S. Army, 21 May 1990, Fort Bragg, N.C.; Flanagan, *Battle for Panama*, pp. 192–93; Donnelly, Roth, and Baker, *Operation Just Cause,* pp. 222–23.

[57] Rpt, Patton, Operation Just Cause, 15 Mar 1990. Quotes from Interv, Wright with Taylor, 21 May 1990. Interv, Wright with Kutschera, 8 Jan 1990; Hedegaard, Task Force Wolf in Operation Just Cause.

Air Assault, Tinajitas, **by Al Sprague, 1990**

Soon thereafter, as the Black Hawks entered the same airspace in what one pilot would call the "hottest" of the 82d's landing zones, the paratrooper who had minutes before been marveling at the skyscrapers of Panama City began hearing sounds like popcorn popping, "you know, the tic-tic-tic-tic thing." The sounds, he quickly realized, were AK47 rounds "bouncing off our helicopter." In several instances, the fire was accurate and effective, as crew members and infantrymen received wounds from bullets coming through the floor, prompting numerous after action comments about the need to outfit assault helicopters with Kevlar floor blankets. The worst injury occurred aboard the lead Black Hawk, where the officer in charge of the formation, Capt. Thomas Muir, had been shot in the head and was bleeding profusely. When word of his wound was radioed to the other pilots, one, 1st Lt. Lisa M. Kutschera, "realized in my gut" what before she had known only "in my head that 'yeah we can get killed doing this.' But that was when it really, really hit me that . . . yeah, we can die out here." Some Black Hawks, unable to land, hovered over Landing Zone Leopard, causing some paratroopers vacating the aircraft to jump from a height of thirty feet into eight-foot-high grass. In the haste to exit the helicopters and find safety on the ground, several troops inadvertently left their gear behind. The Black Hawks then pulled away, having to do a steep climb to avoid power lines strung over two hundred feet above the ground and across the designated departure routes. "We couldn't really fly under them because there was another set of wires—shorter wires—underneath them," Kutschera later related. "So we basically had to climb and make a good target out of ourselves to get out of there." The two Cobra escorts would follow suit, but not before they fired a half-dozen 2.75-inch rockets into the hilltop garrison, helping to suppress the steady volume of enemy fire. The Black Hawk with Captain Muir flew to Howard Air Force Base, where, after medical treatment, the officer survived his near-fatal wound. The other crews returned to Torrijos-Tocumen to examine their aircraft, several of which had been hit. When the damage was not so extensive as to prohibit flying, the pilots began preparing for further missions.[58]

Back at Landing Zone Leopard, the battalion had suffered its first fatality, Pfc. Martin D. Denson, who had been shot while exiting a Black Hawk. Seriously wounded, he was then hit by mortar fragments and later died. Enemy mortar and automatic weapons fire from the PDF garrison and from civilian buildings around the landing zone kept the other paratroopers pinned down, even as they tried to maneuver to a location from which they could get their bearings. The Panamanian defenders "had our number," one soldier was convinced, and were prevented from inflicting heavier casualties only by the actions of the Apaches and Cobras, which "suppressed those guys long enough for us to get in alive." Once all three companies were on the ground, they began working their way up a hill near the target, moving as quickly as they could through the unexpectedly steep and difficult terrain, thus forcing the PDF mortar crews to adjust their weapons continuously. The original plan for fixing and flanking the Las Tinajitas garrison gave way to one in which each company just did what it could to reach high ground. In the process, a second U.S. soldier at Las Tinajitas, Pfc. Jerry Scott Daves, received a fatal wound. While a medic was treating him on the hillside for a gunshot to the head, a mortar round went off next to them, killing Daves and wounding the medic. Rescue helicopters arriving to evacuate the wounded had to be waved off because of the heavy volume of enemy fire. After an exhausting climb, the Americans took a hilltop across from the PDF garrison. Two hours later, when they reached the garrison itself, the defenders had fled. At 1433, Colonel Marable declared the area secured, although, as at Panama Viejo, sniper fire continued for the remainder of the day.[59]

When the fighting at Las Tinajitas and Panama Viejo was later analyzed, several obvious similarities emerged. Perhaps the principal one had to do with the vulnerability of the helicopters during a daylight airborne assault. Too many Black Hawks had been shot up, although none had been shot down. No crew members had been killed, but a number had suffered wounds, as had several of the troops aboard. The attack helicopters had fared no better. Designed for night fighting, both overwatch Apaches at Las Tinajitas had been struck by machine gun fire as they flew close to enemy positions

---

[58] Donnelly, Roth, and Baker, *Operation Just Cause,* pp. 223–25. First quoted word from Flanagan, *Battle for Panama,* p. 193, and see also pp. 192, 194. Second and third quotes from Interv, Wright with Taylor, 21 May 1990. Remaining quotes from Interv, Wright with Kutschera, 8 Jan 1990. Flanagan writes that a SOUTHCOM spokesperson later stated that two of the Black Hawks flown into the hot landing zones at Las Tinajitas were piloted and commanded by female aviators. Flanagan further observes that the 7th Infantry Division's aviation brigade had twenty females, four of whom were pilots (p. 195). On a related note, Task Force Wildcat's Capt. Linda Bray received widespread attention for commanding her MP company in a combat assault.

[59] Quotes from Interv, Wright with Taylor, 21 May 1990. Rpt, Patton, Operation Just Cause, 15 Mar 1990; Donnelly, Roth, and Baker, *Operation Just Cause,* pp. 225–26; Flanagan, *Battle for Panama,* p. 194.

after dawn. The Kiowa, too, had been hit. As for the troops, they initially found themselves helpless to suppress enemy fire, either while en route to their objectives or after they were on the ground in landing zones that in daylight were well known to the enemy and, save for the tall grass, offered little concealment. Also, in both operations, the tropical heat, especially in the sunlight, played a detrimental role. At least a half dozen of the seventeen nonfatal U.S. infantry casualties at Las Tinajitas resulted from heat exhaustion or heat stroke. According to one account, "In spite of the superb physical condition of the paratroopers, a number of them fell victim to heat exhaustion during the attack up the hill, so steep that the paratroopers often fell to their hands and knees to climb through the thick and tall elephant grass. Heat, lack of water, and fatigue slowed the troops."[60] Finally, there seemed to be a consensus that, had the PDF defensive effort at Panama Viejo and Las Tinajitas been more determined and more professional, American casualties would have been much higher, and success in each of the operations might have taken much longer to achieve.

Compared to Panama Viejo and Las Tinajitas, the air assault of the 4th Battalion, 325th Infantry, into Fort Cimarrón seemed almost anticlimactic. The first airlift did not even leave Torrijos-Tocumen until 1205 on 20 December. Six helicopters took Company A into Landing Zone Cougar west of the fort, and six took Company B into Landing Zone Tiger to the south. A second airlift brought the remaining paratroopers to the two landing zones, neither of which received enemy fire. Indeed, the fort was empty. What units had been there before H-hour had fled to the north, made their way to Panama City, or taken casualties in the battle at the Pacora River bridge.[61]

With the successful execution of the three air assaults on Panama Viejo, Las Tinajitas, and Fort Cimarrón, all PDF units in the Panama City area had been neutralized. By the evening of 20 December, a number of enemy personnel who had not surrendered continued to harass U.S. military positions around the capital, and some of Nix's troops were called into the city to deal with a hostage situation at the Marriott Hotel. But the bulk of Noriega's military and Dignity Battalions had scattered, hoping to elude capture until the situation stabilized and their future status became clearer. The PDF had ceased to exist as an organized entity. The centerpiece of the fighting in and around Panama City, the Comandancia, had fallen without any significant body of PDF reinforcements reaching the scene. The plan to isolate the PDF headquarters had worked, thanks to the efforts of soldiers, sailors, marines, aviators, and special operations forces executing their attacks on over a dozen targets in and around the city. There was also no repeat of 3 October, when during the coup attempt against Noriega, PDF units from Rio Hato had flown to his rescue. A battalion of U.S. Rangers had made sure of that.

[60] For the heat problem, see Donnelly, Roth, and Baker, *Operation Just Cause,* pp. 225–26. Quote from Flanagan, *Battle for Panama,* pp. 194–95. Operation JUST CAUSE took place a little over six years after the U.S. invasion of Grenada in 1983. During the latter operation, the heavy loads carried by U.S. soldiers had also led to numerous cases of heat exhaustion. While the subject received much attention at the time, the problem had not been remedied, despite Johnson's intention to go in "light" as a means of making his troops somewhat less vulnerable to the heat.

[61] Rpt, Patton, Operation JUST CAUSE, 15 Mar 1990.

9

Task Force Red-R at Rio Hato

☆ ☆ ☆

In using the concentric circle template to describe U.S. combat operations on the Pacific side of Panama during the first day of Operation Just Cause, the Comandancia was the bull's-eye, while the first three rings contained enemy units based in and near downtown Panama City, close enough to the PDF headquarters to interfere in the fighting there if they chose to do so. The fourth and final ring, however, targeted the PDF's 6th Infantry Company (Mechanized) and 7th Infantry Company (known as *macho de monte*), both based at Rio Hato, a bit over sixty miles southwest of the capital.[1] The rationale for linking these distant units to the Comandancia harkened back to the October coup attempt, when troops flown to the capital from Rio Hato played a decisive role in defeating the rebel forces and maintaining Noriega in power. To prevent a similar maneuver during Just Cause, the two infantry companies at Rio Hato would have to be neutralized.

Plans, Preparations, and Deployment

In the Elaborate Maze and Blue Spoon contingency plans written between March 1988 and October 1989, the list of D-day objectives that would likely involve combat operations between U.S. and Panamanian forces targeted hostile units and command, control, and communications centers in the vicinity of Panama City and Colón. Special Forces would place PDF units at Rio Hato under surveillance, and JTF-Panama would stand ready to launch an attack if necessary. Most planners assumed, however, that the seizure of enemy bases that far from the capital would occur in the course of follow-on or on-order missions. When, after 3 October, the PDF 6th and 7th Companies became a primary H-hour target, U.S. Rangers previously assigned to the Comandancia mission found themselves charged instead with conducting an airborne assault to "seize Rio Hato airfield and eliminate PDF resistance in sector." As part of the proposed operation, U.S. Navy SEALs would search nearby beach houses belonging to Noriega and frequented by his cronies.[2]

[1] *Macho de monte,* often translated as "mountain men," was in fact a Spanish-language nickname for a Baird's tapir, a jungle animal the elite PDF company had adopted as its mascot.

[2] The mission statement for the Rangers is from Briefing Slides, 75th Ranger Rgt, Operation Just Cause, n.d. In the Southern Command Blue Spoon operation order dated 30 October

Of the numerous hard targets at Rio Hato, the most prominent was a commercial-military airfield with its 1.5-mile runway laid out south to northwest. The southern end of the facility was about a thousand yards inland from the Gulf of Panama, with dense jungle covering the land approach from that direction. The barracks and headquarters for both PDF companies were located just west of the jungle and southwest of the runway; also in the area were the 6th Company's motor pool; a communications center; an engineer platoon and motor pool; the Herrera-Ruiz Military Institute, a training school for noncommissioned officers; and a medical dispensary. South of the runway and jungle were the beach houses, while slightly north of the runway's midpoint was the Inter-American Highway, which ran east to west and bisected the airstrip. Northwest of that was an ammunition supply point.[3]

Because Noriega had relocated some elements of the 6th and 7th Companies to bases in and around Panama City after the coup attempt, U.S. intelligence estimates of the number of armed defenders the Rangers would encounter at Rio Hato varied from three hundred to five hundred, including up to fifty engineers. Not included in these numbers were two hundred fifty cadets believed to be at the noncommissioned officer school, the assumption being that, lacking weapons, they were unlikely to join the fight. In contrast, both PDF infantry companies had well-stocked arsenals. The *macho de monte* reportedly possessed 2 .50-caliber machine guns, 9 bazookas, 2 recoilless rifles, a couple hundred rocket-propelled grenades, 20 mortars, and 15 motorcycles. The mortars and motorcycles could enable the 7th Company to disperse and then bring indirect fire onto the attacking force; they could also shell the airfield, thereby disrupting the landing of U.S. aircraft scheduled to arrive after the initial fighting abated.

According to one account, the arsenal of the 6th Company was "even more formidable." In addition to machine guns, RPGs, recoilless rifles, and mortars, it also included three V300 and over a dozen V150 Cadillac Gage armored vehicles and three or more ZPU4 antiaircraft weapons, only one of which U.S. military intelligence could account for going into December. Furthermore, the SOUTHCOM intelligence directorate could not confirm rumors that Panamanian troops at Rio Hato possessed Soviet-built surface-to-air missiles. Such missiles and the ZPU4s were capable of knocking ranger transport planes out of the sky, while each four-barreled antiaircraft gun could wreak havoc among American soldiers parachuting onto the airfield, the only practical drop zone for the objective. Planners hoped that darkness and surprise would minimize this enemy advantage; even so, the casualty estimate for friendly forces was frightening, with some figures ranging as high as 30 percent. Because of the role both PDF companies had played in rescuing Noriega on 3 October, BLUE SPOON planners also assumed that the soldiers

1989, the Rangers were tasked to "neutralize the PDF," and, specifically, to "seize and secure" the Rio Hato airfield. CINCSO OPORD 1–90 (BLUE SPOON), 30 Oct 1989.

[3] The list of hard targets at Rio Hato is in Briefing Slides, 75th Ranger Rgt, Operation JUST CAUSE, n.d.; Flanagan, *Battle for Panama*, p. 141; Donnelly, Roth, and Baker, *Operation Just Cause*, p. 336.

were still loyal to the dictator and would likely mount a spirited defense of their positions.[4]

Given these considerations, Colonel Kernan, the commander of the 75th Ranger Regiment, assigned the Rio Hato mission to his 2d and 3d Battalions (minus). Taken together, they formed the fighting core of Task Force Red-R. The concept of operations called for both units to fly from the United States to Panama, where, at H-hour, numerous aircraft—AC–130s, AH–6s, AH–64s, and F–117As—would pound enemy targets for three minutes, after which the Rangers would begin their drop (*Map 15*). Once on the ground, Companies A and B of the 2d Battalion, commanded by Lt. Col. Alan Maestas, would seize the 6th and 7th Companies' barracks. The battalion's Company C would form a reserve force. Meanwhile, Lt. Col. Joseph F. Hunt's 3d Battalion, minus the company taking part in the Torrijos-Tocumen operation, would isolate the airfield; seize the camp headquarters, the communications center, and the motor pool; maneuver to sever the Inter-American Highway; and secure the ammunition storage area. The fighting would probably be intense but not prolonged. Despite the anticipated resistance, planners expected the Rangers to secure the airfield within an hour, allowing follow-on U.S. aircraft carrying personnel and equipment to begin landing.[5]

Rounding out Task Force Red-R, each of the two ranger battalions had a subordinate task force organized around it. In the case of the 2d Battalion, this included a broadcast team from the 4th Psychological Operations Group, whereas the 3d Battalion had two such teams, civil affairs personnel from the 96th Civil Affairs Battalion, and an Air Force combat control team for preparing the airfield and directing the landings of follow-on U.S. aircraft. For command and control, Colonel Kernan would accompany the Rio Hato contingent of Rangers and, with a small headquarters component, monitor events on the ground while reporting to the JSOTF commander, General Downing. Included in the elements attached to his regimental headquarters were an AC–130, a number of special operations aircraft, another Air Force combat control team, and personnel from a joint medical augmentation unit.[6]

Once the Rangers received the Rio Hato mission, the 2d and 3d Battalions began drawing up plans and making other preparations. These included a series of exercises geared to the objective and culminating in the "first-ever regimental-level" rehearsal in mid-December in Florida. Also involved in the

---

[4] Estimates of the strength of the Panama Defense Forces at Rio Hato are in Briefing Slides, Committed PDF Forces, n.d.; Flanagan, *Battle for Panama*, p. 142; Donnelly, Roth, and Baker, *Operation Just Cause*, p. 338; McConnell, *Just Cause*, pp. 86–87. According to Lt. Gen. Carl W. Stiner, the JTF-South commander, "These 6th and 7th Companies demonstrated in the 3 October coup that they probably had the highest degree of loyalty and readiness of any that we saw react to that." Interv, Robert K. Wright Jr. with Lt Gen Carl W. Stiner, U.S. Army, 2, 7, 27 Mar and 11 Jun 1990, Fort Bragg, N.C.

[5] Briefing Slides, 75th Ranger Rgt, Operation JUST CAUSE, n.d., describe the basic concept of operations. Additional details are found in Flanagan, *Battle for Panama*, p. 141; Donnelly, Roth, and Baker, *Operation Just Cause*, pp. 339–40; Jerry L. Thigpen, *The Praetorian STARShip: The Untold Story of the Combat Talon* (Maxwell Air Force Base, Ala.: Air University Press, 2001), p. 321.

[6] Briefing Slides, 75th Ranger Rgt, Operation JUST CAUSE, n.d.

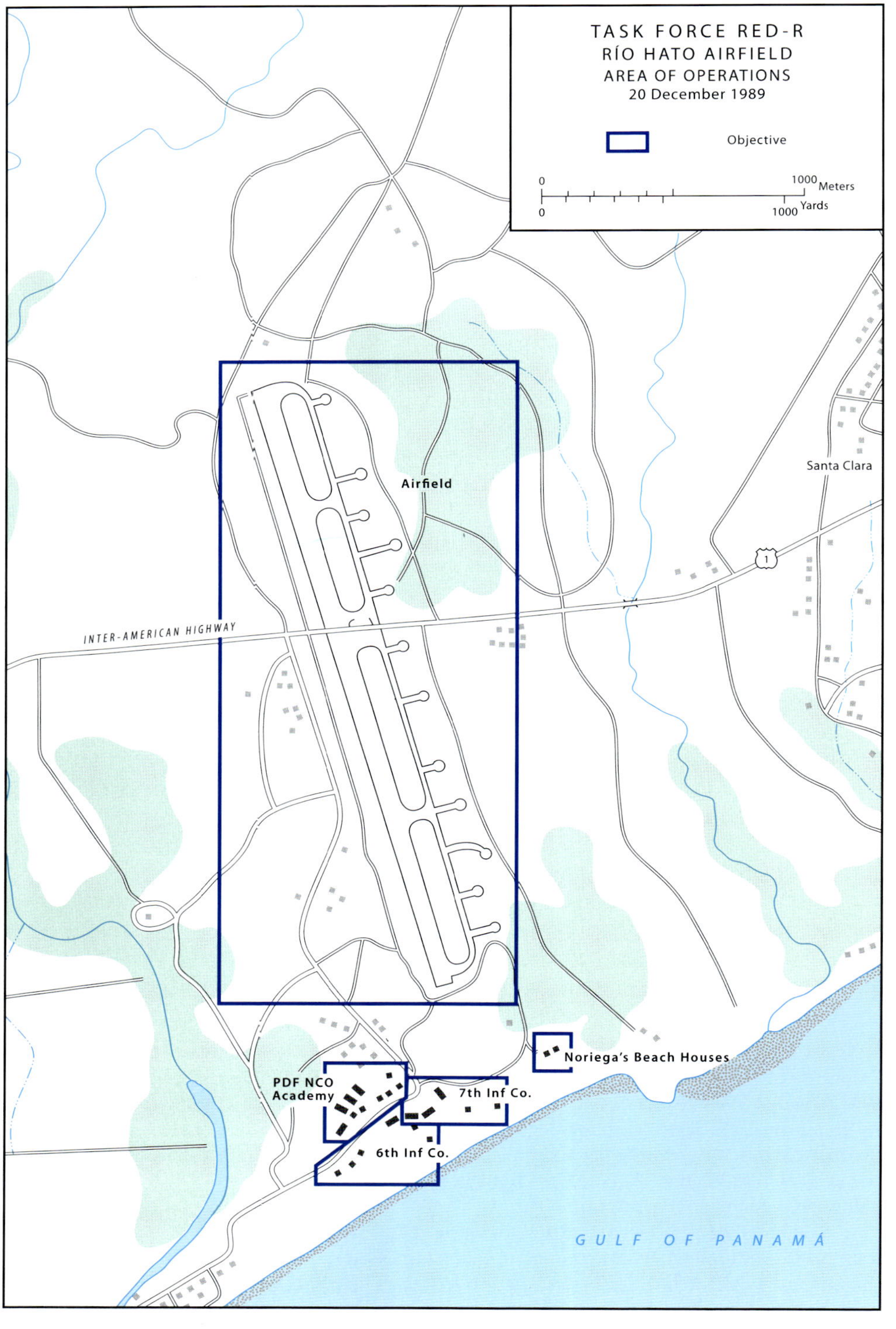
TASK FORCE RED-R
RÍO HATO AIRFIELD
AREA OF OPERATIONS
20 December 1989
Objective
0
1000 Meters
0
1000 Yards
Airfield
Santa Clara
1
INTER-AMERICAN HIGHWAY
Noriega's Beach Houses
PDF NCO Academy
7th Inf Co.
6th Inf Co.
GULF OF PANAMÁ

Map 15

practice run were an armada of Air Force troop and equipment transports and other Special Operations Forces, necessitating a series of coordination visits, the employment of a complex communications setup, and the exchange of liaison officers among the participating units and headquarters. Battalion-level rehearsals preceded the big event by three days, after which the 2d Battalion at Fort Lewis flew to Lawson Army Airfield at Fort Benning.

> The 2d and 3d Battalions were then cross-loaded on their transport aircraft. They would be parachuting into Hurlburt Field on the Florida panhandle, where the Air Force stations most of its special operations aircraft, for their assault. Waiting for them at Hurlburt were two companies of the 101st Airborne Division (Air Assault), who were to play the part of the PDF, including the areas where they would be positioned around the airfield. The Rangers' task was to secure the airfield within thirty minutes and then receive airlanded reinforcements.[7]

With only one exception, the mockup awaiting the Rangers was extremely accurate, and the code names of specific objectives mimicked the terms used in the actual contingency plan. The airborne companies from the 101st "gave no quarter," according to one account, but the Rangers still managed to accomplish their primary tasks within the required thirty-minute time limit. Generals Lindsay and Stiner observed the rehearsal, which both characterized as successful. Lindsay went so far as to declare "it was as fine an example of an airfield seizure as he had seen." At the other end of the chain, at least one private in a ranger weapons squad offered another assessment: "The dry runs we had in Florida benefited Head Quarters [*sic*] and the brass more than it did the platoons or squads." Another Ranger offered a more detailed critique. "During this exercise," he wrote several years later, "we did something that I had never seen before; each of the 13 chalks of C–130s came from a different direction as to confuse the enemy, and hopefully negate the known ADA [air defense artillery] threat. Well, from my point of view, this didn't work very well because it added more chaos to the already difficult task of assembly."[8]

Whatever the feelings of the participants, the eight days of rehearsals ended with all three ranger battalions returning to their home bases by Saturday, 16 December. Many of the troops were given some time off to enjoy the holiday season and lost little time in joining the weekend's round of parties and celebrations. Others simply sought to spend some quiet time with their families or to do some traveling. Whatever one did to relax from the strains and stresses of the previous week, the respite would be short-lived. Within

---

[7] Donnelly, Roth, and Baker, *Operation Just Cause*, p. 329.

[8] While the regimental rehearsal is mentioned in numerous sources, the narrative in this chapter, including all quotes but the last two, is mostly based on the succinct account in ibid., pp. 329–31. Last two quotes are from material written by Rangers who participated in the Rio Hato operations and posted on http://www.suasponte.com/m_riohato.htm. See also Joint After Action Report, Operation Just Cause, U.S. Southern Command, 14 Feb 1991. Briefing Slides, 75th Ranger Rgt, Operation Just Cause, n.d., say that the Rangers rehearsed the operation plan on 9–16 December. This included three days of battalion-level rehearsals, followed by the regimental-level exercise.

twenty-four hours of Lieutenant Paz being killed, the Rangers in the 2d and 3d Battalions found themselves again being placed on a state of alert.[9]

Kernan was the first to be notified over a secure telephone line at his Fort Benning home Sunday night. He, in turn, contacted his three battalion commanders and called together his battle staff. Officers from the 3d Battalion based at Benning met with company commanders Sunday night in a "special room for reading and discussing operational plans." When the formal warning order arrived, the staff made copies and, after receiving permission four hours later, disseminated them down to platoon level. This allowed the company commanders and their platoon leaders to review their specific targets and individual assignments and to finalize the detailed tactical plan. Around midnight, the officers finished tweaking their company-level and platoon-level operation orders. As several of the men tried to get some sleep, they realized that, despite the night's frenetic activity, they were not going anywhere for at least another two days.[10]

At Fort Lewis, the 2d Battalion's activities mirrored those of the 3d, although with the three-hour time difference between the coasts, they occurred earlier in the evening. Moreover, as Colonel Maestas had his staff begin recalling all personnel, he knew his battalion would have to return to Fort Benning by the next day. In both locations, Rangers returned to their barracks, most after only a day's reunion with spouses, families, and friends.

By Monday morning, most of the key staff and commanders had learned that they were not being assembled for another exercise. Many of the rank and file had figured that much out as well, despite the efforts of some officers to preserve operations security by portraying the alert as a continuation of the previous week's rehearsal or as a separate readiness exercise unrelated to Panama. (One intelligence officer, after telling Rangers at Fort Lewis that they were going on an exercise to seize and secure an airfield at Fort Bliss, Texas, actually passed out maps of that facility to support his deception; few recipients took the ploy seriously.) Sunday's news broadcasts prominently reported Lieutenant Paz's death as the lead story, usually accompanied by reports of the detention of the U.S. Navy lieutenant and his wife and the brutal treatment to which they had been subjected. Even though President Bush had made no public response, speculation circulated that this time the Noriega regime had gone too far. For the Fort Lewis Rangers, the sight of materiel being readied for the airlift to Benning offered further evidence of what was coming. Not only was there more equipment than usually required for an exercise, but pallets containing ammunition suggested a real mission. The same could be said about the kinds of weapons in view, such as the M21 antitank mines, which now appeared with other ammunition crammed into the vehicles to be flown to Benning and then to Panama.[11]

---

[9] The summary of postrehearsal activities is from ranger material on http://www.suasponte.com/m_riohato.htm.

[10] Flanagan, *Battle for Panama*, p. 134. Quote from Donnelly, Roth, and Baker, *Operation Just Cause*, p. 324, and see also pp. 323, 325–38.

[11] *Seattle Post-Intelligencer*, 19 Jan 1990. Material taken from firsthand accounts of ranger experiences posted on http://www.suasponte.com/m_riohato.htm.

On Monday, the Rangers at Fort Lewis traveled to nearby McChord Air Force Base and, that afternoon, boarded C–141s for their flight to Georgia. That night, they arrived at their remote marshaling base at Lawson Army Airfield, where they set up a tent city in the inclement weather that prevailed throughout the southeastern United States. The rain that had accompanied their departure from Washington, one Ranger joked, seemed to have followed them. Once accommodations for the night had been pitched, Maestas called his men together and formally told them that they would be mounting an airborne assault on Rio Hato. After going over the principal mission-related tasks as he knew them to be at that moment, he released the battalion in hopes that the men could get some sleep before the issue of updated operation orders and ammunition began early on the eighteenth.[12]

As the Rangers headed back to their tents, they did not realize that most other U.S. troops taking part in JUST CAUSE, namely those stationed at Fort Bragg and in Panama, would have another day's wait before being informed that they would be going into combat. The rangers' advance notification was not exactly an unqualified blessing for either the officers or the men of the 2d and 3d Battalions. If it allowed them more time to go over the plans and maps, to learn their specific tasks, and to outfit themselves for the flight and fight ahead, it also allowed them more time to think and to dwell on personal and professional matters and the dangers awaiting them. Among the former concerns, there was one Ranger who, leaving his expectant wife, knew that his first child would be born while he was in Panama. Another regretted that he and his wife had been arguing when he received the alert; now the contentious point, whatever it was, might fester until his return, if he did return. And another thought about the plans he and his wife had made for their first wedding anniversary, just a couple of days away.

When thoughts turned from personal affairs to the mission at hand, one sentiment common to all was apprehension. Commanders and enlisted men confessed to being scared. As with the paratroopers from the 82d, many worried about the enemy's ability to shoot down U.S. troop transports or to kill individual Rangers as they parachuted to the ground, helpless to defend themselves. While planners had estimated significant ranger casualties at Rio Hato, their figures came nowhere near the exaggerated 70 percent alleged in rumors now spreading among the anxious soldiers. To minimize casualties, the Rangers would jump at five hundred feet above the objective, but such combat jumps entailed their own risks. If a parachute failed, there would be little time to employ a reserve chute (thus little reason to wear one). Other malfunctions could mean a hard landing that could inflict lasting physical damage or at least leave one vulnerable to enemy fire or incapable of fulfilling an assigned role.[13]

Thoughts like these lingered in many rangers' minds as they engaged in routine preparations before boarding the troop transports headed for Panama later Tuesday afternoon. Along with checking their weapons, equipment,

[12] Ibid.

[13] The personal thoughts of several Rangers between the alert notification and their actual deployment are from Flanagan, *Battle for Panama*, especially pp. 137–39; Donnelly, Roth, and Baker, *Operation Just Cause*, especially pp. 323–33; firsthand accounts of ranger experiences posted on http://www.suasponte.com/m_riohato.htm.

and gear, the men drew the ammunition they needed. To their surprise, they found that they could obtain almost anything they wanted and in amounts that vastly exceeded normal exercise loads. "It was like an ammo super store," one Ranger remembered. Another compared the experience to being in "a big candy store." Most Rangers filled their rucksacks with anything they could get their hands on, but mostly extra M16 magazines and grenades. A few took a more practical approach, taking only what they thought they would actually use. As one soldier later stated with macabre logic, if the fight at Rio Hato was going to be so intense as to require "that much ammo, there are going to be enough dead Rangers around that there'll be ammo all over the place. It's not the Battle of the Bulge."[14]

Also to consider was how much ammunition an individual could realistically carry. By the time some of the soldiers finished stuffing extra magazines, grenades, and other items into their rucksacks, the containers weighed near or over a hundred pounds. One Ranger noted that his rucksack was so heavy that he could only move it through the C–130 he boarded with two hands. Another's was so overweight that he had to have two jumpmasters carry it to the plane. An overloaded rucksack could also cause problems during the jump, not to mention the difficulties of carrying it around in Panama's tropical heat.[15]

Tuesday afternoon, both battalions prepared to board the waiting C–130s. That they would not fly to Rio Hato in C–141s like those available to the 1st Battalion going into Torrijos-Tocumen harkened back to General Downing's determination that the Starlifters were better suited to the longer drop zone at the Panama City objective. As a result, the Rangers going into Rio Hato had to cram themselves into the smaller and slower Hercules.[16] Before boarding, however, there were some last-minute formalities that had to be observed, as the troops, wrapped in wool blankets, stood in the cold, rain, "slosh and mud," fortified to some extent by coffee, soup, and cocoa provided by the ranger support element. Colonel Kernan and the regimental sergeant major walked through the assembly area, with the ranger commander stopping whenever he could to deliver a confidence-building talk to the troops. One Ranger simply remembered that Kernan gave "some kind of hoo-ah speech, [and] I think the RGMT Chaplain mumbled something." Others recalled the colonel's remarks as "inspirational" and "great for morale," especially since they knew Kernan would be jumping with them.[17]

Another preboarding procedure involved receiving an update on the operation plan and the latest intelligence. As platoon leaders briefed the material, some seemed clearly upset about something. The source of their displeasure soon became apparent: the aerial fire support they had been told to expect just prior to the ranger airborne assault had included F–117A

[14] First and second quotes from firsthand accounts of ranger experiences posted on http://www.suasponte.com/m_riohato.htm. Last quote from Donnelly, Roth, and Baker, *Operation Just Cause*, p. 332.

[15] Ibid.

[16] Interv, Wright with Stiner, 2, 7, 27 Mar and 11 Jun 1990.

[17] Material and quotes from firsthand accounts of ranger experiences posted on http://www.suasponte.com/m_riohato.htm. See also Donnelly, Roth, and Baker, *Operation Just Cause*, p. 333.

Stealth fighters dropping two 2,000-pound bombs on the PDF reservation at Rio Hato. The resulting destruction was likely to be devastating to the enemy, leaving few in any condition to mount an effective defense. Now, however, on the eve of the operation, unit commanders received word that the bombs would not be dropped. Confused and frustrated, the Rangers had one question: why? To one platoon leader, the answer was obvious, although he refrained from voicing it at the time: the decision was the result of "rotten politics."[18]

In actuality, the 2,000-pound bombs would be dropped but near, not on, the enemy barracks. And the platoon leader was partly right. The decision, if not rotten, had indeed been political, made by the highest-ranking civilian and military personnel in the Pentagon and approved by the president. It was also one of the most controversial decisions surrounding Operation JUST CAUSE on two counts: the use of the F–117s in the first place and the determination not to bomb the target directly.

Discussion surrounding the decisions had been lengthy and at times heated. Once Rio Hato became a primary target, U.S. planners devised a plan of attack that, from the outset, sought to minimize the threat to exposed Rangers dropping onto the airfield. What concerned General Stiner as much as the ZPU4 air defense systems was the "small arms fire in mass" the two PDF companies could train on the Rangers as they descended onto the objective. One means to prevent the Panamanian forces from mounting a deadly defense was to subject them to a massive aerial barrage just prior to the ranger assault. In the SOUTHCOM and JTF-South versions of the operation plan approved by the Joint Chiefs in early November, the preassault fires were assigned to

F–117A Stealth fighter (photo taken one year after Operation JUST CAUSE)

[18] Flanagan, *Battle for Panama*, p. 135.

General Kempf

attack helicopters, AC–130 gunships, and F–117A fighters. A week later, when Powell and Pentagon operations officers briefed the secretary of defense on the plan, Cheney was skeptical about the inclusion of the Stealth fighters, each of which cost $100 million and none of which had yet been used in combat. Employing such an expensive and highly technological asset against the Panamanian military might appear as though the Air Force was just manufacturing an opportunity to see how the plane would perform on a real but low-risk, even "trivial" mission. Thus Cheney's query of Powell and the others, "Why the hell do you want to use the 117?"[19]

Powell's answer was simple and direct: General Stiner, with the backing of General Downing and Colonel Kernan, had requested the fighters. To minimize ranger casualties at Rio Hato, they wanted an aircraft from the Air Force that could deliver accurate fire with "maximum shock effect." Apache helicopters could meet these requirements, but the two AH–64s available for action at Rio Hato were to be used against the ZPU4s and other targets. Stiner considered using F–111 fighters with 750-pound bombs to create the desired psychological impact, but Lt. Gen. Peter T. Kempf, the Twelfth Air Force commander whose planes would be used in Just Cause, recommended his service's new F–117As because they were more accurate. Equipped for precise targeting and designed to penetrate enemy radar undetected, the Stealth fighters could deliver a devastating payload on an unsuspecting foe. Stiner expressed concern that the Stealth was still a secret weapon, but Kempf responded that "for something as critical as this, it could probably be brought out." There was more to Kempf's offer: a choice as to whether the planes would drop 500-pound bombs or 2,000-pound bombs. Stiner decided in favor of the latter, not because of their size but because of their accuracy:

> The reason for recommending the 2,000-pounder over the 500[-pounder] is because of the accuracy it has because of higher technology in the guidance fins. Whereas the 500-pounder does not have that. So the Air Force came back with that. They sent a squadron commander and his operations officer to see me—I was in Panama at that particular time—and he laid out for me

[19] First quote from Interv, Wright with Stiner, 2, 7, 27 Mar and 11 Jun 1990. Cole, *Operation Just Cause*, p. 31. Second quoted word from Flanagan, *Battle for Panama*, p. 143. Last quote from Woodward, *The Commanders*, p. 140.

> the accuracy data of I believe it was eighty-three 2,000-pound bombs that had been dropped versus something like a couple of hundred or maybe 250 of the 500-pound bombs. It was very clear to me that the 2,000-pound had the best record, and so that's what we elected to go with.[20]

After listening to Powell, Cheney kept the F–117As in BLUE SPOON, at least for the time being. Two of the fighters would be needed for Rio Hato: one for the PDF 6th Company, one for the 7th Company. Four others, operating in two pairs, would be used elsewhere: one pair at Noriega's ranch at La Escondida in northern Panama, the other pair at the dictator's beach house at Boquete in the westernmost part of the country.[21]

Although Stiner's proposed use of the F–117As had the support of Downing and Kernan, the three "violently disagreed" over just where at Rio Hato the two planes should drop their bombs, with Stiner arguing that the 2,000-pound bombs should fall close to—but not on—the enemy compounds, and Downing and Kernan arguing for direct hits. Driving Stiner's thinking was the guidance he had received from the National Command Authority—and which was reflected in the rules of engagement—to keep enemy casualties to a minimum. A high death toll could antagonize even anti-Noriega Panamanians and undermine the nation-building and stability operations that would inevitably follow the fighting and the change in government. In an interview after JUST CAUSE, Stiner candidly discussed his concerns. The "last thing we wanted," he told a reporter, "was a Beirut Barracks situation on our hands." Such a slaughter, he reiterated, would have stiffened PDF resistance throughout Panama and left a lasting bitterness, resulting both in increased U.S. casualties during the combat phase and delays in reconstructing the country afterward. What Stiner proposed, therefore, was using the 2,000-pound bombs as humongous "stun grenades." Assuming the U.S. surprise attack would catch the PDF infantrymen at Rio Hato in their barracks, the bombs' powerful explosions about 150 yards away would so confuse and disorient the enemy troops as to preclude their mounting an effective defense. The dazed soldiers might even give up when U.S. psychological operations teams broadcast their surrender appeals. In this scenario, the highly touted accuracy of the Stealth fighters would be a critical factor in placing the bombs where they could have the greatest impact in neutralizing the two infantry companies without causing an undesirable level of casualties.[22]

Downing, like Stiner, registered his concern over the havoc the PDF companies could wreak with their small arms and machine guns during the ranger descent. "They had something like fifteen armored cars there with .50-calibers on them," he cautioned. "They could have set them up on that

---

[20] First and third quotes from Interv, Wright with Stiner, 2, 7, 27 Mar and 11 Jun 1990. Second quote from Clancy with Stiner and Koltz, *Shadow Warriors*, p. 325.

[21] Clancy with Stiner and Koltz, *Shadow Warriors*, p. 325; Woodward, *The Commanders*, p. 140; Cole, *Operation Just Cause*, pp. 31–32.

[22] Interv, John Partin with Maj Gen Wayne A. Downing, U.S. Army, 6 Jan 1990, Howard Air Force Base, Panama. First quote from Flanagan, *Battle for Panama*, p. 142. Second quote from Interv, Wright with Stiner, 2, 7, 27 Mar and 11 Jun 1990. Last quote from Donnelly, Roth, and Baker, *Operation Just Cause*, pp. 340–41. Cole, *Operation Just Cause*, p. 31.

airfield and that would really have gone bad for us." Kernan shared this concern. "We were going to take a lot of casualties if these people were able to marshal their forces as quickly as they had already proven they could." His preference was to eliminate this worst-case scenario by destroying the two PDF compounds. He and Downing lobbied for this approach without success in the month and a half leading up to JUST CAUSE. On the eve of battle, the operation plan continued to call for the offset bombing option.[23]

On Monday, 18 December, the day after President Bush decided to invade Panama, Secretary Cheney conducted another review of the BLUE SPOON operation order with Powell and others, asking detailed questions and again revisiting the need for the F–117As. Powell reiterated Stiner's position and then contacted Thurman in Panama for his views. The SOUTHCOM commander later recounted his position: "If we could give [the defenders] the biggest compression shock of their lives, they would quit fighting. . . . The most accurate weapon system delivery in the United States Air Force is a 117 with a 2,000-pound laser-guided bomb . . . so why not use it?" In light of Thurman's "strong argument," the secretary decided to leave that mission intact. But neither Cheney nor Powell thought the fighters necessary for the La Escondida and Boquete missions, so they scratched the aerial attacks planned for both locations. After the meeting, Powell called in General Kelly, the Joint Staff's operations officer, and began grilling him on the plan—sometimes in a "confrontational way," Kelly remembered with some irritation. When the discussion turned to the F–117As, the J–3 cited "weapons effects experts" as predicting that the bombs were not likely to cause much damage. Stiner, however, was not convinced, contacting Kelly at the last minute to warn that, if the bombs landed too close to the PDF barracks, they might do more damage than the experts anticipated. Powell had reached a similar conclusion and ordered the bombs to be dropped 200 yards away from one compound and 250 yards from the other. The next day, Tuesday, less than twenty-four hours before JUST CAUSE was scheduled to begin, the president approved these decisions, saying of the F–117A, "If that's the best plane, use it." On another matter related to Rio Hato, the latest intelligence indicated that Noriega was not going to be in the beach house near the airfield, so the Navy SEAL mission was canceled and the seizure of the beach house transferred to the Ranger company slated to be the Task Force Red-R reserve during the first phases of the assault.[24]

---

[23] Interv, Partin with Downing, 6 Jan 1990. Downing quotes from Flanagan, *Battle for Panama*, p. 142. Kernan quote from Donnelly, Roth, and Baker, *Operation Just Cause*, pp. 340–41.

[24] First quote from Briefing, Gen Maxwell R. Thurman at SAMS, 23 Sep 1992. Second quote from Cole, *Operation Just Cause*, pp. 31–32. Remaining quotes from Woodward, *The Commanders*, pp. 176–77. One handwritten submission for the ranger regiment's after action report claimed that the "decision to bomb or not bomb 6th & 7th Company barracks and the beach house at Rio Hato was bandied about until H-Hour, both in tactical terms and the political implications. . . . Once the SEALs fell out adjustment to AH–6 plan had to be made and plans for the employment of AH–64 Apaches coordinated." The lesson learned was that "we must be flexible enough to react to those decisions and aggressive enough to rework the fire sup-

Whatever negative feelings the Rangers expressed to their chain of command in reaction to word that the Rio Hato barracks would not be bombed, there was little time to brood over the decision. The plan for the airborne assault had to be studied one more time, the rules of engagement reiterated, and the mission-essential task list reviewed. As one platoon leader related, the rules of engagement emphasized the minimum use of force and the avoidance of collateral damage, but, with no large groups of civilians located near the Rio Hato facilities, the rules were not as restrictive as they were for other units taking part in JUST CAUSE. As for the task list, he noted that it focused solely on how to engage the enemy force in combat operations. (To his later regret, it contained nothing about the stability operations that might follow.) Time ultimately forced an end to these final preparations, and the troops began drawing their parachutes and rigging in the ceaseless rain. When they finished, they took their overweight rucksacks and boarded the warm but humid C–130s where, crammed together, they attempted to get some rest. The ones who could not sleep tried their best not to dwell on the unknown that awaited them. Most were unsuccessful.[25]

At 1800 on the nineteenth, the aircraft began taking off from Lawson Army Airfield for their seven-hour flight to Panama. The C–130 airstream consisted of thirteen troop-carrying transports and two planes carrying the rangers' heavy equipment. Also in the air were the two F–117A Stealth fighters (which took off from the Tonopah Test Range in Nevada) and the KC–10A aircraft needed to refuel them. Timing was critical. If the Rangers arrived over the target before the F–117As dropped their bombs, the consequences could be bloody. En route, however, an even greater concern surfaced when Colonel Kernan, like other U.S. officers that night, received word that the operation had been compromised: the Panamanians knew an invasion was imminent. From the additional intelligence updates radioed to the colonel's C–130, Kernan learned that U.S. commanders had changed H-hour at some targets to 0045 and that the PDF commander of the 7th Company, besides alerting his counterpart in the 6th Company of the inbound American forces, had ordered his men to block the airfield runway with vehicles. The rangers' plan assumed enemy troops would be sleeping when the Stealth fighters dropped their bombs. What would happen if, after being notified about the impending attack, the 6th and 7th Companies had left their barracks, dispersed, and set up a formidable defense around the airfield? It was a chilling prospect.[26]

THE ATTACK

At 0100, the F–117As arrived over Rio Hato precisely on time. Each carried a 2,000-pound Mark 84 bomb with a delayed fuse so that the bomb,

---

port plans to support the maneuver in light of those decisions." Lessons Learned Worksheet, 75th Ranger Rgt, Operation JUST CAUSE, n.d., sub: Fire Support.

[25] Firsthand accounts of ranger experiences posted on http://www.suasponte.com/m_rio hato.htm; Presentation, Capt David Haight, author's History of Modern Warfare course, n.d. [early to mid-1990s], U.S. Army Command and General Staff College.

[26] Chronology, 75th Ranger Rgt, n.d.

in General Stiner's words, "would go into the ground, and that would shield a great majority of the blast, but it would throw up a lot of dirt and debris, and create the kind of commotion and diversion that we wanted." The first pilot was to drop his Mark 84 near the 7th Infantry Company's barracks, but a combination of high winds and possibly incorrect or confusing targeting information resulted in the bomb landing near the 6th Company barracks instead. The second pilot used the first plane to orient his attack, causing his bomb to fall over three hundred yards away from the target. Even so, both Kernan and Stiner later expressed satisfaction with the 117s' performance. Although prior knowledge of the U.S. invasion had caused most Panamanian forces at Rio Hato to leave their barracks and take up defensive positions around the airfield, the blast from the two bombs still caused a certain amount of confusion, prompting some of the enemy to discard their weapons and flee. Overhead, as the approaching Rangers prepared for their combat jump, they could hear the explosions through the open doors of the C–130s. Cheers went up, as did morale.[27]

The Rangers may have heard noises of other attack aircraft as well. In the minutes between the bomb drop and the troop jump, a Spectre gunship, four Little Birds, and two Apaches opened fire on targets in and around the Rio Hato runway. A ranger liaison officer aboard the Spectre had briefed the crew, setting the "priority for target engagement." The helicopter personnel had also received advance word on what weapon systems to hit. The fusillade lasted only ninety seconds but proved both psychologically and physically effective, with at least one—perhaps two—of three ZPU4 antiaircraft platforms and a number of smaller defensive positions being destroyed.[28]

As the C–130s carrying the Rangers approached the drop zone, the preparatory fires stopped, with the helicopters and Spectre remaining on station to identify and engage any enemy formations or weapons that challenged the assault force. Meanwhile, in the troop transports, the green lights came on, and

---

[27] Quote from Interv, Wright with Stiner, 2, 7, 27 Mar and 11 Jun 1990. Cole, *Operation Just Cause*, pp. 38–39; U.S. Special Operations Command, *History, 1987–2007*, 20th anniversary ed. (MacDill Air Force Base, Fla.: U.S. Special Operations Command, n.d.), p. 36; F–117A: Operation Just Cause (Panama), at http://www.f-117a.com/Panama.html; Donnelly, Roth, and Baker, *Operation Just Cause*, p. 340. In April 1990, the F–117As became the object of controversy when the Pentagon disclosed that what Secretary Cheney had described as a "pinpoint accuracy" attack had been anything but that. On 10 April, Cheney ordered an inquiry into "why the Air Force failed to inform him that one of its Stealth fighters missed its target during the Panama invasion." Three months later, an official report was released to the press, charging that a high-ranking Air Force general had failed to pass the information of the F–117As' performance to the secretary. A smaller controversy also surrounded where the bombs were supposed to be dropped in the first place. General Stiner said he ordered them dropped 165 yards from each of the PDF barracks. General Powell said he believed they were to be dropped 275 yards away, and the Air Force said it ordered the pilots to drop them about 55 yards from each target. Quote from *New York Times*, 11 Apr 1990, and see also 4 Apr 1990, 2 Jul 1990. *Boston Globe*, 3 Jul 1990.

[28] Quote from HHC 3/75 Narrative on Operation Just Cause, n.d. Donnelly, Roth, and Baker, *Operation Just Cause*, p. 341; U.S. Special Operations Command, *History, 1987–2007*, p. 36. The following narrative of the battle for Rio Hato owes much to Duty Officer's Log, 7th Ranger Rgt, Rio Hato, 20 Dec 1989.

the Rangers with their unwieldy rucksacks headed for the exit. As most waited their turn, they could hear "ticking," the sounds of bullets striking the aircraft. (Eleven of the thirteen troop carriers and both heavy equipment transports took hits from ground fire.)[29] Questions and concerns multiplied. How many PDF gunners were there? What would be the volume of enemy fire during the twelve-second descent to the ground? How accurate would it be? Would I land near the runway, on flat ground but exposed; or would I land in the tall grass nearby; in a tree or power line; on a fence, building, or garden; or, worst of all, in the Gulf of Panama? Another fear factor was the drop from only five hundred feet. If a parachute failed to function properly, there would be almost no time to take corrective action.

Enemy ground fire, in fact, turned out to be extensive and intense during the jump, something Stiner later attributed to the operation being compromised. As the first to jump, the Rangers from the 2d Battalion exited over the southern end of the airfield. As they floated to the ground, they could see enemy tracer rounds zipping past them. The 3d Battalion encountered the same sight as its members descended toward the northern end of the runway. At least one Ranger was wounded in midair. There was nothing they could do to stop the bullets, so they tried to concentrate on where they would land and how quickly they could shed their parachutes and arm themselves. The plight of Colonel Kernan did not set a promising example. As he neared the ground, his parachute became entangled in some power lines near the airfield. He freed himself just as the chute caught fire, shorting out electricity in the area around him. The resulting darkness helped obscure Rangers still in the air from Panamanian fire but also made it more difficult for them to calculate their landing site. As for Kernan, once freed from the wires his drop to earth turned out to be only six inches. "Lightest landing at Rio Hato," he would quip.[30]

Others survived similar harrowing predicaments. One Ranger specialist, for example, tried to exit his C–130 carrying a 90-mm. recoilless rifle in a weapons container never tested as "jump worthy." When he moved forward toward the plane's open side door, the container failed to clear it, causing him to bounce back into the troops lined up behind him. He then exited sideways, but in such a way as to descend "inverted" toward the drop zone, with the remaining Rangers pouring out behind him in a "tight group." Like Kernan, the specialist was lucky enough to land safely. Thirty-six other Rangers were not so fortunate. As in the parachute jumps at Torrijos-Tocumen, there would be bruised or broken feet, heels, ankles, legs, and arms. Some of the injured tried to carry on; others

---

[29] U.S. Special Operations Command, *History, 1987–2007*, p. 36; Donnelly, Roth, and Baker, *Operation Just Cause*, p. 342.

[30] *New York Times*, 27 Feb 1990. Quote from Donnelly, Roth, and Baker, *Operation Just Cause*, p. 344, and see also p. 342. Flanagan, *Battle for Panama*, p. 145. The *New York Times* article, which quoted Stiner extensively, noted that the general's account of the intense shoot-out at Rio Hato "varied sharply from the Pentagon's initial after-action reports. The Pentagon spokesman, Pete Williams, said in late December that the Rangers were not fired on as they parachuted into Rio Hato because the defenders had been stunned by two 2,000-pound bombs dropped by Stealth F–117 fighter-bombers." Rangers who in 1983 had jumped into enemy fire in Grenada said that the shooting was much more intense at Rio Hato.

could not and had to make their way or be carried to the medics for treatment. The worst case was Pfc. Patrick McElrath, who broke his back on hitting the ground. When a medic reached him, McElrath could not feel his legs. Further examination would reveal that he had severed his spinal cord. Heavy rucksacks also caused some Rangers to land hard or to hit their heads upon landing, in one instance knocking a soldier unconscious. Another Ranger recalled that upon hitting the drop zone, "my ruck was so heavy I had to roll into it on the ground then roll up into my feet, it was all I could do to move forward." In a related experience, a Ranger recounted, "My ruck weighed 100 pounds and I landed on the tarmac and hit my elbow pretty hard on the asphalt. I got out my gun, laid behind the ruck which was full of 10 lbs of C-4, 4 grenades, 400 rounds or so of 5.56mm and a claymore. Why am I using this for cover? Am I stupid or what?" Despite these difficulties and the unceasing hostile fire, no Rangers were killed by enemy bullets during the jump. One, however, Pfc. John M. Price, died when his static line broke. He was found, according to one account, with "his reserve parachute open on his lap, a victim of the low jump altitude. There is no time for a reserve parachute when people were landing."[31]

On the ground, the Rangers confronted a potent, but chaotic and disjointed defense, seemingly devoid of organization and leadership. PDF armored vehicles and trucks, often with their headlights flashing off and on, sped back and forth the length of the runway and the roads running parallel to it, the occupants firing wildly and mostly inaccurately at the ranger drop zones and the grass and jungle next to them. Other enemy soldiers manned positions on and around the airfield and aimed small-arms fire into the same areas. As feared, the 6th and 7th Companies had not only left their barracks, but those soldiers who had not fled the area seemed determined to fight, at least until the Americans could mass their forces. What this meant for the Rangers was that many of them would find themselves in a firefight well before they could assemble and launch attacks on their assigned targets. In some cases, an individual Ranger who had just touched down confronted several enemy soldiers or vehicles. In other cases, a Ranger making his way to his assembly area would pick up one, two, three, or four of his comrades en route, form an ad hoc fire team, and engage PDF defenders in lethal skirmishes. As Kernan later described these initial contacts with enemy soldiers, "They were everywhere. . . . It was a 360-degree firefight. They were places where we hoped they would not be." As another Ranger put it, "There were little fire fights all over the airfield." In all this confusion, the Spectre gunship flying overhead, with its crew systematically selecting and firing on targets of opportunity, seemed a godsend.[32]

[31] First, second, and third quoted words from Handwritten accounts, 3d Squad, Co A, 3d Bn, 75th Ranger Rgt, n.d. Sixth and seventh quotes from firsthand accounts of ranger experiences posted on http://www.suasponte.com/m_riohato.htm. Fifth quote from www.combatreform2.com/combatjump.htm. Almost immediately after the battle at Rio Hato, a story circulated that a Ranger standing in the doorway of a C–130 preparing to jump had been shot in the head and killed. The story later proved false, although one Ranger was apparently wounded while still inside his aircraft. U.S. Special Operations Command, *History, 1987–2007*, p. 36.

[32] Flanagan, *Battle for Panama*, p. 146. First quote from Donnelly, Roth, and Baker, *Operation Just Cause*, p. 348. *Seattle Post-Intelligencer*, 19 Jan 1990, p. 1; Mir Bahmanyar,

For S. Sgt. David DeBaere of the 2d Battalion, contact with the enemy came before he reached the ground. Jumping at the southern end of the airfield, he drifted into the jungle area at the edge of the runway. There, his parachute caught on a tree, leaving him dangling helplessly just five feet over a PDF bunker containing at least one enemy soldier firing his weapon. When DeBaere tried to climb down the tree, a branch snapped. Alerted by the sound, the Panamanian looked upward and let loose a burst from his automatic rifle through the thatched roof covering the bunker. The bullets missed DeBaere, who responded by dropping a fragmentation grenade through the roof. He encountered no further fire as he touched ground at the base of the tree.[33]

S. Sgt. Louis Olivera, a jumpmaster like DeBaere, was not so fortunate. He, too, landed in a tree, but, after managing to get to the ground without incident, his luck changed. As he made his way to the west edge of the airfield, enemy fire forced him to take cover in a ditch. He jumped in without looking first and immediately felt a sharp pain in his shoulder and chest. There were at least two enemy soldiers from the 7th Company in the ditch with him, and they had shot him twice before he was even aware of their presence. As they stripped him of his patches, one of the men shot him in the head at point-blank range with a pistol. They then wrapped a black bandana reading "*macho de monte*" around his rifle and left him for dead. Around 0600, Olivera regained consciousness and, although bleeding profusely, was able to radio for help. One of the medics who responded described the wounded Ranger as having "two sucking chest wounds, and I could see about one inch into his skull. A bullet had penetrated his helmet and stopped in his forehead. He was barely alive." The Kevlar helmet, in fact, had slowed down the bullet enough to prevent it from entering the brain. Olivera recovered, and his comrades absorbed the lesson, "Kevlar helmets are great."[34]

S. Sgt. Richard J. Hoerner likewise had a terrifying, life-threatening experience. A victim of poor timing, Hoerner landed in the middle of a road. Looking up, he saw a PDF truck speeding directly toward him. Once the occupants saw him, they opened up with "*full automatic AK-47 fire.*" Hoerner rolled out of the way of the oncoming vehicle, but the truck drove through his still-inflated parachute, which got tangled in the transport's drive shaft. "The vehicle was going so fast," Hoerner recalled, "it slammed me to my back and started to drag me down the road, going north toward the cadet barracks." After about a hundred feet, he was able to cut himself free from his parachute. The truck sped off, leaving him literarily in its dust. He then retraced the route of his ordeal to pick up his helmet, weapon, and the rest of the gear

Shadow Warriors: A History of the US Army Rangers (Oxford, U.K.: Osprey Publishing, 2005), pp. 128–29. Second quote from firsthand accounts of ranger experiences posted on http://www.suasponte.com/m_riohato.htm.

[33] *Seattle Post-Intelligencer*, 19 Jan 1990, p. 1.

[34] The account of Olivera's ordeal is from ibid. Quote describing Olivera's condition from firsthand accounts of ranger experiences posted on http://www.suasponte.com/m_riohato.htm. Last quote from Donnelly, Roth, and Baker, *Operation Just Cause*, pp. 347–48.

from which he had been so abruptly separated. After that, he headed for his assembly area.[35]

Soon after Hoerner hit the drop zone, 1st Sgt. Joseph L. Mattison from the 3d Battalion also touched down and, like Hoerner, encountered a PDF truck. The defenders inside were firing their weapons but not directly at the sergeant, which gave him time to grab his LAW. As the vehicle stopped near him, he fired the antitank weapon, knocking his target out of commission. Soon thereafter, Mattison was joined by a staff sergeant, who also had a light antiarmor weapon. Mattison instructed the man to prepare his LAW, just as a second enemy vehicle approached. In quick succession, the second vehicle "smashed into the rear of the truck" that had just been destroyed, Mattison opened fire with his M16, the staff sergeant fired his LAW but missed, and the second vehicle sped away. Hours later, after dawn, Mattison returned to the site to inspect the truck he had put out of action. Although he did not realize it at the time, the damaged vehicle was the same one responsible for Sergeant Hoerner's harrowing experience.[36]

As Rangers trying to find their assembly points coped with confusion and diverse dangers, one enemy position stood out as disproportionately threatening: a .50-caliber machine gun and its crew ensconced atop a two-story stone archway and tower that served as a gatehouse to the airfield. On the rangers' battle maps for Rio Hato, the structure was Objective Green, a location where a team from the 3d Battalion's Company A was to rendezvous and launch its operations. First to arrive were S. Sgt. Wayne Newberry and 1st Lt. Loren Ramos, who had landed in tall grass just over a hundred yards from the objective. Under fire, they began crawling through the grass toward the tower, pausing occasionally to empty an M16 magazine or two at the enemy. Not until they drew much closer to the gate could they see the machine gun, which was creating havoc across the airfield. Fortunately for Newberry and Ramos, a wall that provided cover for the gun prevented its barrel from being depressed enough to hit either of them.

As both men reached the archway, two other Rangers joined them. Together, the four-man team mounted an attack, their initial action being to throw grenades into the structure's entrance on the ground floor. Next, Ramos and Newberry moved onto a road running past the entrance but were almost run down by a V150 speeding by them. All four Rangers opened fire on the vehicle, but it continued on unscathed, making a turn onto the Inter-American Highway. Refocusing their attention on the machine gun nest, the Rangers lobbed in another grenade, only to have an enemy soldier inside respond with one of his own. Five minutes later, two defenders tried to escape, perhaps thinking that their grenade had done its work. Their mistake proved fatal, as the Rangers shot and killed both men. According to Newberry, "They just dropped and ran. . . . But they didn't make it." In clearing the gatehouse room by room, the team came across the machine gun, still hot from being fired.

---

[35] Handwritten accounts, 3d Squad, Co A, 3d Bn, 75th Ranger Rgt, n.d. Quotes from Flanagan, *Battle for Panama*, p. 146. Emphasis in the first quote is in the original.

[36] Flanagan, *Battle for Panama*, p. 148; Handwritten statement, 1st Sgt Joseph L. Mattison, 2 Jan 1990.

Rangers at the entryway to Rio Hato airfield after capturing a .50-caliber machine gun

Both Ramos and Newberry later received the Bronze Star for removing this major impediment to the ranger plan of attack.[37]

Like the gateway, the airfield's control tower was another critical objective, and it came under attack soon after the Rangers hit the ground. One specialist in particular, after engaging an enemy vehicle with his squad automatic weapon, moved toward the tower, where he discovered another Ranger entangled in concertina wire and under fire from the tower's main room overlooking the airfield. In the specialist's own words, "I shot up the top floor and quickly got him out of the wire." The specialist then chanced upon an ad hoc ranger fire team composed of three sergeants and a specialist who had maneuvered to the tower in order to secure it. After shooting the lock off the entrance, the men entered and discovered a stairway leading to the top of the structure. From above, a Panamanian defender began shouting at them in a mix of Spanish and English. The team responded with entreaties and threats, after which the lone defender determined it was in his best interest to surrender. The Rangers then ascended the stairs, stopping to clear each level before reaching the control room at the top. There, one of the sergeants broke out the largest of the windows and placed three chem lights in the frame as a signal that the facility was in U.S. hands.[38]

In other action near the gateway and the control tower, elements from Company A, 3d Battalion, fanned out north of the Inter-American Highway to "screen the Americans from possible counterattacks." In this maneuver, they employed six motorcycles that had been dropped by the two C–130s

[37] The bulk of this account of the assault on the gatehouse and the quote are from Donnelly, Roth, and Baker, *Operation Just Cause*, pp. 344–46. See also Flanagan, *Battle for Panama*, p. 149; U.S. Special Operations Command, *History, 1987–2007*, p. 37.

[38] Flanagan, *Battle for Panama*, pp. 146–47.

carrying heavy equipment for the assault. Also in the drop were four jeeps, which elements from the battalion's Company B used to establish blocking positions on the east side of the airfield near the highway. From these positions, the Rangers could interdict military and civilian traffic along the major thoroughfare, engaging the former and forcing the latter to turn back, sometimes with the assistance of a few warning shots.[39]

Within an hour after the beginning of the Ranger airborne assault, PDF resistance on the airfield began to subside. This allowed most of Company B, 3d Battalion, to begin removing obstacles from the runway in preparation for the scheduled landing of U.S. planes carrying additional soldiers and supplies. The Panamanians had employed barrels and barbed wire to block the landing path, but the largest obstacles were two trucks that had been driven onto the runway and then had their tires deflated. There were also obstructions the Rangers themselves had created, namely hundreds of abandoned parachutes that littered the premises and would have to be policed up. Kernan had originally hoped to have the airfield ready by H plus 35 minutes to receive incoming planes, but the intensity of the initial PDF resistance and the time needed to clear away the obstacles added an hour to that estimate. One consequence of the delay was that the Little Bird helicopters used in support of the ranger assault were beginning to run low on fuel and ammunition. A team assigned to set up a forward area rearming and refueling point had parachuted in with the Rangers, but their equipment, dropped by a heavy-lift C–130, failed to work. The helicopters would have to wait until backup equipment arrived on two MC–130 Combat Talons, scheduled to be among the first five planes to land at Rio Hato. Around 0215, the AH–6s began setting down on beaches to the east of the objective until they could refuel and rearm.[40]

The first Combat Talon landed around 0230 and, after disgorging the follow-on Ranger force, gun-jeeps, and other vehicles, taxied to the southeastern end of the field to set up the rearming and refueling point. A second Talon followed suit. When it arrived at the refueling site,

> the lead Talon already had deployed the fuel hose and had set up a fuel pumping system that was connected to the aircraft's single-point refueling manifold. All available crew members were cleared off headsets to assist the two ranger munitions specialists in transporting the 180-pound rocket containers approximately 100 yards to the helicopter refueling and rearming point. As soon as the FARP was operational (approximately 15 minutes), [A]H–6 helicopter gunships began arriving for fuel and ammo servicing.

While the refueling point was being established, the three other incoming planes—each a special operations low-level C–130E—landed. Two unloaded their cargo of troops and equipment. The third "was reconfigured internally to a medevac [medical evacuation] configuration and remained in position ready to extract wounded personnel from the airfield."[41]

---

[39] U.S. Special Operations Command, *History, 1987–2007*, p. 37.

[40] Ibid.; Flanagan, *Battle for Panama*, p. 144; Thigpen, *The Praetorian STARShip*, p. 327.

[41] Thigpen, *The Praetorian STARShip*, p. 328.

The rearming and refueling points were not the only functional elements becoming operational on the battlefield. Command, control, and communications facilities were being set up at various locations. Within the 3d Battalion's Headquarters and Headquarters Company were the "supply personnel, cooks, S–5 personnel, and Spanish linguists" who constituted the enemy prisoner-of-war (EPW) team. This diverse group moved around the airfield from one secure area to another, finally organizing a control point at which they processed over two hundred prisoners. Also present were medical personnel who would establish the battalion's casualty collection point run by the battalion surgeon and headquarters company first sergeant.[42]

As the reinforcement and resupply planes landed at Rio Hato, U.S. forces began coming together to carry out their planned attacks. The fight to secure the airfield had taken longer than expected, over an hour, as many Rangers had experienced difficulty negotiating the elephant grass in the dark, while others, relying on their own initiative and skills, had jerry-rigged ad hoc fire teams to deal with the tenacious enemy resistance. The resulting delays, however, did not necessarily indicate an unfavorable situation. To the contrary, the presence of a sizable number of PDF forces at the airfield, combined with the large numbers of uniformed Panamanians seen fleeing the scene, could mean that few hostile elements remained in the company compounds and other targets listed in the Task Force Red-R battle plans.

Once a sufficient number of troops from the two ranger battalions had gathered at their assigned assembly areas, the units were finally able to begin moving to their primary objectives. For elements of Company A, 2d Battalion, that meant clearing and securing the Herrera-Ruiz Military Institute, the noncommissioned officer training school in the vicinity of both the PDF 6th Company and 7th Company compounds. The assault began with a fire team

The gate to Herrera-Ruiz Military Institute at Rio Hato

[42] HHC 3/75 Narrative on Operation Just Cause, n.d.

from one ranger platoon breaching a chain-link fence and moving up to a communications building near the school. At the entryway to the building, the group stopped and threw a grenade into the main room, with the explosion temporarily knocking out power. After determining the target was "cold," the platoon leader received word to continue to the training center itself.[43]

Although the school was supposed to be unoccupied, the Rangers had come under fire from that general location, so in the words of the company commander, "We proceeded against it with fairly good caution." As a fire team approached the training center, it saw several individuals "running through the courtyard and going into doors." Once in position outside the institute, the team called three times for anyone inside to surrender. When there was no response, the men began the clearing process. A Ranger threw a grenade into the first room he came upon, after which the team sprayed the area with small-arms fire. To their surprise, a number of men, all cadets at the school, came "flying up" from under their bunks, their hands in the air. The grenade had wounded eight in the room, but "quick thinking & reaction prevented [the] students from getting killed." Finding one cadet who spoke English, the Rangers followed him to other rooms occupied by the students. At each, he assured his classmates that, if they gave up, they would not be shot. In all, 167 unarmed cadets surrendered. Escorted into the school's courtyard, all were searched and the wounded received medical treatment. They were then transported to the prisoner collection point in groups of twenty-five. Just before the school was pronounced secured, an armed PDF instructor who had fled into the jungle also surrendered.[44]

Near to Company A, about sixty Rangers from the 2d Battalion's Company B departed their assembly area around 0230 to clear the PDF 7th Company compound, labeled Objective Cat. The group was at 60 percent strength, a cause for some concern, but, to await additional troops would have meant further delay, so the men moved out. If the *macho de monte* succumbed without much resistance, Company B would continue to the PDF 6th Company compound and clear it as well. If the 7th Company put up a fight, 2d Battalion's Company C, the reserve force, would assume responsibility for subduing the 6th Company.

To gain access to Objective Cat, the Rangers of Company B had to walk with their heavy rucksacks for thirty minutes through approximately two hundred yards of unexpectedly thick jungle, negotiate an eight-foot-high chain-link fence, and "use demolition charges to blow holes in the wall surrounding the compound." That done, they advanced on the PDF buildings with one platoon taking the left side of the barracks, another the right. A weapons squad was to provide fire support from four M60 machine guns and a sniper, but three of the four machine gun crews were forced to drop out en route to the compound, and, according to one participant, there was initially no covering fire from the unit's remaining M60 machine gun. The four M60s, he observed, had "either burned in or did not work." Some point after the attack began, however, the remaining M60 and the sniper were able to provide

43 Handwritten notes, Sgt Darren Smith [2d Plt, Co A, 3d Bn], 1 Jan 1990.

44 All quotes except first from ibid. First quote from Flanagan, *Battle for Panama*, pp. 149–50. U.S. Special Operations Command, *History, 1987–2007*, p. 37; Donnelly, Roth, and Baker, *Operation Just Cause*, p. 349.

effective suppressive fire, while AH–6s peppered the back of the compound from the air.

In maneuvering against the barracks, the same Ranger noted that "MOUT training was no good. In training you are supposed to enter from the top. How would you do that?" With no way to get to the rooftops safely, the attackers accepted the reality of the situation: clearing the structures would take place from the bottom up—an adjustment that other U.S. forces in Panama had to make as well. A squad entered the first building in the compound, a two-story structure, by throwing grenades through the front door and then firing three-round bursts to clear each room. The Rangers found no enemy soldiers. The two platoons employed similar tactics for each succeeding barracks, as they leapfrogged squads from building to building. From a roofed shed near the last building, a defender fired at the advancing Americans, but, dazed and bleeding from his nose and ears—probably from the concussive effects of the Americans' weapons—he offered no resistance once two Rangers kicked down the door and confronted him. After that, around 0430, the compound was reported cleared, and the Rangers established a defensive perimeter while preparing to attack the 6th Company compound. Up to that point, there had been no friendly fatalities and the enemy had not put up much of a defense. One Ranger, though, while not disappointed with the relative ease with which the operation had been completed, was unhappy on another count. "We were led to believe that over 400 elite Panamanians would be here—that's what Regimental Intel said. Now I was really pissed because they had scared the s——t out of me earlier."[45]

Near the 7th Company compound was a complex containing a motor pool and fuel tanks. While Company A was focused on the school, and Company B was concentrating on the *macho de monte* barracks, an ad hoc team of three Rangers who had run into each other after the parachute drop decided to move against the smaller cluster of buildings. Two of the soldiers were from the 3d Battalion, the third from the 2d. "The reason we opted to attempt [the takedown]," one of them later explained, "was due to the fact a secondary objective was to block any armor movement north from 6th and 7th company [area of operations] to reinforce drop zone or escape." They quickly devised a plan to move first toward a fuel tank northeast of the motor pool and to use grenades to destroy a row of vehicles parked near the container. Before the three men could implement their plan, however, they were joined by six more Rangers, including a platoon leader who took charge of the operation. To avoid secondary explosions from the vehicles' gas tanks, he rejected the idea of the grenade attack. That notwithstanding, the men still managed to secure the vehicles "without incident." One Ranger then entered the motor pool, where he shot and killed an armed Panamanian trying to exit from the other end of the building. There was no further shooting at the site, although from their new location, the men spied two additional buildings, one to the south, the

[45] First quote from U.S. Special Operations Command, *History, 1987–2007*, pp. 36–37. Flanagan, *Battle for Panama*, pp. 150, 155. Last quote from firsthand accounts of ranger experiences posted on http://www.suasponte.com/m_riohato.htm. Remaining quotes and firsthand account of a Ranger who took part in clearing the compound from Bahmanyar, *Shadow Warriors*, pp. 128–30.

other to the west. Preparatory fire on both buildings included M203 high-explosive rounds, with a LAW also being employed against the building to the west. A fire team then used grenades and small arms to clear the southern building, taking one prisoner in the process. The team used the same tactics to secure the building to the west but found no Panamanians inside. Meanwhile, the Rangers remaining in the motor pool began to set up "a defense vehicular ambush" site.[46]

At 0400, even before Objective Cat was officially cleared, Company B moved out from its perimeter around the *macho de monte* barracks to begin securing the 6th Company compound, Objective Lion. Again there was a walk through the jungle, this time about fifty yards, before reaching the target, after which the assault began. As the ranger squads started the systematic clearing of each building in Lion, some came under enemy fire from Panamanians sitting in ambush positions around the compound. Overcoming this opposition, the Rangers completed their task, after which they formed a "wagon wheel" defensive perimeter. Only then did many of them learn that one of their number, Spec. Philip Lear, had been killed, and another, a Sergeant Howard, severely wounded. The news hit hard, triggering a mixture of emotions. One Ranger remarked later, "In retrospect I think that if they had dropped the bomb on the barracks Lear would not have gotten killed. It pisses me off to this day." Another was walking when "six feet away I saw SSG Howard on the ground, shot through the chest, and Lear dead. It didn't really hit me right away but then—damn—tunnel vision—this can't be happening!" Still another Ranger, after viewing Lear's body later in the day, lamented, "He was gone forever," and then remembered how "just days before I helped Lear tow his car with my Ford Bronco. He was with his fiancé that day, I remember how kind she seemed and how they appeared to adore one another. That was all

Outside the PDF 6th Infantry Company headquarters, a sign in Spanish reads "Loyalty or Death."

[46] Flanagan, *Battle for Panama*, pp. 151–52.

over now." In the midst of this sad news, the battalion commander reported to Colonel Kernan at 0628 that Objective Lion was secured.[47]

Knowing that many Panamanian defenders—perhaps up to three hundred—had fled the area of operations, elements of the 2d Battalion moved during daylight into the town of Rio Hato, where the families of many of the PDF troops lived. The Americans assumed that most males of military age in the vicinity were enemy soldiers. With that in mind, according to one Ranger, "We rounded up all males between 14 and 80, kicked in doors, flex-cuffed individuals and took them to [battalion headquarters] or Regimental HQ—they were handling all EPWs."[48]

Company B's success in clearing both PDF company compounds at Rio Hato meant that Company C, 2d Battalion, was no longer needed as a reserve force; it was free to move along the coastline to seize Noriega's beach house and neighboring guest houses. The operation took place early on the morning of the twentieth, with the Rangers expecting serious resistance. Even so, they were under strict orders not to destroy the houses. Thus, as a first order of business, the company established observation and supporting positions on the high ground overlooking the objective. When Company C actually began to advance toward the houses, the men came under only brief and ineffective fire from a small group of Noriega's bodyguards, who fled soon thereafter (to be picked up at a ranger blocking position in the vicinity). At dawn, the 3d Platoon, in the words of one Ranger, "got to clear the nicest of all the mansions," but only after using a LAW—"a large key," he said with a smile—to blow open the steel front doors. As the soldiers cleared the rooms inside without incident, they found the kitchen where "there was food still hot on the stove and beer that was just opened." There was also a "conference room [that] had shelves of movies and documentaries. It was a big room, the kind you would find in a big business." Once the beach and guest houses were secured, elements from the company established a perimeter along the high ground. The men had suffered no casualties and had captured no enemy soldiers.[49]

Meanwhile, north of the Inter-American Highway, Rangers were trying to overcome the last pockets of PDF defenders. The Panamanians were conducting a systematic withdrawal, "fighting from building to building through a small built-up area." During one engagement, U.S. forces called for aerial fire support. Two AH–6s soon arrived and opened up on the targets they had been given. The pilots did not realize that, in a flanking maneuver, a group of Rangers had advanced into a tree line near the targets. When one of the pilots saw the Rangers' movement, he was convinced they were enemy

[47] U.S. Special Operations Command, *History, 1987–2007*, p. 37; Bahmanyar, *Shadow Warriors*, p. 130; Flanagan, *Battle for Panama*, p. 155. Quotes from firsthand accounts of ranger experiences posted on http://www.suasponte.com/m_riohato.htm.

[48] U.S. Special Operations Command, *History, 1987–2007*, p. 37. Quote from Bahmanyar, *Shadow Warriors*, p. 130.

[49] On not destroying the beach houses, see Daily Staff Jnl, HHC, 75th Ranger Rgt, 20 Dec 1989. First and second quotes from Donnelly, Roth, and Baker, *Operation Just Cause*, p. 349. Third and fourth quotes from and firsthand accounts of ranger experiences posted on http://www.suasponte.com/m_riohato.htm.

forces and fired on them. Four were wounded and two—S. Sgt. Larry Barnard and Pfc. Roy Brown Jr.—were killed.[50]

By midmorning on 20 December, the Rangers at Rio Hato had accomplished their principal missions. The airfield was in friendly hands, and U.S. aircraft were landing with reinforcements and much needed supplies. The most prominent PDF units and facilities in the vicinity had been overwhelmed. As was the case in other battle zones throughout Panama, the Ranger 2d and 3d Battalions continued to receive sporadic mortar and small-arms fire throughout the day, but enemy forces could no longer mount any organized and significant resistance.

As the day passed, individual ranger units sought access to the supplies that were now on the ground, a process impeded by the fact that few vehicles capable of transporting water, food, and other commodities had been dropped from the heavy-equipment C–130s at H-hour. Trucks arrived on later transports, but, until then, the Rangers at Rio Hato, like their counterparts at Torrijos-Tocumen, relied on their own initiatives to make distribution. In one case, two sergeants from the 2d Battalion's Company A presented their executive officer with an air-conditioned Mazda RX–7 they had found; the lieutenant used the car to carry filled canteens to Rangers in desperate need of water in the tropical heat. A red station wagon that could only move in reverse was also "commandeered" to ease the transportation burden. While these and other Rangers coped with the resupply problem, still others received orders to find equipment that had been lost during the initial drop or to recover parachutes and help clean up the airfield's runway. Policing the battlefield took some time. "There was plenty of damage. Cars were charred and anti aircraft guns lay blown apart." In a grimmer task, medics and those assisting them drove around in "a beat up jeep with the chaplain and body bags," collecting the enemy dead. "Bodies, bodies everywhere," one Ranger recalled. "I felt empty coping with the loss." The chaplain would give dead Panamanians their last rites before they were placed in the body bags, which were then put on the hood of the jeep and taken to a consolidation point. In all, thirty-four defenders were known to have been killed; the number of wounded was unknown. Three hundred and sixty-two prisoners, along with forty-three civilian detainees, had been turned over to the prisoner collection point for processing. As for the Rangers, they had suffered four deaths, two from friendly fire, and twenty-seven men wounded in action. Increasing the casualty figures were the three dozen injuries they had sustained during the parachute jump.[51]

With Panamanian military installations in and around Panama City and Rio Hato under American control, the Panama Defense Forces could no longer mount an effective counterattack against U.S. troops on the Pacific side of the canal. There were, to be sure, isolated PDF units headquartered outside

---

[50] Quote from U.S. Special Operations Command, *History, 1987–2007*, pp. 37–38. The names of the Rangers killed in the friendly fire incident are from Bahmanyar, *Shadow Warriors*, p. 138.

[51] The material on supply issues is from Flanagan, *Battle for Panama*, pp. 154–55. The account of policing the battlefield and the quotes are from the firsthand accounts of ranger experiences posted on http://www.suasponte.com/m_riohato.htm. The casualty and prisoner-of-war numbers are from Briefing Slides, 75th Ranger Rgt, Operation Just Cause, n.d.

the canal area as far away as David in the west and Darién Province in the east. But these garrisons contained no major combat units and were slated to be subdued in follow-on operations. Thus, by midday on 20 December, U.S. forces had completed their priority missions on one side of Panama. While they had been doing so, Task Force Atlantic was attempting to duplicate the feat on the other side of the country.

10

Task Force Atlantic: D-day in Area of Operations North

☆ ☆ ☆

Given the priority assigned to the mission of removing the leadership of the Panama Defense Forces and neutralizing the organization's most potentially dangerous elements, the preponderance of Just Cause D-day operations targeted the numerous enemy units and installations on the Pacific side of the Panama Canal. Several task forces had engaged in this effort, and in most cases they had accomplished their missions before the end of the first day, a feat attributed to extensive planning, detailed and realistic rehearsals, the use of decisive force, some element of surprise, the professionalism of the U.S. military, and the mismatch between American forces and their Panamanian adversary. The result was that enemy resistance on the Panama City side of the country had been defeated or isolated without an inordinate number of casualties (friendly, enemy, or civilian) or amount of damage to the capital city.

In contrast, combat troops responsible for operations on the Atlantic side of the country belonged to only one task force, Task Force Atlantic, an organization initially established under JTF-Panama during Operation Nimrod Dancer. Headquartered at Fort Sherman, a U.S. installation situated on the west side of Limón Bay across from the city of Colón, the task force had fifteen Just Cause missions, including nine H-hour targets. As if this did not present enough difficulties, General Stiner had labeled Task Force Atlantic an economy of force: superior firepower would have to compensate for what was at best a one-to-one friendly-to-enemy troop ratio, well below the desired ratio of three to one, if not four to one or even five to one, of attacker to defender. Alluding to the mandated limits on the number of combat personnel available to him, the task force commander told his subordinates shortly before H-hour, "You know, you guys are on your own; we've got no reserve. . . . I can't help you if you get in trouble. You've got to fight this one through on your own. I've got me and my S3 with a rifle, and that's the reserve of the task force."[1]

[1] Quotes from Interv, author with Brig Gen Keith Kellogg and Lt Col Gregory Gardner, both U.S. Army, 9 Apr 1992, Fort Bragg, N.C. Intervs, author with Gardner, U.S. Army, 7 Apr 1992, Fort Bragg; with Lt Col Johnny Brooks, U.S. Army, 11 May 1993, Fort Leavenworth, Kans. Both Brooks and Gardner refer to the economy-of-force status.

Perceptions, Planning, and Preparations

The speaker was Col. Keith Kellogg, commander of the 3d Brigade, 7th Infantry Division (Light). He had arrived in Panama in mid-October, part of the scheduled rotation in which he replaced Col. David R. E. Hale, the 7th's 1st Brigade commander, as the head of Task Force Atlantic. Significantly, in view of what would occur during Just Cause, Kellogg perceived the situation in Panama very differently from his predecessor. Whereas both men were experienced and regarded themselves as "warriors," Hale was more contemplative and readily understood that political considerations inherent in the crisis would at times trump traditional military thinking and necessitate some constraints on the actions of his troops. In contrast, the more orthodox Kellogg was seen as "likely to shoot from the hip and sort it out later." As he later reminisced, "I thought that we were too restrictive because you're asking your young soldiers to do too much thinking instead of being soldiers, and I thought you could make it much more black and white and take the onus away from their having to make real hard decisions. That's my job." Shortly after arriving in Panama, the colonel met with General Thurman and came away reassured that "we weren't going to take any crap from the Panamanians" and that the SOUTHCOM commander "wasn't afraid to get into a nose-to-nose, hard-ass confrontation with them," even if it meant a shooting incident that might trigger a war. Based on the four-star's guidance, Kellogg was determined to place Task Force Atlantic on a war footing as a prelude to what he saw as the "disaster waiting to happen." In brief, he anticipated adopting "a much more aggressive approach to doing business down here as a task force."[2]

In the weeks that followed, Kellogg's task force "turned up the heat quite a bit" on the Panama Defense Forces through the zealous implementation of Sand Fleas and contingency readiness exercises. Yet, while the colonel believed this approach to be in keeping with Thurman's intentions, his immediate superior was another general, Marc A. Cisneros, the JTF-Panama commander. At first, Kellogg did not see this arrangement as cause for concern. Cisneros, as the SOUTHCOM operations officer under General Woerner, had been known to favor a tougher stance in dealing with the Panama Defense Forces than his boss had thought prudent. Moreover, once in command at Fort Clayton, Cisneros had moved quickly to introduce several measures designed to put additional pressure on the Panamanian military. Yet, there were limits to how far even he was willing to go, and Kellogg, he soon concluded, was too often inclined to cross those boundaries. The colonel and his staff for their part came to perceive Cisneros as someone who "would kind of talk a good game, but when you actually stopped somebody [the Panamanian military], he was on that line in a heartbeat saying, 'Kellogg,

[2] Quote describing Kellogg's persona from Interv, author with Brooks, 11 May 1993. All other quotes from Interv, author with Kellogg and Gardner, 9 Apr 1992. Although Kellogg and Hale had little face-to-face time in mid-October while conducting the Task Force Atlantic transition, the two colonels and their staffs had met in Panama for about three days in August to discuss the upcoming handoff. Interv, author with Gardner, 7 Apr 1992. In discussions about the personality differences and leadership styles of Hale and Kellogg, at least two officers who knew each of the colonels well (Brooks and Gardner) concluded that Hale was best suited for working under Woerner and Loeffke and that Kellogg was more attuned to Thurman's approach. As Gardner observed, "Of the two of them, the right guy was in the right place at the right time."

what are you doing over there. . . . You're getting out of control,' and that kind of thing." In one such case, Cisneros had ordered Kellogg to dismantle a roadblock the colonel had set up on his own authority to control traffic on a main avenue near Coco Solo following a drive-by shooting in the area. In another, Cisneros had rejected a proposal from Fort Sherman to have Task Force Atlantic elements conduct a night assault exercise onto Fort Amador on the Pacific side of the canal area. "That's out of your area," the general is reported to have told his subordinate commander.

As a result of these and other such cautions, Kellogg and his staff came to regard Cisneros as a "political general" like Woerner and Loeffke, someone who did not want "to make waves." For his part, Cisneros developed very strong reservations about the ability of Kellogg and those around him to recognize the nuances of the current crisis and, should hostilities come, to keep from overreacting in such a way as to cause unnecessary casualties and physical damage. Still, despite the general's wariness and his attempts on occasion to rein in the Task Force Atlantic commander, in the weeks preceding JUST CAUSE Kellogg generally enjoyed a high degree of autonomy in his day-to-day operations, in part because of his location on the other side of the isthmus, well removed from Fort Clayton.[3]

While U.S. troops on the Atlantic side carried out their various assignments, Kellogg and his immediate subordinates found themselves caught up in another critical activity. Their arrival in mid-October ensured that they would be deeply involved in the planning effort that Thurman had accelerated immediately after the failed coup. Back in 1988, both Kellogg, then the 7th Infantry Division's operations officer, and Lt. Col. Gregory Gardner, who was now the colonel's operations officer in Task Force Atlantic, had been involved in the early planning directed by Woerner, and both had been instrumental in determining what role their division would play, initially under ELABORATE MAZE, then under two of its successors, the POST TIME and BLUE SPOON operation orders in the PRAYER BOOK. At the time, the 7th was the only U.S.-based division "to support the plans with all three brigades," a commitment that meant it would be designated JTF-Panama's Army Forces component. By September 1988, the division's set of contingency plans for the Panama crisis had reached a "steady state," in Gardner's words, as he moved into his next assignment, the S–3 position with the 3d Brigade, the unit that had come under Kellogg's command the previous month.[4]

A year later, as Kellogg and Gardner settled in at Fort Sherman, they learned that the "old plan was out."[5] Thurman had named General Stiner his war fighter, and planners at the XVIII Airborne Corps, by bringing in the 82d

---

[3] On the pre–JUST CAUSE relationship between Kellogg and Cisneros, see Interv, author with Kellogg and Gardner, 9 Apr 1992; Interv, author with Gardner, 7 Apr 1992, which contains the quotes. In his interview, Gardner, who was Kellogg's operations officer, conceded that Cisneros' views of the crisis were probably right but that Task Force Atlantic under Kellogg simply had its "warfighting focus."

[4] On Kellogg's and Gardner's participation in contingency planning during 1988, see Interv, author with Kellogg and Gardner, 9 Apr 1992; Interv, author with Gardner, 7 Apr 1992, which contains the second quote; Interv, author with Maj Harry Tomlin, U.S. Army, 22 Mar 1991, Fort Leavenworth, Kans. First quote from 7th Infantry Division (Light) Input, Overview: Planning and Pre–JUST CAUSE Deployment, to JTF-Panama AAR, Operation JUST CAUSE, 3 Feb 1990.

[5] Interv, author with Kellogg and Gardner, 9 Apr 1992.

Airborne Division for H-hour combat operations, had effectively relegated the two light-fighter brigades still in the United States to follow-on missions. Furthermore, Stiner's JTF-South, not the 7th Infantry Division, would be dual-hatted as the Army Forces component in the event of war. What this meant for Kellogg was that he and his subordinates would have to help revise the contingency plans for those units under the control of the one brigade headquarters from the 7th that would be in Panama at the outset of hostilities, Task Force Atlantic.

Kellogg had only one week to complete the initial revisions, presumably so that the Southern Command and the XVIII Airborne Corps could incorporate them into CINCSO Operation Order 1–90 and JTF-South OPLAN 90–2 that would be presented to the Joint Chiefs of Staff in early November. Kellogg met this deadline, which culminated in a briefing he gave to Stiner. Following that, work on updating the plan continued throughout November and into mid-December, with the colonel meeting frequently with his headquarters staff, as well as with the commanders and staff officers of the security enhancement battalion under his control and of whatever battalion was currently rotating through the Jungle Operations Training Center. At the time Kellogg arrived in Panama, the 7th Infantry Division's 2d Battalion, 27th Infantry, commanded by Lt. Col. Jeff Rock, was the security battalion, with its headquarters at Fort Davis. For the next month, Rock's people added missions and refined the ones already assigned to them under Blue Spoon and 90–2. Then, at the end of the third week in November, the battalion redeployed to the United States, to be replaced by another 7th Infantry Division unit, the 4th Battalion, 17th Infantry, commanded by Lt. Col. Johnny Brooks. Brooks' battalion had just come out of the jungle training center where, like those units that had preceded it during Nimrod Dancer and those that would follow it prior to Operation Just Cause, it had maintained a "formal command relationship" with Task Force Atlantic for the purpose of contingency planning.[6]

As plans neared their final form, the JTF-South Operations Plan 90–2 listed several tasks for U.S. forces under Kellogg. These included neutralizing the Panama Defense Forces and capturing arms caches in the Task Force Atlantic area of operations; seizing and securing Fort Espinar, Coco Solo, and Renacer Prison; securing Madden Dam; isolating, neutralizing, and, if necessary, destroying enemy elements at Cerro Tigre and Gamboa and being prepared to do the same to the PDF military regional headquarters in Colón; protecting American lives and facilities; and preventing enemy reinforcements from moving into or out of the area of operations (*Map 16*).[7]

As these tasks were refined, necessary adjustments were made to Task Force Atlantic's Sand Fleas, contingency readiness exercises, and Purple Storms, bringing each closer in line with a specific wartime mission. For example, after the JOTC battalion received the Blue Spoon responsibility for taking Gamboa, the unit began conducting more Sand Fleas in that area, in the process gathering additional intelligence on the objective, especially through

---

[6] Ibid. Quote from Briefing, Lt Col Johnny Brooks, 29 Apr 1992, Fort Lewis, Wash.

[7] JTF-South OPLAN 90–2, 3 Nov 1989. A summary of the Task Force Atlantic mission statement is also in 7th Infantry Division Input, Purpose, to JTF-Panama AAR, 3 Feb 1990.

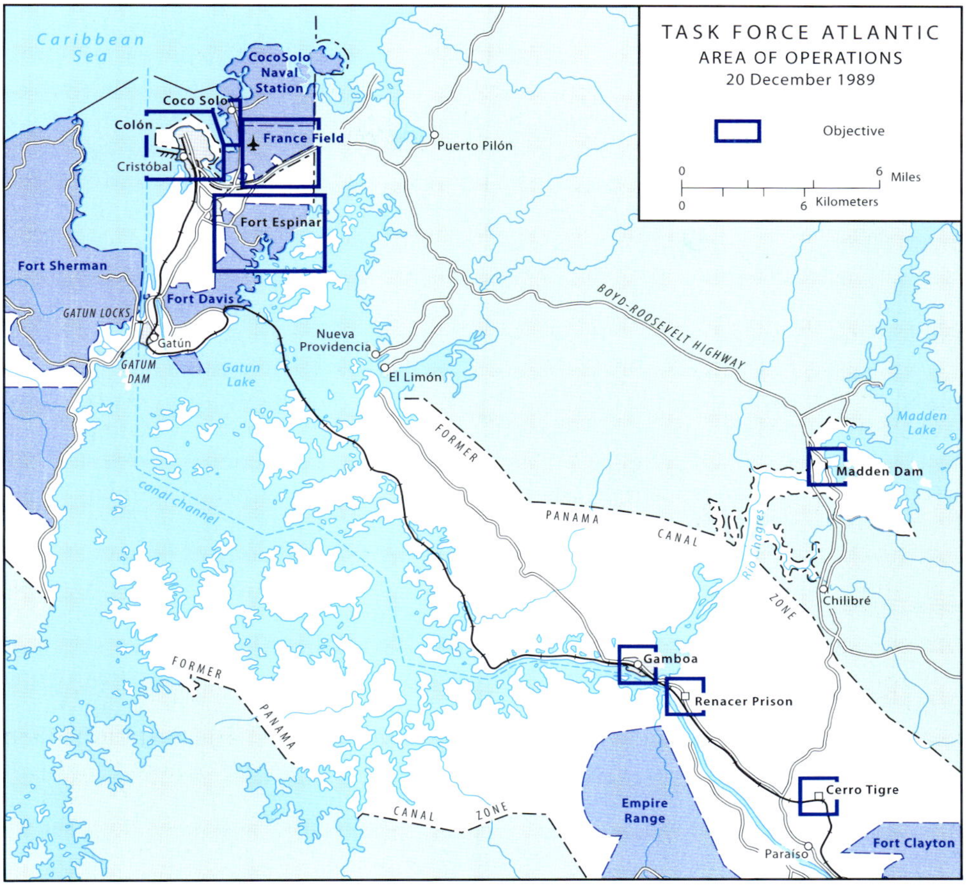

Map 16

contact with friendly Panamanians living there. (One operation, a simulated air assault on the potential target, met with PDF attempts to obstruct it, which caused the U.S. battalion commander to have his soldiers fix bayonets after which the Panamanians moved aside.)[8]

For some targets, acquiring the desired intelligence proved difficult. Colón was such a case: legal issues pertaining to Panamanian sovereignty prohibited positioning U.S. forces where they could observe several objectives in and around the city. Also, restrictions on U.S. troop movements near an ammunition dump in the vicinity of Cerro Tigre limited intelligence gathering at that target. The isolation of Renacer Prison created additional problems. While Brooks' battalion was in the Jungle Operations Training Center, the lieutenant colonel acquired blueprints for the prison and, on a few occasions,

[8] Briefing, Brooks, 29 Apr 1992; Intervs, author with Kellogg and Gardner, 9 Apr 1992; with Gardner, 7 Apr 1992; with Brooks, 11 May 1993. In his briefing at Fort Lewis, Brooks emphasized that, during the period from mid-October through mid-December, U.S. forces under Task Force Atlantic engaged in many live-fire exercises and carried ammunition with them everywhere they went, resulting in an increase in the troops' confidence.

flew as close to it as he could. Not until ten days before Just Cause, however, were elements from the JOTC battalion, now from the 82d Airborne Division, able to walk the terrain near the facility, thanks to the fact that Kellogg's staff had begun plotting "green to green" routes for Task Force Atlantic exercises (meaning that U.S. troops, as allowed under the canal treaty, could move through sovereign Panamanian territory if it lay across the most direct route from one treaty "area of coordination"—in which American forces were allowed—to another). The Sand Flea near the prison and a videotape of the maneuver made by a Navy SEAL team provided some very useful, if belated information. In sum on the eve of Just Cause, Kellogg had adequate intelligence on most but not all of the objectives for which he was responsible.[9]

In addition to linking exercises to a particular unit's contingency mission, planners also divided the Task Force Atlantic area of operations into two parts, AO North and AO South. In the former, the security enhancement battalion would be the principal U.S. combat force. Its main objectives included Coco Solo, Fort Espinar, Colón, France Field (a last-minute addition), and the transisthmian Boyd-Roosevelt Highway. The JOTC battalion would be the principal combat force in AO South, with the unit's main missions including Madden Dam, Gamboa, Cerro Tigre, and Renacer Prison. Other units, including a battalion cobbled together from the cadre in the jungle training center, would execute critical operations and provide essential support both north and south. In addition, Battery B from the 2d Battalion, 62d Air Defense Artillery, would provide Vulcan light antiaircraft guns armed with 20-mm. Gatling cannon capable of firing 1,000 or 3,000 rounds per minute, "the most firepower we had," according to Brooks. As of mid-December, the Task Force Atlantic order of battle consisted of the following forces:

Headquarters and Headquarters Company, 3d Brigade, 7th Infantry Division
4th Battalion, 17th Infantry, 7th Infantry Division
 Company C, 2d Battalion, 27th Infantry (plus)
3d Battalion, 504th Infantry, 82d Airborne Division
Cadre, Jungle Operations Training Center
7th Military Police Company
Battery B, 7th Battalion, 15th Field Artillery
Battery B, 2d Battalion, 62d Air Defense Artillery
Company C, 13th Engineer Battalion
Company C, 7th Medical Battalion
Company C, 707th Maintenance Battalion
Company C, 7th Supply and Transportation Battalion
3d Platoon, Company B, 127th Signal Battalion[10]

[9] Briefing, Brooks, 29 Apr 1992; Intervs, author with Kellogg and Gardner, 9 Apr 1992; with Gardner, 7 Apr 1992; with Brooks, 11 May 1993. The terms *green to green* and *area of coordination* are derived from sections of the canal treaty pertaining to Panamanian sovereignty, the interaction of U.S. and Panamanian armed forces, and the legal restrictions applicable to both militaries.

[10] Interv, author with Gardner, 7 Apr 1992. Quote from Briefing, Brooks, 29 Apr 1992. Donnelly, Roth, and Baker, *Operation Just Cause*, pp. 236–37. The task organization of JTF-South, including Task Force Atlantic, is at http://www.history.army.mil/documents/panama/taskorg.htm.

A U.S. soldier firing a Vulcan air defense system at a ground target

On the night Lieutenant Paz was killed, Colonel Kellogg, on his own authority, put these forces on alert, had ammunition issued to them, and had them secure certain sites. He then notified General Cisneros, who accepted the fait accompli. "So we kind of got ahead of the whole game on our own," Kellogg later acknowledged. Units gathered at the assembly points for their assault missions, with the only unit out of place being the company slated to block access to and from Colón. The problem was partially remedied by putting the unit on trucks and driving the men from Fort Sherman to Fort Davis, thus locating them on the east side of the canal just south of the city. By Sunday morning, Task Force Atlantic was ready to go to war, only to have Cisneros tell Kellogg and the other JTF-Panama commanders to have their troops adopt more "normal" behavior as a means of lulling Panamanians into believing that the United States was not going to retaliate militarily for the death of the marine. For the remainder of Sunday and throughout the next day, Kellogg had his men engage in routine activities, even while remaining in positions dictated by their contingency plans.[11]

By Sunday evening, Kellogg had learned that elements of the 7th Infantry Division at Fort Ord "were moving." That news and the arrival of XVIII Airborne Corps personnel Monday convinced him that he would be ordered to execute Task Force Atlantic's portion of OPLAN 90–2. Later that night, in the JTF-Panama meeting with Cisneros at Fort Clayton, the information Kellogg received confirmed his conjecture. Shortly before the meeting convened, the colonel sat on a table outside the Simón Bolívar conference room rubbing his hands together. It was a long-standing habit of his, but, to the casual observer, the image was of a man eager to settle the crisis in Panama once and for all.[12]

[11] Quotes from Interv, author with Kellogg and Gardner, 9 Apr 1992. Intervs, author with Brooks, 11 May 1993; with Maj Christopher J. Rizzo, U.S. Army, 15 May 1997, Fort Leavenworth, Kans.

[12] Interv, author with Kellogg and Gardner, 9 Apr 1992. The author was the uninitiated observer. Over three years later, during the interview with Johnny Brooks, the lieutenant colo-

After the meeting, Kellogg contacted Maj. Gen. Carmen Cavezza, his division commander at Fort Ord, to inform him of the impending invasion, only to be told that Stiner had already telephoned the general with the news. Kellogg then placed a secure telephone call to Col. Linwood Burney, the 2d Brigade commander at Ord, asking that Burney assign one of his companies to help Task Force Atlantic "finish the job" with respect to its D-day missions. Burney not only agreed to the request but dispatched one of his best units, which arrived in the country Wednesday morning. Following the phone conversation and some other business at Clayton, Kellogg flew back the length of the canal to Fort Sherman, arriving at 0400 Tuesday. Knowing that he would not be getting much sleep for some time to come, he allowed himself the luxury of not being awakened until 1000.[13]

The remainder of the day was taken up with meetings and a host of last-minute preparations. After again flying to Fort Clayton, this time for the conference at 1600 with Generals Thurman and Stiner, Kellogg returned to the Atlantic side and assembled his battalion and company commanders at 1830. "Gentlemen," he began, "we have an H-hour, and that H-hour is at 0100, 20 December." Reacting to the announcement, one company commander remembered thinking, "That's tonight! That's six hours from now!" In the course of the briefing, Kellogg covered in detail the rules of engagement Stiner had distributed, and he emphasized the role that the professionalism of the U.S. military, together with Task Force Atlantic's firepower, would play in compensating for the doctrinally inauspicious one-to-one ratio between the attacking U.S. forces and the defending Panamanians. The room was tense, although Kellogg could later reminisce, "The joke was 'We're going to knock on the door with an AT4.'" He also relayed to the group Thurman's sentiment that, if any civilians were killed in the fighting to come, the operation would be considered a failure.[14]

Following the meeting, the colonel's field commanders returned to their headquarters with only a few hours to brief their subordinates, issue the necessary operation orders, procure whatever ammunition had yet to be distributed along with other essential items such as glint tape, conduct inspections, and in general prepare the soldiers for battles only a few hours away. There were also the usual personal matters many of the troops tried to squeeze in, such as writing letters or making videotapes to be sent to their families. Then, having done all they could in the time allotted, the units involved began departing for their assault positions. Each soldier, according to one platoon leader,

---

nel, in emphasizing a point he was making about Kellogg, rubbed his hands together the way the colonel had the night of 18 December. Brooks went on to say that, while the habit certainly reflected Kellogg's assertiveness, one should not read too much into it.

[13] Ibid.

[14] Last quote from ibid. Remaining quotes from Interv, author with Rizzo, 15 May 1997. Evan A. Huelfer, "The Battle for Coco Solo Panama 1989," *Infantry* 90 (January-April 2000). The article by Huelfer, who served as a platoon leader in Company C, 4th Battalion, 17th Infantry, is a highly detailed account of the company's activities at Coco Solo and can be found on several Internet sites.

> carried a basic load of 210 rounds in seven magazines for his individual weapon, plus two or more hand grenades. Some carried additional rounds for machine guns, demolition materials, or antitank rockets. Nobody seemed to mind toting extra ordnance. Every soldier wore a protective vest and kevlar helmet and carried at least four quarts of water. Leaders also had night vision devices, binoculars, radios, and flashlights. Altogether, the typical combat load exceeded 50 pounds without even counting the rucksack.[15]

After briefing his own troops at 2100, Capt. Christopher J. Rizzo, the 4th Battalion's Company C commander, had the additional task of orienting a platoon that would be attached to him from the airborne battalion at the Jungle Operations Training Center. The platoon would assist in the attack on the PDF naval infantry company at Coco Solo, Rizzo's primary mission at H-hour. He had briefed the paratroopers the night before, but now he had to provide them with ammunition, teach them how to fire the AT4s without being injured from the missiles' backblast, and get them into position. (Despite the captain's warnings, two of the paratroopers suffered backblast burns during the actual operation.) The fact that the platoon did not arrive at its assembly point until shortly before midnight compressed the time Rizzo could spend with the men. Further complicating matters for him, eight American families living in a housing area across from his objective had to be evacuated and moved to safer residences nearby, a process that began shortly after midnight. The procedure of going door-to-door posed some risk to operations security. As Rizzo's immediate superior Colonel Brooks worried, "Somebody turns a light on, then you go to the next house and somebody turns a light on, and you go to the next house." To his relief, the PDF infantrymen did not detect the evacuation in progress. At least they did not let on if they did.[16]

By 0030, the various elements of Task Force Atlantic were either on the move toward their assault positions or, if already there, in the process of setting up their larger weapons and getting into their tactical formations. At Fort Sherman, Kellogg had assembled a "roving team" that included himself, Gardner, a sergeant major, the brigade's intelligence officer, a fire support officer, a liaison officer, an operations sergeant, and, for protection, two military police vehicles. Once JUST CAUSE kicked off, Kellogg would leave two captains in charge of the tactical operations center at Sherman while he took his team to observe the assault on Renacer Prison in AO South, returning to AO North at dawn. In the latter area of operations, Brooks oversaw the final disposition of his forces. In a move to demonstrate the sense of trust he held in his company commanders, he sent out several field grade officers from his staff and from the JOTC battalion's staff to be with the units under his control that had H-hour objectives (*Map 17*). The officers' purpose was not to micromanage the tactical operations but simply to provide Brooks with information from the front, thus obviating the usual practice of his having to be in frequent and direct communication with company

---

[15] Quote from Huelfer, "The Battle for Coco Solo." "Conquest At Coco Solo," in *Soldiers in Panama: Stories of Operation Just Cause* (Washington, D.C.: Office of the Chief of Public Affairs, Department of the Army, 1990), p. 26.

[16] Interv, author with Rizzo, 15 May 1997. Quote from Interv, author with Brooks, 11 May 1993. "Conquest At Coco Solo," p. 26.

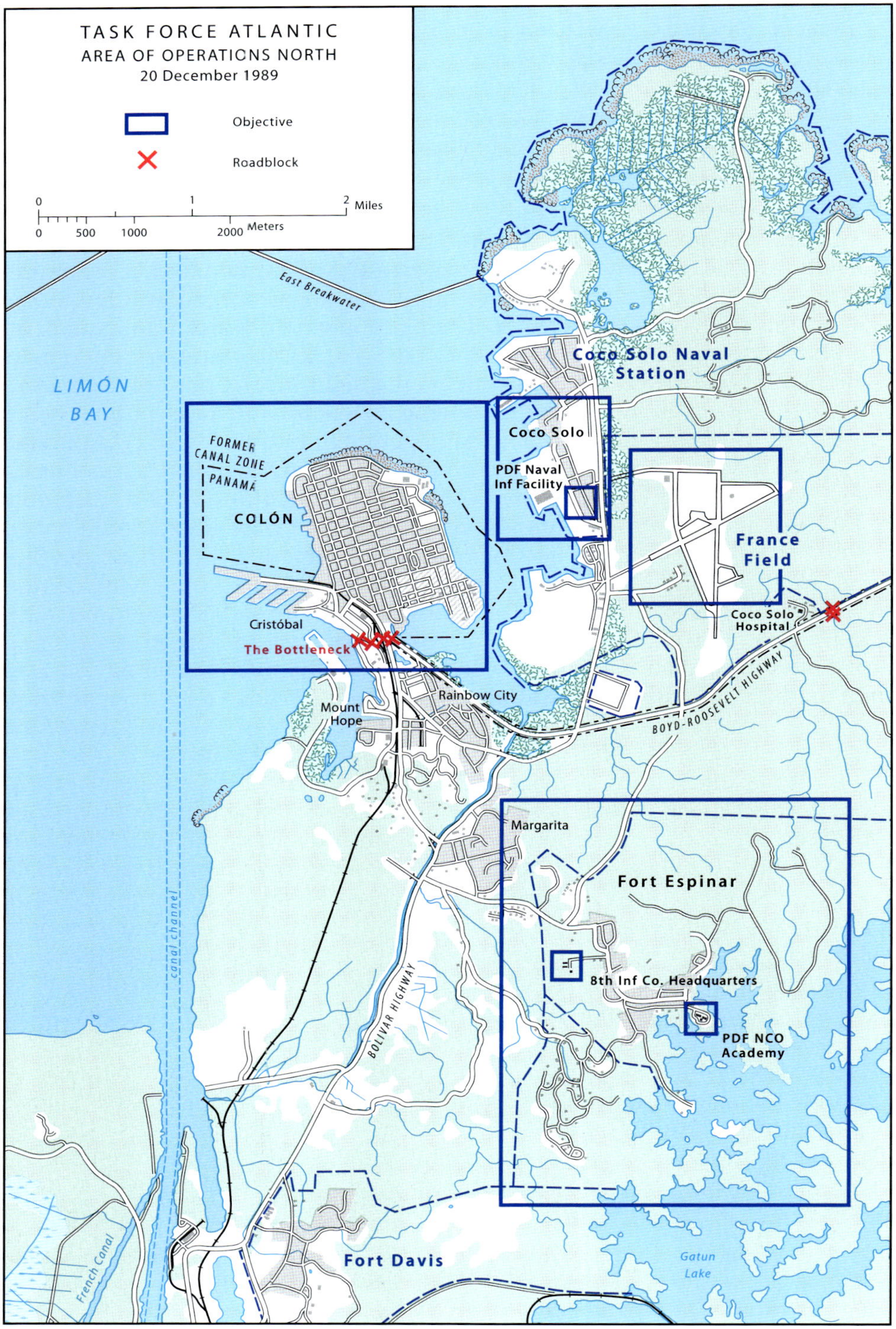
TASK FORCE ATLANTIC
AREA OF OPERATIONS NORTH
20 December 1989
Objective
Roadblock
0 1 2 Miles
0 500 1000 2000 Meters
East Breakwater
LIMÓN BAY
FORMER CANAL ZONE
PANAMÁ
COLÓN
Cristóbal
The Bottleneck
Coco Solo Naval Station
Coco Solo
PDF Naval Inf Facility
France Field
Coco Solo Hospital
BOYD-ROOSEVELT HIGHWAY
Rainbow City
Mount Hope
Margarita
Fort Espinar
8th Inf Co. Headquarters
PDF NCO Academy
BOLIVAR HIGHWAY
canal channel
French Canal
Fort Davis
Gatun Lake

Map 17

commanders, who in his opinion should be left alone as much as possible to conduct their missions. As he described his feelings on the issue, "what they don't need is for me to be screaming and yelling at them on the radio." As for Brooks himself, he would set up his field headquarters at Coco Solo, arguably the most dangerous H-hour mission in AO North.[17]

As with the other task force commanders in Just Cause, Kellogg was aware by the evening of the nineteenth that the operation had been compromised. Yet, as was the case with most other JTF-South components, the 0100 H-hour for Task Force Atlantic remained unchanged. That seemed to pose no problem for Kellogg's troops as they entered their final hour of preparations. The colonel had talked with his commanders about what would happen if something went wrong prior to H-hour, but although some units were having to scramble more than others to reach or prepare their final assault positions, on the whole the scheduled deployments seemed to be proceeding well. Then, about twenty minutes before 0100, a gunshot could be clearly heard from the vicinity of Coco Solo. The war on the Atlantic side had begun, suddenly and prematurely.

Coco Solo

Coco Solo, a former U.S. Navy base, is located east of the Colón peninsula, a half mile across Manzanillo Bay. In December 1989, near the shoreline running north to south, there were two large, almost identical, three-story buildings, each with three wings. Behind them, along the bay, were a large parking lot and a dock area. In part, the complex was a "joint" military facility, with U.S. troops stationed in the northern building and in a gymnasium adjacent to it, and the PDF's 1st Naval Infantry Company—also known as a

Coco Solo

[17] First quoted words from Interv, author with Gardner, 7 Apr 1992. Second quote from Interv, author with Brooks, 11 May 1993.

marine company—in the southern building, only two hundred yards away.[18] As if that were not complicated enough, the northern structure also served as an American junior and senior high school, while the PDF building contained a Chinese restaurant and, in its northern and southern wings, a garment factory. At the southern tip of the PDF facility was a Soviet Lada car factory, across from which was the housing area mentioned earlier. Roads ran in front of the two main buildings, behind them, and between them (*Map 18*).

In late November, when Brook's 4th Battalion had moved out of the Jungle Operations Training Center to become the security battalion at Fort Sherman, Captain Rizzo had learned that in the event of hostilities his company would have the mission of neutralizing the marine company at Coco Solo and any patrol boats and other enemy craft docked in the bay behind the headquarters building. In response, Rizzo had moved the troops of Company C into an unused wing of the American high school. At the same time, his predecessor as company commander gave him the Coco Solo battle book, together with some valuable information and guidance that helped Rizzo begin drafting a tactical plan for accomplishing the mission. A week later, the captain gave his superiors the first of three briefings on the course of action he and his staff had developed.[19]

The concept of operations relied on Company C's three infantry platoons; the attached platoon of paratroopers from Company C, 3d Battalion, 504th Infantry, from the jungle school; two squads from the 7th Infantry Division's 2d Battalion, 62d Air Defense Artillery, each with a 20-mm. Vulcan; a platoon from the 549th Military Police Company at Fort Davis; and a signal detachment from the 127th Signal Battalion. Before H-hour, these units would move into their assigned positions. Rizzo's 1st Platoon would cover the dock area and be responsible for seizing any PDF boats present at the onset of the fighting. The 3d Platoon would take up a blocking position to the south of the naval infantry company headquarters and in front of the southern housing area. The military police would establish four checkpoints at key approaches to the complex, while one of the two Vulcan squads would cover the northern flank, including a second housing area and the waterways leading out of the

[18] The term *joint* is in quotation marks because, in U.S. military doctrine in 1989 (and of this writing), it applied to the interaction of two or more American services, with the term *combined*—today *multinational*—used to denote the interaction of U.S. forces with foreign forces. In the context of the Panama Canal treaties, the term *joint* meant the interaction of U.S. and Panamanian forces.

[19] Interv, author with Rizzo, 15 May 1997. In his briefing at Fort Lewis, Washington, in 1992, Colonel Brooks stated that the original mission had called for the PDF patrol boats to be destroyed. Feeling that, in light of the restoration effort that would follow combat operations in Panama, it might be better to capture the boats, he raised the issue with Kellogg, who agreed. The wording in the mission was therefore changed to *neutralize*. Briefing, Brooks, 29 Apr 1992. As for the guidance Rizzo received for preparing his plan, Huelfer indicates that Brooks, the captain's immediate superior, had "five major concerns regarding operations at Coco Solo": "He could not allow the boats to escape with their firepower intact. He had to protect U.S. citizens in the area. He had to minimize collateral damage to private property. He had to prevent small groups of armed PDF soldiers from escaping the initial battle. He had to prevent the PDF from using their heavy machineguns to influence the battle." See Huelfer, "The Battle for Coco Solo."

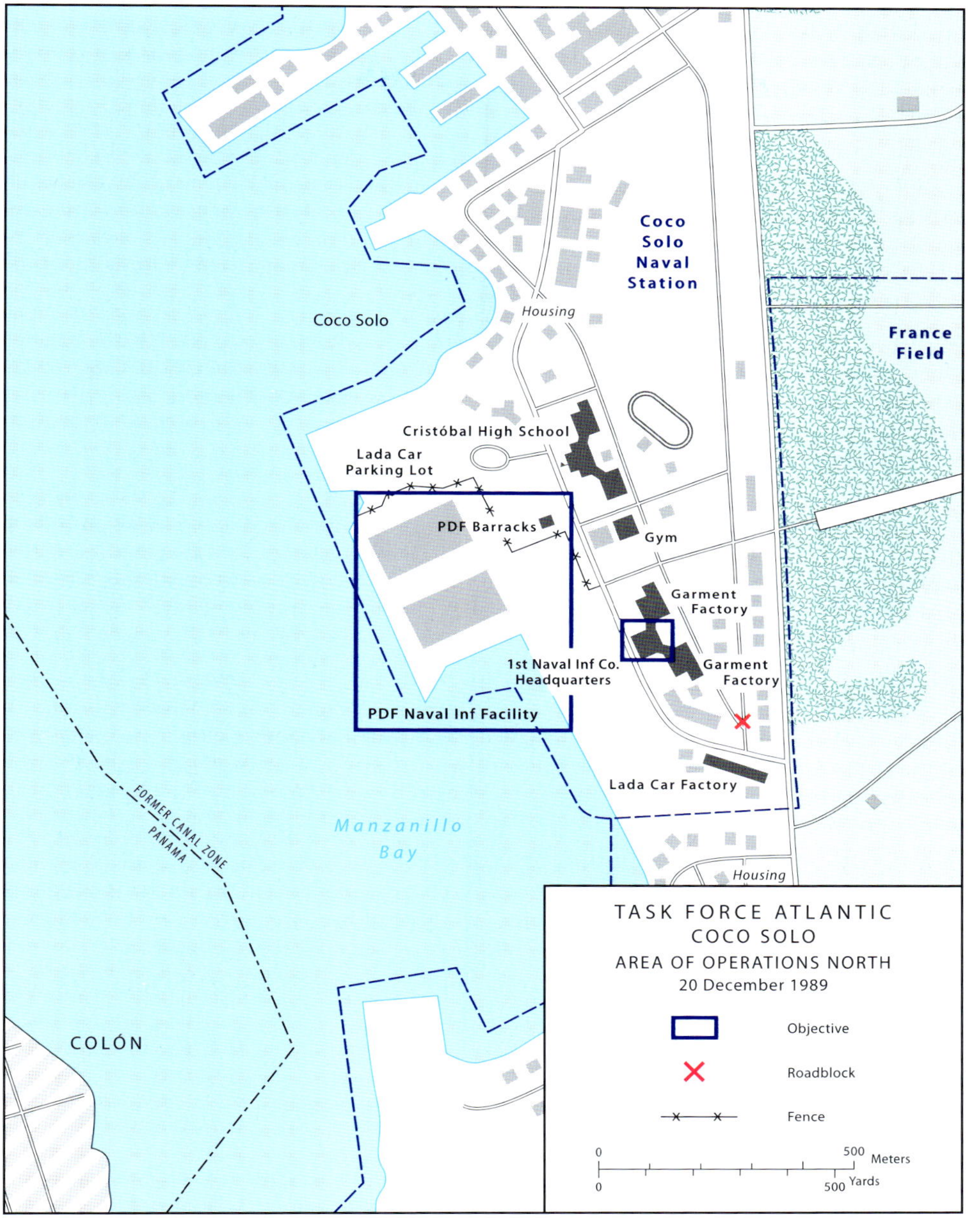

Map 18

bay. Finally, the paratroopers accompanied by the other Vulcan squad would take up firing positions about a hundred yards from the front of the main PDF building, while soldiers in Rizzo's 2d Platoon prepared to enter the north wing of the headquarters via the third floor and then work their way down to enemy positions on the lower levels. The 2d Platoon would not move, however, until the Panamanians had been given a chance to surrender. As an incentive for the defenders to do so, the U.S. troops in front of the building would open fire at 0100 with the Vulcan, a combination of AT4 and LAW antitank weapons, M60 machine guns, and a variety of small arms. The barrage would be intense but brief, after which a loudspeaker team would broadcast a surrender demand. If the enemy rejected the appeal, the cycle of firing and broadcasts would be repeated. Only if the second attempt proved futile would the building be breached.[20]

If the PDF marines did not capitulate after the displays of U.S. firepower, Kellogg and his subordinates anticipated intense combat. "We thought this would be one of our tough fights," the colonel recalled, "because the naval infantry company had been openly aggressive during the entire time we were down there." The aggressive posture to which he referred ranged from a daily ritual of obscene gestures and verbal abuse to the actual pointing of weapons at U.S. soldiers. What Kellogg did not say was that, in the course of Sand Fleas geared to rehearsing the plan, Rizzo was having one of his platoons take up positions in front of the marine company headquarters two to three times a week. For twenty minutes, the men would lie there with their weapons, then leave. On occasion, the Company C commander would also "roll the Vulcan out" and position it in such a way as "to augment the infantry platoon's intimidating posture. At first, these demonstrations unsettled the PDF soldiers but after a while only served to lull them into a false sense of security." In short, both sides in Coco Solo were acting aggressively; the crisis had reached the point where such behavior was no longer considered unusual.[21]

Even so, going into mid-December, Company C still lacked a good deal of information concerning its potential opponent. The order of battle Task Force Atlantic possessed for the naval infantry company estimated a unit of up to two hundred men, the same number of troops that Rizzo would have to pit against them. (The company actually carried 130 men on its rolls.) There was, however, no updated intelligence as to how many marines were present, nor was it known for sure where their living quarters were located in the PDF building. Through the Sand Fleas and other exercises, Brooks and Rizzo tried to acquire some of this needed information. Company C's platoon leaders also ate regularly at the Chinese restaurant located in the PDF building, simultaneously dining and observing. Unfortunately, in light of subsequent events, they did not realize that a large Chinese family was living on the

[20] Rizzo's plan is covered in Intervs, author with Rizzo, 15 May 1997; with Brooks, 11 May 1993; Briefing, Brooks, 29 Apr 1992; "Conquest At Coco Solo," p. 25; Donnelly, Roth, and Baker, *Operation Just Cause*, pp. 242–43. According to Huelfer in "The Battle for Coco Solo," an AC–130 was scheduled to be overhead to provide Company C additional fire support, but it did not arrive until "it was too late to be useful." The company's mortar section also proved of limited use because of rules of engagement prohibiting indirect fire without the brigade commander's authorization.

[21] First quote from Donnelly, Roth, and Baker, *Operation Just Cause*, p. 243. "Conquest At Coco Solo," p. 25. Second quote from Interv, author with Rizzo, 15 May 1997. Third quote from Huelfer, "The Battle for Coco Solo." In his interview, Rizzo also noted that his 2d Platoon used the school in which it was quartered to rehearse the clearing operations it might have to conduct in the near identical PDF building to the south.

premises. In another undertaking just two days before JUST CAUSE, a SEAL team in the area accompanied by some of Rizzo's people sailed by Coco Solo, taking photographs that revealed the location of the naval infantry company's living quarters.[22]

All this paid off once Rizzo received word that the operation plan would be executed. Shortly past midnight on 20 December, after he had accompanied the platoon from the 82d Airborne Division to its position in front of the PDF company headquarters, he returned to his command post in the courtyard at the school and contacted his other platoon leaders to make sure they were in place. He also passed along the units' various call signs to the 4th Battalion operations officer at Brooks' command post in the gymnasium and was going through several final checks when, unexpectedly, a round went off behind him. As he recalled, the shot "stirred things up in the naval infantry company headquarters," as several Panamanians ran out the back door of the building and headed for the docked patrol boats. With H-hour still over a quarter hour away, the key questions of the moment concerned not only what had happened but what to do about it.[23]

Years later, the whole episode and its ramifications would remain shrouded in controversy. What was clear within minutes after the shot was fired, however, was that a U.S. military policeman, Sfc. Charlie Gray, had shot and killed a Panamanian in the vicinity of the Coco Solo complex. Gray was the platoon sergeant of the 549th Military Police Company's 3d Platoon. About thirty minutes before H-hour, he was in the process of leading four other military policemen to a position—the last of the four MP checkpoints called for in Rizzo's plan—from which they could seal off one of the entrances to the PDF headquarters. Near the location was a wooded area, and, as Gray and the others approached, they came across a group of Panamanians. According to Gray's own account, the men had a radio and were "giving away our position." The sergeant informed his platoon leader of the situation and, in return, purportedly received an order "to take that position and to silence that radio at all costs." As the other military policemen provided cover, Gray approached a guard shack in which a man wearing a PDF uniform and holding a Motorola radio was standing. When Gray ordered him to surrender, the man instead tried to grab the sergeant's M16 rifle. Hand-to-hand fighting ensued, at the conclusion of which Gray shot and killed the individual. According to the sergeant, the victim was in the process of unholstering and cocking a revolver. Soon thereafter, the other Panamanians in the wooded area emerged with their hands raised.[24]

To have the H-hour attacks in AO North compromised by over a quarter hour created no small degree of consternation at Brooks' command post, where word of the shooting was radioed back to Kellogg. In the days that followed, both officers and a number of their subordinates voiced the opinion that the incident could have been avoided. They attributed its occurrence to several causes, beginning with the military policemen involved. Kellogg later

[22] Intervs, author with Rizzo, 15 May 1997; with Brooks, 11 May 1993.
[23] Interv, author with Rizzo, 15 May 1997.
[24] Gray's account is from Donnelly, Roth, and Baker, *Operation Just Cause*, pp. 240–41.

characterized them as USARSO "traffic cops"; in his opinion, they lacked the experience and skills necessary for wartime operations. Had a combat-trained military police unit from the 7th Infantry Division been present instead, he clearly implied, the incident would not have taken place.

Reinforcing this point, Kellogg, Gardner, and Rizzo cited previous tension between elements of Brooks' battalion and the Panama-based military police. The colonel and his S–3 went further, identifying Gray specifically as a "loose cannon" whose "flaky" behavior had caused problems for Task Force Atlantic in at least two episodes prior to JUST CAUSE. As for the encounter with the Panamanians in the wooded area, Kellogg's view was that Gray should not have been there. According to one staff officer, Gray "didn't have an order. He didn't have a mission to take anything down, simply to block." Another officer went so far as to label the shooting a murder, although a criminal investigation conducted by the Army found no grounds on which to bring charges against the first sergeant. Beyond dispute, however, was the fact that the shooting "had certainly jeopardized the entire operation tactically," since together "with the recently slain PDF soldier, surprise became the second casualty of the night." For this reason, Brooks recounted, "things got moving rather expeditiously."[25]

When Brooks' operations officer radioed Kellogg about the incident, the colonel was in a helicopter observing preparations for the U.S. assault on Renacer Prison in AO South. Particularly troubling was the news relayed to him about the PDF marines, who, following the premature shot, were trying to escape via the military boats in the dock area. The craft sported 20-mm. chain guns and .50-caliber machine guns, weapons capable of wreaking havoc on Company C and other U.S. forces should the enemy be allowed to board the boats unchallenged. To prevent this, Captain Rizzo was urgently requesting permission through Brooks to begin the operation, despite what Kellogg had stated emphatically earlier that evening: H-hour would stand at 0100 regardless of what might happen before then. Unequivocal as that declaration had seemed at the time, it did not survive Gray's actions and Brooks' subsequent entreaty, "We've got to do it now!" Kellogg did not hesitate to reverse himself. Without informing JTF-South in advance, he quickly authorized an immediate attack, even though he and his battalion commander knew that other Task Force Atlantic forces in AO North were still en route to their assault positions.[26]

As soon as Rizzo received permission to open fire, he turned to the gunner on the Vulcan positioned in front of the PDF company headquarters and tapped him on the shoulder. When the first 20-mm. round hit the target, the explosion provided the cue for the U.S. forces in front of the building to begin firing with their AT4s, LAWs, M60s, and various small arms. The barrage initially peppered the top floor of the building and then worked its way down. In the process, the Vulcan fired about eleven hundred rounds, the paratroopers

[25] For Task Force Atlantic's perspective on the shooting incident and the fourth quote, see ibid. Interv, author with Rizzo, 15 May 1997. First quote from Interv, author with Kellogg and Gardner, 9 Apr 1992. Second and third quoted words from Interv, author with Gardner, 7 Apr 1992. Sixth quote from Huelfer, "The Battle for Coco Solo." Last quote from Briefing, Brooks, 29 Apr 1992.

[26] Interv, author with Kellogg and Gardner, 9 Apr 1992. Quote from Briefing, Brooks, 29 Apr 1992. Donnelly, Roth, and Baker, *Operation Just Cause*, p. 245.

nearly sixty antitank rockets. Then, according to plan, the shooting stopped, and a loudspeaker team broadcast its first appeal for the PDF defenders to surrender. Brooks had thought the shock of the Vulcan and antitank weapons would result in a quick capitulation, but, much to his surprise, "There was absolutely no response," save some scattered and poorly aimed gunfire from within the building. So the second barrage began, as Brooks thought to himself, "I really don't want to do this." (Two years later, Brooks would learn that many of the Panamanian defenders were in the back of the building where they could not hear the U.S. broadcasts.)[27]

While most of what one officer called the "deafening roar" of battle emanated from U.S. positions in front of the naval infantry company headquarters, the "main fight" at Coco Solo, according to Rizzo, took place in the dock area behind the building. Rizzo's 1st Platoon was responsible for that sector, and, around 2330, ten to fifteen minutes before the shooting began, soldiers in the unit had begun infiltrating into the adjacent Lada parking lot. Separating the parking lot and the docks was a chain-link fence. Once the platoon had assembled, the men split into two elements: a small support group led by the platoon sergeant and a larger assault team of eighteen men headed by the platoon leader. The plan called for the support group to seize and secure a boathouse on the other side of the fence, in the northeast corner of the docks, and, next to it, a PDF barracks. From these positions, the sergeant and his men could provide covering fire for the assault element as it worked its way single-file in a duck crawl along a seawall on the western edge of the lot. The twelve-inch-wide seawall ended where it intersected with the chain-link fence, with the fence running slightly past the wall and into the bay. Once the assault force had swung around the end of the fence, the men would be on the dock and in a position to capture the nine PDF boats that were moored there. Prebattle estimates predicted the presence of no more than a half-dozen defenders who would have to be dealt with quickly given the relative lack of cover available to any U.S. soldier on the south side of the fence.[28]

Sgt. Rick Mowatt was near the front of the assault force that began maneuvering along the seawall. He had not yet reached the fence when, following the shot fired by Sergeant Gray, he heard the motors on several boats start up and saw several Panamanians leaving the naval infantry company building through its back door. Mowatt hurried to get the men behind him around the fence and onto the dock. Once there, however, they had to wait four or five minutes until the details of their situation were communicated up the chain of command to Kellogg and until the colonel's permission to open fire was relayed back down the chain to them. During that time, they were completely exposed except for the cover afforded by a landing craft and the dry dock constructed of railroad ties to which it was moored. While the men marked time, two large PDF gunboats began moving southward from

[27] Interv, author with Rizzo, 15 May 1997. Quotes from Briefing, Brooks, 29 Apr 1992. Huelfer, "The Battle for Coco Solo."

[28] First quoted words from Huelfer, "The Battle for Coco Solo." Second quoted words from Interv, author with Rizzo, 15 May 1997. Donnelly, Roth, and Baker, *Operation Just Cause*, pp. 243, 245.

the docks into the bay, away from the 1st Platoon and beyond the range of its weapons.[29]

Once the platoon received permission to engage the enemy, the support group blew a hole in the fence that separated it from the small PDF barracks, advanced on the structure, and, when there was no response to an appeal to surrender, tossed grenades inside what turned out to be an empty building. In the boathouse next door, two Panamanians were killed when grenades from the support group set off a butane gas explosion. Two other defenders took cover under trucks on the dock and then directed small-arms fire against the attackers. Their resistance and the increasing volume of enemy fire coming from the remaining PDF patrol boats caused the 1st Platoon leader, 1st Lt. Walter Burke, to order Mowatt forward with one member of his team to employ their LAWs against the Panamanians. Joining Mowatt and his partner were two men from the platoon's antitank section who also brought several LAWs with them. The first rocket they fired hit a mast on one of the boats, but at the cost of revealing the small group's vulnerable position. Thus, as Mowatt prepared to follow up with a second LAW, the Panamanians brought the four men under withering automatic weapons fire, hitting two, one in the thigh, the other in the hand and ankle. Despite the casualties, Burke, Mowatt and his partner, a radio man, an M60 machine gunner, the two wounded men, and the medic attending them managed to stake out a forward position about ten yards south of the landing craft that the remainder of the assault force was continuing to use for cover. In both positions, the soldiers composing 1st Platoon's main element found themselves outgunned, under heavy enemy fire, and, because of a problem with their radio, out of touch with Captain Rizzo for about fifteen minutes.

The ensuing firefight lasted over an hour. Once Rizzo became aware of the 1st Platoon's predicament, he ordered the Vulcan that was situated on the northern flank of the battle area to move down and into a position from which it could fire on the PDF gunboats. Meanwhile, on the dock, Burke sought to get his men dispersed and moving toward the enemy, something many of the young soldiers were reluctant to do given the volume of hostile fire. The platoon's noncommissioned officers decided to set the example and crept forward, prompting many of the men to follow them. Their movement, however, was tentative, except for Mowatt, who moved out in front of the advancing troops and fired four antitank missiles at the enemy positions, hitting two boats and unmistakably seeing two of the enemy fall. This suppressive fire allowed those behind him to advance farther, although their maneuvering only accelerated after the relocated Vulcan and two M60 machine guns began pouring rounds onto PDF positions. In the process, the tactical advantage shifted to the Americans; indeed, according to one account, "The additional firepower had all but silenced resistance on the docks and allowed 1st Platoon to close onto its objective." By that time, the Panamanians had begun to run short of ammunition, and U.S. troops had seized two boats. When a Spanish-speaking American soldier gave the remaining defenders an ultimatum to

[29] According to Huelfer, "The Battle for Coco Solo," one of the two vessels that escaped was later destroyed by an AC–130 gunship.

surrender in thirty seconds, they did so, to 1st Platoon's surprise and relief. One source concluded that, in all, "about twenty-five PDF had taken part in the fight—several times as many as the soldiers were told to expect." Twelve marines surrendered on the dock; others were captured hiding in buildings nearby. Mowatt, who later received a Bronze Star, attributed 1st Platoon's success to luck and the unheralded but heroic actions of the men who were with him.[30]

In the battle for the dock area, the opening exchange of shots between the 1st Platoon and the Panamanians trying to escape occurred shortly before the Vulcan positioned on the other side of the naval infantry company headquarters fired its first round. Rizzo's 2d Platoon, assigned the task of breaching the main PDF building, later corroborated this. "You hear these little potshots, then you hear the [M]60s, and then the Vulcan," said one of the unit's fire team leaders. The sounds of the premature battle raised the level of anxiety as the 2d Platoon hustled out of the gymnasium, where it had been preparing for its part in the assault, and into the noise and surreal sights—it "was just like a light show out there"—caused by the massive firepower demonstration the Vulcan, the paratroopers, and Company C's 3d Platoon had unleashed on the enemy. Before the men crossed the eighty yards from the gym to the northern wing of their objective, the platoon's 1st Squad leader tossed a smoke grenade to obscure his soldiers' movement. The smoke provided concealment but also attracted enemy fire. Even so, with PDF tracer rounds seemingly everywhere, visible even through the smoke, the entire platoon made it safely through a door by the Chinese restaurant and to a staircase just inside the naval infantry company headquarters.[31]

The platoon worked its way by squads up the stairs to the second floor of the north wing where, without a blueprint showing the location of rooms that would have to be cleared, the troops began searching for enemy defenders. As a private in the 2d Squad approached one area, a Chinese woman opened a door, saw the soldier with his camouflage face paint and ragpot helmet, and fainted. What the private and his comrades quickly discovered was a three-room complex that served as living quarters for seventeen Chinese. After ascertaining that the residents were unarmed, members of the 3d Squad removed them to a safer area and stood guard over them.[32] Meanwhile, the other two squads moved up to the third floor of the north wing and then

[30] The most detailed account of the 1st Platoon's actions at Coco Solo is in Huelfer, "The Battle for Coco Solo," which contains the first quote; Donnelly, Roth, and Baker, *Operation Just Cause*, p. 253, which contains the second quote, and see also pp. 245–46, 248–49. Huelfer also notes that, once the 1st Platoon's advance on the PDF positions gained momentum, the Vulcan rounds came within the "danger close" range to U.S. troops, causing the platoon executive officer to run over to the Vulcan crew and order it to stop shooting.

[31] Unless otherwise noted, this account of the actions of the 2d Platoon is based on Donnelly, Roth, and Baker, *Operation Just Cause*, p. 246, which contains the first quote, and see also pp. 247, 249–53; Huelfer, "The Battle for Coco Solo," which contains the second quote.

[32] At some point, the Chinese were removed from the building and relocated across the street in the housing area that had been vacated. To gain access to the empty houses, Brooks ordered a few of his men to "kick some doors down." The next day, the facility engineer for Panama on the Atlantic side protested both the damage to the doors and Brooks' decision to allow the Chinese to live inside, however temporarily. Briefing, Brooks, 29 Apr 1992.

maneuvered toward the main area of the building, the one that housed the Panamanian company. Arriving there, the plan next called for them to work their way back down to the second floor and, after clearing it, to the first.

The troops first had to gain access to the main wing through a locked steel door surrounded by bolts of fabric from the building's garment factory. To accomplish this, a private put a three-pound charge of C4 plastic explosives on the door. It was a mistake. "He screwed up," Brooks conceded. "He used too big a charge." The subsequent explosion not only blew the door open, but also set the fabric ablaze, a fire that quickly spread to the rafters and began to burn out of control. The 3d Squad had to be pulled out of the clearing operation to fight the flames, while the smoke and the fumes made several soldiers nauseous. As for the private who set the charge, Brooks would later smile and say, "We made him a sergeant a year later."[33]

After spending only the couple of minutes needed to clear the third floor of the building's main section, the two squads from 2d Platoon still committed to the fight prepared their descent to the second floor by rolling a few grenades down the stairs to eliminate or discourage any enemy personnel on that level. The troops then negotiated the stairs, reaching the next floor below without incident. In clearing each room on the second floor, they used the standard procedure of first throwing in a fragmentation grenade, then firing a burst from an M16 rifle. The sweep produced no significant results, since by that time, all the PDF defenders were gathered together on the first floor.

What followed, with both sides offering different accounts, reflected the fog and friction inherent in most combat operations. According to several U.S. soldiers who took up positions by the main stairway leading to the first floor, one or more Panamanians below were yelling up at them. Darkness and smoke prevented individuals on either side from seeing their adversaries, while the distance that separated the two forces, together with the noise of the battle outside and of soldiers inside moving and shouting to one another, made what was being said almost impossible to decipher. To some of the Spanish-speaking troops on the American side, the words coming from below sounded like taunts and demands for a U.S. surrender; others believed that the defenders were themselves trying to give up. Another possibility was that the shouting contained a mixture of defiant epithets and desperate appeals. Not being certain as to the Panamanians' intentions, the 2d Platoon leader, 2d Lt. Daniel K. Kirk, was unwilling to put his men at risk. He instead ordered one of his squad leaders to throw a grenade down the stairs. Other grenades followed before the U.S. soldiers moved down the stairway to confront an enemy that they knew was waiting for them.

The battle the 2d Platoon anticipated did not take place. Rather, in a rear area of the first floor, they found the PDF company commander, Capt. Amadis Jimenez, and about twenty-three of his men waiting to surrender. In fact, Jimenez later related, without adequate weapons to defend his marines from what he regarded as an unnecessarily excessive American onslaught, he had been trying to give up for some time. But his shouts to the "very nervous" U.S. servicemen upstairs, both in Spanish and in English, had been acknowledged

---

[33] Interv, author with Brooks, 11 May 1993.

with grenades, the explosions from which had wounded several in his company. The Americans, he added, had also thrown grenades out the back windows of the building to discourage the Panamanians cornered below from trying to escape over the dock area, as some had done earlier. Now that U.S. forces had taken positions just down the hallway from Jimenez, the captain's efforts to negotiate a surrender entailed additional risks. Through a translator, Lieutenant Kirk told Jimenez to lead his men out the back door. The captain agreed, but when he set foot outside, someone tossed a grenade in his direction, fragments from which hit one of his men in the eye. The Panamanians scurried back into the building, and Jimenez heatedly asked the translator, "Why do you ask me to surrender and then throw grenades at me?" In reply he received an apology and an explanation that the U.S. troops were indeed nervous. Again, Jimenez was directed to leave the building with his men, and again he complied. Some further confusion ensued, but Kirk was finally able to complete the surrender. The naval infantry company building was considered secured by 0240, a time much later than Brooks had anticipated. In the course of the operation, the 2d Platoon took fifteen prisoners, while suffering no casualties.[34]

While the 1st Platoon was fighting on the dock and the 2d Platoon was clearing the marine headquarters, Captain Rizzo's 3d Platoon was busy in its flanking positions to the south. The platoon participated in the initial firepower demonstration against the PDF company, but only for about one minute before the platoon leader told the men to stop shooting. "I was sure there were no more targets in our sector by that time," he later explained, "and I was concerned about the many civilians in our area." Some of those civilians surfaced around 0330, when U.S. troops heard voices in a V-shaped factory across the street. A Spanish-speaking soldier investigated and found eleven workers, whom he was able to coax into leaving their hiding place. Later, after dawn, the platoon's 2d Squad searched the building and the Lada factory next to it. In the latter, the men found three PDF members who had donned civilian shirts. The three surrendered immediately.[35]

As the fighting in Coco Solo tapered off, the fire in the naval infantry company building still needed to be extinguished. Once the Chinese residents and Captain Jimenez and his men had been escorted out of the building and the rooms cleared, the platoon from the 82d Airborne Division was told to take over the efforts to put out the blaze, which by that time had destroyed much of the third floor. As the paratroopers emptied one fire extinguisher after another, they could not stop the flames from spreading to all three wings and all three floors. About an hour after dawn, a fire truck from the vicinity of Colón arrived, but it was too late to be effective, especially after the roof of the building collapsed. In this instance, the attempt to avoid collateral property damage had failed, in contrast to the success of the mission itself. By 0700, when U.S. interrogators arrived to question the prisoners, Company C had

[34] Quotes from Donnelly, Roth, and Baker, *Operation Just Cause*, pp. 251–52. Brooks' feeling that the process of clearing the building had taken more time than he had anticipated comes from Briefing, Brooks, 29 Apr 1992. Captain Jimenez's account of what transpired at the PDF company headquarters is from Interv, Dolores De Mena with Capt Amadis Jimenez, 13 Nov 1991, Fort Clayton, Panama.

[35] Huelfer, "The Battle for Coco Solo."

captured twenty-seven PDF troops, killed nearly a dozen (and perhaps more on the two vessels that had escaped), and wounded many others. Sixty-five civilians had been detained, but all were released by noon. In the course of the battle, there had been no U.S. fatalities and only a handful of wounded soldiers. The preparation, discipline, and firepower of the American troops had made the victory possible, as had the failure of the defenders to bring some of their heavy weapons, especially mounted machine guns, into play.

Fort Espinar

Two miles south of Coco Solo, between the Boyd-Roosevelt Highway and Gatun Lake, was Fort Espinar. Until the 1980s, it had been Fort Gulick, an American base that was home to the Army's School of the Americas, a training facility for U.S. and Latin American military and police personnel. Under the canal treaties, the fort was renamed and transformed into a shared U.S.-Panamanian installation, with the Panama Defense Forces billeting its 8th Infantry Company on the premises. In 1984, the School of the Americas was moved to Fort Benning, Georgia, and the building it had occupied turned into a PDF academy for noncommissioned officers. By early 1989, U.S. military units had vacated the post, leaving behind various facilities, including an American housing area for dependents, a school, and a commissary.

With the execution of Operation Nimrod Dancer in May 1989 and the consequent creation of the Task Force Atlantic headquarters at Fort Sherman, a decision was made to station one of the companies from the 7th Infantry Division's security enhancement battalion on Fort Espinar. While the canal treaties allowed for such an arrangement, a breakdown in coordination led to a confrontation and near firefight on 18 May between the PDF 8th Company and the camouflaged and well-armed light fighters as the latter approached the fort's main gate. After several tense hours, the situation was defused. Once ensconced in the former U.S. officers' club at Espinar, the company commander followed orders received through his chain of command to begin reasserting American treaty rights in Panama, including the reoccupation of a building on the premises that the United States had leased to the Panamanians but on which the Panama Defense Forces had stopped paying rent.[36]

The movement of the U.S. infantry company onto Espinar in May and the reoccupation of the leased building in September bracketed Colonel Hale's tenure as the Task Force Atlantic commander. When Colonel Kellogg replaced Hale in mid-October, the combined impact of the new commander's more aggressive style and the increased tensions following the failed coup against Noriega was felt throughout Task Force Atlantic's area of responsibility. With respect to Fort Espinar, one immediate change Kellogg insisted on was to forgo any personal contact with the 8th Company commander. Hale had maintained contact with the Panamanian officer, often speaking with

---

[36] For an account of U.S. military activities on Fort Espinar during Operation Nimrod Dancer, see Yates, *U.S. Military Intervention in Panama, June 1987–December 1989*, pp. 188–89, 235–36.

him personally. Kellogg broke the connection with a man he regarded as the enemy. As Gardner phrased the new arrangement, the colonel's "attitude was basically, 'Screw you, Jack.'"[37]

In mid-November, with the handoff of the Task Force Atlantic security enhancement force to Brooks' 4th Battalion, 17th Infantry, the unit's Company A, commanded by Capt. Michael F. Beech, moved onto Espinar. As with Captain Rizzo at Coco Solo, Beech inherited and updated his predecessor's contingency plans for the fort, the principal missions of which consisted of neutralizing the 8th Company and the noncommissioned officer academy. Because some of the barracks for the U.S. and Panamanian companies were located only 218 yards apart, Beech's troops were able to study their adversary very closely through the use of binoculars and videotapes. The fact that U.S. intelligence had wiretapped the PDF headquarters also proved invaluable in acquiring information essential to the planning process. One vital piece of intelligence concerned the 8th Company commander, Maj. Luis Guardia; the officer was perceived as a professional soldier who required professional conduct from his men as well. If hostilities broke out, Brooks and Beech feared, U.S. forces at Fort Espinar would be in for "a difficult fight." From well-fortified defensive positions, the one hundred fifty to two hundred Panamanians estimated to be in Guardia's company could inflict serious casualties on an attacking force.[38]

To ensure that the outcome of a fight would be favorable, Company A fine-tuned its contingency plan so that, in keeping with Kellogg's guidance, the firepower and skills of American troops would overwhelm the enemy in such a way as to minimize casualties on both sides, protect civilians living in the housing area, and limit physical damage to property. More specifically, the plan called for the 3d Platoon from the 7th Infantry Division's 7th Military Police Company to seal off the installation. Company A's 2d and 3d Platoons, billeted just a block and a half from the 8th Company, would neutralize the unit, the bulk of which was housed in two large, multistory buildings surrounded by a chain-link fence built atop a two-foot-high brick wall and with dense jungle outside the perimeter to the east. The company's 1st Platoon would provide fire support, as would an M60 machine gun team, the company's antitank section, and a Vulcan crew. As for the noncommissioned officer academy at the other end of the fort, a composite platoon of about twenty-nine paratroopers from the Headquarters Company of the 3d Battalion, 504th Infantry—"cooks, clerks, you name it," as Brooks would describe the hodgepodge—had the mission of neutralizing what was thought to be a platoon-size force at the school. The students, U.S. planners believed, did not pose a serious threat, but, if left unchecked, they would be in a position to disrupt the main operation against the infantry company. Furthermore, as Capt. Matthew Halder, commander of the Headquarters Company, observed, "They could easily get a hostage situation going," given the number of Americans living in the housing area on post. As with the paratroopers employed at Coco Solo, the makeshift platoon

[37] Interv, author with Gardner, 7 Apr 1992.

[38] Quote from Briefing, Brooks, 29 Apr 1992. Donnelly, Roth, and Baker, *Operation Just Cause*, p. 260; Department of the Army (DA) Field Manual (FM) 7–10, *The Infantry Rifle Company*, December 1990, ch. 2.

responsible for the academy would be transported by truck to Fort Espinar from the Jungle Operations Training Center.[39]

Again, as was the case with U.S. units throughout Panama, as the plan neared its final form, the Sand Fleas, Purple Storms, and contingency readiness exercises for Company A were tailored more and more to the wartime mission. Squad leaders, even though they would not be informed of the exact plan until just hours before JUST CAUSE, repeatedly led their troops on patrols over the routes they would most likely be using in the event of hostilities. Checkpoints they established near the commissary and an elementary school, as well as patrols in the housing area, allowed soldiers assigned those sectors to become familiar with the surroundings. As for the Vulcan crew from the 2d Battalion, 62d Air Defense Artillery, every night the men positioned the mounted weapon "on a baseball field in plain view of the PDF company headquarters." Also, around Thanksgiving, Company A began to reconnoiter the routes it would take through the jungle to get to its assault position near the two main 8th Company buildings. According to the wiretap intercepts, all of these maneuvers were having the desired effect of lulling the Panamanians into a false sense of security: "The Americans are at it again. They're doing this just to harass us," were the kinds of statements that frequently appeared on the transcripts provided Kellogg. Or as one PDF officer put it, "People would always cry, 'The wolf is coming,' but the wolf never came." Only on one occasion, about a week before JUST CAUSE when the Panamanians began placing several barriers near Company A's assault position, were Kellogg and everyone down the chain to Beech and his staff worried that the contingency plan had been compromised. They would not find out until much later that the PDF actions were solely a response to the U.S. exercises, nothing else.[40]

The night Lieutenant Paz was killed, Captain Beech, like other commanders in Task Force Atlantic, placed his company on alert, only to stand down the next day, on orders passed to Colonel Kellogg by JTF-Panama. Then, on Tuesday, the captain climbed into a van with Brooks, Rizzo, and the Company B commander and headed to Fort Sherman to hear Kellogg's announcement that Operation JUST CAUSE would begin in a matter of hours. When the colonel finished his talk, Beech returned to Fort Espinar to prepare his soldiers, briefing them at 2100 at the officers' club. After further preparations, the three platoons of Company A moved to a building closer to the PDF 8th Company.

---

[39] First quote from Briefing, Brooks, 29 Apr 1992. Second quote from Donnelly, Roth, and Baker, *Operation Just Cause*, p. 261, and see also p. 262. DA FM 7–10, *The Infantry Rifle Company*, December 1990, ch. 2. Donnelly, Roth, and Baker raise again the subject of the "mutual disdain" that had developed between the military police of USARSO's 549th Military Police Company and the military police from the 7th Infantry Division. Concerning plans for executing the contingency mission on Fort Espinar, the USARSO military police were originally slated to seal off the installation, only to have Task Force Atlantic planners replace them with the 7th's military police because the Panama-based personnel were "considered too close for comfort to the Panamanians they would have to fight," in that "they patrolled together with PDF troops."

[40] Briefing, Brooks, 29 Apr 1992. All but last quote from Donnelly, Roth, and Baker, *Operation Just Cause*, p. 260, and see also p. 261. Last quote from Interv, De Mena with Jimenez, 13 Nov 1991.

From there, the two platoons that would approach the objective through the jungle departed under Beech's personal leadership to begin their two-hour trek along a trail they had dubbed Route Stampede.

Meanwhile, the Vulcan team waited at the officers' club for word to advance to its firing position on the baseball field, some seven minutes away. Around 2300, the composite platoon of paratroopers arrived at the club, also to wait in the darkness before moving out. The platoon and Captain Halder departed just before 0040, carrying a .50-caliber machine gun and tripod en route to an assault position across from the noncommissioned officer academy. Within minutes of leaving the club, the men heard the Vulcan firing at Coco Solo, an indication that something had almost certainly gone wrong. Picking up the pace of their advance, they moved through the post's housing area and reached Gulick Lodge, the three-story building from which they would engage the Panamanian students, all assumed to be senior noncommissioned officers. At this moment, while Halder was attempting to get the platoon's three squads and the .50-caliber machine in position near the academy, one of his men took it on himself to open fire on the building. Though the act was unauthorized, the defenders responded in kind, and a ten-minute firefight ensued, well before H-hour.[41]

The shooting at Coco Solo and, closer to their position, at the PDF academy caught the two Company A platoons moving through the jungle toward the 8th Company compound far short of their assault positions. Given the priority of their mission, the men began sprinting over the remainder of Route Stampede, a risky move given the dark and the precarious terrain. As one soldier recalled, "We had to stay, like, right behind the guy in front of us just so you could know where you're going." When they finally emerged from the jungle, they realized they had missed their objective by at least a hundred yards. They went back into the densely wooded area, made the needed adjustment, and then waited a few minutes until 0100, the original H-hour. To their relief, the elements providing fire support—mainly the Vulcan—had not yet opened up on the PDF buildings. Thus, despite the chaos at the noncommissioned officer academy a short distance away, Company A would begin its attack on the 8th Company on schedule.[42]

The men in the assault force did not realize at the time just how tenuous the restraint shown by the soldiers assigned to provide supporting fires had been. With Captain Beech accompanying the 2d and 3d Platoons and with his decision to "leave the senior NCOs with their rifle platoons," control of the fire support elements had fallen, not to the first sergeant as planned, but to 1st Lt. Daniel E. Evans, Company A's executive officer. His task was to employ suppressive fires as needed to prevent Panamanian forces from exiting the 8th Company buildings into their prepared combat positions and to protect the assault force as it breached the chain-link fence and maneuvered against the enemy. To these ends, Evans had moved his people and their weapons into position well before H-hour, but once the battle at the academy began, he came under intense pressure from the fire support elements, especially from

[41] Donnelly, Roth, and Baker, *Operation Just Cause*, pp. 262–63.

[42] Ibid., pp. 263–64.

the Vulcan team, to authorize them to begin shooting. The lieutenant resisted. From his vantage point, he could see that the Panamanian infantrymen inside the two buildings were not deploying for combat. Furthermore, given the darkness, the defenders probably did not even know the Americans were in place to attack them. Unless either or both of those conditions changed, there was no reason to second-guess Captain Beech's clearly stated intent to launch the attack at H-hour. So Evans held firm. The first Vulcan round did not hit its target, a guard shack inside the fenced area, until the precise time designated in the plan.[43]

Once the Vulcan began firing, the barrage that followed was comparable to the one leveled at the Coco Solo naval infantry company. AT4s and LAWs quickly joined in, as did the M60 machine gun and various small arms. Said one member of the 1st Platoon, the PDF headquarters building closest to his position soon resembled "Swiss cheese." Yet, intense as the shooting was, the defenders managed to return the fire with small arms and rocket-propelled grenades, one of which accounted for the only U.S. casualty at the site—a sergeant who was wounded. The support element's suppressive fire served its purpose, though, as the assault platoons were able to climb a steep slope to the PDF compound and breach the fence with a bangalore torpedo. They then waited for the scheduled cease-fire call from the U.S. side, to be followed by a psychological operations team issuing a surrender message. The broadcast, as it turned out, had an immediate and positive effect: the defenders had been outgunned, had fired back almost blindly, had taken casualties, and were ready to call it quits. Major Guardia, whom Brooks and Beech feared would mount a spirited defense, had fled to Colón well before H-hour. The forty or so soldiers left behind under the command of a first sergeant had a formidable arsenal at their disposal, but as Lieutenant Evans observed, "It looked like they were ready to fight but they changed their minds," probably the result of the firepower demonstration featuring what Brooks called "the magnificent Vulcan." After the surrender, the assault platoons cleared the two buildings, while the prisoners were taken to the fort's front entrance secured by the 7th Infantry Division military police and then removed through the back gate.[44]

The battle at the academy had gone as well as the other task force missions. The platoon of paratroopers had used antitank weapons against the defending force, compelling thirty-five noncommissioned officers to surrender before the fight was an hour old. U.S. troops then waited until dawn, when all three infantry platoons from Company A arrived. Having just finished clearing the

---

[43] Briefing, Brooks, 29 Apr 1992. Quote and additional material on the fight are from Daniel E. Evans, Actions at Fort Espinar, Republic of Panama, 20 December 1989, n.d., copy at http://www.benning.army.mil/library/content/Virtual/Donovanpapers/other/STUP5/EvansDaniel%20E.%20CPT.pdf. DA FM 7–10, *The Infantry Rifle Company,* December 1990, ch. 2. In a vignette designed to demonstrate the "utility of the commander's intent," the field manual uses the example of Evans' "proper decision" to withhold fire until 0100.

[44] Last quote from Briefing, Brooks, 29 Apr 1992. Other quotes from Donnelly, Roth, and Baker, *Operation Just Cause*, p. 265, and see also p. 264. In contrast to Donnelly, Roth, and Baker, who place the number of prisoners from the 8th Company at "about 40," DA FM 7–10, *The Infantry Rifle Company,* December 1990, ch. 2, states that eighty-four enemy soldiers surrendered after the short fight.

A PDF 8th Infantry Company building at Fort Espinar after the firefight

8th Company buildings, the company began to do the same at the school, aware that there might be holdouts in some of the rooms. The Americans broke into three-man teams, using grenades when necessary to ensure that a room was safe. In the process, a procedural blunder proved costly when nine light fighters found themselves together in the same room, and one of them threw a grenade through an opening into what he thought was an adjacent room. The grenade hit a wall and bounced back amid the tightly bunched men, with fragments from the explosion wounding everyone present, six seriously.[45]

News of the casualties, the most suffered in any one incident involving Brooks' battalion, dampened spirits in what had otherwise been a successful, if far from perfect, operation. Around 0500, Beech radioed Brooks to report that Company A and the units attached to it had accomplished their mission. Both the PDF 8th Company infantrymen and the academy students had been neutralized. Beech then sent troops into the American housing area to check on the residents. As Brooks later related, the "Americans stayed in their quarters. . . . It must have been frightening because the Vulcan was blowing up right next to their houses. The fighting was going on right there, and explosions. . . . The backblast of the AT-4s and the percussion caused by the Vulcans knocked the glass out in the housing the people lived in that we had moved them out of."[46] The situation was similar to that at Fort Amador, where the battle against the PDF 5th Company had taken place in the front yards of some of the American dependents living on post. In

[45] Donnelly, Roth, and Baker, *Operation Just Cause*, pp. 266–67; Briefing, Brooks, 29 Apr 1992; Interv, author with Brooks, 11 May 1993.

[46] Donnelly, Roth, and Baker, *Operation Just Cause*, pp. 266–67. Quote from Interv, author with Brooks, 11 May 1993.

both cases, there would be lasting psychological repercussions for many of the families involved.

THE COLÓN BOTTLENECK

By dawn on 20 December, most of Task Force Atlantic's D-day objectives in AO North had been accomplished, with the two targets considered the most dangerous—Coco Solo and Fort Espinar—well under control. Two lesser missions had been completed as well. At H-hour, U.S. forces had seized France Field, an airfield southeast of Coco Solo, while, even farther south, other soldiers from the Headquarters and Headquarters Company, 4th Battalion, 17th Infantry, had positioned four school buses across a wide bridge to stop traffic along the Boyd-Roosevelt Highway. At the latter location, another platoon of paratroopers, this one from Company D, 3d Battalion, 504th Infantry, provided additional firepower to keep enemy reinforcements from reaching Colón and to help cover Army engineers as they reinforced the roadblock with concertina wire and 55-gallon drums. While the engineers were hurrying to complete their task, Panamanian policemen arrived in a van and tried to talk their way through the roadblock, to no avail. At the time, there were only eight paratroopers on the scene, and they were concerned that the men in the van would radio for reinforcements to help outflank the school buses and other barriers. The Americans therefore told the police that, if they did not leave the area, they would be killed. Despite the death threats and some warning shots fired by the paratroopers, the Panamanians returned to their vehicle but remained where they were. Soon thereafter, two police cars "came racing down the sides of the road, and they had their high lights on and they were flashing their lights on and off." At that point the Americans opened fire in earnest, disabling the van and one of the cars and killing three of the policemen. An hour later, after the full airborne platoon had arrived, the paratroopers engaged a squad-size PDF element that was approaching the buses, again inflicting casualties. After that, the situation at the roadblock quieted down.[47]

There was one other AO North objective, the primary one: Colón. Task Force Atlantic was charged with securing the bottleneck leading into and out of the built-up area and neutralizing specific targets inside the city itself.[48] Company B from the 4th Battalion, 17th Infantry, drew the mission but, because of legal prohibitions on the movement of U.S. troops in Colón, the unit had not had the opportunity to conduct anything approximating an on-site rehearsal—no Sand Fleas, Purple Storms, or contingency readiness exercises in the city. There were photographs, of course, and a map, and troops from the company had been able to conduct an exercise in an area within sight of their objective. Still, most of

[47] Donnelly, Roth, and Baker, *Operation Just Cause*, pp. 254–55. Quote from Interv, Robert K. Wright Jr. with 2d Plt, Co D, 3d Bn, 504th Inf, 31 Dec 1989, Madden Dam, Panama; Michael Ryan, C-645 essay [on securing France Field] (Student paper, U.S. Command and General Staff College, Fort Leavenworth, Kans.).

[48] Unless otherwise noted, this account of the Colón bottleneck is based on Donnelly, Roth, and Baker, *Operation Just Cause*, pp. 254–59; Briefing, Brooks, 29 Apr 1992.

the company, commanded by Capt. Doug Thorp, had not seen the actual target prior to D-day. That night, to make matters worse, the initial elements of the assault force were in five-ton trucks en route from Fort Davis to the bottleneck when the shooting at Coco Solo began. By the time the column reached Colón, the PDF defenders were ready. As soon as light fighters from Thorp's 2d and 3d Platoons dismounted, they found themselves under fire. Tall elephant grass in the area obscured enemy positions, while the lights of the city illuminated Company B's presence all too well. As the two sides discharged their weapons at one another, Spec. William Gibbs from the 3d Platoon was killed, the battalion's only fatality that night. Another U.S. soldier was wounded.

As the shooting continued, the company's 1st Platoon and a squad of engineers managed to establish a roadblock at the bottleneck, using school buses and concertina, much as the paratroopers had done on the Boyd-Roosevelt Highway. A platoon of military police from the 549th Military Police Company guarded the position from HMMWVs mounted with M60 machine guns and a 2½-ton truck with a .50-caliber machine gun. Meanwhile, the 2d and 3d Platoons seized objectives on the east and west flanks of the roadblock. Once they established their positions, they began shooting out streetlights within range. Around 0230, a Panamanian bus tried to run the roadblock, and the 3d Platoon received authorization to shoot at only its tires. The order proved providential, as the vehicle was full of civilians, including babies and children, many of whom would have certainly been killed or wounded had the soldiers tried to destroy the bus.

Further shooting occurred when U.S. troops fired on two PDF boats that approached their positions from Manzanillo Bay. Enemy snipers in the area also proved a menace throughout the night. In response, Captain Thorp requested that Task Force Atlantic's artillery employ illumination rounds to light up the area presumed to conceal the shooters. Antisniper teams also helped bring the threat under control. By 0500, the bottleneck was secured. Twelve of the enemy were known to have been killed or wounded, but there had also been casualties among the numerous civilians who had tried to flee the fighting. With Panama's second largest city isolated by Company B's blocking positions, OPLAN 90–2 called for Kellogg's task force to be "prepared to isolate, neutralize, and/or destroy PDF military region headquarters in COLON" and to protect other facilities there.[49] For Task Force Atlantic, this translated into entering the city on D-day, and the command's plans reflected this assumption. All that was needed was word from JTF-South to proceed. But on 20 December, the word never came. In fact, the opposite happened. Kellogg received an order not to go into the city. For reasons that were not immediately clear, the planned sweep of Colón had been put on hold.

[49] JTF-South OPLAN 90–2, 3 Nov 1989.

11

Task Force Atlantic: D-day in Area of Operations South

☆ ☆ ☆

While Lt. Col. Johnny Brooks' 4th Battalion, 17th Infantry, and the units attached to it were engaging their targets in the northern part of Task Force Atlantic's area of operations, the bulk of the 82d Airborne Division's 3d Battalion, 504th Infantry, commanded by Lt. Col. Lynn D. Moore, was playing the lead role in D-day missions throughout the southern portion. In the ten days leading up to Just Cause, the battalion was still in the midst of a three-week rotation in USARSO's Jungle Operations Training Center. Since the school's founding in 1968, a variety of light infantry, airborne, ranger, and Marine battalions had found themselves subjected to rigorous, sometimes dangerous, courses in survival skills, land navigation, and jungle warfare.[1] The school accommodated one battalion at a time, with the schedule of training dates for specific units being determined months in advance. With the execution of Operation Nimrod Dancer, the battalions entering the facility not only confronted the challenging course but also acquired, under the operational control of Task Force Atlantic, a number of highly classified responsibilities related to the crisis. These included planning for contingency exercises and possible Blue Spoon wartime operations; executing the contingency exercises and incorporating the feedback from them into further iterations of the war plans; and, in the event of hostilities, carrying out combat missions. All this generated no small amount of suspense when a battalion began its jungle training and its commander and staff learned, many for the first time, of their additional assignments. If war came, would it be while their battalion was at the school? On 19 December, as word of the impending conflict made its way down through the ranks of U.S. troop units, the suspense ended.

Preparations, Deployment, and Prewar Activities

That the 3d Battalion found itself in the Jungle Operations Training Center on that date was something of a fluke. Originally, Moore's men were not

[1] The jungle facility could, in fact, trace its history back to a provisional headquarters that in 1951 laid the groundwork for the Army's Jungle Training School at Fort Sherman. The training center was established after the Department of the Army directed U.S. Army, Caribbean, to take on the mission of "keeping the art of jungle warfare alive in the Army." For a brief overview of the JOTC, see http://www.forum.militaryltd.com/jungle-warfare/m475-jungle-warfare.htm.

scheduled to receive jungle training in Panama until February. But, in October, Col. Jack P. Nix Jr. took command of the 82d's 1st Brigade, the 3d Battalion's immediate higher headquarters. After looking over the congested training requirements for the units under him, the colonel made some adjustments, in one instance plugging the 3d into a gap created by a Marine battalion that had dropped out of the mid-December JOTC rotation. Among the paratroopers affected by the decision, news that they would be spending the Christmas and New Year's holidays in rough tropical terrain evoked little enthusiasm.[2]

In November, Moore flew to Panama to survey accommodations for his men at Fort Sherman and, more important, to obtain information on the battalion's crisis-related responsibilities, about which he had only a general notion based on what he had learned from another airborne battalion that had engaged in contingency planning while in Panama. With respect to the site survey, he quickly realized that the presence at Sherman of Task Force Atlantic's headquarters and a company from the command's security battalion would limit the number of troops and equipment he could deploy from his own unit. There simply were not enough billets and available space at the fort to accommodate NIMROD DANCER personnel and a full-strength JOTC battalion simultaneously.

Even more discouraging for Moore was the stone wall he hit while talking to Colonel Kellogg's people about the contingency role the 3d Battalion would play. When he asked them about existing plans and battalion-level missions, he was informed that all of that "was too highly classified for us to know." Thus, the missions "were withheld at the brigade level." When Moore returned to Fort Bragg, he assembled his key commanders and staff officers and told them, "We have never been closer to going to war," and "there is a great deal of planning going on right now at the highest levels." This came as a revelation to most in the group, who had no indication that there were any crisis-related missions attached to the jungle-training rotation. As one company executive officer later wrote, "Even the least experienced among us began to suspect that our holiday trip might include some activities not regularly scheduled for units deploying to Jungle Warfare School."[3]

With details of the combat missions 3d Battalion might have to execute in Panama squirreled away in unobtainable, compartmentalized plans, Moore and his staff tried to determine what would be appropriate predeployment preparations for the paratroopers—preparations relevant both to the Panama crisis and to possible contingencies elsewhere in the world. They concluded that some level of training in military operations on urbanized terrain would be useful. Valuable, too, would be company-level "movement to contact" exercises, designed, as one source explained, "to gain or regain contact with the enemy and develop the situation in order to conduct either a hasty or deliberate attack." At the platoon and company levels, Moore's staff scheduled night live-fire attacks. Various exercises at

[2] Interv, Robert K. Wright Jr. with Lt Col Lynn D. Moore, U.S. Army, 29 May 1990, Fort Bragg, N.C.

[3] First and second quotes from ibid. Remaining quotes from Clarence E. Briggs III, *Operation Just Cause, Panama, December 1989: A Soldier's Eyewitness Account* (Harrisburg, Pa.: Stackpole Books, 1990), p. 8.

Bragg were also to incorporate medical evacuation procedures, "an often neglected part of training."[4]

In hindsight, Moore would contend that, "just by chance," the pre-Panama training had been right on target, but not flawless. The one problem cited most often pertained to room-clearing procedures. Existing Army doctrine, as found in Field Manual 90–10–1, *An Infantryman's Guide to Urban Combat*, provided useful information on the best task-organization and movement techniques to use but was not precise on the "backblast area required for antitank weapons used for breaching walls and doors." In a related issue, realism at the MOUT site had suffered when the Rangers responsible for overseeing the battalion's training there would not allow the use of fragmentation grenades in the live-fire exercises. As a result, soldiers went to Panama assuming that "ceilings, walls, and floors, usually constructed of wood or plaster, would keep the fragments from exiting the room in which the explosion occurred." In JUST CAUSE, some of the paratroopers would discover the error of this assumption firsthand. Similarly, they would learn during combat that, in clearing operations, the suppressive fire from an attack unit's automatic weapons, when directed into a room in which a grenade had just exploded, could not be accurately pinpointed, thus posing a risk to friendly soldiers entering the room immediately after the blast.[5]

As the 3d Battalion trained at Bragg, Moore and others had to decide which elements of the unit would actually deploy to Panama. As the colonel later told an interviewer, "Normally the entire Fort Sherman facility is dedicated to the [JOTC] battalion that's down there."[6] Since the presence of NIMROD DANCER units had altered that arrangement, making it impossible to accommodate an additional battalion and its equipment in the remaining space, determining who and what to take from the 3d Battalion became a priority issue. Further complicating the selection process was news that the Air Force would only have one C–141 and one KC–10 Extender available to transport the soldiers and equipment from Pope Air Force Base to Panama.

Starting from the position that "we really had to go heavy with infantry force," meaning his three rifle companies, Moore then sorted out who else would be required. In the shuffle, several officers on his battalion staff would be left behind, as would the psychological operations personnel who normally

---

[4] Interv, Wright with Moore, 29 May 1990. Quotes from Briggs, *Operation Just Cause*, pp. 8–9.

[5] DA FM 90–10–1, *An Infantryman's Guide to Urban Combat*, September 1982. First quote from Interv, Wright with Moore, 29 May 1990. Second quote from Briggs, *Operation Just Cause*, p. 11. Remaining quotes from Operations, nos. 9 and 10, in After Action Review, Operation JUST CAUSE, 3d Bn (Abn), 504th Inf, n.d. Regarding the problems with the use of fragmentary grenades when clearing rooms, the after action review recommended that future military training for urban operations demonstrate the effect of grenade fragments on various construction materials, that fragmentary grenades in combat be used when only enemy forces were known to be in a room, and that units' "basic loads should include a number of stun or concussion grenades" for room clearing. As for suppressive fire with automatic weapons employed in room-clearing tasks, the review recommended that the weapons could be more accurately employed if fired in a semiautomatic mode.

[6] Interv, Wright with Moore, 29 May 1990.

would have deployed. In contrast, a squad of riggers made the cut because its expertise was essential in support of the battalion's airborne undertakings. Other deploying "slices" included his antitank company, a squad of air defense personnel, some military police, some personnel from the 3d Battalion, 319th Field Artillery (but no artillerymen or howitzers), and, in terms of firepower, all of the line companies' individual and crew-served weapons, including M16 rifles, M203 grenade launchers, M249 squad automatic weapons, M60 and .50-caliber machine guns, six sniper systems, and 60-mm. and 81-mm. mortars. The availability of the KC–10 transport also allowed Moore to take four HMMWVs with him.[7]

The battalion began deploying to Panama on Saturday, 9 December, and finished Monday evening after the last of the unit's 633 officers and soldiers, along with their equipment, arrived at Fort Sherman. Given the limited availability of Air Force transports, the first two flights of personnel out of Pope were aboard commercial L–1011 TriStar airliners, which flew to Howard Air Force Base. As one of the officers on the lead flight noted, "Instead of jumpmasters and safeties screaming commands and passing out barf bags and earplugs, we had stewardesses serving us food and drink while the movie 'Batman' played on the screen." The trip from Howard across the isthmus to Fort Sherman was not nearly so pleasant, given the combination of tropical heat and uncomfortable buses. Meanwhile, back at Pope, the remainder of the battalion crammed aboard the available C–141. On Sunday, after several hours in the crowded, smelly, noisy, and stewardless aircraft, the men parachuted onto Fort Sherman.[8]

Operations security dictated that key personnel in the 3d Battalion did not arrive together. Thus, when Moore received his first briefing from the Task Force Atlantic staff immediately after he reached Fort Sherman, he was accompanied by only his operations officer and two company commanders; the other staff members and troops were still en route. As Moore tried to digest the crisis-related information denied to him during his November visit, he and the small group with him realized that, as a priority, jungle training was going to fall well behind the need to update the battalion's portion of OPLAN 90–2 and to conduct and evaluate contingency readiness exercises and Sand Fleas. This realization, in turn, caused some second-guessing about those elements of the battalion that, of necessity, had been left behind at Fort Bragg. Contingency planning alone would require the presence of several staff officers who had not made the trip, while Moore's soldiers, should they find themselves in combat, would not have with them resources they would need for transportation, communication, and other critical battlefield functions. Fortunately for the battalion, Colonel Kellogg's headquarters, having worked closely with the three JOTC battalions that had preceded Moore's, had already designed procedures for facilitating this latest transition. For many, even most, of the shortages plaguing Moore in personnel, administrative support, and

[7] Ibid.

[8] Operations Summary and Operations Log Extract, in After Action Review, Operation Just Cause, 3d Bn (Abn), 504th Inf, n.d. Quote from Briggs, *Operation Just Cause*, pp. 12–13.

equipment, Task Force Atlantic stood ready to step in and compensate for the deficiencies.[9]

Moore described the battalion's first week in Panama as one of "almost continuous planning," as his staff tried to balance revising the contingency plan and organizing contingency readiness exercises mandated by JTF-Panama. Of necessity, updating the combat plan received the highest priority, with Kellogg's brigade staff "cheerfully" providing "material, information, and assistance," as well as use of the task force's planning area, which was larger and in other ways superior to that reserved for the battalion. As Moore's officers studied the plan they inherited, pouring over the target folders prepared by Kellogg's intelligence section, the main problem they identified was having to take what had been written for the light infantry and Marine battalions that had recently gone through the jungle training center, retain the essence of those arrangements, and then tailor them for the quite different organization of an airborne battalion. For starters, Moore's battalion had an antitank unit, Company D, that the light infantry and Marine battalions lacked. As the colonel observed, "Our TOE [table of organization and equipment] would not support what the Marine TOE might." Besides the organizational differences, an airborne battalion also had several different weapons, the .50-caliber machine gun being the most obvious. The existing plan made no provision for .50-caliber ammunition and, since the battalion had left behind at Fort Bragg much of what it might need in an actual fight, the paratroopers would have only two thousand rounds available to them for combat operations. According to the 3d Battalion after action review, "the lack of .50 caliber ammunition remained a problem." In contrast, adjustments made in the plan to incorporate the airborne's antitank company had a positive effect, according to Moore, because they "freed up more of my infantry for the more infantry-intensive targets."[10]

As the planners worked around the clock modifying the operation plan, they also had to prepare orders for the battalion's first company-level contingency readiness exercise scheduled for Tuesday night, 12 December. On Saturday, when Moore first learned of the exercise and its objective, Madden Dam, he had only his Company B and Company D commanders with him. The latter drew the assignment. If 90–2 were actually executed while the 3d Battalion was in the jungle course, the antitank company was slated to release one of its platoons to Colonel Brooks' force in and around Colón. The company therefore would not have its full wartime complement of troops. Since securing Madden Dam and its hydroelectric plant would require fewer men than some of the other Task Force Atlantic targets, Company D (minus) became the logical choice for that D-day mission and thus for conducting contingency readiness exercises near the facility. The unit received its formal orders for the exercise Sunday night, briefed and rehearsed the mission on Monday, and marched to the site Tuesday. The men encountered no trouble en route and returned to Fort Sherman later that day.

---

[9] Interv, Wright with Moore, 29 May 1990; Operations, nos. 1 and 2, in After Action Review, Operation Just Cause, 3d Bn (Abn), 504th Inf, n.d.

[10] All quotes except the second to the last are from Interv, Wright with Moore, 29 May 1990. Second to last quote from Operations, no. 3, in After Action Review, Operation Just Cause, 3d Bn (Abn), 504th Inf, n.d.

Despite the uneventful excursion, one fact, apparent to all, was that exercises mandated by JTF-Panama were "not a walk in the woods"—the troops went out carrying live ammunition and loaded weapons.[11]

During the briefing Saturday at which the Company D commander learned that Madden Dam was both his rehearsal and wartime objective, his Company B colleague, by virtue of being the only infantry company commander present at the time, was able to choose which one of the remaining missions he wanted. To Moore's satisfaction, the captain picked Cerro Tigre, a key PDF logistical base and the largest, most complex target on the list. Since an exercise aimed at that objective was not scheduled until Wednesday, the soldiers of Company B had time Tuesday to log in a full day of jungle training: highlights included briefings on poisonous frogs and other indigenous species that could kill them, as well as instructions on how to construct temporary shelters, or hooches. On Wednesday morning, 13 December, the troops trained in guerrilla tactics, then turned their attention to the Cerro Tigre exercise, the formal order for which they had received fewer than twenty-four hours before.[12]

The Sand Flea was not very complex but did entail certain risks.[13] The company would embark on naval LCMs, move southward through the Panama Canal to a point near Cerro Tigre, conduct an amphibious landing, and secure a friendly signal station and power plant near the PDF logistical base. Having accomplished this, the men would then walk to a golf course just south of the target to be extracted by helicopter. The paratroopers would be maneuvering in clear view of Panamanian guards and other PDF personnel in the area. Moreover, the mission statement, besides calling for the assertion of U.S. rights and the rehearsal of "a portion of the TF Atlantic contingency plan," also wanted the paratroopers to "elicit a response from the PDF." The "underlying implication" of this extremely vague directive was open to interpretation. In the opinion of Company B's officers and staff, it meant nothing less than "to get shot at first by the PDF." While the staff officers at Fort Clayton responsible for writing the statement probably had something less lethal in mind, their inexact handiwork generated "a frank and fruitful discussion" among the participants in the Sand Flea. Because the paratroopers would be carrying live ammunition, they conducted a careful review of the rules of engagement under which they would be operating.

Commanded by Capt. Bryan Dyer, the airborne troops had been trained as warriors to fight and win on a conventional battlefield. As with many such U.S. units rotating into and out of Panama during 1988 and 1989, the men of Company B initially reacted to the political considerations of the politico-military crisis with uncertainty, some apprehension, and no small degree of

[11] Operations Log Extract, in After Action Review, Operation JUST CAUSE, 3d Bn (Abn), 504th Inf, n.d. Quote from Interv, Wright with Moore, 29 May 1990.

[12] Interv, Wright with Moore, 29 May 1990; Briggs, *Operation Just Cause*, pp. 18–20, 24; Operations Log Extract, in After Action Review, Operation JUST CAUSE, 3d Bn (Abn), 504th Inf, n.d.

[13] Unless otherwise noted, the following account of the Cerro Tigre exercise and the quotes are from Briggs, *Operation Just Cause*, pp. 24–39. Some sources refer to the exercise as a contingency readiness exercise, others as a Sand Flea. In many cases, the line separating the two was hazy, and the terminology employed less than precise.

discontent.[14] This was especially true of the rules of engagement they received on their arrival (*Table 3*). As 1st Lt. Clarence E. Briggs III, Company B's executive officer, phrased the problem, "Most of the rules and laws were beyond an infantryman's scope of training." Very soon, he and those around him were asking the same questions that had troubled marines going onto the Arraiján Tank Farm in April 1988, or light fighters and mechanized soldiers running convoys at the outset of NIMROD DANCER a year later. When was it appropriate for the soldiers to open fire? What was considered a threat? How were soldiers to translate such general, even ambiguous, terms as *minimum force*, *self-defense*, *selected marksmen*, and *appears to be*?

> There were so many things that needed to be clarified. We could lock (place the magazine in the magazine well), but not load our weapons (chamber a round) during [a] Sandflea unless someone locked and loaded and pointed a weapon at us. The problem, of course, is that by then it is too late to defend yourself. Furthermore, designating selected marksmen beforehand is not feasible, as you cannot possibly know who will be shot at and when and where it will occur. Also, how do you tell who is an "innocent civilian"? Many of the Dignity Battalion soldiers wore civilian clothes.

Again, as with other units over the nearly two-year buildup of U.S. troops in Panama, several of Company B's leaders determined that they would not allow what they regarded as an unorthodox situation and confusing rules of engagement to paralyze their decision making in the field or to put them and their soldiers at undue risk. Thus, there was tough talk at first of permitting some of the troops, if finding themselves in dangerous circumstances, to lock and load their weapons before the rules suggested that such action was warranted. On further reflection, though, the officers "finally decided that we would just have to wait and handle the situation as best we could when and if it arose."

The live ammunition Company B took on the Cerro Tigre Sand Flea included rounds for M16 rifles and M60 machine guns, AT4 antitank rockets and LAWs, claymore mines, and grenades. As Briggs recorded, "It was the first time any of us had carried an infantryman's basic load, and it was heavy," seemingly even more so in the tropical heat. Once the soldiers had their supplies, there was a "forced hydration ceremony" (each man had to consume at least one quart of water), after which the troops boarded the LCMs that would convey them through the canal to their objective. Hovering overhead, Colonel Moore observed the activity from his seat in an OH–58.

While no confrontation with the PDF developed during the Sand Flea, the paratroopers experienced the inevitable fog and friction inherent in military undertakings. The maps they had of the area proved outdated, and the jungle less penetrable than anticipated. Intelligence reports concerning an unused railroad line were refuted by an onrushing train that drove the company's ambulatory 1st Platoon off the tracks and into the jungle. Finally, at the golf course extraction site, a squad leader and one of his fire teams failed to board

[14] For the controversial issues surrounding rules of engagement for U.S. forces in the Panama crisis from early 1988 to December 1989, see the relevant sections in Yates, *U.S. Military Intervention in Panama, June 1987–December 1989*.

Table 3—Rules of Engagement for Cerro Tigre Sand Flea

Under the Laws of War You Must:

1. Treat civilians and detainees humanely.
2. Respect civilians and their property.
3. Avoid forbidden targets, tactics, and techniques.
4. Prevent and report to your superiors any crime committed under the laws of war.

Rules for Leaders

1. A commander will take all steps necessary and appropriate for his unit's self-defense.
2. Use only the minimum force necessary to control the situation.
3. If possible when returning fire, use selected marksmen.
4. Without endangering your unit or risking the success of the mission, take measures to minimize risk to civilians.
5. Riot-control agents may only be used when authorized by the division commander or his designated representative.
6. Upon cease-fire, take necessary measures to maintain control and assist any injured.

Rules for Individuals

1. If fired upon, fire back if necessary in self-defense.
2. If it appears that you are about to be fired upon, fire if necessary in self-defense.
3. When returning fire, aim directly at its source; do not spray your fire into a general area.
4. Cease fire when the threat is over.
5. Allow anyone who is trying to surrender to do so.
6. Treat innocent civilians with respect.

Source: Clarence E. Briggs III, *Operation Just Cause, Panama, December 1989: A Soldier's Eyewitness Account* (Harrisburg, Pa.: Stackpole Books, 1990), pp. 27–28. The same list with typographical errors is in An. F, to After Action Review, Operations Order for Sand Flea to Cerro Tigre, n.d.

the waiting trucks (a last-minute substitution for the anticipated helicopters, which were unavailable). The absence of the men was not noticed until the convoy had returned to Fort Sherman. There followed a scene marked by confusion and considerable swearing, after which Moore, Dyer, and the platoon leader made the 2½-hour truck ride back to Cerro Tigre and retrieved the errant team. The furor notwithstanding, the troops had accomplished their mission, which consisted of securing the antenna site and electrical substation southwest of Cerro Tigre. In the process, their march took them near enough to the logistical complex to observe its layout, the terrain in the area, and, in the event of combat, the best place to breach the fence surrounding it.[15]

On the day Company B conducted its Sand Flea, the 3d Battalion's remaining rifle companies, A and C, received orders for their contingency exercises at Gamboa and Renacer Prison, respectively.[16] Moore later observed in an interview that each company's assignment matched well with the personality of its commander. Company C's captain had shown himself in the past to be "very aggressive," exactly what would be needed for the "real shootout" anticipated at the well-guarded prison in the event of actual combat. The commander of Company A, in contrast, was "very steady," according to Moore, exactly the kind of leader needed for Gamboa, "the only target that we had that had American families on the target. So we really had to show a great deal of restraint there, because no matter what we did throughout the rest of the country, if we harmed a single American or damaged a single American home, then it was all for naught."[17]

The exercises for Gamboa and Renacer Prison were described in the same mission statement and would occur on the same day. A team from Company A would fly to Gamboa via four Chinook helicopters, mount a mock air assault onto a landing zone dubbed Sword, seize and secure a U.S. school in the vicinity, and then head to a beach labeled Normandy for extraction by an LCM. Meanwhile, a team from Company C would arrive by LCM at a site designated Omaha Beach, conduct a "route reconnaissance" near the prison, secure an objective identified as Iron, and then move to a Panama Canal Commission golf course for an airmobile extraction using the Chinooks. Should either exercise turn into a confrontation with armed Panamanians, Team A had 81-mm. mortars at its disposal, while two attack aircraft—an AH–1 and an OA–37—would provide any necessary fire support for Team C. The exercises received the code name BLUE DEVIL and, similar to Cerro Tigre, called for "highly visible" maneuvers to "elicit the maximum response possible from the PDF." By this time, the controversy surrounding such wording had subsided. The troops knew that they were not to engage in confrontational acts or take unnecessary risks. To the contrary, their guidance expressly prohibited various provocative measures. They were not, for example, to block traffic along roads located in their objectives. As the commander's intent portion of their operation order further elaborated,

[15] See also Donnelly, Roth, and Baker, *Operation Just Cause*, p. 281.

[16] The 3d Battalion's official chronology indicates that the activity at Cerro Tigre took place on Wednesday, 13 December, while Briggs' account suggests that it was on Thursday, the fourteenth.

[17] Interv, Wright with Moore, 29 May 1990.

> Because of our visibility, we will be subject to the perusal of the Panamanian people and press. Absolutely nothing must occur that would present other than a 100% professional appearance. Because we will be operating in close proximity to PDF units, all leaders must be thoroughly familiar with maneuver rights under the Canal Treaty and each soldier must know the [rules of engagement] especially with regard to the use of deadly force. I will not have a soldier killed because he was unsure of his rights of free passage and self defense.[18]

The two companies accomplished their missions, although, in the case of Company C at Renacer Prison, not without incident. At that site, the troops disembarked from the LCMs at Omaha Beach only to be greeted by the prison's Panamanian guards with their weapons at the ready, locked and loaded. The PDF officer in charge, a lieutenant colonel, approached Company C's commander, Capt. Derek Johnson, and informed him that under the canal treaties the U.S. force had no right to be in the area without having received prior authorization. According to Moore, Johnson responded, "Okay, I'm coordinating it right now. We're going through here in five minutes." Following some animated discussion among the Panamanians, the PDF officer proposed that the U.S. troops wait fifteen minutes before proceeding. Johnson accepted the compromise, and, at the appointed time, his men divided into two elements and reconnoitered the area. As Moore later noted with pride, his company commander had "stood toe to toe with the P.D.F. lieutenant colonel and backed him down." All the while, the Sand Flea was being videotaped by a SEAL team nearby, thus providing Task Force Atlantic with invaluable information on the location of the PDF guard posts and the layout of the prison. What they could not determine was the number of prisoners inside the complex. Still, Johnson was pleased: "A company commander couldn't ask for anything better except the names of the prison guards."[19]

In summing up the 3d Battalion's first week in Panama, Moore talked about having "almost a 24-hour cycle here of planning the CRE, executing the CRE, doing the after-action review on that, adjusting the contingency plan to fit our lessons-learned, and then executing 24 hours later the next CRE." The exercises culminated on Saturday, 16 December, with something of a milestone: authorized personnel within the 3d Battalion received the OPLAN 90–2 operation orders they might have to execute. In almost every case, the actual battle plans approximated what Moore's four principal companies had rehearsed during the Sand Fleas. Company A would conduct an air assault into Gamboa to protect American citizens and property and to seize police stations and the FUFEM (*fuerzas femininas*) Barracks, a PDF women's training facility. Company B would conduct air assaults to seize the Cerro Tigre logistical center, secure an ordnance school and a nearby intersection, and block enemy use of that portion of the transisthmian Boyd-Roosevelt Highway passing near the objective. Company C (minus) would conduct an air and LCM assault to seize Renacer Prison and rescue the political prisoners there, including one or two American "ecologists"

[18] After Action Review, Operations Order for Sand Flea to Gamboa/Renacer Prison, n.d.

[19] Second quote from Interv, Wright with Moore, 29 May 1990. Interv, author with Lt Col Gregory Gardner, U.S. Army, 7 Apr 1992, Fort Bragg, N.C.; Briggs, *Operation Just Cause*, p. 29. Other quotes from Donnelly, Roth, and Baker, *Operation Just Cause*, pp. 270–71.

(presumably intelligence operatives), together with a number of Panamanian officers who had taken part in coup attempts against Noriega in March 1988 and October 1989. Finally, Company D (minus) would conduct a road march to secure the Madden Dam and to observe another portion of the transisthmian highway (*Map 19*).[20]

Moore's staff and company commanders barely had time to peruse the contingency plans when word reached Fort Sherman of the Paz shooting. More details soon followed, some describing troop movements by Panamanian forces at several key locations. For example, reports had V300 Cadillac Gage personnel carriers taking up positions around the Comandancia in Panama City, while other PDF elements in the capital were setting up roadblocks, including one at the Bridge of the Americas, and moving to defend other military and police facilities. At some point, so reports ran, the entire defense force, as well as some Dignity Battalions, had increased their alert status. On the Atlantic side, one group of DigBats was known to have been issued arms and told to guard a Panama Canal Commission building in Colón. More disconcerting, the PDF 8th Company at Fort Espinar supposedly had been ordered to "take out" any U.S. soldiers attempting to block the gates there. All of this led to a corresponding increase in Task Force Atlantic's own alert posture, which included "frantically" issuing ammunition to the troops. Equally unsettling, the transportation assets needed to convey Kellogg's forces to their combat objectives began arriving at Fort Sherman, just in case the death of Lieutenant Paz escalated into a shooting war.[21]

Tension eased somewhat on Sunday and Monday, although the monitoring of Panamanian radio and television stations revealed that the regime apparently expected some kind of U.S. retaliation for Saturday night. To increase its readiness posture, the 3d Battalion canceled further training at the jungle school (although some of Company B's officers "went out on the land navigation course just for kicks"), and company commanders began "scouring the area" in search of buildings that at least resembled each company's actual targets. When a replica could not be found, long strips of engineer tape stuck to the ground sufficed to approximate a target's blueprint. Rehearsals followed, lasting into Tuesday. In the meantime, on Monday night, Moore was in the room with other Task Force Atlantic officers when Kellogg, who had called the meeting, wrote the day-time group of H-hour on a piece of paper, an unofficial heads up regarding the conflict just over one day away. As a result of the news, Moore told his company commanders to order an earlier than usual "lights out" and later surprised several of them "with how irate I was when they weren't asleep at 10:00 o'clock that night." That his troop leaders

---

[20] Quote from Interv, Wright with Moore, 29 May 1990. Operations Summary and Operations Log Extract, in After Action Review, Operation Just Cause, 3d Bn (Abn), 504th Inf, n.d.; JTF-South OPLAN 90–2, 3 Nov 1989; McConnell, *Just Cause*, p. 166. While several sources mention the presence of one or two American prisoners at Renacer, McConnell's account is among those that suggest that Moore and other U.S. planners believed them to be intelligence operatives.

[21] Interv, Wright with Moore, 29 May 1990. Quoted words from Briggs, *Operation Just Cause*, pp. 41–42. Operations Log Extract, in After Action Review, Operation Just Cause, 3d Bn (Abn), 504th Inf, n.d.

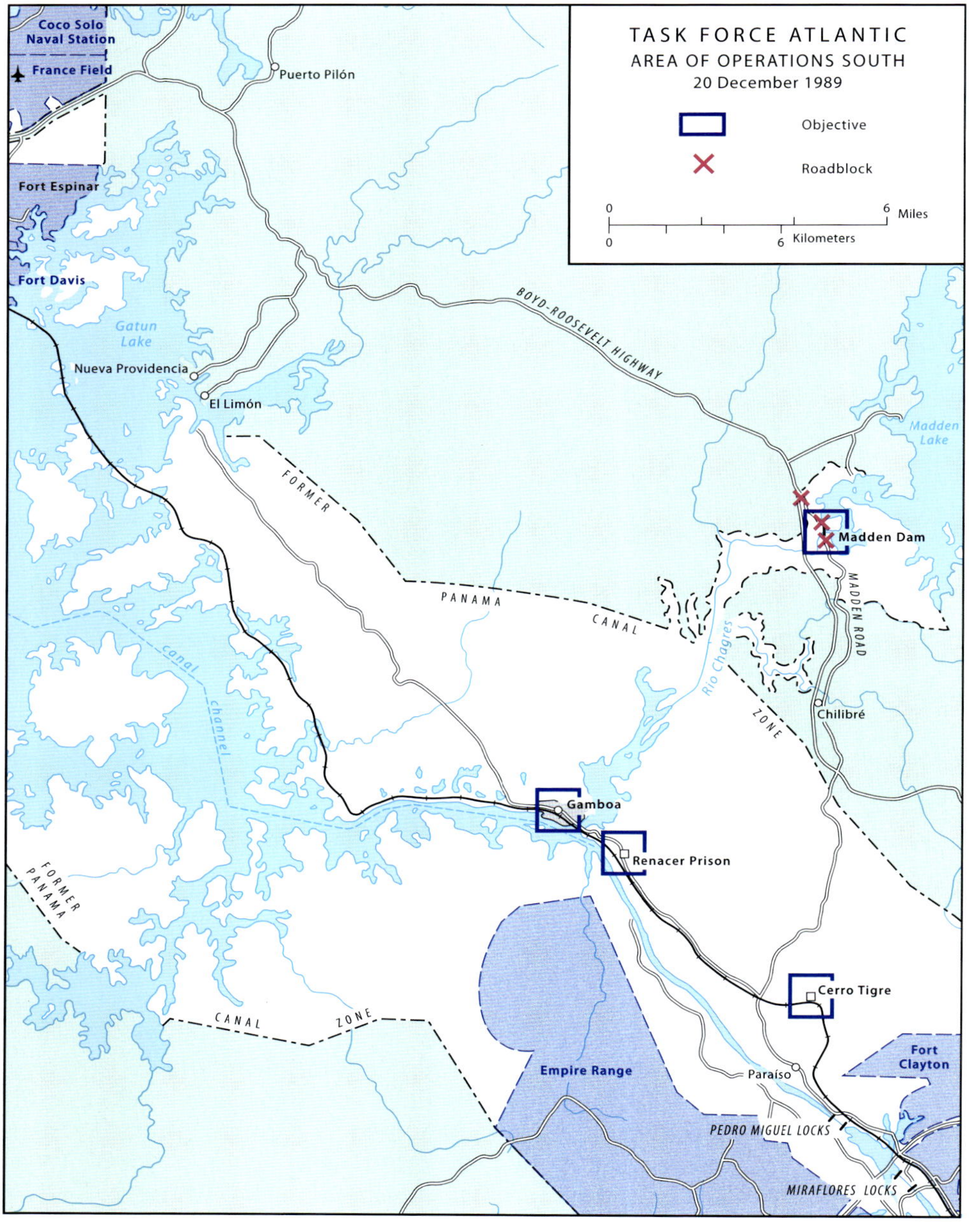

Map 19

get sufficient sleep was more important to Moore than having them continue the war-gaming they were doing when he issued his directive.[22]

[22] First quote from Briggs, *Operation Just Cause*, pp. 42–43. Other quotes from Interv, Wright with Moore, 29 May 1990. As an example of the makeshift rehearsals, one account states that Captain Dyer, the Company B commander, by "a stroke of sheer luck" found buildings at Fort Sherman that "closely resembled" those at the logistical complex at Cerro Tigre.

The official announcement of impending hostilities was given to 3d Battalion at 1800 the next evening and to lower echelons at 2100. By then, the battalion had completed a physical training run, and speculation that war was imminent was already filtering down through the ranks. One clue that something was afoot had come earlier that morning when troops had awakened to find an array of transport and attack helicopters, some rigged with armaments, lined in formation on the field outside the battalion's barracks. The likelihood that they were about to go into combat led a number of soldiers to call their families and sweethearts in the United States, often revealing in the course of their conversations sensitive operational details over unsecured telephones easily monitored by Panamanian officials, thanks in part to the fact that Noriega's brother-in-law controlled the telephone exchanges. To stop these security violations, Moore banned further communications with the outside world, "although some damage had already been done." The remainder of the day was spent fine-tuning and rehearsing the contingency plans and then going through the same standing operating procedures that occupied almost all U.S. combat units on each side of the canal as they prepared to depart for their objectives. Also in common with other U.S. troops about to take part in Just Cause, morale was high, albeit dampened somewhat by the news, received around 2300, that the Panamanian military knew the Americans were coming.[23]

Madden Dam

The leaders and soldiers of Company D, Moore's antitank company, had little chance to analyze the information provided them during the 2100 meeting. Only an hour later, they were boarding trucks that, once transported across the Panama Canal, would take them to Fort Davis, their staging area and departure point for the 1½-hour trip to Madden Dam, fifty miles to the southeast. Possible traffic in the canal prompted the decision to leave Sherman hurriedly; in being ferried across the waterway, the convoy did not want to "get stuck behind some ship going through the [Gatun] Locks." Even if the transit took place without incident, the company still had to negotiate the overland ride from Davis to the objective. The roads to the dam were primitive, and, although there were no enemy combat units at the facility itself, the Panama Defense Forces had small elements armed with AK47s and .38-caliber pistols manning two checkpoints and a customs station along the route. There was also the PDF 1st Infantry Company at Las Tinajitas, but it would not be able to get reinforcements to the dam for at least ninety minutes, if it was able to respond at all given the air assault planned for it.

Donnelly, Roth, and Baker, *Operation Just Cause*, p. 282. Another account tells how Company C used trash cans and Meals, Ready to Eat cases to build a mock-up of its target, the Renacer Prison. See CALL, "Renacer Prison Raid," in Newsletter 91–1, *Rehearsals* (1991), at http://www.globalsecurity.org/military/library/report/call/call_91-1_sec6.htm.

[23] Operations Log Extract, in After Action Review, Operation Just Cause, 3d Bn (Abn), 504th Inf, n.d. Quote from Briggs, *Operation Just Cause*, pp. 44, 48.

Madden Dam

Another factor that could possibly disrupt Company D's timely deployment was the rules of engagement. Prior to 0100, the company commander, Capt. John Campbell, would not have the right under the pre–H-hour restrictions imposed on him to take hostile action against any enemy troops he encountered unless those troops fired on his own soldiers first. Besides the legal distinction between precombat and combat phases, there was another compelling reason for the constraint. If Campbell were to initiate the use of lethal force prematurely—the noise of which would be heard for miles—he would risk compromising the other missions planned for Task Force Atlantic's southern area of operations. Moore's advice to the captain: if you are stopped, lie your way through. Colonel Kellogg had emphasized the same point at his meeting with the battalion and company commanders. Campbell was to tell any armed Panamanians he met along the route that he and his troops were on their way to Howard Air Force Base to catch flights back to the United States. As an alternative to that cover story, Campbell could simply say that he was lost. Right after the meeting, the captain, realizing the vulnerability of his troops, requested protective air support, but Moore turned him down: "We don't . . . divide air cap for the convoys. You're on your own. We'll be on strip alert; call us if you have a problem."[24]

Madden Dam, as Campbell knew, was considered "a vital facility to the operation of the Panama Canal." For this reason, it was regarded in at least

[24] Donnelly, Roth, and Baker, *Operation Just Cause*, pp. 286–88. Quotes from Interv, Wright with Moore, 29 May 1990. During Moore's interview with Wright, when Moore said that he had told Campbell to call back to Fort Sherman for air support if there was a problem, the written transcript indicated "laughter" on the part of Wright or Moore or both. As for the estimates concerning enemy forces along Campbell's march route and at the dam, see After Action Review, Operations Order for Sand Flea to Madden Dam, n.d. Although the Sand Flea occurred just one week before Just Cause, there had been no reason to change the estimates.

one account as "Task Force Atlantic's only strategically significant objective, the others being enemy positions or areas where American civilians were in danger." The dam provided power to U.S. military facilities south of it. Moreover, the water it retained in Madden Lake, if released through an act of sabotage, would quickly submerge the nearby town of Gamboa. Such an event would be catastrophic, but U.S. planners saw the threat as remote. Indeed, they expected Campbell to secure the dam and block the highway north of it with less than his full company, one of his platoons having been assigned to the Coco Solo mission. In partial compensation, the captain received as attachments to his force two squads of engineers, a mortar section from the battalion's Company C, and two teams of military police from the 549th Military Police Company.[25]

The Company D convoy consisted of a military police HMMWV, a command and control HMMWV, and three 2½-ton trucks, one of which was mounted with a .50-caliber machine gun. The vehicles and troops departed Fort Davis on time at 2330, with Campbell's plan being to keep their speed around thirty-five miles per hour so as not to draw unwanted attention. The military police lieutenant attached to Company D, however, had other ideas, and he would be at the head of the convoy. According to Campbell, the young officer thought the captain's imposed speed limit too slow and too dangerous. "Sir, that won't work," he is quoted as saying. "There's only one speed, that's hauling ass. That's the only safe way to go." Campbell argued to the contrary, but once the convoy set out, he found himself chasing after the junior officer's lead HMMWV. As a result, the assault force arrived in the vicinity of the dam twenty minutes before H-hour. By that time, Moore was hovering over AO South in the OH–58, monitoring his battalion's movements over the helicopter's single radio. When he heard that Campbell was prematurely approaching Madden Dam, he ordered him to find the nearest cover and go to ground until H-hour.[26]

Campbell complied but stayed in place for only fifteen minutes. At 0055, he decided that his troops had waited long enough in the "hiding position" he had chosen at an intersection on the Boyd-Roosevelt Highway near the dam and the PDF customs station. "I'm not going to just sit here," he reportedly declared. Having made this decision, he and the engineers moved down the road toward the dam, their advance covered by one of his platoons armed with two .50-caliber machine guns. The only hostile force they expected to encounter was a single guard who, as a result of previous contacts, was known to be friendly. Still, approaching the armed Panamanian was "one of the tightest things" Campbell had to do, since he could not answer one crucial question: would an overriding sense of patriotism and honor compel the guard at the last minute to offer some resistance? To everyone's relief, the soldier

---

[25] First quote from After Action Review, Operations Order for Sand Flea to Madden Dam, n.d. Interv, Wright with Moore, 29 May 1990. Second quote from Donnelly, Roth, and Baker, *Operation Just Cause*, p. 286, and see also pp. 287–88.

[26] Military police lieutenant quote from Donnelly, Roth, and Baker, *Operation Just Cause*, pp. 288–89. Interv, Wright with Moore, 29 May 1990.

readily surrendered his weapon and opened the gate to the dam. About fifteen minutes later, Campbell's troops had secured the facility.[27]

While the engineers moved into the dam and power plants, Campbell fortified his position, placing a platoon on the roof of the electrical station, another at the roadblock north of the dam, a mortar section with two 60-mm. mortars south of the installation, and, somewhat later, a team of scouts near Madden Road and Boyd-Roosevelt Highway where they could monitor traffic along those routes. With these locations secured, he took eight men from his 1st Platoon and moved to the PDF customs station. The two guards there decided not to offer even token resistance, surrendering without firing a shot. The remainder of the night proved uneventful. Members of the 4th Platoon spied a couple of Panamanians with AK47s, but the men fled once they saw American troops. In another minor incident, U.S. soldiers apprehended an old man foraging for food. When they brought him to Campbell, the company commander could only laugh. "What the hell are we going to do with this man?" The answer was to give him some Meals, Ready to Eat and send him on his way. After dawn, Campbell decided to seize the smaller of the two PDF stations in the vicinity. As he and ten of his men in two military police vehicles approached the building, they saw a group of armed Panamanians surrounded by a group of children. The Americans held their fire, which allowed the enemy to abandon the kids and escape out the back of the station into a wooded area. Campbell had to settle for confiscating the weapons left behind and returning to Madden Dam. The rest of the day passed without incident.[28]

Cerro Tigre

Because there were no shooting incidents at Madden Dam, the battalion's after action report noted that it was the only objective at which there was no contact with the enemy. The degree of resistance encountered at the other targets varied greatly. In the case of Cerro Tigre, it turned out to be considerably less than anticipated. On the eve of the mission, the officers in Company B "fully expected to go into a hot landing zone (LZ) and to fight our way up to the objective." The enemy force, some estimated, could be as large as a company of soldiers.[29]

Unlike the Sand Flea in which Company B had deployed in LCMs, the real attack on the PDF logistical center would be via an air assault, employing four transport helicopters from U.S. Army, South's 1st Battalion, 228th Aviation; two UH–1s would take the 1st Platoon to the objective, while two CH–47s would follow close behind with the rest of the company. Initially, planners had located the landing zone for the Hueys inside the enemy complex, with the troops on board then establishing a secure perimeter for the remaining soldiers who would land thirty seconds later. But when word circulated that

[27] First and third quotes from Interv, Wright with Moore, 29 May 1990. Second quote from Donnelly, Roth, and Baker, *Operation Just* Cause, pp. 288–89.

[28] Donnelly, Roth, and Baker, *Operation Just Cause*, pp. 289–90.

[29] Operations Summary, in After Action Review, Operation Just Cause, 3d Bn (Abn), 504th Inf, n.d. Quote from Briggs, *Operation Just Cause*, p. 49.

the mission had been compromised, a last-minute decision shifted the landing zone to a golf course near the site, the same greens from which the company had been extracted after its Sand Flea. Supporting the troops from the air would be a Cobra attack helicopter from Task Force Hawk. Originally, it was to hold its fire until Captain Dyer requested support, but, once the breakdown in operations security shattered hopes that Company B might be able to "sneak in without causing suspicion or compromise," there was no reason to stop the Cobra from firing on PDF guard positions—one of which boasted a ZPU4 Soviet antiaircraft gun—even before Dyer's infantry had landed on the golf course.[30]

The transport helicopters for Company B arrived at Fort Sherman around midnight. While platoon leaders briefed their superiors, the troops received their final inspections, moved to the pickup zone, and began boarding the Hueys and Chinooks. Then, the waiting began, with its toll on one's nerves and self-confidence. The principal fears, of course, stemmed from wondering how one would react in combat and whether one would survive the experience intact. As Briggs later wrote in an allusion to the rules of engagement, "The principles of military necessity, proportionality, and avoidance of unnecessary suffering were brushed aside. It all boiled down to 'us or them,' and it sure as hell was not going to be us. We could worry about the legal ramifications later, assuming that we lived through this. In the meantime we would go in with both barrels blazing, hitting them with everything we had to gain total and immediate victory."[31]

The two Hueys and the first Chinook took off on time with their passengers. When the second Chinook tried to lift off, however, it broke down. The troops quickly exited and made their way to a backup CH–47, but the loading process delayed a timely departure. Further disrupting matters, the backup pilot was uncertain of the route he was supposed to take to Cerro Tigre, the result being that he flew past the objective and almost all the way to Panama City before discovering the error, returning to the correct landing zone, and disgorging his passengers nearly ten minutes late. As for the troops who arrived at the golf course on schedule, they encountered enemy fire on their approach, but 2.75-inch rockets from the Cobra and machine gun fire from door gunners on the Hueys quickly silenced the PDF guards doing the shooting.[32]

---

[30] Interv, Wright with Moore, 29 May 1990; Douglas I. Smith, Army Aviation in Operation JUST CAUSE (Study project, U.S. Army War College, Carlisle Barracks, Pa., 1992), p. 53. Smith was the commander of USARSO's aviation battalion. Donnelly, Roth, and Baker, *Operation Just Cause*, p. 282. Quote from Briggs, *Operation Just Cause*, pp. 48–49.

[31] Donnelly, Roth, and Baker, *Operation Just Cause*, p. 282; Briggs, *Operation Just Cause*, pp. 49–50. Briggs also notes the concern over friendly fire, especially given reports that a number of AC–130 gunships would be in the area providing fire support. To lessen the chances that the soldiers of Company B would be mistakenly identified from the air as the enemy, the soldiers "brought along glint tape" that the AC–130 crew could see from the air.

[32] Interv, Wright with Moore, 29 May 1990; Briggs, *Operation Just Cause*, p. 53. According to Moore, only the second Chinook arrived late at the golf course. According to Captain Dyer, both Chinooks arrived late, thus leaving the 1st Platoon that had already exited the Hueys alone and vulnerable. "I had a platoon on the ground for five minutes by themselves," Dyer is reported as saying. Donnelly, Roth, and Baker, *Operation Just Cause*, p. 283.

In Colonel Moore's opinion, Captain Dyer had a very sound plan. After engineers had breached the fence surrounding the target, the full company would move into the logistical complex and begin a three-platoon operation to clear the buildings in each of three areas into which the planners had divided the compound. For the breaching operations, the officers had ruled out claymore mines, grenades, and antitank weapons as being ineffective against a chain-link fence. Instead, the engineers would use C4 explosives. "They made a belt charge," Briggs explained, "by attaching a coat hanger to the top of an empty bandoleer. They could attach it to the fence to facilitate vertical separation of the wire when the charge exploded." As things turned out, the innovative rigging went for naught. As the first elements moved toward the fence, a startled PDF guard fled. Dyer's men then discovered that a gate in the barrier had been left open, enabling the entire company to walk into the compound unimpeded.[33]

While Company B's 2d Platoon initially stayed in place as a reserve, the 3d Platoon moved to the western part of the complex, with the 1st Platoon going to the south. As they began clearing the guard shacks, warehouses, and other buildings, the few enemy soldiers they came upon fled into the jungle. At one guardhouse, however, the 3d Platoon saw enough enemy activity to warrant taking the structure by force. Small-arms fire and an AT4 were enough to do the job, with the hostile troops fleeing like the others. This success, however, came close to exacting a very high price. To the surprise of the 1st Platoon and the company's executive officer situated in a nearby warehouse, many of the rounds being fired at the guardhouse ricocheted into their area, at first creating the fear that an enemy counterattack was under way. In the warehouse, in fact, Briggs had laid his grenades in front of him, switched off his rifle's safety, and was preparing to open fire on a group of men he saw moving in his direction. Only when he heard them swearing in English and saw their rucksacks did he lower his weapon. They were soldiers from the 3d Platoon still maneuvering against the guardhouse. As the situation stabilized following this close call, Dyer brought in his 2d Platoon to join the clearing process. By 0600, Company B controlled the logistical site. No enemy soldiers had been killed and no wounded had been found. As for the U.S. force, two men had been injured from the use of M203 grenade launchers. One of the soldiers had thought he was firing a buckshot round but had chambered a high explosive grenade instead. Given the short distance over which the grenade traveled, fragments from its blast wounded the shooter. Afterward, the two casualties succeeded in convincing some of their superiors that M203 fragmentary grenades were not best suited for clearing rooms, unless a unit had actually trained with such ammunition, which Company B had not.[34]

Once the logistical compound was secured, the company's 1st Platoon concentrated on the ordnance school just over a half-mile to the north. The troops in the attacking party rode most of the way to the facility on two PDF trucks taken from the Cerro Tigre motor pool and then dismounted and made

[33] Interv, Wright with Moore, 29 May 1990. Quote from Briggs, *Operation Just Cause*, pp. 49, 54.

[34] Donnelly, Roth, and Baker, *Operation Just Cause*, pp. 283–84; Briggs, *Operation Just Cause*, pp. 55–56.

the final approach along a road through the jungle. The Panamanians at the school, having anticipated the route, had prepared some defensive measures. The fact that the Americans were exhausted and moving along the jungle path too close to one another only increased their vulnerability. Suddenly, there was an explosion, freezing the men in their tracks except for the soldier who took some of the shrapnel in his arm. "I'm hit! I'm hit!" he cried out. Surprised by the blast and unnerved by the soldier's screaming, 1st Platoon's initial attempts to return fire on a guardhouse some fifty yards away proved ineffective, with the rounds going high. Finally, a corporal from a squad positioned across a road from the target made a run for a berm in front of the guardhouse. The rest of the squad quickly followed, bringing with them a PDF .30-caliber machine gun they had also appropriated from Cerro Tigre. They shot up the structure but, by then, the guard who had set off the explosion had escaped.[35]

As the squad moved toward the objective, the men discovered that they were in the middle of an area filled with wired booby traps. Without engineers to disarm the devices, the platoon carefully crept forward, marking the locations. As Briggs recalled, "I examined every twig and patch of dirt before me, imagining that something was going to explode between my legs at any moment and leave me squirming on the ground with stubs of raw hamburger." Once the men made it through to the school's barracks area, they began the clearing operations, generally starting the process by lobbing M203 grenades through windows, following with shotgun fire, and then with personnel entering the premises. As they went from building to building, they met no resistance. By early morning, the ordnance school was secured.[36]

Once Cerro Tigre and the ordnance school were under U.S. control, Company B began regrouping, redistributing ammunition, sorting through captured PDF equipment and weapons, and setting up defensive positions against a possible (if unlikely) enemy counterattack. In the meantime, some elements of the company returned to Fort Sherman where, they learned, their numbers and firepower were needed to assist in the delayed sweep through Colón.[37]

Gamboa

Southwest of Madden Dam and northwest of Cerro Tigre, the town of Gamboa was home to three PDF facilities: a headquarters said to contain three platoons of forest rangers, a police station containing two platoons, and a counterintelligence training center for a company of FUFEM soldiers. The town also contained about one hundred seventy American citizens, many of whom were Panama Canal Commission or Department of Defense school system employees. The mission for Capt. Peter Boylan's Company A was to neutralize the enemy forces while protecting the Americans, a responsibility

[35] Quote from Donnelly, Roth, and Baker, *Operation Just Cause*, pp. 284–85. Briggs, *Operation Just Cause*, pp. 67–69; Interv, Wright with Moore, 29 May 1990.

[36] Interv, Wright with Moore, 29 May 1990. Quote from Briggs, *Operation Just Cause*, pp. 69–70. Donnelly, Roth, and Baker, *Operation Just Cause,* pp. 285–86.

[37] For details on the Colón sweep, see the next chapter.

that all knew would require restraints on the use of deadly force in ways that could put U.S. troops at risk from hostile fire. To accomplish the mission, Boylan assigned the forest ranger headquarters to his 1st Platoon, the police station to his 2d, and the FUFEM to the 3d. The 2d Platoon also had the task of establishing a blocking position on a bridge over the Chagres River. Renacer Prison, another 3d Battalion objective, was located on the other side of the bridge.[38]

As in the case of Cerro Tigre, there was a last-minute change to the planned air assault on Gamboa once word reached Moore's headquarters that the enemy knew the time hostilities would commence. The original plan called for the helicopters involved—one Huey and two Chinooks—to put down at two landing zones: one, McGrath soccer field in the center of the city, near the forest ranger headquarters, and the other near the female counterintelligence facility. Once news of the breach in operations security reached Fort Sherman, the ranking officers at Task Force Atlantic had to assume that both landing zones at Gamboa would be hot. To preclude having Company A's divided force trying to land under lethal fire in two different locations, a decision was made to put the whole force down at McGrath field, labeled Landing Zone Vulture, on the grounds that all three enemy targets were accessible from the field and that a full company could more easily overwhelm any enemy resistance at that location. The pilots of the helicopters factored in this new information, and, at 0040, Company A was in the air en route to its objectives. The Huey took the lead, "carrying the 3d Platoon leader, his radio operator, and three gunnery teams, each consisting of an M60 gunner, an assistant gunner, and a squad leader carrying an M203 grenade launcher." The preponderance of automatic weapons in the Huey reflected the critical role the soldiers aboard would play in securing the landing zone for the two Chinooks behind them. The Chinooks transported the 1st and 2d Platoons and what remained of the 3d Platoon, with Boylan and his company headquarters among the passengers on the second CH–47.[39]

The three helicopters followed the Chagres River until they reached Gamboa, then veered over the golf course, and headed for McGrath field. As feared, the Panamanian defenders opened fire as the first paratroopers jumped from the aircraft. Despite the "tremendous volley," only one U.S. soldier was hit, a thigh wound that he did not even realize he had sustained; he just thought he had banged his leg getting out of the Chinook. The three M60 teams quickly established their firing positions, with each machine gun crew targeting one of the company's three objectives. The forest ranger headquarters was the closest and, because of the Americans living around it, the most problematic. No LAWs or AT4s could be employed against the facility for fear of hurting or killing the people Boylan and his force had arrived to protect.[40]

---

[38] Donnelly, Roth, and Baker, *Operation Just Cause*, pp. 276–77; Interv, Wright with Moore, 29 May 1990; Briggs, *Operation Just Cause*, pp. 62–63.

[39] Interv, Wright with Moore, 29 May 1990. Quote from Donnelly, Roth, and Baker, *Operation Just Cause,* pp. 278–79.

[40] Donnelly, Roth, and Baker, *Operation Just Cause,* pp. 278–79. Quote from Briggs, *Operation Just Cause*, pp. 63–64, and see also p. 65. The soldier who was wounded in the thigh, Sgt. Christopher Thompson, was evacuated to a hospital. After learning that the medical staff

As the 1st Platoon exited its Chinook, the men became aware of another problem concerning the forest ranger headquarters: it "was located on a hilltop up a 70-degree slope." They had scouted the area only a few days before during the Blue Devil Sand Flea, but the route they had taken to observe the facility actually concealed the steep incline they now discovered directly in front of them. Moore later made an additional admission: "Maybe if we had looked more closely at the 1:50,000 map we might have seen it." At H-hour on 20 December, however, the 1st Platoon could only curse its misfortune and accept the inevitable, as the men began climbing the hill. In the heat and humidity of Panama, even at night, this was no simple feat, requiring the soldiers to crawl upward "with aching lungs and burning muscles." Reaching the top of the hill, they discovered a fence, the breaching of which sapped more of their energy. Once through this last obstacle, they approached the PDF headquarters, positioning themselves at several locations around the building, including on property or against residential housing that was assumed to be Panamanian. At one house, a U.S. soldier who had moved just inside its breezeway was confronted by a man walking onto the patio, loudly asking what was happening. The man turned out to be an American citizen who, once informed of the danger he faced, went back inside. The soldier, for his part, quickly realized that his continued presence there could draw enemy fire toward the house and its friendly inhabitants.[41]

Once the 1st Platoon was in position, it opened fire on the headquarters, using small arms, machine guns, and grenade launchers in a prelude to storming the building. One minute later, the troops launched their attack. The defenders quickly fled out the back and into the area planners had declared a no-fire zone because of the American housing it contained. As Moore related, one U.S. machine gunner "had to make the choice between shooting the fleeing P.D.F. members and not firing towards American buildings, and he held fire." When the 1st Platoon entered the headquarters and began clearing the rooms, it used "small arms with a couple of grenades," Boylan noted.[42]

By the time the forest ranger headquarters was secured, the 2d and 3d Platoons had already achieved their objectives. The police station south of McGrath field was the first to fall, although not easily, as U.S. troops coming under fire had to respond with great care, not because there were any American citizens in the area, but because a stray round into a gasoline station next to the police building might set off an unwanted conflagration. As with the forest rangers, the PDF police fled—in this case into the jungle—after the 2d Platoon began its final assault. Once inside the station, the clearing operation uncovered a variety of weapons, including assault rifles, bayonets, shotguns, and pistols. With the building secured, two teams from the platoon moved to the Gamboa bridge to establish the blocking position that would help the

---

there planned to send him back to the United States for an operation, he left and rejoined his unit.

[41] First and third quotes from Briggs, *Operation Just Cause*, p. 63. Second quote from Interv, Wright with Moore, 29 May 1990.

[42] Briggs, *Operation Just Cause*, pp. 63–64. First quote from Interv, Wright with Moore, 29 May 1990. Second quote from Donnelly, Roth, and Baker, *Operation Just Cause,* p. 279.

battalion's Company C isolate Renacer Prison on the other side of the Chagres. By 0230, the 2d Platoon controlled its area of operations.[43]

A half hour later, the final target, the counterintelligence training center, had been taken by the 3d Platoon. Moore recounted that, before JUST CAUSE, the idea of U.S. troops having to fight female personnel elicited smiles from many of the planners. The reaction was quite different from the soldiers who actually had to conduct the attack. Well armed and highly trained, the women waged a brief firefight, suffering several casualties during the shooting before they fled or surrendered. By 0300, the 3d Platoon had finished clearing the two main buildings of the school.[44]

After the successful execution of its JUST CAUSE mission, Company A reported four soldiers wounded but no fatalities. Also no American citizens had been harmed during the fighting. As for the Panamanian defenders, blood trails and the accounts of eyewitnesses indicated that several had been wounded, perhaps a few killed. None had surrendered prior to the fighting; they had not been offered the chance. Given the concentration of American citizens in the battle zone, Moore and Boylan did not want to give the enemy an opportunity to take hostages. Hence, they dispensed with the planned PSYOP broadcasts as too time-consuming and went straight into the attack. In looking back at how those actions had fared, Moore was pleased with the performance of Boylan's company at Gamboa, emphasizing the point that light infantry had demonstrated that it was capable of conducting surgical operations. Referring to the forest ranger headquarters specifically, he noted, "I don't think we had a single bullet hole in the surrounding quarters."[45]

RENACER PRISON

The 3d Battalion's fourth major H-hour objective, Renacer Prison, was considered the unit's most dangerous mission. After the confrontation with the Panamanian guards that had occurred during Company C's Sand Flea exercise there, Colonel Moore and Captain Johnson both anticipated that seizing the facility and securing the prisoners inside would probably involve the toughest fight the battalion would face that night.

The layout of the prison and the terrain in the vicinity did nothing to ease their concerns. The prison contained some twenty cinderblock or wood buildings, including the main prison, a headquarters, a barracks, office facilities, a chapel, and storage areas. The prison yard measured about a third the size of a football field and was surrounded by a high chain-link fence topped with razor wire. Outside the fence at the northeast corner was a single guard tower, the most conspicuous structure in the installation. As for the terrain surrounding the compound, it was also formidable. As described in one account, the prison "spread across the cleared top of a steep jungle bluff above the confluence of the Canal and the Chagres River near the town of

[43] Donnelly, Roth, and Baker, *Operation Just Cause*, pp. 279–80.

[44] Ibid., p. 280; Interv, Wright with Moore, 29 May 1990.

[45] Interv, Wright with Moore, 29 May 1990.

Gamboa." Scattered within the landscape were high-voltage power lines, light poles, towers, the prison fence, and a railroad bridge built a hundred feet above the Chagres. Mounting a combined air and sea assault on the target would not be easy, one reason being the restrictions in the canal treaties that had prohibited U.S. forces from practicing air assaults in the area. Complicating matters further was a lengthy list of unknowns. While, to Moore's previously noted satisfaction, the Sand Flea and a four-day reconnaissance by a SEAL team had allowed Company C to study the prison and identify the location of the guards' quarters and stations, the number of PDF defenders was still undetermined, with estimates ranging from twenty to as high as seventy-five. The number of prisoners, placed at sixty, also remained a guess, and no one knew for sure where they were located inside the complex. Most of the guards, analysts believed, carried only small arms, including AK47s and T65s (a Taiwanese assault rifle resembling the U.S. M16). But these defenders, as the Sand Flea had also demonstrated, could be reinforced quickly by police units in Gamboa, which was one reason for Company A's mission to establish a blocking position on the bridge between the town and the prison.[46]

The plan for Renacer at H-hour was to insert into the prison yard two groups of paratroopers from Company C's 2d Platoon. Originally, Chinooks were to be employed for that part of the operation, but, given the small size of the landing zone, two Hueys were used instead.[47] Each carried eleven soldiers, the total being called Team Oswalt after the 2d Platoon leader, 2d Lt. Christopher F. Oswalt. A third UH–1 would land a team of ten scouts at the front gate to seal off that entrance from any enemy reinforcements. In this first phase of the operation, a Cobra attack helicopter and two Kiowas—including the one with Moore aboard—would provide suppressive fire. The right door gunners on the Hueys would join the effort as well, backed by their passengers armed with squad automatic weapons and grenade launchers. To make sure the rounds struck only places known to be occupied by the guards, there would be no firing from the left side of the Hueys, the side that would be facing those buildings most likely housing the prisoners.

While the helicopters were flying the lead elements into the prison compound, the bulk of the Company C attack force would be landing via two LCMs, the part of the operation that had been rehearsed during the Sand Flea. The remainder of the 2d Platoon, armed with five M60s and twenty AT4 antitank weapons, would be the first troops to disembark. Once off their landing craft, the troops would climb the ridge in front of them and set up a position from which to deliver additional suppressive fire. Since Team Oswalt would be inside the fence by that time, the prison yard was to be considered a no-fire zone, meaning that the fire support element would concentrate on targets such as the headquarters building and guard quarters outside the fence. Meanwhile, paratroopers from the 3d Platoon would disembark from the

[46] Donnelly, Roth, and Baker, *Operation Just Cause,* pp. 268, 270–73; "Raid At Renacer," in *Soldiers in Panama: Stories of Operation Just Cause* (Washington, D.C.: Office of the Chief of Public Affairs, Department of the Army, 1990), p. 6; Interv, Wright with Moore, 29 May 1990; CALL, "Renacer Prison Raid." Quote from McConnell, *Just Cause*, pp. 165–66.

[47] The information regarding the switch from Chinooks to Hueys is from Briefing, Lt Col Johnny Brooks, 29 Apr 1992, Fort Lewis, Wash.

second LCM. As Oswalt and his men concentrated on seizing the main prison buildings within the compound and securing the prisoners, the 3d Platoon would assault and clear prison facilities outside the perimeter fence.

Around 2100 on the nineteenth, Captain Johnson approached the soldiers from his 3d and 2d Platoons (minus Team Oswalt) as they were loading onto the LCMs. When he told them that they were not going on another exercise but into combat, their reaction, as he later described it, was "calm." "There was no fidgeting around or scurrying for cover or Hail Marys or anything. There was a job to do and they were going to do it." Back in the briefing room, the interaction between Moore and some of the helicopter pilots assigned to the mission was not so dispassionate. At issue was the part of the plan that called for the Cobra and two Kiowas to make a last-minute reconnaissance over the prison. Once word arrived that Operation JUST CAUSE had been compromised, the pilots argued for canceling the recon mission: the PDF defenders would be waiting, and, if the flyover itself did not draw fire, it would certainly alert the enemy that American forces were on the way. Moore refused the request, and a heated argument ensued, especially with the Cobra pilot who, the battalion commander bemoaned, demonstrated "not quite the attitude that I wanted him to have." Moore reportedly told the aviator that "he was going to fly the mission the way it was briefed, or I'd find somebody else to fly it. . . . We had some guys on the LCM who were hanging out a lot more than he was." Given that Moore outranked the pilots, his view prevailed. In the opening phase of the attack, however, he would have an additional reason to express his displeasure with the Cobra pilot.[48]

The LCMs left Fort Sherman around midnight. The helicopters were in the air by 0030, with the two Kiowas leading the Hueys along the Chagres River to the objective. The weather, while far from threatening, could have been better, as clouds and dense fog obscured the waterways below and much of the terrain on each side. Once the Huey pilots arrived in the vicinity of the target, they had to fly across the canal, then make a 180-degree turn before approaching the prison. To avoid the power lines and the prison fence, the UH–1s, in Moore's words, had "to come in at a very high angle and then shoot straight down" on about a 40-degree glide slope.[49]

Fifteen seconds before H-hour, as the Hueys were in the process of performing this difficult maneuver, a sniper on one of the Kiowas opened fire on the prison's guard tower. Moore, in the other Kiowa, fired his M16 at the target as well, while the Huey door gunners poured M60 rounds into the buildings where the guards were presumably located. The few soldiers aboard armed with SAWs and M203 grenade launchers joined in, although those with the M203s quickly found that pressure from the helicopters' rotor wash pushed the rounds downward, causing them to fall well short of their targets. When the Panamanian defenders returned fire, the numerous tracer rounds from both sides reminded Moore of firefights he had seen in Vietnam. As for the pilot in the lead UH–1, CWO Michael Loats, he could only marvel

[48] Johnson quotes from Donnelly, Roth, and Baker, *Operation Just Cause,* p. 270, and see also pp. 268, 272–73. Moore quotes from Interv, Wright with Moore, 29 May 1990. Johnson's 1st Platoon, it should be recalled, was assigned to a mission in AO North.

[49] Interv, Wright with Moore, 29 May 1990.

AH–1 Cobra helicopter

at surviving the shooting unscathed. "How we never got hit, I don't know," he recounted. "All we saw were tracers in front, on the side, and behind us." The Cobra arrived late, according to an aggravated Moore, who also complained that the fire from its 20-mm. Gatling gun, once it finally became engaged, was erratic. "I think we were fortunate in that regard that we didn't take any of our own rounds." The effect of the Cobra's fire support was later debated, but not the sequence of events that followed its arrival on station: the Hueys departed the prison yard, the twenty-two soldiers they had deposited there from the 2d Platoon began their assault on the main prison building, and the troops from the LCMs began moving against the outer areas of the prison complex.[50]

To get Team Oswalt inside the main building, an engineer and his security man crawled forward under fire to place three one-pound blocks of C4 explosives on the doorway.[51] The security man was hit in the arm by grenade fragments, but the engineer managed to set the charge, pull the pin on the igniter, and then crawl away through broken glass and gravel. The explosion rocked the compound as the locked door vanished, leaving a hole through which the remainder of Oswalt's soldiers passed. Men from the team's 3d Squad headed for the cellblock, where the irregular lumps they spied barely visible on the floor of the darkened area turned out to be mattresses the prisoners had

[50] Last quote from ibid. Other quotes from CALL, "Renacer Prison Raid." Donnelly, Roth, and Baker, *Operation Just Cause,* pp. 272–74; Briggs, *Operation Just Cause*, pp. 59–61; McConnell, *Just Cause*, pp. 170–72.

[51] The account of the clearing operations is common to the following sources, with some differences of detail that this author has tried to reconcile: Interv, Wright with Moore, 29 May 1990; Flanagan, *Battle for Panama,* pp. 131–32; McConnell, *Just Cause*, pp. 173–74; "Raid At Renacer," in *Soldiers in Panama*; Donnelly, Roth, and Baker, *Operation Just Cause,* pp. 274–76; Briggs, *Operation Just Cause,* p. 61. Citations will be provided for only quoted material or additional sources.

pulled over themselves. Meanwhile, the 2d Squad seized a nearby recreation building without incident.

Outside the compound, Company C's 3d Platoon led by 2d Lt. Chuck Broadus made its way up the ridge from the LCMs' docking area and passed through the fire support positions set up by those 2d Platoon elements that had also been sealifted to the objective. En route, one of Broadus' men saw two armed Panamanians and pointed them out to an M60 machine gunner, who promptly shot and killed them. The platoon continued its advance, only to run unexpectedly into a ten-foot-high wire fence hidden by the overhang between the office and the headquarters buildings. Broadus called for claymore mines and grenades to blow a hole in the obstruction, but none of the blasts had any effect. Finally, two of his men crawled to the fence and cut a hole in the wire with their bayonets. Within minutes, the remaining soldiers had passed through and reached their objectives.

For squad leader Sgt. Kevin Schleben, that was the headquarters building on the southern side of the complex. When he reached the main corridor, he found himself overcome by a cloud of chemical stun gas. He and his men went back outside and then returned wearing their protective masks. In the process of clearing the building, Schleben discovered a blood trail on the floor and followed it out the back door. There, just a few yards from him, were two armed Panamanians, one wounded in the leg. Neither saw him, as both seemed to be preparing to fire on three American soldiers moving toward the prison's office block. Schleben yelled, "Down!" causing the two enemy guards to turn toward him. As they did, he opened fire with his M16, killing both of them. As Broadus later stated, "They could have shot four or five of us. He went between us and them. He could have taken our fire or their fire."[52] For his quick actions that may have prevented several American casualties, Schleben received the Bronze Star.

Around 0200, Company C reported that the main prison building and the prisoners were secured.[53] Clearing operations continued, however, for the next three to four hours. By and large, these were uneventful, save for the anxiety generated by entering forbidding structures at night, each one a possible hideout for enemy forces not yet ready to concede their inevitable defeat. The U.S. troops, despite their fears, displayed remarkable discipline, as demonstrated by one episode in a home located near the prison. The house was a family dwelling, a duplex, and after clearing one apartment and while preparing to enter the other, the squad of soldiers heard a woman's voice yelling, "Don't shoot." The men held their fire as a PDF officer, his wife, and their child emerged, unharmed but visibly shaken. The thought of "what might have been" left its impact on the soldiers as well. One specialist who was present could not get over feeling, "I came close to killing that woman and her baby."[54]

---

[52] Flanagan, *Battle for Panama,* p. 132.

[53] Operations Log Extract, in After Action Review, Operation Just Cause, 3d Bn (Abn), 504th Inf, n.d.

[54] Flanagan, *Battle for Panama,* p. 132.

Fourteen die-hard Panamanians did not surrender until later that morning. They had holed up in their quarters, taking potshots at U.S. soldiers who ventured into view. To silence their weapons, Captain Johnson ordered antitank rounds fired into the building, after which the shooting stopped. When a Company C element finally entered the quarters, it found the soldiers under their beds and ready to give up. Their surrender brought the total number of enemy prisoners taken by Company C to twenty-two, six of whom were wounded. Five of the defenders had not fared so well, having been killed in the firefights. American casualties numbered four wounded, none killed. There were also sixty-five former Renacer prisoners now under U.S. protection, the number including the two Americans known to have been incarcerated there. There were also five Panamanians who had participated in coup attempts against Noriega. Accounts differ, but at some point between 0430 and 0600, the entire prison complex and surrounding area were quiet and deemed to be secured.

Aside from his problems with the Cobra pilot, Moore was pleased with the outcome of the raid at Renacer. In interviews and briefings afterward, he would cite several aspects of the operation in which Company C had excelled, usually emphasizing the novel way in which the unit had cleared the complex's nearly twenty buildings—from the inside out. The concept, referred to in the battalion as "star burst," came from an article Moore had written for *Infantry* magazine on night attacks. In the author's words, it called for the assault force to "make the breach, penetrate to the center of the target, and attack outwards from there." As executed at Renacer, this tactic, he claimed, had ensured that U.S. troops, once inside the cellblock, would not be firing on the prisoners Company C had been sent to liberate.[55]

With the mutual success of operations conducted by Brooks' 4th Battalion, 7th Infantry, and Moore's 3d Battalion, 504th Infantry, Task Force Atlantic had achieved by midday on the twentieth almost all of its H-hour objectives—Colón being the major exception—in both halves of its area of operations. In the process, a number of insights had been gained, some unique to the task force's experience, some in common with other U.S. units in Operation JUST CAUSE. General observations included the value of the prewar exercises conducted in Panama, particularly as they pertained to "crowd control, urban patrols, check points operations, and other security missions." There was also praise for the predeployment training in urban operations, although existing MOUT doctrine was criticized for providing "no nonlethal solutions or proportional response techniques for conditions that include civilians in close proximity to the target, or for situations where the enemy is indistinguishable from friendly civilians." On a more specific level, elements of Task Force Atlantic touted the importance of flak vests, the superiority of shotguns (captured from the Panamanians) over grenades and M16s for clearing rooms, the effectiveness of LAWs in use against buildings, and the reliability of the PRC–126 radios.[56]

---

[55] Interv, Wright with Moore, 29 May 1990.

[56] Quotes from Daniel E. Evans, Actions at Fort Espinar, Republic of Panama, 20 December 1989, n.d., copy at http://www.benning.army.mil/library/content/Virtual/Donovanpapers/other/STUP5/EvansDaniel%20E.%20CPT.pdf. Michael F. Beech, 'Quasi-War': Training Infantry

For the remainder of 20 December and in the days that followed, U.S. units on the Atlantic side of the canal would consolidate their initial gains while providing on-order responses to countless other matters that would arise. Some of these follow-on operations would take only a few hours to execute, others a day or two, and still others a week or more. Some would be routine, others quite dangerous. Meanwhile, U.S. forces on the scene would also have to accommodate themselves to the transition from strictly combat to stability operations. This, too, would often prove a difficult proposition.

Small Units for Operations Other Than War (Student paper, School of Advanced Military Studies, Fort Leavenworth, Kans., Academic Year 1995–1996).

12

Loose Ends

20 December 1989–3 January 1990

☆ ☆ ☆

By the evening of 20 December, U.S. forces had achieved most of the D-day combat objectives set forth in the JTF-South Operation Plan 90–2. The Comandancia had fallen to Task Force Gator after a fierce fight, while the remainder of Task Force Bayonet had neutralized organized resistance by the Panama Defense Forces and Dignity Battalions stationed or operating near the PDF headquarters. Farther out, in the suburbs and outlying areas of the capital and in towns close by, Task Force Red-T had captured the country's main commercial-military airfield complex, while Task Force Semper Fi, Special Operations Forces, and elements of the 82d Airborne Division had dealt with other enemy units that could have reinforced their colleagues at Torrijos-Tocumen or in downtown Panama City. Meanwhile, on the northern side of the canal area, troops from Task Force Atlantic had secured Coco Solo, Renacer Prison, Madden Dam, Gamboa, and Cerro Tigre.

Despite these impressive accomplishments and the speed with which they had been achieved, the outcome of some D-day and other high-priority missions remained uncertain. For example, Noriega was still at large, with none of the U.S. search teams possessing timely intelligence on his location. Complicating matters, while Panamanian citizens displayed unexpected eagerness to turn in their deposed dictator, the plethora of information they proffered often overwhelmed U.S. forces; much of it also proved unverifiable or inaccurate. Meanwhile, the securing of Colón, Panama's second largest city and the neutralization of PDF elements there had been put on hold. In addition to this unfinished business, Generals Thurman and Stiner also had to deal with follow-on operations, some planned, such as the reduction of PDF units and facilities in the provinces west and east of the canal; some arising from unforeseen emergencies, such as the rescue of hostages at the Marriott Hotel in Panama City. Equally in need of urgent attention was the widespread looting and criminal activity the chaos of the invasion had spawned on both sides of the isthmus. Finally, as the fighting died down, units in Panama City received directives to secure vital installations, such as embassies, warehouses, commercial facilities, transformer stations, and so on—over seventy-two locations in all, most to be guarded by five- to six-man teams. Thus, while an American victory had been assured by late afternoon on the twentieth, there was still much to accomplish militarily before the success of U.S. arms could be translated into a meaningful political achievement.

Hostages at the Marriott Hotel and Elsewhere

One reason that Thurman and the XVIII Airborne Corps had objected to General Woerner's contingency plan for a gradual buildup of U.S. forces in Panama was that the extensive time needed to complete the projected troop movements would allow enemy forces to exploit the vulnerability of American citizens living in the country. Supporting this argument were U.S. intelligence reports of PDF plans to take civilian hostages in the event of hostilities. The best way to negate this threat, Thurman and Stiner believed, was to mount a massive surprise strike with such speed and power that there would be little time for the enemy to do anything other than surrender or run. While the reasoning behind this line of thought seemed compelling, it was hardly foolproof. As the planners of 1–90 and 90–2 clearly understood, the element of surprise in such a large-scale operation as an invasion of Panama could not be taken for granted. Further, no matter how sudden and devastating the initial strikes, they could not guarantee that the enemy would not have sufficient time to take hostages, either by design or as a target of opportunity. Indeed, in the wee hours of 20 December, several unsuspecting passengers on the commercial airliner that had arrived late at Torrijos International Airport found that, only minutes later, they had become PDF captives merely by walking into the terminal to collect their baggage. Then there was also the case of the Americans staying at the Caesar Park Marriott Hotel in Panama City. They fared no better when, soon after the shooting began, a group of men dressed in civilian clothes and carrying assault rifles entered the building, gathered a number of the hotel's

The Marriott Hotel (*tall building at center foreground*) in Panama City

occupants in the lobby, and announced, "We're being invaded, so we're taking hostages."[1]

Details of what happened inside the Marriott from this initial encounter to midafternoon remain confusing, even conflicting. Among the hotel occupants were American citizens employed by various news organizations. Eastern Airlines pilots and crew members were also guests. There were over a hundred other foreign nationals as well. Some registrants managed to hide from the various intruders who came and went until well after dawn. Some of the hostages—initial reports ranged from twelve to sixteen—were forcefully removed and taken away in trucks. Others returned to their rooms between visitations. Electrical power was out but the telephones were still working, allowing reporters to call their home offices and transmit raw information that, in turn, was relayed to Washington. Soon, NBC was reporting that two news producers, one from CBS, the other from ABC, had been abducted. Off-camera, network executives were telling U.S. officials that they wanted their people in Panama, and especially in the Marriott, protected. According to telephone calls received from one of the Eastern Airlines pilots, there were twenty-nine American hostages still in the hotel.[2]

The Crisis Action Team in the Pentagon sifted through this information. Then, at 1500 on the twentieth, the team received a message from Thurman, listing the Marriott among the general's top three priorities at that moment. President Bush shared this sense of urgency, a fact that Secretary of Defense Cheney directed Colin Powell to transmit back to the Southern Command. Up to that point, Powell had not tried to influence U.S. military operations in Panama. Now, given the concerns of the president, he had to intervene, making it clear to the SOUTHCOM commander that the situation at the hotel had to be resolved as soon as possible. In a conversation later in the day, Powell went further: U.S. troops had to rescue the occupants and secure the Marriott by midnight. Thurman discussed the edict with Stiner, who passed word down the JTF-South chain of command to the task force he chose for the mission. In making that choice, he would have preferred to tap Special Operations Forces, but they were busily engaged in a variety of sensitive missions, including the search for Noriega. Next in line, given their location and their operational status, were elements of the 82d Airborne Division. As to the selection of the specific airborne unit for the hostage rescue mission, Stiner left that decision to the Task Force Pacific commander, Maj. Gen. James H. Johnson Jr., and the 82d's assistant division commander for operations, Brig. Gen. Joseph W. Kinzer. After conferring, the two officers selected the 2d Battalion, 504th Infantry, at Panama Viejo. In recalling his thinking at the time, Johnson indicated that he was not clear as to whether the Americans at the Marriott were in fact hostages or just trapped by the military situation in the capital. In either case, they had to be rescued.[3]

---

[1] Donnelly, Roth, and Baker, *Operation Just Cause*, p. 230.

[2] Ibid., pp. 230–31; *Washington Post*, 21 Dec 1989; *Chicago Sun-Times*, 21 Dec 1989; Associated Press, 20 Dec 1989; *New York Times*, 22 Dec 1989; Cole, *Operation Just Cause*, p. 49.

[3] Cole, *Operation Just Cause*, p. 49; Interv, Dolores De Mena with Brig Gen Joseph W. Kinzer, U.S. Army, 20 Apr 1992, Fort Clayton, Panama; Donnelly, Roth, and Baker, *Operation Just Cause*, p. 231. In his SAMS briefing on 23 September 1992, Thurman said that military

The 2d Battalion commander, Lt. Col. Harry Axson, had spent the better part of the afternoon at Panama Viejo with his operations officer, Maj. Jonathan Chase, and other key staff members, drawing up plans to have the battalion's Company B mount an air assault on Paitilla airfield, where the Navy SEALs had suffered heavy casualties at H-hour. Around 1730, Axson received a call from his immediate superior, 1st Brigade commander Col. Jack P. Nix Jr. "I'm flying in. I want to see you," Nix said. "We're getting ready to go out of here," Axson replied, referring to the Paitilla mission. Nix's response was crisp: "No you're not. Your mission's been changed." The colonel arrived some time later aboard an OH–58 and got straight to the point. Axson would put together a rescue party that, before midnight, would seize and secure the Marriott, free the twenty-nine hostages being held there, and evacuate them to the Torrijos-Tocumen airport complex. Nix told the lieutenant colonel that he would have an AC–130 to provide fire support. The battalion mortar platoon would be on hand to add more, if needed. After Nix departed, Axson and his staff immediately began planning for the new mission. The time was 2000. The 2d Battalion had just learned how quickly priorities could change at the tactical level as the result of directives passed down from officials in the highest reaches of the U.S. government.[4]

The first question Axson had for his staff was, "Where is the Marriott?"[5] The battalion intelligence officer turned and pointed across the Bay of Panama toward the downtown area of the city where the hotel's large marquee was clearly visible from a distance that Axson estimated to be about 2½ miles. The assembled group then located the building on a map and determined the best route Company B should take in getting to its new objective. Here Axson was aided by Capt. Charles Durr, an officer who had accompanied Nix to Panama Viejo. Six months earlier, Durr had finished a tour in Panama, during which he had gained more than a passing familiarity with the well-traveled streets between the hotel and the old city's stone ruins, the latter a tourist attraction now serving temporarily as the 2d Battalion's operations center. The captain not only recommended a route for the rescue force but, at Axson's request, agreed to accompany the lead element in case it needed to make a detour as a means of avoiding a skirmish or any other obstacle that might slow down its advance. To improve the chances for unimpeded movement, Axson had his tactical air controller, Air Force Capt. John Wittington, coordinate with

reporting of JUST CAUSE went up the chain of command and not down and that Powell did not question him or interfere in the operation. The Marriott Hotel was clearly an exception to that generalization.

[4] Quotes from Interv, author with Lt Col Harry Axson, U.S. Army, and Maj Jonathan Chase, U.S. Army, 8 Apr 1992, Fort Bragg, N.C. Interv, Robert K. Wright Jr. with 1st Lt Christopher Miller, U.S. Army, 15 Jun 1990, Fort Bragg, N.C. Miller was Axson's mortar platoon leader.

[5] Quote from Interv, author with Axson and Chase, 8 Apr 1992. Unless otherwise noted, the account of the rescue mission that follows is based primarily on this interview but also on the following four sources: Interv, Wright with Maj Jonathan Chase and 1st Lt Robert Cejka, U.S. Army, 4 Jun 1990, Fort Bragg, N.C.; Flanagan, *Battle for Panama*, pp. 181–91; Donnelly, Roth, and Baker, *Operation Just Cause*, pp. 229–32; "The Day Of The Bronze Stars," in *Soldiers in Panama: Stories of Operation Just Cause* (Washington, D.C.: Office of the Chief of Public Affairs, Department of the Army, 1990), pp. 23–24.

the AC–130 crew, in part by numbering on two maps—one in Wittington's possession, the other in the plane—the buildings the rescue party would pass en route to the hotel. Should the gunship's sophisticated imagery devices detect a concealed enemy position on a rooftop, the AC–130 crew had only to pinpoint the site by relaying to the paratroopers the building's assigned number.

Axson and Chase told Company B's commander, Capt. Stephen C. Phelps, that his unit was to depart at 2200, just enough time for Phelps to accomplish the mission provided there were no significant delays. Not until 2130, however, was the rescue plan far enough along to brief the captain, who then had to go over it with his platoon and squad leaders, a process one later described as "very compressed." In its final form, the rescue mission called for Company B to maneuver to the hotel on foot with its 1st Platoon and engineer squad in the lead, accompanied by Chase and Durr. The 2d Platoon would follow, while behind it would be Axson's command post, to include the lieutenant colonel, key staff members, and his sergeant major. The 3d Platoon would come next and bringing up the rear would be a confiscated van with two specialists—one the driver—in the front seats and a sergeant in the back. The van would serve as a crude ambulance, picking up any casualties and evacuating them back to Panama Viejo for medical attention. Last-minute preparations ate up additional minutes. As Chase put it, "It just took a little time to get everybody together, and get the word out and the route and all that." As a result, "we left at 2215."[6]

Fewer than a hundred yards outside the battalion perimeter at Panama Viejo, the rescue team passed an open garage with the lights on. When Captain Durr ordered the lights shot out, three or four men hiding inside the structure opened fire. Both sides exchanged shots, after which the hostile force fled carrying one of its own. Company B then resumed its advance at an accelerated pace. There had been no U.S. casualties thus far, but that was about to change. About five hundred yards past the garage, the 1st Platoon turned a corner and heard the loud, grinding sound of shifting gears—the noise suggested a large vehicle—but the men could see nothing in the dark. Several nerve-racking seconds passed before a truck suddenly emerged from a side street and headed straight for the gap between the 1st and 2d Platoons. In the front seat, a passenger was firing a pistol at the soldiers. He also threw a grenade, but it sailed over its target, falling harmlessly over a nearby seawall and into the bay. As the truck sped among the troops, the side suddenly rolled up "like a Venetian blind," revealing five men, two with AK47s and three firing machine guns mounted on tripods bolted to the floor. For a brief moment, the paratroopers could only absorb the blow. They did not have a good angle with which to return fire. Nor, given the seawall at their backs, did they have much room to maneuver out of the way. One paratrooper jumped over the wall into the water twenty feet below; three others, including Chase, were wounded.

---

[6] Last two quotes from Interv, author with Axson and Chase, 8 Apr 1992. First quote from Maj Robert Haycock, NEO—A Platoon Leader's Experience: Operation Just Cause/Marriott Hotel NEO (Student paper, U.S. Army Command and General Staff College, Fort Leavenworth, Kans., 27 May 1999). Also, two of the official JUST CAUSE chronologies place Company B's time of departure at 2230 vice 2215.

As the truck barreled toward Axson's command element, members of the first two platoons were finally able to begin shooting, their bullets setting the vehicle on fire. At almost the same time, a paratrooper crouching behind the colonel stood up, walked to the middle of the street, and fired a round from an M203 grenade launcher, hitting the passenger and taking off one of his arms. With the truck still bearing down on the soldier, he fired a second round that went through the windshield on the driver's side. The vehicle swerved and crashed into a building. Inside, the paratroopers found the driver and four of the men in back dead; the passenger in the front seat and the fifth man in back were wounded. Stiner later identified all the occupants as members of UESAT.[7] As for the three American casualties, Chase's wounds were to his face, while the other two men's were superficial. The van following the column picked up the three and headed back to Panama Viejo, receiving sporadic fire en route.

With the truck destroyed, Company B again resumed its advance, at which point the AC–130 overhead informed Axson's air control officer that a dozen or so enemy soldiers had set up an ambush position on a bridge leading to the hotel. Phelps' men and the gunship had the firepower to deal with the threat, but the Marriott was the objective and the clock was ticking. Axson quickly decided to maneuver around the ambush, so he radioed Captain Durr with instructions to select a detour. The captain picked an alternate route, and word was passed to the troops. News of the change, however, failed to reach the sergeant and two specialists in the van-cum-ambulance who, after dropping off the wounded paratroopers and providing a brief update to Colonel Nix at Panama Viejo, were speeding back to rejoin the rescue column. Sticking to what they believed was the correct route, the three soon approached the defended bridge, at which time the Panamanians began shooting at them, blowing out all the vehicle's windows and wounding the specialist in the passenger seat. The driver turned and headed away from the ambush, only to find that an enemy vehicle was following him. In what one author has called a "running street battle, Mafia movie style," the sergeant in the van, a former Marine sniper, exchanged shots with the pursuers, even as the van swerved wildly on and off the road, not as an evasive maneuver but as a result of the driver having had his eye glasses shattered.[8] The fight ended when the near-blind specialist lost the Panamanians by making a sharp turn onto a side street. Once certain that the way was clear, he returned to Panama Viejo to get medical help for the wounded specialist and to pick up additional ammunition and a new pair of glasses. When the sergeant with him began preparing for a second attempt to rejoin Company B, Colonel Nix stepped in and vetoed the move. The unit had already reached the Marriott and was preparing to enter the building.

Across the bay, as Axson, Phelps, and their men were on the last leg of the approach to the hotel, they had heard the shooting at the bridge, guessed what had happened, and hoped the occupants of the van were unharmed. About ten minutes later, the rescue party again encountered problems of its own. A communication from the AC–130 alerted Axson to four enemy snipers on the

---

[7] Interv, Wright with Lt Gen Carl W. Stiner, U.S. Army, 2, 7, 27 Mar and 11 Jun 1990, Fort Bragg, N.C.

[8] Flanagan, *Battle for Panama,* p. 187.

roof of a building a block or so from the Marriott. The 1st Platoon had just made a left turn away from the building in question and the 2d Platoon was preparing to do so. Axson surmised that, once the entire column had its back to the snipers, they would begin shooting. He therefore directed the gunship crew to open fire on the enemy position, telling them to use 40-mm. rounds. In response, he was asked to move the 2d Platoon eight feet off the sidewalk and away from the building. The AC–130 then fired three rounds onto the roof, scattering the enemy snipers and removing the last obstacle to the paratroopers' advance. A minute or so later, the gunship reported two armed men atop the Marriott itself, but Axson, fearing for the safety of the hotel's occupants, told the crew not to fire.

When Company B reached the hotel, a few of Captain Phelps' men approached the double doors in front and tried to push them open. The doors would not budge, so the paratroopers broke the glass in a nearby window, entered the Marriott, and opened the doors from the inside. The soldiers found the doors were not locked, rather they opened outward not inward. Stifling a laugh, Captain Phelps had one of his three platoons secure the entrance while the other two began clearing the hotel floor by floor. In the meantime, Axson told Captain Durr to man the hotel's desk and to telephone every room in the building. If the call was answered, the occupant was told to prop the room's door open with a pillow and then lie face down inside until U.S. troops arrived to provide an escort to safety. During the process, Axson set up a command post on the fifteenth floor, while Phelps remained in the lobby.

Shortly after midnight, Axson had accounted for twenty-eight of the twenty-nine American hostages—the twenty-ninth, a reporter he was told, had already fled the Marriott. The rescue team had also gathered up over a hundred other people, including hotel employees and guests from several countries. Because Company B was still the target of an occasional drive-by shooting or sniper round, the assembled civilians were processed in the hotel's windowless discotheque. There had been no fatalities during the room-to-room search, although one of the guests had suffered a heart attack. He and some of the paratroopers were examined by the battalion's surgeon, Capt. John Marriott (not connected to the hotel chain). Of the two armed Panamanians detected by the AC–130, one had been wounded and the other had escaped. Axson tried to report this situation to Colonel Nix but failed to make contact. He then called Fort Clayton and talked with General Kinzer, who promised an immediate resupply of ammunition and medicine.

Later that morning, around 1000, the 2d Battalion's Company D, commanded by Capt. Gregory L. Sawyer, arrived at the Marriott in an evacuation convoy composed of four HMMWVs armed with .50-caliber machine guns, one Sheridan light tank, two cargo trucks, and two Marriott "silver-side" trucks his men had hot-wired and commandeered hours before they received the rescue mission.[9] Company D's journey to the hotel had been uneventful until the last leg of its trip from Panama Viejo. It was just

[9] The account of Company D's trip from Panama Viejo to the Marriott and back again is from Haycock, NEO—A Platoon Leader's Experience. See also Interv, Wright with Cos C and D, 2d Bn, 504th Inf, 8 May 1990, Fort Bragg, N.C.

as the Marriott came into view, about a half mile from the lead vehicles, that well-concealed enemy snipers opened fire on the paratroopers. At first the shots were sporadic and ineffectual, but the firing soon increased in volume and accuracy. In response, the convoy's soldiers, increasingly fearful and exasperated, searched in vain for targets. Then, according to 1st Lt. Robert Haycock, a Company D platoon leader, two people suddenly

> appeared out of the shadows and shrubs. These individuals appeared to be pointing a large shoulder-fired weapon at the convoy. The reaction was instinctive and deadly. Nearly every weapon system in the convoy . . . traversed to the right and opened fire on what we believed to be a PDF carrying an RPG or LAW. The men were cut to shreds by SAW and M16 fire and although we knew the men were dead, the medic (Doc Massa) ran over and confirmed this fact. The men were not PDF, but instead were reporters who happened to be at the wrong place, at the wrong time, doing the wrong thing, to the wrong people—an unfortunate accident.[10]

The tragedy was nearly repeated as the convoy increased its speed and approached the entrance to the Marriott. According to Haycock, "We were again surprised by two figures running out of the bushes at us. . . . Every soldier again traversed the guns to acquire the target. Fortunately, I saw that one of the figures had long blonde hair. Recognizing that PDF were generally dark complexioned, I made the assumption that this was not a PDF and called for a check fire." The order saved the lives of the two people, both of whom were, again, reporters. The troops grabbed them and threw them into separate vehicles. The blond reporter landed in Haycock's HMMWV, where a sergeant slapped a helmet on her head, and the lieutenant sat on her "to keep her down and out of harm's way."

Between the incidents with the two pairs of reporters, Company D received fire from inside the Marriott. This time, there was no doubt as to the location of the shooter; muzzle flashes were visible from one of the windows several stories up. In the convoy, the paratroopers knew that, once the building was secured, Company B intended to occupy the bottom and top floors of the Marriott, nothing in-between. Haycock drew the logical conclusion: the shooter was a PDF sniper. "I guess everyone else must have thought the same thing," the platoon leader recalled, "because before the leadership could say a word, nearly every crew-served and individual weapon system in the convoy opened up on the window of the hotel." For good measure, Haycock's driver fired a LAW, putting it through the window in what Haycock considered a near impossible shot. Only later did he learn that the shooter was not Panamanian, but a sergeant in Company B, who despite the LAW blast and two .50-caliber bullets in his abdomen survived the experience, if only barely.

Once the convoy stopped in front of the hotel, close to a hundred civilians rushed for the two Marriott trucks, "all trying to jump onto the silver-sides at

---

[10] Haycock writes in his paper that some of his men later told him that only one reporter had been shot and killed, but he insists that it was two, in part "because my medic went over to assess the casualties."

the same time, each carrying luggage!" Because of the suitcases and bags, the trucks filled to capacity well before all passengers were inside. To impose some semblance of order, a platoon sergeant from Company D told the civilians in less-than-friendly terms that they could jettison their luggage and stay on the trucks or stay behind with their bags. As Haycock described what followed, "Luggage began to fly out of the trucks," after which the loading resumed. The process was punctuated by sniper fire throughout, and, despite the flak vests made available to the passengers, one Panamanian civilian received a superficial wound. Once the convoy pulled away from the hotel, however, the shooting subsided, and the trip back to Panama Viejo was without incident. Once there, many of the foreign nationals boarded CH–47 helicopters, which took them to the Torrijos International Airport. By noon, the affair was as good as over. Still unaccounted for were the hostages who had been taken from the Marriott at gunpoint, but the Panamanians responsible for their abduction released them incrementally at various locations throughout the day.

In an unrelated but equally publicized hostage development, eleven Smithsonian Institution scientists and research assistants (and the four-year-old daughter of one), whom thirty Panamanian soldiers had taken captive in the San Blas Islands on the first day of hostilities, also gained their freedom. Moved to a jungle on the mainland, the hostages were released on 21 December and flown via U.S. helicopters from a Kuna Indian village to Howard Air Force Base. Another hostage situation did not have a favorable outcome. Raymond Dragseth, a teacher at the Defense Department's Panama Canal College, was dragged from his apartment by armed Panamanians while his wife and daughter looked on terrified. Despite the family's continuous efforts to have U.S. military police and soldiers rescue Dragseth, his body was discovered in a shallow grave on 28 December, bringing to three the number of American civilians killed in Operation Just Cause. In the end, changes authorized by Thurman to the Blue Spoon and 90–2 contingency plan in hopes of precluding hostage situations had proved effective but not flawless. Left to the realm of speculation, however, was how much better or worse things might have been had the changes not been made.[11]

The Move into Colón

For Colonel Axson, Major Chase, Captain Phelps, and the other soldiers and airmen who took part in seizing and securing the Marriott Hotel, the rescue operation was an unexpected mission that began hours after the 2d Battalion had achieved its 90–2 D-day objectives. That same day, on the other side of the isthmus, elements of Task Force Atlantic had yet to begin moving into Colón to conduct what they had thought would be their main D-day assignment, a sweep of the city. During the first four hours of the invasion, Company B from the 7th Infantry Division's 4th Battalion, 17th Infantry, together with engineers and military police, had fought to establish and secure

[11] On the Smithsonian hostages and the Dragseth incident, see Donnelly, Roth, and Baker, *Operation Just Cause*, pp. 232–34.

a roadblock across the narrow strip at the southern base of the city, known as the Bottleneck. During the skirmishing, one U.S. soldier had been killed and another wounded. A dozen enemy defenders had been killed, as had several civilians caught in the cross fires. Now, with the land approach to Colón sealed off, American troops prepared to enter the isolated city and neutralize enemy elements there, particularly those located at a DENI station and the PDF headquarters for Military Zone 2 (*Map 20*). Task Force Atlantic expected to begin the sweep that day, but, as the hours passed, Colonel Kellogg received no order to do so.

General Cisneros had much to do with that turn of events. Once General Stiner had arrived in Panama and taken command of JTF-South, the JTF-Panama commander officially became his deputy. That position, however, left Cisneros with little to do in the operations center once the XVIII Airborne Corps began running the war, so he made plans to spend D-day in the field, initially with units from Task Force Bayonet and then from Task Force Atlantic, two commands that, while going under Stiner's operational control, had since May been an integral part of JTF-Panama. After personally checking on the well being of Panama's new president who had accepted the hospitality of a safe house at Fort Clayton, Cisneros went to Fort Amador out of concern for how the American families living there would fare during the U.S. air assault against the PDF's 5th Company and other hostile elements.

Entering the installation about thirty minutes before H-hour, the general made his presence felt almost immediately by ordering the fort sealed off early because of the attempts of some armed Panamanians to escape.[12] After daylight, with the heaviest fighting at Amador winding down, he visited Task Force Wildcat's area of operations around Ancon Hill and then headed for the Atlantic side to check on the situation there. Prior to his departure, he received reports about Colón being sealed off at the Bottleneck. Knowing that the next step would be to send U.S. troops into the city, Cisneros contacted Col. Thomas H. Needham, the JTF-South operations officer, asking that no such action be authorized before he, Cisneros, had a chance to evaluate the situation firsthand. Needham relayed the request to Stiner, who agreed to it. At some point, Cisneros also discussed his recommendation with Maj. Gen. Carmen Cavezza, commanding general of the 7th Infantry Division (Light), who after his arrival in Panama from Fort Ord, assumed command of Task Force Atlantic on the night of the twentieth. (Because the 7th's other two infantry brigades were deploying to Panama for operations on the Pacific side of the country, Cavezza set up his operations center in the officers' club at Albrook Air Station.)[13]

---

[12] See Chapter 5.

[13] Intervs, author with Maj Gen Marc A. Cisneros, U.S. Army, 30 Jan and 29 Jun 1990, Fort Clayton, Panama; author, Wright, and Joe D. Huddleston with Lt Gen Carmen Cavezza, U.S. Army, 30 Apr 1992, Fort Lewis, Wash.; Panel Comments, Lt Gen Marc Cisneros (Ret.), 19 Jan 2011, U.S. Army Center of Military History, Washington, D.C. With a helicopter at his disposal, Cavezza would make daily visits, most of them relatively brief, to Fort Sherman, where Kellogg otherwise remained the senior commander on the scene.

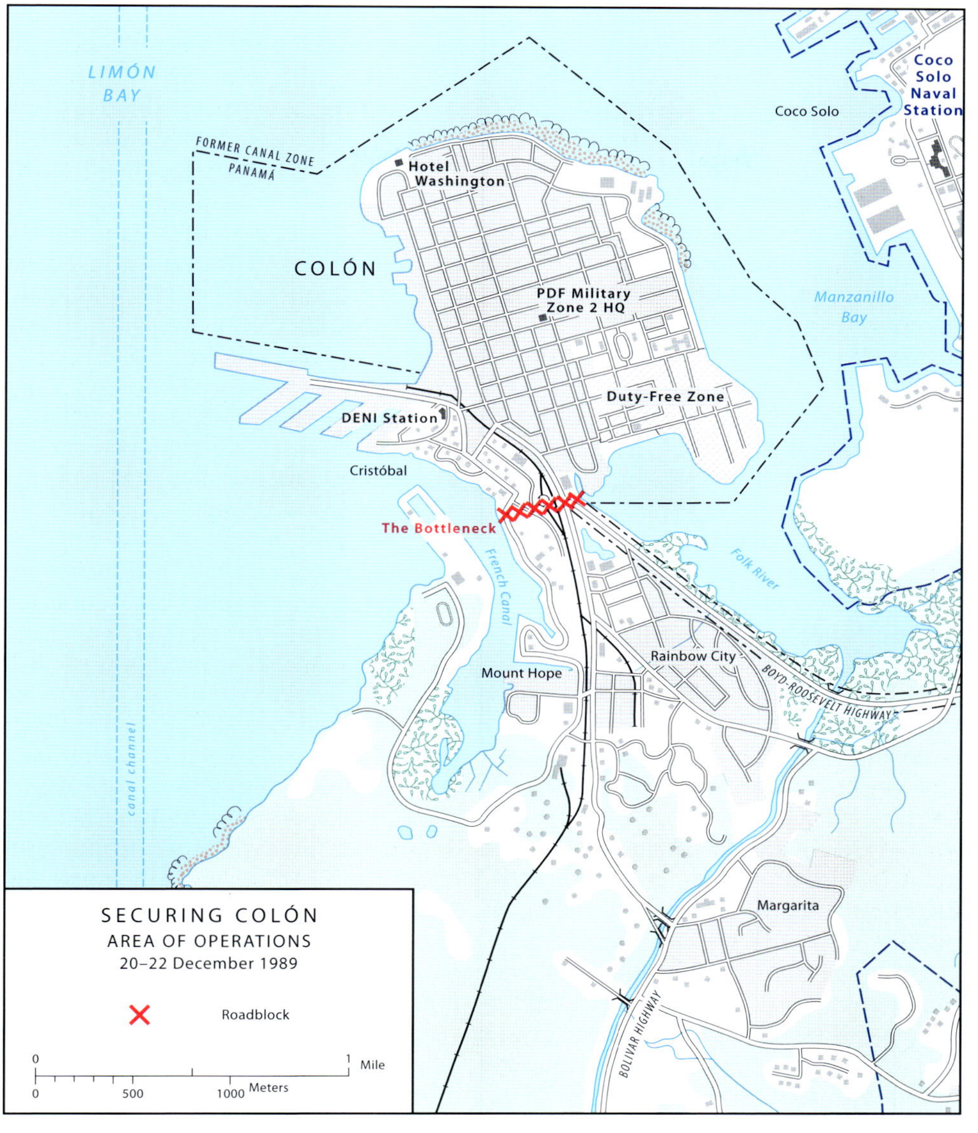

Map 20

Cisneros' desire to put the sweep of Colón on hold arose from two principal concerns. The first was the damage the city could suffer once the U.S. troops entered. "If we'd gone in there," he said a month later, "we'd have had a massive burning, and a massive amount of refugees, and we would be in a dire world of hurt right now." Cavezza was in agreement for the same reasons. "I did not want to go into Colón if there was any risk of burning that place down. We would have had a heck of a mess on our hands. There was just one way out of that city. It was a tinderbox. . . . Not only that, we would have had a bunch of civilians killed in the fire. What would we do with all those civilians?

And in my mind, there was not anything militarily essential that we had to do right then."[14]

Cisneros' second concern was Colonel Kellogg, whom he described as "antsy to attack." Since becoming commander of Task Force Atlantic in October, Kellogg had been Cisneros' subordinate, and, over the two months leading up to Just Cause, the general had concluded that, on more than one occasion, the colonel's behavior had been more aggressive and provocative than the circumstances required. Cisneros, in recounting his uneasiness regarding Colón on 20 December, alluded to his doubts about Kellogg in remarks regarding the "warrior . . . mentality of attack, attack, attack," even though in many cases "it was not necessary."

> You need to have the discipline of your soldiers so that if one person fires from a building doesn't mean that you destroy the whole building. . . . you can tell your soldiers to take cover, and if they take cover, they're not going to get hurt. And then they start working, make an assessment, carefully make sure how many families are in there, try to work them [the snipers] out. Then, if they keep firing and doing things, then you make the appropriate judgments on it. But you don't just automatically react to destroy things just because one sniper fires at you from someplace.[15]

Cisneros was correct in thinking that the Task Force Atlantic commander wanted to seize Colón sometime during the night of 20–21 December. His soldiers were "itching to go," Kellogg informed Thurman when the latter told him to postpone the operation. Later, Kellogg would state flatly, "We wanted to go in the night of D-day, because we'd finally powered up enough forces by securing all the objectives." But he also recognized the logic of the position taken by Cisneros, Cavezza, and Stiner. Planners had anticipated stiff resistance in the city from PDF and Dignity Battalion elements, the latter of which included unemployed, angry, and armed stevedores and port workers. In light of this prognosis, there was no point in fighting a battle on urban terrain, given the inherent risks of high casualties, both military and civilian, and extensive property damage, unless it was absolutely necessary. Holding back to allow the situation to develop "was the right call," Kellogg acknowledged. Not that he lacked a counterargument for moving into the city sooner. U.S. forces at the Bottleneck were still coming under fire. Moreover, extensive looting, which threatened the country's economic recovery, had begun the afternoon of the twentieth in Colón's business district and duty-free zone and was continuing unchecked except when merchants could employ armed guards to protect their goods. Finally, the civilian inhabitants, understandably shaken to the point of panic, were beginning to flee despite U.S. loudspeaker broadcasts urging them to stay in their dwellings. As they arrived at the Bottleneck, each person and vehicle had to be checked for weapons and drugs. The consequent congestion

[14] First quote from Interv, author with Cisneros, 30 Jan 1990. Second quote from Interv, author, Wright, and Huddleston with Cavezza, 30 Apr 1992.

[15] Interv, author with Cisneros, 30 Jan 1990.

tied up U.S. troops, raised the level of danger, and interfered with Company B's efforts to focus on the planned sweep.[16]

As Cisneros traveled north to make an on-the-scene evaluation, Lt. Col. Johnny Brooks, the commander of the 4th Battalion, 17th Infantry, from Kellogg's brigade, was spending much of the afternoon at Coco Solo finalizing plans for going into Colón. After receiving guidance from his boss, Brooks, his operations officer, and the 4th's command sergeant major fashioned a foray that called for the battalion's Companies A and C, accompanied by Navy SEALs, to board LCM–8s and cross Manzanillo Bay under cover fire from a Navy patrol boat and a Spectre gunship. (Brooks later learned to his chagrin that the AC–130 he had counted on would not be available.) As the landing craft approached Colón, the SEALs would swim ashore and emplace infrared chemical lights that would mark a beachhead for the infantry. Once the soldiers landed, they would raid the Military Zone 2 headquarters, subdue its defenders, and then withdraw. Simultaneously, Company B at the Bottleneck, together with Company B from the battalion of the 82d Airborne Division located on the Atlantic side, would move into the city and mount an assault on the DENI station, which would be subjected to preparatory fires from 105-mm. howitzers provided by the 82d's Battery B, 7th Battalion, 15th Field Artillery. The operation was set to begin at 0300 the next morning. Brooks briefed the plan at 1800 on the twentieth, after which the participating units began making preparations and moving into position for the attack.[17]

Not until 2130 did Brooks receive word that the operation had been put on hold. No explanation accompanied the news, but, during the afternoon of the next day, he found out why: Cisneros had arrived and was negotiating the surrender of the city's Panamanian defenders. As Cisneros himself would relate the story, he suspected even before reaching Colón that many enemy soldiers and policemen, if given a chance to lay down their arms, would choose not to fight for the city. Once he reached Task Force Atlantic's area of operations, he stopped at Fort Davis, where the prisoners from Coco Solo were being held, and sought out the ranking PDF officer, who turned out to be Capt. Amadis Jimenez, the naval infantry company commander. Cisneros talked to the captain, who was married to the daughter of a man prominent in the anti-Noriega movement. Records provided to Cisneros also indicated that Jimenez was a highly professional officer. The general asked the captain to help him contact ranking PDF officers in Colón. At first, Jimenez responded with his name, rank, and serial number, but later that day Cisneros convinced him that his assistance would help prevent a conflagration in the city. To this end, Jimenez produced a directory of PDF telephone numbers and, following Cisneros' instructions, placed a call to the Military Zone 2 headquarters. When a sergeant answered, Jimenez handed the phone to Cisneros, who was told that most of the garrison, including the ranking officers, had fled the premises, leaving only a few men behind to burn documents. That the zonal commander,

[16] First and third quotes from Interv, author with Brig Gen Keith Kellogg and Lt Col Gregory Gardner, both U.S. Army, 9 Apr 1992, Fort Bragg, N.C. Second quote from Donnelly, Roth, and Baker, *Operation Just Cause*, pp. 292–93, 295.

[17] Donnelly, Roth, and Baker, *Operation Just Cause*, pp. 295–96; Briefing, Lt Col Johnny Brooks, 29 Apr 1992, Fort Lewis, Wash.

Maj. Mario del Cid, had deserted his post did not surprise Cisneros. But del Cid was still in the city, as were other PDF officers and several hundred enlisted men. Cisneros therefore delivered a blunt message to the sergeant: the war was over, thus any resistance to a U.S. movement into Colón would be foolish and futile, not to mention deadly. That being the case, any Panamanian military personnel still at the zonal headquarters should come to the Bottleneck and surrender. The sergeant responded that the men remaining with him were prepared to comply but only after receiving certain assurances. According to Cisneros, "They were scared and I just had to convince them that I was not going to shoot them or torture them." The general was patient and persuasive, and in the end the men at Military Zone 2 headquarters turned themselves in.[18]

With the help of Jimenez and the PDF sergeant, Cisneros was able to determine the whereabouts in Colón of other Panamanian military leaders and to reach them by telephone, in each instance making a pitch similar to the one that had already proved so effective. Surrender appeals from U.S. loudspeaker teams and in broadcasts over U.S.-controlled radio and television stations reinforced the message. During the evening of 20 December and throughout the next day, the success of Cisneros' initiative could be seen at the Bottleneck, as Red Cross vans arrived carrying enemy soldiers. As midnight approached on the twenty-first, 127 Panamanians had surrendered at the roadblocks. At least a hundred more would follow. One consequence of the surrender process was that Thurman postponed for a second time the planned move into the city.[19]

When Johnny Brooks learned of Cisneros' role in postponing the initial sweep planned for Colón, the delay did not upset him. To the contrary, he had been considering contacting General Cavezza personally—even if that meant circumventing the chain of command—with an appeal to put the operation on hold until the defenders had been given ample opportunity to surrender. At the time, he did not realize Cavezza had already reached the same conclusion. Later, however, he would sound very much like the general in describing Colón as a tinderbox. As Brooks phrased it, just one tracer round could have set the whole city ablaze. Also influencing his thinking was an aspect of the initial plan, dictated by higher headquarters, that troubled him: following the proposed raid on the Military Zone 2 headquarters, the U.S. troops involved were to withdraw from both the objective and the city itself. In his words, "I was going to get kids killed, and in getting kids killed . . . I was going to have to go back in that town later on and do it all over again. My personal opinion was, if we're going to go in, let's go in one time and do it. And let's don't burn

[18] Intervs, author with Cisneros, 30 Jan and 29 Jun 1990; De Mena with Capt Amadis Jimenez, 13 Nov 1991, Fort Clayton, Panama. Quote from Interv, De Mena with Cisneros, 12 Jun 1990, Fort Clayton, Panama.

[19] Intervs, author with Cisneros, 30 Jan and 29 Jun 1990; De Mena with Cisneros, 12 Jun 1990; author with Gardner, 7 Apr 1992, Fort Bragg, N.C.; with Lt Col Johnny Brooks, U.S. Army, 11 May 1993, Fort Leavenworth, Kans.; Briefing, Brooks, 29 Apr 1992; Donnelly, Roth, and Baker, *Operation Just Cause*, pp. 296–97; 7th Infantry Division (Light), at http://www.globalsecurity.org/military/library/report/1990/90-9/9091ch1.htm.

the town down. Let's wait them out and let them surrender, which is in essence what happened."[20]

The surrender of Panamanian soldiers in large numbers on the twentieth and twenty-first militated for further postponement, and, because a revised version of the plan for securing Colón included a change that overcame one of Brook's concerns—once U.S. troops moved into the city, they would stay, not withdraw—the battalion commander's desire to avoid an unnecessary firefight held firm. It was a sentiment all of his company commanders shared, if with a slightly different emphasis. While Cisneros, Cavezza, Kellogg, and Brooks had to balance protecting the city with protecting the U.S. troops assigned to take it, the company commanders concentrated almost exclusively on the dangers their soldiers would confront in urban combat. As one expressed his anxiety, "I didn't know if I had the manpower to go another six to eight blocks under heavy fire." Capt. Christopher J. Rizzo, Brooks' Company C commander, echoed the sentiment. "Now, that's a mission you don't want, if it's in a MOUT environment with any kind of folks who are hostile to you." Yet, though the emphasis of their reasoning differed ever so slightly, Brooks and his subordinates welcomed the decisions to postpone sending the 4th Battalion into Colón. In retrospect, the lieutenant colonel only wished that someone had informed him sooner about what Cisneros and Jimenez were doing. Overall, though, the play of events on 20 and 21 December increased Brooks' respect for Cisneros.[21]

Despite the success of the Cisneros-Jimenez initiative, at least two significant problems remained. One history of Operation JUST CAUSE concluded that, with the surrender of the large number of enemy forces in Colón during the first two days, "The PDF had abandoned the city." Many of the enemy, of course, had decided not to fight, but there were still an estimated four hundred who had not yet come out, and U.S. troops at the Bottleneck were still receiving sniper fire. The second problem was the looting, which continued unabated. Kellogg had personally visited the duty-free zone, a target of the looters containing an estimated $2 million worth of goods, and, on the twenty-first, his operations officer, Lt. Col. Gregory Gardner, had passed near the area en route to Coco Solo. Seeing several people carrying off whatever of value they could put their hands on, Gardner ordered the military policemen with him to fire over their heads. He then walked around a corner and saw a thousand or so civilians engaged in the same illegal activity. The military policemen repeated their warning shots—the rules of engagement and common sense prohibited them from aiming at the civilians—and dispersed the crowd, only to have it return as Gardner drove off. These encounters left an impression. At some point sooner than later, Task Force Atlantic would have to do something to stop the damage being inflicted on Panama's economy.[22]

---

[20] Interv, author with Brooks, 11 May 1993.

[21] First quote from Donnelly, Roth, and Baker, *Operation Just Cause*, p. 296. Interv, author with Brooks, 11 May 1993. Second quote from Interv, author with Maj Christopher J. Rizzo, U.S. Army, 15 May 1997, Fort Leavenworth, Kans.

[22] Quote from Donnelly, Roth, and Baker, *Operation Just Cause*, p. 297. Intervs, author with Kellogg and Gardner, 9 Apr 1992; with Gardner, 7 Apr 1992. According to McConnell,

The matter came to a head on Friday, 22 December. Kellogg and Brooks were talking at Fort Sherman when the brigade's intelligence officer walked in and told them of a telephone call the operations center had just received from the chief of security in the free zone. The civilian guards being paid to protect merchandise from the looters were running out of ammunition. Could Colonel Kellogg help him? The colonel's first reaction was to swear, but he then told the S–2 to assure the man that U.S. troops would get there but not to tell him when. In fact, for the third consecutive day, an advance into the city had been scheduled, this one to take place during the wee hours of Saturday morning. But both Kellogg and Brooks realized that something had to be done sooner. To allow the pillaging of Colón's stores and warehouses to continue would be, in Brooks' opinion, "a real serious black eye" for the United States. Brooks had recently driven by the duty-free zone and seen that there was a back way that U.S. troops could use to reach the area safely. When, following the telephone call, he reported his observation to Kellogg, the colonel, in Brooks' words, "shot up like a bullet," ran out of his office, and shouted, "Get me a landing craft, now!" Sensing a "window of opportunity" to get troops into the free zone "through the back door," Kellogg on his own authority ordered the entire sweep operation to begin that afternoon, around 1430, meaning in broad daylight, ten hours earlier than scheduled in the most recent orders he had received.[23]

The two-day delay in moving into Colón had allowed Task Force Atlantic to acquire additional intelligence and to adjust its plans to accommodate unanticipated developments, such as the surrender of large numbers of Panamanian troops. Yet, even though details of the sweep had been revised, its objectives still included the Military Zone 2 headquarters and the DENI station, with the duty-free zone being added to the list. Estimates of organized enemy resistance were revised downward, with the most serious threat now seen as coming from the DENI station, where remnants of the Panama Defense Forces and Dignity Battalions were rumored to be holed up. Company A of the 4th Battalion would lead the amphibious assault from Coco Solo, but rather than raiding the Zone 2 headquarters as originally planned, it would instead move into the duty-free zone. Colonel Gardner arranged for a civilian security guard to be waiting for the soldiers after they had disembarked from the landing craft to show them the route. Following Company A would be the battalion's Company C, accompanied by Company C from the 7th Infantry Division's 2d Battalion, 2d Brigade, which had flown into Howard Air Force Base the previous day as part of the Just Cause follow-on force. These two units would land and head for the Military Zone 2 headquarters. Finally, the two Company Bs on the scene, one from the 7th and the other from the 82d, would move with the 82d's artillery battery from the Bottleneck to the DENI station. Supporting the operation would be Navy patrol boats, Navy SEALs, and attack helicopters. Given the suddenness of Colonel Kellogg's decision to

---

Just Cause, pp. 256–57, criminals freed from jail and armed by Dignity Battalion members were among the looters, as were the DigBats themselves.

[23] Donnelly, Roth, and Baker, *Operation Just Cause*, p. 297. Last two quotes from Interv, Kellogg and Gardner, 9 Apr 1992. Remaining quotes from Interv, author with Brooks, 11 May 1993.

launch the undertaking several hours early, the units involved had little time—in some cases only a half hour—to get to their lines of departure. As they were making their last-minute preparations, they received verbal reminders that, despite the numbers of enemy troops who had already surrendered, Colón still contained many weapons in the hands of people who were willing to use them against the Americans.[24]

The amphibious operation began as planned, with Company A crossing Manzanillo Bay first. When it reached the other side, however, the security guard who was to meet the men was nowhere in sight. Knowing that the duty-free zone was only a couple of blocks away, Capt. Michael Beech, the company commander, did not linger. His troops moved out, and, to their surprise, the Panamanians lining the street and positioned on the rooftops along the route began cheering wildly and waving flags. Unfortunately, the unit also received some sniper fire, which caused the men to hit the ground and "pop smoke" grenades. There were no casualties. When the troops reached the main gate of the free zone, they found a large group of armed Panamanians in civilian attire waiting for them. Not knowing anything about the men or their intentions, the company's lead element ordered the group to lie on the ground and relinquish its weapons. Beech and others soon learned that the men were merchants and security guards who, in trying to ward off looters, had themselves looted many of their weapons from the PDF Military Zone 2 building. Discussions ensued between Maj. Tom Ryan, Brooks' operations officer, and the zone's security chief, during which an agreement was reached in which the Americans promised to provide additional arms, namely assault rifles and pistols, to the security force. Kellogg, who was monitoring the deliberations from the Coco Solo side of the bay, felt uncomfortable with the deal and let Brooks know his misgivings through a continuous series of radio communications. Kellogg also wondered why U.S. troops were not shooting, the idea being that a barrage of warning shots would quickly clear the duty-free zone of looters. Brooks, who had been on one of the LCMs in the amphibious assault, repeatedly assured his boss that Beech and Ryan had the situation under control. Finally, in frustration, the battalion commander simply said, "Roger. Out." According to Gardner, who was with Kellogg, the colonel completely understood his subordinate's abrupt end to their exchange. By 1500, the free zone was considered secure.[25]

Brooks himself had come ashore with the two companies responsible for seizing the Military Zone 2 headquarters. In contrast to Company A, whose Panamanian guide had failed to appear, Brooks' contingent had not arranged for a guide but, in an opportune development, found one waiting for them at the beachhead. The man, by appearances a wealthy Panamanian, wore a white armband and headband and held a megaphone in his hand. Worried that the arrival of the U.S. troops would result in unnecessary damage to the city, he offered to lead the units to their objective. Brooks talked to the man at length, concluded that he was sincere and his intentions were good, and then

[24] The account of the move into Colón, unless otherwise noted, is based on Donnelly, Roth, and Baker, *Operation Just Cause*, pp. 297–305.

[25] The account of Colonel Kellogg "badgering" Brooks via radio messages is from Interv, author with Brooks, 11 May 1993.

put him and his bullhorn at the head of the troop column. As the soldiers made their way down Colón's 11th Street, the volunteer guide called on any armed Panamanians hiding in the area to throw out their weapons. He also directed the throngs of excited observers along the route to go indoors so as not to slow down the Americans. Both appeals went largely unheeded. As was the case with their comrades in Company A, U.S. troops advancing on the zonal headquarters could not believe the exuberant welcome they were receiving. For the 4th Battalion's Company C, which had fought fiercely two nights before in Coco Solo, the cheers—punctuated by a sniper round here and there—seemed incongruous.

The Military Zone 2 headquarters consisted of several buildings located within a square-city-block and surrounded by a high masonry wall with spikes on the top. When the two rifle companies arrived at the objective, they began to clear the complex systematically. Only one enemy soldier was present, and, when he tried to fire his weapon, he was killed immediately. Otherwise, the clearing process would have been uneventful had it not been for a telephone call warning the troops that the complex was rigged with explosives that could be set off by a remote detonator. By then, Kellogg had arrived, only to hasten out of the headquarters with the rest of his force. An explosives team began a search of the buildings, but, when it found no explosives, the clearing operation resumed, after which the headquarters was left under U.S. guard.[26]

Thus far, armed resistance in Colón had been unorganized and sporadic. But the DENI station, which some U.S. officers considered the primary threat, had yet to be taken. That phase of the operation did not kick off until right before sunset, with a platoon from the 4th Battalion's Company B advancing into the container yard that separated the Bottleneck from the objective. The platoon cleared the way for units that would follow by cutting openings through fences and looking for possible threats along the route. The airborne Company B then began its movement to the target, leaving the Bottleneck at 1830. The artillery battery passed through the vacated area an hour later. On the paratroopers' right flank, the remaining platoons from the light fighters' Company B provided security. It also attracted most of the attention, as crowds cheered the soldiers as they passed by. This fortuitous development allowed the paratroopers and the artillerymen behind them to approach the DENI largely unimpeded. Once they reached the station, a white cement building, Capt. Bryan Dyer, the commander of Company B, 3d Battalion, 504th Infantry, put his fire support platoon at the southern end of the structure, and his two other platoons on the west side. The artillerymen positioned themselves to the east and, on Dyer's order, brought up two 105-mm. howitzers. The northern end of the building was not part of the PDF installation but housed the Panama Canal Commission's Atlantic headquarters. Because it was not believed to harbor enemy troops, it was not targeted by the U.S. attackers.

---

[26] According to Donnelly, Roth, and Baker, *Operation Just Cause,* p. 299, the telephone call may have been triggered by events of the previous day when the headquarters commander had driven a truck loaded with explosives to the front gate. He had left the vehicle there "with a time fuze wired to it," but someone in Colón had driven the truck away before the U.S. troops arrived.

As soon as the cannons were ready to fire, the artillerymen opened up. According to Kellogg, by then, any hostile forces inside the station had had ample opportunity to surrender. Determined not to take casualties but to avoid widespread destruction, he authorized the artillery to be used in a direct-fire mode, the same way Task Force Black Devil had employed its howitzer at Fort Amador. The artillery blasts signaled U.S. infantrymen to open up with their M16s and squad automatic weapons. The fusillade lasted for twenty minutes; eighteen howitzer rounds were fired in the process. The roof at the southern end of the building was ablaze and the walls pockmarked from the hundreds of rounds striking them. Yet, despite the amount of ordnance expended, there was no collateral damage. The Panama Canal Commission facility remained untouched. Cisneros, who by this time had returned to the Pacific side, considered the use of the artillery as overkill—"now we have a quagmire of damage and repair to do"—but the barrage had the result Kellogg desired. A white flag appeared at the DENI station, as three Panamanian soldiers surrendered. One of the captives said that the rest of the defenders had fled out the back door once the artillery opened fire. There had been no casualties on either side so far. But when Company B closed in on the station to begin clearing operations, the men came under "uncontrolled fire," not from the enemy but from the platoon of the 549th Military Police Company, a friendly unit, that had provided security for the artillery battery in its movement from the Bottleneck. No one in Company B was hit—"Luckily they were bad shots," Dyer remarked sardonically—but the episode was unnerving and served to intensify the distrust and animosity between the infantry and the military policemen in evidence two days earlier during the battle at Coco Solo.[27]

With the reduction of the DENI station, the forces under Task Force Atlantic exercised control over their three principal objectives in Colón. Yet, significant as this achievement was, it did not mean that the city itself had been secured. That would require Kellogg's troops to fan out, establish command posts, and mount patrols, in the process seizing PDF and Dignity Battalion members still at large and restoring law and order and public services. On the night of 22 December, the initial actions demanded by these requirements had not been taken, which was unfortunate for at least one reason: public relations—not in Panama, but in the United States. When Colonel Kellogg returned to Fort Sherman that night around 2300, he was curious as to what was being reported on American television. Watching ABC's *Nightline,* the colonel found himself viewing a live interview in which General Thurman was saying that Colón was now under U.S. control. Getting to a communications radio, Kellogg immediately contacted Brooks with the urgent appeal, "Johnny, whatever you do, get me Colón by six in the morning and for God's sake, don't get in a firefight as you do it, because General Thurman just told the entire world we own the city." By sunrise, the troops had fanned out over the city and were beginning their patrols. There was, however, still much to be done.[28]

---

[27] First quote from Interv, author with Cisneros, 29 Jun 1990. Remaining quotes from Donnelly, Roth, and Baker, *Operation Just Cause*, p. 302.

[28] Four years after Operation Just Cause, Brooks recalled receiving the message to take Colón by morning and to do so without a firefight. He remembered receiving it from Colonel Gardner, however, not Kellogg. Interestingly, a typed version of the Task Force Atlantic and 7th

As American forces began policing Colón, several questions required an immediate answer. One concerned what rules of engagement would be in effect now that combat operations had abated. Another related to how the troops should interact with the Panamanian people, large numbers of whom were expressing their approval of Operation JUST CAUSE and the fall of Noriega's regime. Civilians mingled with the soldiers, wanted to talk with them, and brought them food and refreshments. Captain Rizzo was offered even more than that. As he passed one house in the northern part of the city, a man handed him a briefcase containing $10,000. It was Rizzo's for the taking if he would personally protect the man's house. The captain promptly returned the briefcase. In the ensuing days, as his men and those in other companies tried to strike a balance between providing security and exhibiting friendly behavior, the realization set in that they had entered a new phase in which stability operations, for which few of them had been trained, now constituted their principal mission.[29]

COUNTERATTACK AT THE DNTT

The move into Colón on Friday afternoon and evening completed the principal combat missions contained in the operation order for JUST CAUSE. It was not, however, the only fighting to take place on the twenty-second. On the other side of the canal, in Panama City, a much more fierce battle erupted at the PDF's Directorate of Traffic and Transportation. The large building and the complex around it had been taken on D-day by Capt. Donald S. Currie's Company C, 5th Battalion, 87th Infantry, 193d Brigade. The target had been one of several in the Ancon Hill and Balboa area of operations assigned to Task Force Wildcat, commanded by Lt. Col. William Huff III. Two days later, Wildcat was still responsible for securing most of that area, although a portion of it had been turned over to units from the 82d Airborne Division. Furthermore, Huff had "unofficially" lost operational control over two of the units he had on D-day: Company A from the mechanized battalion, which JTF-South had commandeered to secure the legislative assembly building, and a platoon of military police, which, according to the colonel, "just kind of filtered away." Despite the depletion of his force, he still had his three infantry companies, his headquarters company, and some other forces with which to carry out his follow-on missions.[30]

Infantry Division (Light) log for 18 December 1989 through early January 1990 contained the following entry: "221945U Dec 89. . . . All of Colon taken to include MZ2 HQ." Interv, author with Brooks, 11 May 1993; Chronology of Key Events for Operation 'Just Cause,' 161400U Jan 1990. Quote from Donnelly, Roth, and Baker, *Operation Just Cause*, p. 303.

[29] Interv, author with Rizzo, 15 May 1997.

[30] See Chapter 6 for a narrative of Task Force Wildcat's operations on D-day. Quotes from Interv, Wright with Col William H. Huff III and Senior Officers of the 5th Bn, 87th Inf, 193d Inf Bde, 5 Jan 1990, Fort Clayton, Panama. Unless otherwise noted, the account that follows of the PDF counterattack on 22 December is based on this interview and Interv, author with Huff, 29 Jan 1990, Fort Clayton, Panama.

On the twenty-first, those missions included rounding up enemy forces still on the loose, collecting documents and other intelligence, and finding arms caches. Throughout the day, Huff's soldiers received assistance from the local populace. In one case, a cooperative Panamanian led a patrol to a large collection of weapons. Many civilians seemed eager to help ferret out PDF members and political officials who had been active in the now-toppled regime or at least to point out where these people lived. Because some of the fugitives were on wanted lists issued by the Southern Command, U.S. forces received permission to raid several of the identified dwellings. In assessing the situation, one U.S. officer declared that, in general, "the immediate threat was kind of gone from the area," although he conceded that his troops could not always determine "who was the bad guy, who was the nice guy. So we were still picking up what we thought were EPWs [enemy prisoners of war] on the street and processing them and getting information from them." At one point, DigBat members tried to assemble on a road near the DNTT, but, after U.S. sniper fire discouraged the effort, "they were not a threat anymore," in Captain Currie's assessment.

On the morning of the twenty-second, after a night of planning, Huff's task force prepared to conduct a sweep of the Ancon-Balboa area, to begin at 0600. (*See Map 9.*) The intention was to canvass locations Task Force Wildcat had not searched on Thursday and to pick up Panamanian males of military age. The battalion's Company A at the engineer compound and the Headquarters Company securing Quarry Heights would both move to cordon off the area: "They made a big circle around our little piece of the pie," Maj. James Woods, the task force operations officer, observed. Then Company C, minus a platoon left behind to maintain security at the DNTT, would conduct a sweep of a sector bounded by Roosevelt Avenue and Gaillard Highway, while Company B covered the largely residential area to the south—inhabited mostly by Panama Canal Commission employees—running from Balboa Street to the high school being used as a refugee center. The operation began on schedule and was still in progress later that morning when, around 1100, reports began reaching various commanders that something was happening around the Palace of Justice and the Ancon DENI station. Fifteen minutes later, these reports indicated shots were being fired at the DNTT. What became known as the counterattack was under way.

Huff would later emphasize the confusion surrounding events of that day, starting with the arrival at the DNTT of Col. Al Cornell, a former defense attaché at the U.S. Embassy in Panama who had returned temporarily during JUST CAUSE to assist the new government. Cornell had twenty Panamanians in tow and told Huff that President Endara was establishing a new police force, that recruitment was to begin that morning in the large parking lot in front of the transportation building, and that he, Cornell, was taking charge. All this information caught Huff off guard. The night before, Colonel Snell, himself in the dark about much of what was planned at higher levels for Friday morning, had told his Task Force Wildcat commander to expect a former PDF lieutenant who would be coming to pay recruits wanting to join an organization that would replace the now-defunct Panama Defense Forces. Cornell and his entourage, it was obvious, were contemplating a much larger and complex undertaking. As one of Huff's staff officers explained, "We had no idea what was going on and why they were there." Having received no word about a change to the command arrangements, Huff's response to Cornell's assertion that he, Cornell, was taking charge was to tell the full colonel to leave. As Huff described what followed, "So there was a lot of, 'Oh, no, you're not,' 'Oh, yes, I am.' Now, all of a sudden people are starting to come in, stream in. 'Well, do we disarm them now, do we put these . . .' 'No, no, these are good guys,' and we went back and forth and back and forth. But the bottom line is, we finally set up a system," which amounted to a process in which the arriving recruits, several hundred former PDF members in all, would receive a two-tiered interrogation as part of their necessary vetting. Those who passed would be paid and enlisted in the new organization. That the Panamanians pouring into the parking lot had been the enemy only two days before did not sit well with the platoon providing security at the DNTT. "My guys were, needless to say, not polite," Captain Currie recalled. "They were very apprehensive that it would directly jeopardize our security around the place."[31]

As the registration process was getting under way, Currie's tactical operations center received a report from his platoon at the Ancon DENI that "something was going on." Currie informed Huff, who drove over to the facility to look into the matter personally, arriving around the time, as he put it, "when things were kicking off . . . when the rounds started firing." In trying to make sense of the ensuing chaos, Huff and others concluded that shots were being fired at three separate locations: the Palace of Justice, somewhere near the Ancon DENI where Huff was, and the DNTT. At the first of these locations, Panama's new First Vice President Ricardo Arias Calderón had just finished giving a speech to some of the prospective new police recruits. When he and his bodyguards emerged from the building, they came under sniper fire. Scurrying to the automobiles constituting their small convoy, they hastily departed, their drivers showing no concern for speed limits. It was when the motorcade sped past the Ancon DENI that the shots there were fired. By then, the DNTT had come under fire as well; in fact, it seemed to be bearing the brunt of the enemy attack.

When the first burst of fire hit the DNTT, a Panamanian seeking to enroll in the new police force dropped to the ground, dead. About a dozen other Panamanians were wounded. No Americans had been hit, but, in the initial fusillade, none seemed to have been targeted by the assailants who were firing down into the parking lot from a furniture warehouse next to the transportation building. As would be confirmed later, the purpose of the attack was to punish the former PDF members at the scene who, in the eyes of the shooters, were committing treason by signing up for the new force. "They hosed them down pretty good," Huff said of the Panamanians lined up in front of the interrogation and recruiting tables. Initially caught off guard, the U.S. platoon providing security began to shoot back into the warehouse, after which the firefight quickly escalated and a few U.S. soldiers were hit, none fatally. The horde of Panamanian recruits scrambled for cover, some inside the DNTT. The wounded, meanwhile, were tended to until ambulances

[31] The observation that Snell's brigade headquarters knew little about the recruitment drive scheduled for the DNTT (fifth quote) is from Interv, Capt Joseph Nemmers with Capt Robert E. Vikander, 5th Bn, 87th Inf, 193d Inf Bde , 16 Oct 1990, Fort Clayton, Panama. Other quotes from Interv, Wright with Huff and Senior Officers of the 5th Bn, 87th Inf, 193d Inf Bde, 5 Jan 1990.

arrived and evacuated them under fire to Gorgas Army Hospital. About an hour into the fight, enemy snipers started firing 40-mm. grenades into the parking lot from a second position, a housing area across the street that presumably had been cleared earlier that morning in the sweep operation. In response, a company-size element of military police at the DNTT returned fire, shooting indiscriminately toward the houses—they were "totally out of control," according to Currie.

In the midst of this disarray, the three-car convoy containing Arias Calderón, "traveling at about Mach 3," in Huff's words, came racing down the highway that ran past the DNTT. The drivers, thinking they had found a safe haven, pulled abruptly into the parking lot. When the U.S. troops there saw AK47s sticking out of the cars' windows, they almost opened fire. "I thought it was a set up," Currie recalled. Only troop discipline and the realization that one of the automobiles was an armored car used for VIP travel prevented a tragedy. As the vehicles slowed and finally came to a stop, Arias Calderón was identified and hustled with his bodyguards who had not been wounded to the second floor of the DNTT. About thirty minutes after his arrival, the first vice president was "rescued" by another convoy sent to retrieve him.

The close call involving one of Panama's new leaders later prompted Huff to criticize a decision made the previous day by higher headquarters to remove some of the roadblocks Task Force Wildcat had set up around Ancon Hill in the first hours of Just Cause. The reason given him was that the removal would facilitate the movement of motorcades and convoys carrying important Panamanian and U.S. officials as they crisscrossed the capital to assess postcombat conditions. The decision, however, "proved to be quite a problem for us when we got counterattacked at the DNTT," Huff pointed out, alluding to the unannounced arrival of Arias Calderón's speeding motorcade and the disastrous response it nearly precipitated.[32]

Another problem created by removing the roadblocks concerned drive-by shootings, one of which occurred a short distance from the DNTT while the firefight there was in progress. The target was the antitank platoon, attached to Huff's Headquarters Company, that was securing Quarry Heights. When the shooting started on the twenty-second, the platoon maintained the cordon it had established earlier that morning as part of the sweep operation. At one point, a black Mitsubishi Montero, unimpeded by any U.S. checkpoint, suddenly appeared from around a corner, the occupants firing Uzis at several of the troops. There were no casualties, but the Headquarters Company commander, Capt. Robert Jones, noted that one of the platoon's vehicles had been hit three times in the windshield. The rounds did not penetrate; "armored glass, we found out, actually works," he reported. If they had gone through, they certainly would have hit the driver in the chest. The platoon fired at the car with squad automatic weapons, shooting out two windows as the vehicle sped away. Later, it was found on the side of a road; there was blood inside. Huff offered his opinion on the drive-bys: "Literally, a guy could drive up in civilian clothes, point a weapon out the door

---

[32] Seventh quote from Interv, author with Huff, 29 Jan 1990. Huff's comments on the removal of the roadblocks and eighth quote are from Interv, Capt John Hollins with Huff, 20 Jun 1990, Fort Clayton, Panama. Other quotes from Interv, Wright with Huff and Senior Officers of the 5th Bn, 87th Inf, 193d Inf Bde, 5 Jan 1990.

and shoot and keep on going, because there were no roadblocks." On the directive he had received Thursday to remove the checkpoints, in hindsight he realized that "we should have said 'screw that.' We should have, in fact, put roadblocks in and checked every vehicle," he concluded, referring to the day of the counterattack.

Captain Jones recounted another near miss, this one not caused by the removal of the roadblocks but by the general confusion that prevailed during the firefight that morning. When the shooting started, his scouts were in the Balboa train station. To determine what was happening, some went up to the second floor to get a better view. Once upstairs, they heard weapons being fired from the first floor. They hurriedly went back down to find some PSYOP personnel from Fort Bragg firing two M60 machine guns. "What are you shooting at?" the scout platoon leader shouted. "There! There!" was the response. When he looked where the men were pointing, it was not at the warehouse near the DNTT, which the scouts had already identified as the source of most of the enemy activity, but at a detention center that had been set up for enemy prisoners. The facility was guarded by U.S. military police and other friendly forces who, finding themselves under fire from the train station, began shooting back. While Jones was later telling the story, Captain Currie interrupted to point out that Colonel Snell was at the prisoner-of-war center at the time of the friendly fire episode. "He was not pleased about that," Huff added. There were no casualties, but Huff and his senior officers were upset that their scouts, and not the PSYOP personnel, were being blamed for the misdirected fire.

Adding to the mayhem was the fact that there were just too many people, military and civilian, on the streets in Task Force Wildcat's area of operations, many of whom did not have to be there. From the perspective of the Headquarters Company, "a lot of the staffers up at Quarry Heights were driving around in civilian clothes; they were in POVs [privately owned vehicles] with vests on." Often their cars were filled with captured weapons—Uzis, AK47s, M16s, and so forth. "They're driving around in their POVs with this stuff stuck out the window. And we keep telling them, if you do this you're going to get shot at. At which point, one bird colonel said, 'Well then your people are stupid.' My reply was, 'Well then you'll be dead. But then they won't care.'" One of Huff's staff officers complained that "every Tom, Dick, and Harry in USARSO, as well as SOUTHCOM, started getting on the streets and driving around in whatever vehicle . . . or uniform they had, even in civilian clothes." His point was that the infantry battalion commander on the scene, meaning Huff, "didn't know who all these people were." Thus, there was no sure way to monitor the activities of individuals or groups who might be preparing for some hostile activity against fellow Panamanians or U.S. troops. "There were reports of guys in black uniforms running around," but with hundreds of other people out and about, there was little that could be done to identify and isolate them. Once the shooting started Friday morning, some civilians and U.S. military personnel successfully vacated the area, but others, such as Panama Canal Commission employees residing near Ancon Hill, poured into the streets, only creating more distractions as they tried to run or drive to safer locations.[33]

[33] Ninth and tenth quotes from Interv, Nemmers with Vikander, 16 Oct 1990. Other quotes from Interv, Wright with Huff and Senior Officers of the 5th Bn, 87th Inf, 193d Inf Bde, 5 Jan 1990.

How enemy forces were able to mount surprise attacks at the Palace of Justice, the Ancon DENI station, and the DNTT on the twenty-second would be the subject of later analysis and speculation. Once the shooting started, though, Huff had to focus his attention primarily on the firefight at the DNTT, where he sought to devise a response that would neutralize the enemy without exacerbating the chaos and confusion. One step toward that goal was to rely on Company C to implement the bulk of the countermeasures. Of Captain Currie's three rifle platoons, the 3d Platoon was the one at the DNTT. At the time the shooting started, the 1st was at the Ancon DENI having finished its part of the sweep operation, and the 2d was still patrolling its portion. Once the 3d Platoon identified the warehouse as the source of the initial firing, most of the troops concentrated their return fire on that structure, leaving a squad to control the gate into the DNTT parking lot and, to the extent the soldiers could, the crowd already inside. On Currie's orders, the platoon leader also placed some of his men on the roof the DNTT and in the back of the building in the motor pool area, a vulnerable point if the shooters in the warehouse actually tried to retake the transportation complex. Meanwhile, as a result of the sweep, the 2d Platoon had reached a location from which, in Currie's words, "I could maneuver them into position, which turned out to be key." Specifically, after some extensive and time-consuming movement, the platoon would be in a position to move from the south toward the warehouse and the Explansa building next to it—also identified as a source of enemy activity. Rounding out the countermeasures, Huff directed the antitank platoons in Companies A and B to form up for assaults on the targets from the southwest and northeast.

As these units were maneuvering into their preattack positions, the enemy began firing mortar rounds from a location well concealed from Huff's troops, but one that apparently did not allow the shooters a line of sight on the DNTT. Thus, the first round fell close to the Balboa refugee center. The mortarmen made an adjustment, but the second round hit near the warehouse containing their own people. The remainder of about a half-dozen rounds fell behind the DNTT, causing no damage to the compound. In the meantime, one of the antitank platoons picked by Huff had reached an overwatch position vis-à-vis the warehouse, which appeared to be on fire, a result of the enemy mortar round that fell short. The colonel told the platoon to open up with whatever weapons it had, which meant 90-mm. recoilless rifles, LAWs, and TOWs (the only two TOW rounds fired during Just Cause). Some of the ordnance used high-explosive warheads, which intensified the warehouse fire, ensuring that it would soon burn out of control. The column of black smoke produced by the blaze could be seen miles away. A few of the enemy apparently burned to death in the warehouse; others made their escape. Meanwhile, U.S. troops watching the Explansa building could see cars arriving to extract armed men from that location. Some of those leaving the building appeared to be hurt. When the shooting finally stopped, there was no way to determine how many attackers there had been, or how many had been killed or wounded. Two or three days later, military policemen at the DNTT began to smell a terrible odor. When they went looking for the source, they found up to three dozen bodies in the tall grass near the warehouse. Whether the men had been killed in the firefight or in some other exchange with American forces could not be resolved at that point. As for the U.S. defenders, Huff's men had suffered a handful of casualties. A few of the wounds were serious but none were fatal. Of the Panamanian recruits at the DNTT on 22 December, two had been killed and over a dozen wounded.

In assessing the counterattack, Huff and his men quickly reached two important conclusions, neither of which, they conceded, could be proved. First of all, they were almost certain that the attack had been coordinated. They based this assumption on the near-simultaneous shooting incidents at the Palace of Justice, the Ancon DENI, and the DNTT, as well as on the registration of the mortar fire, however erratic, which would have required communication between the mortar crew and someone observing the DNTT. The logical and immediate concern raised by this possibility, of course, was whether other coordinated attacks on U.S. positions would follow. Second, they surmised that the enemy force had used sewer drains near the warehouse to conceal their entry into the structure. Movement via any other approach by men carrying some of the large weapons they employed would most likely have been observed by U.S. troops, even given the large numbers of people inundating the area. In hindsight, there was one fact about the counterattack that was not open to question. Despite reports that enemy diehards might mount another such assault, the fight at the DNTT proved to be the last episode of major combat to occur in Panama City during Just Cause. With the movement into Colón that same day, organized enemy resistance within the canal area and at Rio Hato had been eliminated. There were, to be sure, PDF units and facilities in the outlying provinces that had to be subdued, but operations to that end were already under way.[34]

Ma Bell and Other Operations Outside the Canal Area

During the afternoon of 20 December, when General Cavezza flew into Albrook Air Station and established his command post there, the two infantry brigades from his 7th Infantry Division not already in Panama were either in the air or making the trip from Fort Ord to Travis Air Force Base, California, to board the waiting troop transports. The 2d Brigade—the Division Ready Brigade when the warning order arrived at Ord—was commanded by Col. Linwood Burney. It began landing at Howard around the same time as Cavezza, with the last transport touching down soon after sunrise the next day. The day after that, the twenty-second, the division's 1st Brigade, commanded by Col. David R. E. Hale, began its deployment to Panama.[35]

During Operation Nimrod Dancer, Hale had been the first Task Force Atlantic commander, Kellogg's predecessor. Now, as part of Operation Just Cause, he would operate on the Pacific side of the country, in Panama City, and he would have nearly his entire brigade with him instead of just one battalion. Upon

[34] One point of contention regarding the counterattack was mentioned by Huff and by several of his senior officers. The press reported the action as taking place at Quarry Heights near the SOUTHCOM headquarters. In their interviews, the officers of Task Force Wildcat were determined to correct that reportorial error. Quote from Interv, Wright with Huff and Senior Officers of the 5th Bn, 87th Inf, 193d Inf Bde, 5 Jan 1990.

[35] Chronology of Key Events for Operation 'Just Cause,' 161400U Jan 1990; Flanagan, *Battle for Panama*, pp. 197–98.

arrival, the 1st Brigade became part of Task Force Pacific under General Johnson, the 82d Airborne Division commander, and received the mission of clearing and securing a sector of the Panamanian capital. As for the 2d Brigade and its commander, they had not participated in any of the earlier phases of the Panama crisis. Once deployed for JUST CAUSE, Burney came directly under JTF-South and General Stiner, who told the colonel that his brigade would operate west of the canal as far out as the Costa Rican border. Its mission "included neutralizing the PDF in the area, securing key sites and facilities, protecting U.S. lives and property, restoring law and order, and demonstrating support for the emerging Panamanian government." In the process, Burney would have to integrate the 2d Brigade's activities with the U.S. Special Forces and Rangers who had comparable missions.[36]

With respect to one aspect of the mission, "neutralizing the PDF in the area," there would be an additional participant on the U.S. side: General Cisneros. As Burney's soldiers prepared to move west outside the canal area, they anticipated heavy resistance. It was a prudent expectation. As they saw the situation, they would be facing the diehards of the toppled regime—PDF and Dignity Battalion members who would be heavily armed and determined to fight. More than that, these recalcitrants might take to the jungle and begin a guerrilla war against U.S. forces—the specter of Vietnam still haunted the American military. While this was an unlikely prospect, it could not be dismissed outright, given U.S. intelligence on PDF contingency plans. Cisneros certainly understood this anxiety but disagreed with the assessment that produced it. In his opinion, the remainder of the Panama Defense Forces was not spoiling for a fight. In the interior of Panama, as in the canal area, senior PDF officers had left their headquarters when the invasion began. Few of these men were fanatical loyalists who would endure the hardships of guerrilla warfare, or any other kind of conflict, to prove their devotion to Noriega. Thus, plans to have U.S. forces in the provinces "go in there and fire into the garrisons, and then give them an ultimatum to surrender" would, Cisneros believed, only produce unnecessary casualties. The preferable alternative would be to give those PDF officers who had stayed at their posts a face-saving way to surrender before armed Americans arrived on their doorstep. If offered such an opportunity, Cisneros was convinced, the recipients would take it.[37]

In searching for a way to do "the minimum we needed to do to knock out the headquarters" in the outlying towns and cities, Cisneros did not have to look further than the tack he had taken in Colón. In fact, when he returned to Fort Clayton the evening of 21 December, he brought Captain Jimenez with him, receiving approval from the day-old Endara government to make the officer his assistant. Then, referring to the Colón experience, Cisneros told Jimenez, "Okay, let's try the same thing with the military zone commanders in the provinces." Between the two of them, they had most of the telephone numbers they needed. Jimenez brought with him the directory he had used in Colón, while Cisneros was able to utilize a list of PDF numbers he had requested be put together during one of General Stiner's planning visits. Jimenez suggested that Major Gómez, the

[36] Flanagan, *Battle for Panama,* pp. 197–99.

[37] Intervs, author with Cisneros, 30 Jan and 29 Jun 1990; De Mena with Cisneros, 12 Jun 1990. Quotes from Donnelly, Roth, and Baker, *Operation Just Cause*, p. 351. Cole, *Operation Just Cause,* p. 56.

PDF province chief in Darién, be the first officer they try to contact. Jimenez based his choice on Gómez' solid professional reputation, even though Darién Province was located to the east, near the border with Colombia, an area not yet a target of U.S. operations. But if Gómez cooperated and agreed to contact other officers, many of whom held him in high esteem, the telephone initiative would stand a better chance of success in the west. Cisneros concurred, and Jimenez placed the call.[38]

Gómez answered the phone. By simply being at his duty post, he confirmed Jimenez' assessment of his professionalism. Jimenez explained briefly why he was calling, then handed the phone to Cisneros, who matter-of-factly informed the major that the battle for Panama was over and that there was no need for further bloodshed. He also promised that U.S. forces would not attack PDF members and installations in Darién if Gómez agreed to surrender his troops, collect their weapons, and wait peacefully until the Americans arrived. Gómez accepted Cisneros' assessment and his offer; the major also agreed to contact other zonal commanders with this information and to recommend they follow his example.[39]

While Cisneros was talking with Gómez, Jimenez was already trying to get in touch with other PDF officers outside the canal area. Because Operation Just Cause had played havoc with Panama's telephone exchanges, a number of the many phone calls he placed between 21 and 23 December were routed through a switchboard in Miami. The approach he adopted did not vary significantly from that used in Colón and with Gómez. Once he had made contact with a PDF officer, he relayed just enough information to turn the conversation over to Cisneros. Once on the line, the general employed a mixture of reason, requests, demands, and threats, tailoring the specific blend to each Panamanian with whom he spoke. He appealed to the officers' nationalism, patriotism, and sense of self-preservation. He told them that the regime had been toppled and the Panama Defense Forces throughout the Panama City–Colón corridor defeated. If need be, he stretched the truth, implying that Noriega had been captured. He promised a dignified and peaceful transfer of power and held out the prospect of ranking positions in the new police force if the officers and their men agreed to surrender. He also threatened an apocalyptic employment of U.S. firepower against any PDF cuartel that refused his offer. Most of all, he assured the Panamanians that if they surrendered, they would not be shot. He even arranged for safe-conduct passes signed by him to be distributed widely throughout the country. (For some PDF officers wanted by the United States or the Endara government for serious crimes, the passes, when used, were not always honored.) Once the officers

---

[38] First quote from Interv, author with Cisneros, 29 Jun 1990, and see also 30 Jan 1990. Second quote from Interv, De Mena with Cisneros, 12 Jun 1990. Cisneros' request for the list of PDF phone numbers had been prompted by his inability during the 3 October coup attempt to find the phone number of the Comandancia, where the rebels held Noriega. See also Interv, De Mena with Jimenez, 13 Nov 1991, and Donnelly, Roth, and Baker, *Operation Just Cause*, p. 351. In his interview with Dolores De Mena, Jimenez said that, upon arriving at Fort Clayton, he and Cisneros met with President Endara and his two vice presidents at their safe house. After Endara named Jimenez his liaison officer with Cisneros, the captain and the general went to Cisneros' office, where they would be assisted by Antonio Bonilla and Captain Marquez.

[39] On the talks with Gómez, see Intervs, author with Cisneros, 30 Jan and 29 Jun 1990; Donnelly, Roth, and Baker, *Operation Just Cause*, p. 352.

had accepted his terms for turning over their garrisons, he would forward that information to General Stiner's people (who gave Cisneros the impression that they suspected the PDF leaders were feigning compliance in order to ambush the U.S. troops heading toward them). Perhaps most important, Cisneros made sure the information reached the Special Forces elements responsible for coordinating surrender operations in the provinces and the Rangers and light fighters who would be working with them.

The Rangers at Rio Hato had started these operations west of the canal on the same day Cisneros and Jimenez had begun making their calls from Fort Clayton. At both locations, those involved used a near identical approach, suggesting some degree of communication and coordination beforehand. For the Rangers, the target on the twenty-first was the city of Penonomé, the capital of Coclé Province and home to a PDF prison and the Military Zone 6 headquarters. The commander at each facility was summoned to Rio Hato, with Colonel Kernan arranging their transportation. When the two PDF officers arrived, they received an ultimatum indistinguishable from the one Cisneros was issuing from Fort Clayton. Both men agreed to the surrender terms and returned to Penonomé to set them in motion. Soon thereafter, Company A from the ranger 3d Battalion arrived in the city and secured both installations. At the prison, the Panamanians present were treated to a demonstration of the kind of air assault they would have experienced had they resisted. According to an official history, "Word of this display of force and surrender quickly spread throughout cuartels in the countryside," to good effect.[40]

As for Cisneros, from the outset of his initiative, he had set his sights on capturing Lt. Col. Luis del Cid, Noriega's deputy, as well as the PDF commander of Military Zone 5 in Panama's westernmost province of Chiriquí. "He was a critical one to turn around," the general declared. Del Cid, like his boss, was wanted by the United States for drug trafficking; furthermore, Cisneros considered him one of Noriega's staunchest supporters who might actually fight any U.S. forces approaching his base. It was not by chance, given del Cid's well-established loyalty, that a PDF contingency plan designated Chiriquí as a possible redoubt for Panama's dictator, even though the province was known for harboring a high degree of anti-Noriega sentiment. (A "reliable source" on Saturday, 23 December, reported, incorrectly, that Noriega was in a basement in David, the provincial capital of Chiriquí.) Moreover, the province housed the largest PDF weapons depot in the country. Yet, however strong the reasons for believing that del Cid would not surrender easily, there always remained the possibility that he had already accepted the inevitable and hoped to spare himself. Cisneros needed to find out. That del Cid was also trying to contact him was something Cisneros learned only later. When the two finally did talk on Friday, the general employed all his powers of persuasion, and, after several conversations, del Cid agreed to

[40] Quote from U.S. Special Operations Command, *History, 1987–2007*, 20th anniversary ed. (MacDill Air Force Base, Fla.: U.S. Special Operations Command, n.d.), p. 38. Briefing, Gen Maxwell R. Thurman at the SAMS, 23 Sep 1992. In remarks to the author, Cisneros talked in general terms about his coordination with Special Operations Forces operating in the provinces west of the canal area. Panel Comments, Cisneros, 19 Jan 2011. The word *cuartel* used in this account is a generic term employed throughout Latin America to identify a military base of any kind. With respect to Just Cause, the term appeared repeatedly in U.S. sources referring to PDF bases outside the canal area, as this chapter suggests.

give up. By Friday night, he was flying a white flag over his headquarters in the city of David. On Saturday and again on Sunday, however, he called Cisneros two or three times, nervous and wanting to know why the U.S. forces had not yet arrived. On each occasion, Cisneros told him to be patient, to keep his garrison intact, and to maintain law and order in the province.[41]

The reasons for the delay had to do with the timing of the mission to move into the western interior, the distances that had to be covered, and the several tasks U.S. troops had to carry out en route to David. Planning for the westward movement of Colonel Burney's 2d Brigade did not kick into gear until Thursday, to be followed on Friday—the same day that del Cid agreed to surrender—by talks at Albrook between Burney and Maj. Gilberto Perez, the commander of Company A, 1st Battalion, 7th Special Forces Group (Airborne). Perez had just that morning received a warning order for operations he would have to conduct in the provinces of Herrera, Coclé, Los Santos, and Veraguas. (*See Map 21.*) In these districts, his company would be working in conjunction with Burney's 2d Brigade and the Rangers. The coordinated effort began that day, with Burney deploying one of his three battalions to Coclecito and the other two to Rio Hato, to relieve the Rangers there. Perez also flew into Rio Hato with his unit. As was the case in the Penonomé initiative, the town would serve as the staging area for the operations ahead.[42]

In the process of neutralizing the Panama Defense Forces in the western provinces, each piece of the coordinated U.S. effort became collectively and popularly known as the Ma Bell campaign, a sobriquet coined by a Special Forces officer who realized the importance the telephone would play in what he was about to do. As with the Rangers on Thursday, the approach at the tactical level would be very similar to what Cisneros and Jimenez were doing from the general's office at Fort Clayton. There were, according to Major Perez, two phases to his concept of operations. In the first, he would insert Special Forces elements into the towns of Santiago, Chitré, and Las Tablas, using each town's airfield as the landing zone. From the airport, the Special Forces officer in charge would contact the commander of the nearest PDF garrison by telephone. In the conversation that ensued, the U.S. officer would dictate surrender terms, which usually meant reiterating and reinforcing what the garrison commander or ranking PDF officer on the scene had already accepted in talks with Jimenez and Cisneros: "First, the surrender would be unconditional. Second, all weapons would be placed in the

[41] First quote from Interv, author with Cisneros, 30 Jan 1990, and see also Interv, author with Cisneros, 29 Jun 1990. Second quote from Chronology of Key Events for Operation 'Just Cause,' 161400U Jan 1990. Flanagan, *Battle for Panama,* pp. 215, 218; Donnelly, Roth, and Baker, *Operation Just Cause*, p. 352; Cole, *Operation Just Cause,* p. 56. According to Cole, pp. 56–57, Thurman and Stiner had planned to take David on 25 December in "the last major combat assault of Just Cause." An article entitled "An Overlooked Hero," on Cisneros' role in Operation Just Cause in the *St. Petersburg Times*, 20 December 1999, suggested that Cisneros' knowledge of the plan caused him to dispatch Jimenez in a helicopter to speak with del Cid directly in order to produce a surrender before the impending attack. Years later, Cisneros remarked, "One of my hardest challenges was to convince the XVIII Airborne Corps not to attack David." Panel Comments, Cisneros, 19 Jan 2011.

[42] Chronology of Key Events for Operation 'Just Cause,' 161400U Jan 1990; "Winning the West," in *Soldiers in Panama,* pp. 16–17; Flanagan, *Battle for Panama,* pp. 198, 215.

arms room of the cuartel. Third, the entire garrison of the cuartel would assemble on the parade ground." Once the telephone conversation was over, the Special Forces leader and the PDF garrison commander would meet and fly over the cuartel in a helicopter to ensure that the Panamanian troops were honoring the surrender arrangements. If there was any sign of resistance, an AC–130 assigned to the operation would fire a few rounds near the installation as a warning of the devastation that could follow.[43]

Soon after the surrender had been accepted, the second phase would begin, with the Special Forces moving into the garrison or headquarters, searching the premises, and processing the Panamanians present. Meanwhile, a ranger or light fighter company in waiting would enter the town and begin reestablishing law and order. When necessary, that would include stopping the inhabitants from looting or from taking reprisals against PDF or Dignity Battalion members.

On Saturday afternoon, 23 December, the Ma Bell campaign on the ground formally began, with the ensuing operations hitting only a few snags. The first objective for Perez was the town of Santiago, where Task Force Hawk helicopters flew him and his men to a nearby airfield. Once they landed, the major was unable to contact the cuartel commander, so he took five of his men and flew over the facility. Below they could see Panamanians lined up on the parade ground as instructed, ready to surrender. When the Special Forces team landed, however, someone in the assembled group fired at them. The "misguided youth" was quickly overpowered, and the remainder of the operation went as planned, with the Special Forces company moving into the cuartel and clearing it, while an infantry company from a battalion in Burney's 2d Brigade entered the town. When Perez pulled his people out on Sunday to move on to the next objective, he left a Special Forces detachment to serve as translators for the light fighter company, which remained in place to occupy the PDF garrison and stabilize the area. The company's commander had four missions:

> First, he was to gather intelligence on the location of weapons caches and those personnel of the Panama Defense Forces . . . and dignity battalions who had not surrendered. Second, he was to assist local government officials in reestablishing the civilian infrastructure. Third, he was to assess the public utilities, medical facilities, and law and order capabilities of the area. This information would be used to establish the working priorities for the follow-on civil/military operations. Finally, he was to conduct joint U.S. and Panamanian patrols throughout Santiago in order to reestablish law and order in the community.

The 2d Brigade had follow-on missions planned for the countryside immediately west of Santiago but, according to one report, those were "temporarily prohibited as CINCSO negotiates with new government."[44]

[43] McConnell, *Just Cause,* pp. 291–92; Linda Robinson, *Masters of Chaos: The Secret History of the Special Forces* (New York: Public Affairs, 2004), p. 50. Quote from "Winning the West," in *Soldiers in Panama,* p. 16.

[44] First quote from Flanagan, *Battle for Panama,* pp. 216–17. Second quote from "Winning the West," in *Soldiers in Panama*, pp. 16–17. Last quote from Chronology of Key Events for Operation 'Just Cause,' 161400U Jan 1990.

Map 21

Perez took his next two objectives, Chitré and Las Tablas, without incident on 24 and 25 December, respectively. In each case, the operation followed the script: he contacted the local PDF commander, went to the assigned cuartel, and received the surrender of the troops present. At Las Tablas, a crowd of curious Panamanian civilians assembled outside the base. Perez used their presence to send a message. With the crowd looking on, he assembled the U.S. and Panamanian troops on the parade ground and had the Panamanian flag raised on the compound's flagpole. The gesture, he believed, would convey in dramatic fashion the message that the United States military did not intend to

conquer Panama but to liberate it. He viewed the cooperation he received from the locals during the remainder of his brief stay as proof that the civilians of Las Tablas had understood the symbolism of the flag-raising ceremony.[45]

On Christmas Day, while Perez was busy in Las Tablas, about two hundred fifty Rangers from the 3d Battalion, 75th Ranger Regiment, commanded by Lt. Col. Joseph F. Hunt, and a contingent of about thirty Special Forces led by Maj. Kevin M. Higgins, the on-scene commander during the Pacora River

[45] "Winning the West," in *Soldiers in Panama,* p. 16.

bridge fight, moved on the PDF headquarters in David. Hunt apparently knew about the arrangement Cisneros had made with del Cid, but he was still concerned about "what we were really flying into." Higgins and his men had also received reports that the PDF members in the city would probably not resist the U.S. takeover. Hunt sent the Special Forces into David ahead of the Rangers. If Higgins ran into trouble, two Apache attack helicopters and an AC–130 would be there immediately to provide fire support until Hunt's force landed at the airfield.[46]

The helicopters carrying Higgins and his men set down at a schoolhouse near the PDF cuartel. The operations officer, Capt. Charles Cleveland, knew the area. There was a telephone outside the schoolhouse, he said, that Higgins could use to call del Cid. Hunt had already talked with the PDF lieutenant colonel, informing him that the Special Forces would be arriving soon at his headquarters. The ranger commander also reminded del Cid of the combat power facing him if he should change his mind and make a fight. Now Higgins would close the deal, beginning with his call from the schoolhouse. He dropped the necessary number of quarters into the telephone box, but received no dial tone. He then slammed the phone down, after which an old man sitting on a bench nearby suggested that he use the telephone inside the schoolhouse. When the Special Forces showed signs of knocking down the school's door to gain access, the man produced a set of keys, saying that he was the janitor. The encounter with the out-of-order telephone proved to be the biggest obstacle to Higgins accomplishing his mission. When he contacted del Cid on the school telephone, the PDF officer asked, "What day would you like to come?" Higgins said the Special Forces would be there within the hour.[47]

As with U.S. troops in several other locations, Higgins and his men arrived at del Cid's headquarters to find a crowd of several hundred cheering, flag-waving Panamanian civilians. Del Cid proved accommodating, cooperating with the Americans even as he futilely tried to negotiate an agreement whereby his soldiers would keep their weapons. He also held a press conference during which he swore loyalty to the new government. Then, after he insisted repeatedly that he could play a helpful role in building a new Panama, Higgins suggested that he be flown to a meeting in the capital that would decide much of the country's future. Del Cid agreed and subsequently made the flight. When he landed at Howard Air Force Base, he was promptly turned over to the Drug Enforcement Administration, who put him on an airplane bound for Miami, where he would eventually stand trial on the drug charges U.S. federal grand juries had brought against him in February 1988. Del Cid accused his captors of double-crossing him. None of them felt moved to dispute the allegation or to offer apologies.[48]

While del Cid was still talking with Higgins, Hunt's Rangers flew into David's airfield. In the days to come, they would help pacify the city while

[46] Quote from Donnelly, Roth, and Baker, *Operation Just Cause*, p. 353. Robinson, *Masters of Chaos*, p. 50.

[47] Donnelly, Roth, and Baker, *Operation Just Cause*, p. 353. Quote from Robinson, *Masters of Chaos*, pp. 50–51.

[48] Robinson, *Masters of Chaos*, p. 51; McConnell, *Just Cause*, pp. 291–92; Associated Press, 24 Dec 1989; *Sarasota Herald-Tribune*, 26 Dec 1989.

Higgins' men concentrated on taking over other PDF outposts in the area. In time, elements of Colonel Burney's 5th Battalion, 21st Infantry, commanded by Lt. Col. Robert Cronin, would relieve the Rangers. Major Perez also arrived in David to serve as Cronin's deputy after Higgins and his men had redeployed to Panama City to conduct stability operations there. Together, Cronin and Perez worked to gain acceptance in Chiriquí for the new police force Endara had authorized. Perez also persuaded Burney to redeploy one of his infantry battalions to Rio Hato. From there, Burney could mount light fighter operations designed to canvass the western region, searching for arms caches and rounding up enemy troops on the run. Eventually, the 2d Brigade would be called back to Panama City, and then deployed to the countryside east of the canal, but from late December through early January, Burney's infantrymen working with Special Forces helped secure the area from the capital to David for Panama's new government. As for the Ma Bell campaign, the procedures initiated by Cisneros and Jimenez and further executed in the field by the 7th Infantry Division, Rangers, and Special Forces were credited with getting all nine of Noriega's provincial commanders to surrender, together with around five thousand Panamanian troops. In the process, according to General Thurman, only one shot was fired in anger.[49]

Apprehending the Dictator

Cisneros had made the surrender of del Cid a priority in part because of the demoralizing effect he hoped it would have on Noriega, who was still managing to elude the U.S. forces searching for him. When del Cid did surrender on Christmas Day, the impact of his action was still substantial, but circumstances had changed. On the afternoon of Sunday, 24 December, one day before Higgins and Hunt reached David, Noriega had entered the Apostolic Nunciature, the Vatican's diplomatic mission in Panama City, seeking asylum. Nearly five days after Operation Just Cause had begun, the United States finally knew the dictator's whereabouts. The problem now was how to get him out of the nunciature and into American custody.

As a fugitive, Noriega had moved frequently, usually from the accommodations of one friend or colleague to another. In the meantime, U.S. military personnel were finding themselves overwhelmed with reports concerning the general's location. Moreover, as of 21 December, capturing Noriega was Thurman's second priority, the first being completing the defeat of the Panama Defense Forces. On that same day, the Pentagon suggested placing a bounty of $1 million on the deposed strongman, which was approved. From H-hour on, elite special operations units had led the search for the dictator, often reacting at a moment's notice, regardless of the disruptive effect they might have on the scheduled operations of conventional units in the field. The

[49] *St. Petersburg Times*, 20 Dec 1999; Thurman SAMS briefing, 23 Sep 1992; Flanagan, *Battle for Panama,* pp. 218–19; "Winning the West," in *Soldiers in Panama,* p. 17. At some point after the outlying garrisons had surrendered, Cisneros visited each of the commanders still in place to impress on them their role for maintaining law and order in their areas. Panel Comments, Cisneros, 19 Jan 2012.

day after moving into Colón, for example, Task Force Atlantic had its activities interrupted when, at 0400, some of Kellogg's soldiers answered an unexpected but urgent request to cordon off the Hotel Washington so that Navy SEALs could enter the building to look for Noriega. At other times, special operations and conventional forces worked closely together in the search, as in the case of the Panzergruppen that General Downing cobbled together. There were even conventional units at the tactical level that were drawn into the search on their own simply because they had to run down the numerous leads they were receiving from Panamanian citizens who approached them on the streets or telephoned their headquarters. Also, some of these units, as a result of their D-day missions, found themselves in a position to search Noriega's offices or one of his residences. All the while, the Noriega tracking cell continued to operate out of SOUTHCOM's Tunnel. As Thurman put it, "Got all kinds of spooks—NSA and CIA and all kinds of guys looking for him."[50]

There were several near misses, as search teams often entered a home just after Noriega had departed. "We missed him by one hour here, thirty minutes there," Thurman later declared. Left for speculation was how many of the close calls were caused by the inability of the Southern Command and JTF-South to process in a timely way the enormous amount of information they were receiving and to separate the signals contained in legitimate reports from the noise produced by inaccurate rumors or false accounts, some planted deliberately but most submitted innocently with the best of intentions. Bad luck also played its part. Early in the morning of 24 December, one of Noriega's bodyguards, Ivan Castillo, found an excuse to leave the dictator. Castillo then tried to find a way to turn his boss over to Cisneros, but, despite his assiduous efforts, he either could not find U.S. troops who spoke Spanish, or, when he did, who realized the importance of the information he was imparting. As a result, Castillo did not make contact with Cisneros until 1030. The general sent Captain Jimenez and a sergeant from the Panama-based 29th Military Intelligence Battalion to the location where Castillo had last been with Noriega, but, by the time the two-man team arrived, the elusive quarry was gone. Had Noriega been there, the two had instructions from Cisneros to do nothing other than call and inform him. Just hours later, the object of the manhunt entered the nunciature.[51]

Noriega's decision to seek refuge there was probably based on his sense that U.S. forces were closing in on him. "Finally, we wore him out," Thurman later boasted. Noriega likely realized that, if he continued to move between the residences of various acquaintances, he would eventually be seen and captured. One alternative was to seek asylum in the embassy of a country he considered friendly. But anticipating that Noriega might make such a move, Thurman and Stiner had placed a cordon of U.S. troops around the Cuban,

---

[50] Interv, John Partin with Gen Maxwell R. Thurman, U.S. Army, 12 Apr 1990, Fort Benning, Ga. Quote from Thurman SAMS briefing, 23 Sep 1992. On the sudden operation at the Hotel Washington, see Interv, author with Brooks, 11 May 1993. For a very brief account of how Downing created the Panzergruppen, see Chapter 4. See also Donnelly, Roth, and Baker, *Operation Just Cause*, pp. 110–11; Cole, *Operation Just Cause,* p. 56.

[51] Quote from Thurman SAMS briefing, 23 Sep 1992. Donnelly, Roth, and Baker, *Operation Just Cause*, pp. 111–12; Cole, *Operation Just Cause,* p. 56.

Nicaraguan, and Libyan Embassies, a violation of an international treaty, the former later remarked, after which he defended his action by saying "when in charge, take charge." The two generals also placed some other embassies under surveillance. Noriega may have suspected that the embassies were being watched, or someone may have told him. For whatever reason, he chose not to approach them but to reach out instead to the papal nuncio, Monsignor José Sebastián Laboa, the man who had granted a safe haven to Endara in the aftermath of the violence following Panama's May presidential election. Laboa, seeing a way to help end further bloodshed, agreed to take in the fallen dictator, sending a car to pick him up at a prearranged site—a Dairy Queen in Panama City. The nuncio also made it clear that his offer was good only for as long as it took to arrange for Noriega to be sent to a third country. When the nuncio's car arrived at the Dairy Queen, Noriega jumped in and hid behind the seat as the vehicle sped off. In the meantime, Laboa had been trying to get in touch with Cisneros to inform him about the pickup, with hopes that U.S. forces would intercept the car before it returned. But Cisneros was out with Stiner, observing the progress of various stability operations. When the nuncio finally reached the USARSO commander on the telephone, it was too late. Just as they began their conversation, Laboa looked up and, seeing a pockmark-faced man in running shorts and a T-shirt enter the building, had little time to say more than, "He just walked in." Within a short time, Noriega was joined by an entourage of bodyguards, advisers, other PDF officers, and even their family members.[52]

Cisneros immediately informed Stiner about the phone call, including the nuncio's request to meet and talk. "Go ahead and see Laboa," Stiner replied. "Call me as soon as you have details." The JTF-South commander then contacted Downing and ordered him to send Special Operations Forces to the nunciature and to seal it off. Soon thereafter, Downing had the area secured. In the process, he established his operations center in an elementary school across the street from the two-story stucco building.[53]

---

[52] First and second quotes from Thurman SAMS briefing, 23 Sep 1992. Last quote from Donnelly, Roth, and Baker, *Operation Just Cause*, p. 112. Clancy with Stiner and Koltz, *Shadow Warriors*, pp. 371–72. The use of U.S. troops to cordon off certain foreign embassies in Panama City resulted in two well-publicized incidents. In one on 28 December, the troops stopped the Cuban ambassador and his escort and checked their papers. The escort was an intelligence officer using a diplomatic position as his cover, not an uncommon practice. On instructions from General Stiner, the two men were taken to Panama Viejo but then, again at Stiner's insistence, returned to the ambassador's residence by 1700. The Cuban government protested, but, given the courtesy with which both men had been treated and the short period of the escort's being detained, nothing more came of the incident. More serious, the next day U.S. troops broke into the residence of the Nicaraguan ambassador and found an arms cache. The troops claimed the residence had not been properly marked as being affiliated with the Nicaraguan Embassy, although General Powell would later concede that, attached to the house, a round seal the size of a manhole cover did exactly that. The Nicaraguan government protested, but the United States responded that the arms cache inside the residence violated, and thus overrode, diplomatic immunity. Cole, *Operation Just Cause*, pp. 57, 61–62. The title nuncio refers to a papal legate of the highest rank permanently accredited to a civil government.

[53] Clancy with Stiner and Koltz, *Shadow Warriors*, pp. 372–73.

An M551 Sheridan outside the nunciature during negotiations for Noriega's surrender

After talking to Downing, Stiner called Thurman to give him the news. Thurman was surprised. "I never figured that he'd go into the Papal Nunciature, the Vatican embassy, but happily he did," the general later stated. He notified Lt. Gen. Thomas Kelly, the Joint Staff's operations officer in the Pentagon, about the development, and Kelly informed General Powell. Thurman also spoke to Secretary of Defense Cheney, who was at that moment flying to Panama to deliver a Christmas message to the U.S. armed forces. "He was forty minutes out," Thurman recalled. The general wanted to know who would make the announcement, the White House or himself. Cheney called back from his plane and told Thurman to go ahead and hold a press conference. The secretary was also reported as saying, "Don't let that guy out of the compound." On the diplomatic front, the State Department quickly contacted the Vatican with a strong entreaty that Noriega be denied asylum.

The next day, Christmas, Thurman met Laboa at the front gate of the nunciature to reiterate that the United States wanted Noriega turned over to stand trial. While this marked the beginning of face-to-face talks to obtain that outcome, the man with whom the nuncio would be speaking on a daily basis would not be Thurman but Cisneros. As the commander in chief of the Southern Command, Thurman did not have the time or the inclination to be the lead negotiator. (He also did not have the language skills.) Cisneros, too, was extremely busy. Having worked the telephones to secure the surrender of PDF officers and other personnel in the provinces, he had turned his attention to stability operations aimed at helping the Endara government set up its new police force. Yet, he was the logical choice to conduct the negotiations. Besides his fluency in Spanish, he and Laboa knew one another and shared a mutual respect and trust. A directive from President Bush designating Cisneros as the U.S. negotiator sealed the deal. To assist the general, the State Department sent Michael Kozak, a long-time veteran of the crisis, back to Panama. Either Kozak (who did not speak Spanish) or his deputy would be at Cisneros' side in mapping an effective strategy for getting Noriega to turn himself over to the United States.[54]

The talks lasted ten days and were known almost as much for the events surrounding them as for the results they produced. The Bush administration had to make numerous decisions relating to the negotiations and then adjust to changing circumstances as the talks progressed. One early decision, made on the twenty-fourth, concerned what would happen to Noriega if the nuncio brought him to the front gate and turned him over to the United States. Telephone conversations between the Pentagon and the Justice Department produced an agreement whereby the Drug Enforcement Administration would take custody of the deposed dictator at that point, leaving U.S. forces in Panama to provide security and the transportation needed to get him from the nunciature to Howard Air Force Base and, from there, to Florida.[55]

Also on the first day of Noriega's stay in the nunciature, U.S. officials discussed what could be done, first, to prevent him from providing colleagues who visited him with instructions for carrying on armed resistance against American forces and, second, to keep him or one of his entourage from escaping with a visitor, say in the trunk of a car. The actions recommended as a response to both concerns were complicated and changed over the course of the negotiations. Among the initial decisions, General Kelly informed Brig. Gen. William Hartzog, Thurman's operations officer, that U.S. troops could stop vehicles with diplomatic plates entering and exiting the nunciature, check the occupants' papers, and, if suspicions surfaced that someone might be hiding inside, search the car or van. If, in fact, either Noriega or any one of his associates also wanted by U.S. or Panamanian officials was found inside, the troops could place the person in custody until they received further instructions. Late in the afternoon of the twenty-fourth, the State Department modified those guidelines: all Vatican diplomats and the four top officials in the new Panamanian government were to be allowed unrestricted access to the nunciature. The department also indicated that it would be preferable if the chargé d'affaires at the U.S. Embassy, John Bushnell, informed Thurman of any visit by these people. Thurman quickly objected to the term *preferable*, insisting that he be notified of all such visits without fail, a position that Powell supported. The Deputies Committee of the National Security Council in the White House approved the change two days later on 26 December.[56]

Another issue raised a potentially more lethal threat: the taking of hostages inside the nunciature by Noriega and the men with him, all of whom U.S. authorities assumed were armed. On the twenty-sixth, soon after Cisneros became involved in the negotiations, Laboa gave his oral, then written

[54] Third quote from Cole, *Operation Just Cause,* pp. 57–58. Intervs, author with Cisneros, 29 Jun 1990; Partin with Maj Gen Wayne A. Downing, U.S. Army, 6 Jan 1990, Howard Air Force Base, Panama. Quotes from Thurman SAMS briefing, 23 Sep 1992.

[55] Cole, *Operation Just Cause,* pp. 58–59.

[56] Ibid., p. 59.

permission to have U.S. troops assault the compound if a hostage situation developed. Cisneros and Downing passed the word to Stiner, after which it went up the chain of command, with Powell briefing Cheney, and the latter authorizing American forces to enter the nunciature if they heard shots fired inside. The next day, Acting Secretary of Defense Donald Atwood approved a request from Thurman to employ U.S. snipers in the event of a hostage crisis. According to an official account, Atwood made the following statement: "You are authorized to use the appropriate military force necessary to resolve any hostage situation . . . which you have reason to believe is immediately life threatening. Of course, if time permits, you are expected to consult with the National Command Authorities." Four days later, Atwood revised his guidance in light of certain political considerations. The Vatican had to make a direct request for the troop intervention before U.S. forces could enter the compound. Furthermore, the Vatican would have to acknowledge publicly that it had made the request. These conditions would both reduce the chances of a mistaken assault on the nunciature and, if an attack had to be launched, prevent the Vatican from claiming that the action had been unprovoked or carried out without its knowledge.[57]

Perhaps the most memorable circumstance surrounding the negotiations was the music blaring from loudspeakers the U.S. military placed around the nunciature. Viewed by the news media as a psychological operation aimed at Noriega, the initiative was actually directed at them, at least at first. On Christmas Day, as Thurman was leaving the front gate of the nunciature after his only meeting with Laboa, a reporter on the balcony of a hotel close by yelled out, "Hey General Thurman, how ya doin'? Merry Christmas!" Knowing that, in the negotiations to come, some of the talks would likely take place at the front gate and concerned that reporters might employ directional microphones to eavesdrop on what was being said just below them, Thurman approved emplacement of the speakers after Stiner made the recommendation. The reporters cried censorship, even though the selection of songs—"I Fought the Law and the Law Won," "Voodoo Chile," "You're No Good," "Ring of Fire," "Nowhere to Run To," "In the Jailhouse Now," and others—appeared to have been selected by a desire to harass Noriega, not the press. The fact that the loud music played around the clock, not just when Cisneros and Laboa met at the front gate, also supported the psychological operation rationale. Whatever the case, Washington was soon experiencing adverse fallout from the tactic, as the Bush administration received strongly worded complaints from the Vatican, diplomats, irate Catholics, and Laboa himself. If the music was keeping Noriega awake, it was also having the same effect on the

[57] Quote from ibid., pp. 59–60. Intervs, author with Cisneros, 29 Jun 1990; Partin with Downing, 6 Jan 1990; Clancy with Stiner and Koltz, *Shadow Warriors*, pp. 373–74. In his oral history interview, General Downing referred to the presence of Noriega in the nunciature as being "like a hostage barricade situation, complete with negotiators and everything else." That the negotiations lasted well over a week caused him to add, "you must also practice a high degree of patience." On the possibility of sending U.S. forces into the nunciature, Cisneros later related that Laboa wanted the troops to come in but also hoped that he could deny he had requested such action. The decisions in Washington precluded that tack. Panel Comments, Cisneros, 19 Jan 2011.

nuncio. Moreover, the noise seemed to be making the dictator and his men nervous. Thurman, who readily touted the PSYOP aspect of the music in his communications with Washington, argued in its defense that the racket might make Laboa more eager to rid the nunciature of a man who not only was armed but growing increasingly agitated as well. Cisneros, meantime, thought otherwise and lobbied the State Department representative with him to have the music stopped. After a couple of days, President Bush himself came to regard the ploy as "irritating and petty." On 29 December, under instructions from the National Security Council not to "make things any more difficult or unpleasant for Monsignor Laboa than necessary," Powell directed Thurman to turn off the music. Other ways would have to be found to prevent journalistic eavesdropping and to bring pressure on Noriega.[58]

In the midst of the furor over the music, negotiations continued, with Washington monitoring the process closely. Powell communicated with Thurman several times a day, while Kelly and the Pentagon's Crisis Action Team did the same with Hartzog and his staff. Meanwhile, Cisneros and Laboa began conferring at the front gate. The nuncio argued that diplomatic protocol dictated that only Noriega could make the decision to leave the nunciature. Laboa could use his powers of persuasion to produce that result but would not force Noriega out. As for Cisneros' role, the two agreed that the general should concentrate on persuading individual members of Noriega's entourage to depart, thus gradually isolating the former strongman and using that sense of isolation to convince him to leave. To this end, beginning on 26 December, Cisneros conferred one-on-one at the gate with several of the men who had followed Noriega into the nunciature. Within twelve hours, he had persuaded nine of them to turn themselves over to U.S. authorities, assuring them in the process that they would not be harmed and that the United States did not plan to indict them. General Downing, who was present, praised Cisneros as being "magnificent to work with. He was really, really good." He added, however, that, unwittingly, Cisneros was not being completely candid with the men. "We got them out basically by lying to them," Downing later admitted. What he knew at the time, and what Cisneros did not know, was that Thurman and Stiner had no intention of letting the men go before interrogating them at Fort Clayton. After that, some would be released, others not.[59]

On the twenty-seventh, Cisneros was talking to one of the remaining men, Capt. Asunción Eliécer Gaitán, a UESAT officer, and just about had him ready to walk out, when Downing received a call from either Thurman or Stiner, he could not remember which one, telling him "to knock that s——t

[58] Quotes from Cole, *Operation Just Cause*, pp. 59–60. *Time*, 8 Jan 1990; Panel Comments, Cisneros, 19 Jan 2011; Clancy with Stiner and Koltz, *Shadow Warriors*, pp. 377–78. In his comments on 19 January, Cisneros said that, although he had not been consulted in the matter, in his mind the use of the loudspeakers and the music was not an anti-media ploy that evolved into a psychological operation, but a PSYOP from the very beginning. He only told Laboa that it was aimed at reporters to calm the increasingly irritated nuncio. Stiner, on the other hand, emphasized the need to prevent reporters from overhearing the talks as the basis for his recommendation to Thurman.

[59] Interv, author with Cisneros, 29 Jun 1990. Quotes from Interv, Partin with Downing, 6 Jan 1990. *Los Angeles Times*, 30 Dec 1989.

off." Thurman, it seemed, wanted to change the strategy on which Laboa and Cisneros had agreed. Downing conveyed this "unfortunate" news to Cisneros immediately, thereby terminating the talks with Gaitán. From that point on, the strategy became one of leaving those remaining in Noriega's entourage—with one or two specific exceptions—in the nunciature in order "to let them work against each other."[60]

The question remained as to why Thurman sought the change in the negotiating strategy. The answer most often given is that the SOUTHCOM commander did not trust the nuncio and believed that Cisneros, whom he also suspected, had fallen under Laboa's influence. The nuncio, according to Thurman, advocated the strategy of bringing Noriega's people out one at a time so that, once the entourage had left, Noriega could remain in the nunciature indefinitely. One factor contributing to Thurman's opinion was the presence in the nunciature of a number of Basques from Spain who had been given asylum there several years before. In Thurman's mind, this was proof that Laboa supported terrorism and could not be trusted. Cisneros knew that the Basques' presence was the result of a deal made between the Spanish government and the Vatican; Laboa had not been involved. Once again, Cisneros and his immediate superior were at loggerheads.[61]

Downing, too, was not pleased. "It appeared to us on the scene that the strategy became 'get the Nuncio.' Embarrass the Nuncio and embarrass the Vatican. The plan was we are going to pressure the Catholic Church to give this guy up. My reflection was that the Catholic Church has been around for 2,000 years. They are probably going to be hard to pressure. Maybe what we ought to do is work with the Nuncio and help him." If that approach had been continued, the general believed, "we probably could have ended the thing several days earlier."[62]

The breakthrough that forced Noriega's hand came on the afternoon of 3 January 1990, when a mob of several thousand angry Panamanians marched on the nunciature, carrying anti-Noriega signs and angrily shouting slogans such as "Assassin! Assassin!" and "Down with Noriega!" Stiner moved extra troops into the area, while overhead there were AH–64s and an AC–130, all under orders not to fire indiscriminately into the crowd. In the event that someone in the mob opened fire, the troops were told to locate the gunman and target only him. The message Cisneros sent through Laboa to Noriega, though, deliberately conveyed a different impression. "We will not kill a single Panamanian citizen to defend Noriega. If there's a riot that comes in here and wants to drag Noriega out, we're not going to kill one single Panamanian. It's not worth it, not for us." In Cisneros' opinion, that message convinced the nuncio to tell his unwanted guest, "Your solution is turn yourself over to the Americans or you're going to be faced with a Mussolini experience where the people will come and drag you out and hang you." Laboa added a twist of his own in talking with Noriega, telling him that, come the next day, he intended to lift all diplomatic immunity at the nunciature. Noriega also received a phone

[60] Interv, author with Cisneros, 29 Jun 1990. Quotes from Interv, Partin with Downing, 6 Jan 1990.

[61] Intervs, author with Cisneros, 29 Jun 1990; Partin with Downing, 6 Jan 1990.

[62] Interv, Partin with Downing, 6 Jan 1990.

call—one in a series that Stiner and Downing had arranged—from his mistress, Miss Vicki, telling him, "The decision is in your hands." The combination of pressure and entreaties had the desired effect. A shaken Noriega decided the time had come to give himself up.[63]

During the remainder of the day and into the evening, details of the handover were worked out between the nunciature and U.S. officials. Then, at 2048 that night, Laboa and another priest accompanied Noriega to the front gate. The former general, who carried a Bible and a toothbrush, had shed his T-shirt and running shorts for a military uniform that he insisted he be allowed to wear. Cisneros refused to meet with the man he detested, so Downing was present in his place. After Noriega passed through the gate, U.S. troops handcuffed him over his protests. A helicopter flew him to Howard, where he boarded a C–130 transport. There, Drug Enforcement Administration agents read him his rights, arrested him, and had him exchange his uniform for plain prisoner's coveralls. Less than an hour after leaving the nunciature, he was in the air, en route to Miami. The principal Just Cause objectives pertaining to the defeat of the enemy and the capture of the dictator had now been achieved. There was, however, still much left for U.S. forces to accomplish in Panama.[64]

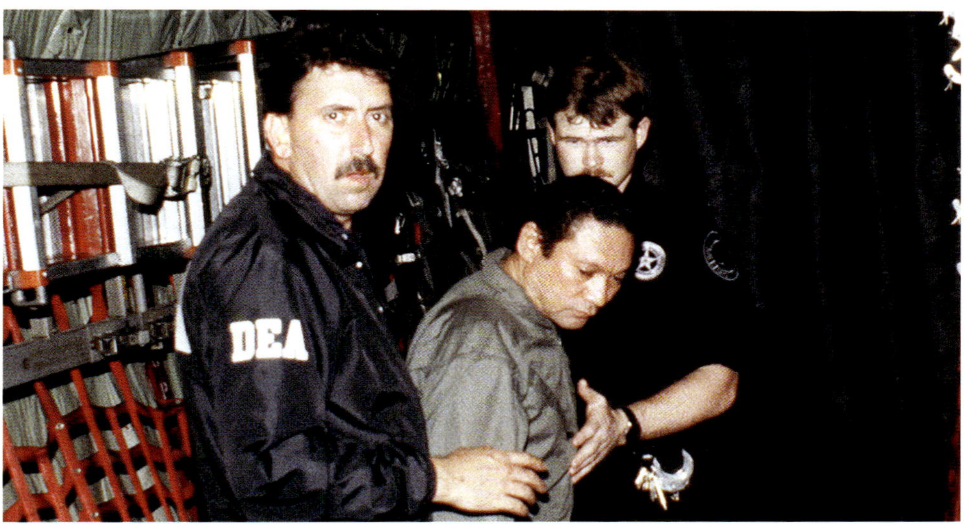

Noriega being placed in custody

[63] Cisneros and Laboa quoted messages from Interv, author with Cisneros, 29 Jun 1990. Cole, *Operation Just Cause,* p. 64. Last quote from Clancy with Stiner and Koltz, *Shadow Warriors*, pp. 379–82.

[64] Cole, *Operation Just Cause,* p. 64; Clancy with Stiner and Koltz, *Shadow Warriors*, p. 382.

13

The Transition to Stability Operations and Nation Building

December 1989–January 1990

☆ ☆ ☆

Throughout the lengthy preinvasion crisis in Panama, U.S. combat units arriving in the country for security enhancement and treaty enforcement duties in 1988 and 1989 had experienced various degrees of frustration in adapting to a situation that required their presence more than their firepower, their restraint rather than their belligerence. Trained for high-tech, conventional battles, these warriors found the bulk of their time in Panama taken up with police-type duties. Even when reasserting American treaty rights—actions that in some cases carried the risk of armed conflict with the Panama Defense Forces—the troops had to observe highly restrictive rules of engagement and rules of confrontation that many soldiers and marines believed nullified their inherent right of self-defense. In their minds, the constraints imposed on them yielded the initiative to their opponent, who would be virtually guaranteed the first shot if a confrontation spiraled out of control.

Following the outbreak of hostilities in late December, many U.S. units engaged in D-day combat operations found themselves making a similar adjustment. During the fighting, they clearly understood that they were to avoid inflicting unnecessary casualties and property damage, but they also knew that the Just Cause rules of engagement explicitly permitted them to take "all necessary and appropriate action" in their own defense.[1] Their adjustment came later, once the shooting began to subside and they had to make an abrupt and, for some, an unexpected and difficult transition from combat to stability operations and nation building. For many of the units, the transition occurred as a result of on-order or follow-on missions derived from JTF-South's Operation Plan 90–2; other units acted in accordance with Promote Liberty, the civil-military operations plan the Joint Chiefs of Staff approved on 21 December, a day after they received it from the Southern Command. Whatever the source, the military tasks that were deemed essential to securing the country, to helping restore law and order, and to supporting the new Panamanian government required American forces to demonstrate several sets of skills—political, diplomatic, law enforcement, social, economic, and humanitarian—for which their war-fighting capabilities and warrior self-image often seemed irrelevant, even counterproductive. Also back in force were much more restrictive rules of engagement. In the slew of interviews

---

[1] JTF-South OPLAN 90–2, 3 Nov 1989.

and after action reports to emerge from JUST CAUSE and PROMOTE LIBERTY, individual soldiers and marines, together with the units in which they served, emphasized that they had not been trained or mentally prepared to conduct stability operations.

The argument could be made that the headquarters and senior military officers responsible for planning and executing the invasion were partly responsible for this lack of preparation. The JTF-South operation plan envisaged military action in Panama taking place in five phases, the third of which called for "Assault Force Operations" and the fourth for "Stabilizing Force Operations." In the latter phase, U.S. objectives were to "restructure the PDF; provide support to civil-military operations; and prepare JTFSO forces for redeployment." Yet, while the plan recognized the significance of stability operations and, in the words of one officer, "acknowledged that a breakdown in law and order was likely," it did not, in his opinion, "address in any detail the actions US forces would have to take to bring order to the situation." In contrast, BLIND LOGIC, the code name for the PRAYER BOOK's civil-military operations plan, did contain specific guidance for short-term and long-term measures to restore and rebuild a postconflict Panama. But, in the second half of 1989, efforts by staff officers at the Southern Command to integrate this document with the CINCSO BLUE SPOON operation order and with what became the airborne corps' OPLAN 90–2 proved largely unsuccessful. Consequently, on the eve of JUST CAUSE, no detailed and seamless campaign plan existed, except as a concept, to facilitate a smooth, timely, and effective transition from armed conflict to peace.[2]

As for the combat units that lacked training and experience in stability operations and nation-building activities, they had little incentive to scrutinize the contents of 90–2's section on Stabilizing Force Operations while they were preparing for a life-and-death struggle with armed Panamanians during the Assault Force Operations phase. The focus was on the battle to come, with most troops either completely unaware or only vaguely cognizant of other, noncombat missions they would be asked to perform. To be sure, units received oral instructions to protect American citizens, innocent Panamanians, and public and private property during the fighting. But with just a few hours left before being placed in harm's way, few recipients of this guidance gave any serious thought to Panama's future beyond the removal of its dictator and the defeat of the enemy force.

[2] First, second, and third quotes from ibid. While an original copy of 90–2 describes a five-phase operation, the After Action Report, Operation JUST CAUSE, 20 Dec 1989–12 Jan 1990, Joint Task Force-South, 19 Mar 1990 (hereafter cited as JTF-South AAR), for Operation JUST CAUSE, forwarded to the Southern Command on 8 March 1990, speaks of operations in only four phases. The "Stabilizing Force Operations," cited as Phase IV in 90–2 is listed as Phase III in the March report. The reason for the discrepancy is that Phase II in 90–2, "In-Place Force Operations," is combined with Phase III, "Assault Force Operations," in the after action report. Remaining quotes from Timothy D. Bloechl, "Operation 'Just Cause': Low Intensity Conflict Stability Operations and Issues of Tactics," *Low Intensity Conflict & Law Enforcement* 2, no. 1 (Summer 1993): 24.

FROM BLIND LOGIC TO PROMOTE LIBERTY

Contingency plans for U.S. civil-military operations in a post-Noriega Panama had their origins in Phase V of the Southern Command's ELABORATE MAZE operation order, written under General Woerner's guidance in March 1988.[3] Initially drafted by two civil affairs (CA) officers on the SOUTHCOM staff, Phase V anticipated what the country would look like at the end of hostilities with the United States: normal government functions would have broken down; various public and private services would have been disrupted; and the Panama Defense Forces, while surviving as an organization, would have been decapitated and shorn of its combat capabilities. In determining the role U.S. armed forces would play in this scenario, planners drew extensively on America's experience in establishing military governments at the end of World War II. From that and other precedents, an operative assumption emerged: the commander in chief of the Southern Command would be in charge of running Panama for thirty days following hostilities. At the end of that period, Phase V read, his "direction and control of selected government functions will be transferred to the US Country Team and/or the [government of Panama]." A second assumption stemmed from the first: the White House would authorize a call-up of Selected Reserve units trained in civil-military operations, especially in the area of civil affairs, to provide qualified American personnel to help run the military government.[4]

In early April 1988, after the Joint Chiefs of Staff directed Woerner to break ELABORATE MAZE into four discrete operation orders, Phase V became KRYSTAL BALL, a code name that was superseded several months later by BLIND LOGIC. During this period, small teams of civil affairs personnel, primarily from the 361st Civil Affairs Brigade, a reserve unit with formal ties to the Southern Command, rotated in and out of Panama on temporary tours of active duty (TTAD) in order to help the four officers in the SOUTHCOM J–5's overwhelmed Civil Affairs Branch write the civil-military operation orders. Their efforts encompassed both the CINCSO version of KRYSTAL BALL and BLIND LOGIC and the immediate supporting plan, the Commander, Civil-Military Operations Task Force (COMCMOTF), version. In August, progress

---

[3] This section draws on the author's *U.S. Military Intervention in Panama, June 1987–December 1989*; and "Panama, 1988–1990: The Disconnect between Combat and Stability Operations," *Military Review* 85 (May-June 2005): 46–52. Also consulted were the works of John T. Fishel, an expert on stability operations and nation building in Panama during JUST CAUSE and PROMOTE LIBERTY. At the time of the crisis, Fishel, a lieutenant colonel in the reserves serving on active duty in SOUTHCOM's J–5 directorate, was intimately involved with various iterations of the planning for BLIND LOGIC. See his *The Fog of Peace*; Fishel, "Planning for Post-Conflict Panama: What It Tells Us About Phase IV Operations," in *Turning Victory Into Success: Military Operations After the Campaign,* ed. Brian M. De Toy (Fort Leavenworth, Kans.: Combat Studies Institute Press, 2004), pp. 169–78; Fishel and Richard D. Downie, "Taking Responsibility for Our Actions? Establishing Order and Stability in Panama," *Military Review* 72 (April 1992): 66–78.

[4] Msg, CINCSO to JCS, 181945Z Mar 1988, sub: USCINCSO/001/ELABORATE MAZE/OPORD 2–88. As for the anticipated call-up of reservists, the president, under Section 673b, Title 10, United States Code, had the authority in 1989 to activate up to two hundred thousand members of the country's Selected Reserve.

on the two orders reached a point at which the planners could set them aside, to be updated as needed.

Then and later, Woerner considered Blind Logic to be "the most important and politically delicate of all the contemplated military actions" contained in the Prayer Book because of the embarrassment contingency plans for a U.S. military government in Panama, a sovereign country, could cause if they became public knowledge.[5] To guard against that possibility, several measures to ensure operations security insulated the planning effort. One of these prohibited each arriving reserve civil affairs team from discussing the evolving plan with the team that it was replacing or with the team that would follow. Similarly, the teams leaving Quarry Heights would be read off the plan, meaning that they would no longer have access to it unless they returned on additional temporary tours. Some of the departing reservists, however, were not only planners but also officers who would be personally involved in overseeing civil-military operations should Blind Logic be executed. In that event, the fact that they might not have seen the plan since their last visit to Panama could prove counterproductive, especially if, in the interim, SOUTHCOM personnel and the other reservists had made extensive changes to the document.

As contingency plans for the Panama crisis were being drafted and revised, Woerner's guidance contained a condition that he regarded as indispensable. In the words of one participant, the general wanted planners to ensure that each of the five Elaborate Maze phases or, later, each of the Prayer Book operation orders, could be executed "independently, concurrently, or in sequence with any other phase" or operation order.[6] The injunction flowed logically from the concept of operations contained in the plans, which reflected Woerner's prudent approach to the crisis. If the United States had to take military action in Panama, the first step, he hoped, could be confined to a gradual buildup of U.S. troops, the psychological effect of which might convince the Panamanian military to oust Noriega. If the buildup alone failed to produce that outcome or if the regime initiated military action that obliged Washington to respond with armed force, the ensuing combat operations would strike only a limited number of targets chosen for their command and control capabilities and their immediate threat to American forces stationed near Panama City. Most PDF units based on the periphery of the capital city would not come under attack, nor would there be an extension of combat operations into distant parts of the canal area or throughout the country at large. By thus restricting the scope of the fighting, but also by hitting the Panama City targets in sequence, the headquarters fighting the war directly under Woerner's command, be it JTF-Panama or a joint task force with the XVIII Airborne Corps in charge, would have sufficient manpower on hand to neutralize the enemy while securing the capital's business areas and neighborhoods.

Under this concept of operations, the execution of the Post Time operation order, which contained the U.S. troop buildup, and the Blue Spoon plan for combat operations would most likely take place sequentially, with the Klondike Key plan for evacuating noncombatants being conducted in either

[5] Fishel, "Planning for Post-Conflict Panama," p. 172, and see also p. 171.

[6] Ibid., p. 170.

phase—if, in fact, Blue Spoon and Klondike Key needed to be implemented at all. If events did result in hostilities, then, ideally, Blue Spoon and Blind Logic would be executed simultaneously, since the need to conduct stability operations—contrary to planning schemes that emphasized sequenced or linear phasing—would surface with the onset of combat operations, not just after the fighting had stopped.

Essential to achieving the capability Woerner stipulated for carrying out the operation orders independently, concurrently, or in sequence were organizational and procedural arrangements that would promote close cooperation and coordination among the various and often dispersed groups of commanders and staff officers involved in drafting the plans. At the Southern Command, the highest level at which U.S. contingency planning for Panama was taking place, officers assigned to a specific top secret and compartmentalized strategic-level Prayer Book order would have to do the best they could, given the obstructions and restrictions under which they worked, to furnish progress reports to colleagues writing each of the other plans. Complementing this lateral flow of information would be a vertical exchange whereby staff directors at Quarry Heights would monitor and provide advice and feedback to subordinate headquarters charged with writing the supporting operational-level plans that sought to convert the general guidance contained in the CINCSO operation orders into tasks and assignments that units at the tactical level could use in writing their more detailed plans for specific missions. In the case of Blue Spoon, the vertical communications network consisted of separate channels running from the SOUTHCOM J–3's plans shop downward to three planning groups on two installations: JTF-Panama at Fort Clayton, special operations staff at Fort Bragg, and, beginning in early 1989, the XVIII Airborne Corps, also at Bragg.

The coordination requirements for Blind Logic were somewhat different. Because the strategic CINCSO version of the plan and the operational-level COMCMOTF supporting plan had both been written at Quarry Heights, they were largely in sync. That fact, together with the security restrictions imposed on the reserve teams once they left Panama, precluded a continuous exchange of information with U.S.-based headquarters specializing in civil-military operations. Instead, what SOUTHCOM planners working on Blind Logic wanted was some form of meaningful interaction with their counterparts drafting the sections on civil-military operations in the CINCSO and joint task force versions of Blue Spoon. In the drafts of both Blind Logic and Blue Spoon, general assumptions concerning the immediate and negative impact of hostilities on Panama were pretty much the same, as planners working separately on each version of each plan considered the possibility of looting, an upswing in criminal activity, vigilantism, die-hard resistance, and even terrorism and insurgency. Thus, at the strategic level, the principal task was simply to fine-tune the language contained in both operation orders. More substantive issues emerged in trying to harmonize the supporting plans. Since Blind Logic called for the establishment of a civil-military operations task force to be run by the SOUTHCOM J–5 director, a brigadier general, questions requiring answers included how that task force would interact with the joint task force executing Blue Spoon, and how civil affairs forces that, in early drafts of JTF-Panama's BLUE SPOON, would be attached to combat units on the eve of hostilities would interact with the civil affairs elements working for the Civil-Military Operations Task Force.[7] Other operational issues demanding coordination involved listing all the units that would be required to conduct stability operations while the war-fighting joint task force was still focused on winning the war; scheduling the timely arrival of those units if they were deploying from the United States; determining the point at which operational control of the troops engaged in stability operations under the BLUE SPOON joint task force would be transferred to the Civil-Military Operations Task Force commander; and assessing what the most effective command and control relationship would be between the war-fighting joint task force and the Civil-Military Operations Task Force at different phases of the military operation. Unless the stability operations called for in BLUE SPOON and BLIND LOGIC could be synchronized during the planning stage, the actual process of providing security for a postconflict Panama while returning it to the status of a functioning state could entail needless delay, hardship for the country's citizens, and prolonged disruption in restoring the country's institutions, basic government services, and social and economic well-being.

Unfortunately, while the need for cooperation and coordination was imperative, the obstacles to achieving those goals were daunting. Because each PRAYER BOOK operation order was compartmentalized, it could only be accessed on a need-to-know basis. This prevented a group writing one plan from examining the progress of a group working on another, even though at Quarry Heights, two groups drafting separate plans were likely to be only a short walk from each other. In the case of BLUE SPOON and BLIND LOGIC, early in 1989 the two plans did reside together in the SOUTHCOM J–3 directorate, the result of a command reorganization, but in May of that year, BLIND LOGIC was returned to the J–5 because of "the relative qualifications of J5 personnel required for its execution and the ongoing relationship between the J5 and the 361st [Civil Affairs Brigade]." Specifically, responsibility for the plan fell to the directorate's Policy and Strategy Division under Lt. Col. John T. Fishel, a reserve officer on an extended tour of active duty. Because that division oversaw a variety of activities not related to the Panama crisis, Woerner supported creation of a BLIND LOGIC "workable planning cell," consisting of Fishel and another reservist. In time, six more officers joined the group, three from the active component's 96th Civil Affairs Battalion and three from civil affairs units in the reserves. Fishel later recorded that, of the six, "three were language qualified . . . while only two of the officers involved had worked on the plan in 1988."[8] Even so, a group now existed to concentrate on BLIND LOGIC, albeit in a separate directorate from those drafting the CINCSO BLUE SPOON.

Besides the security and organizational barriers to coordinated planning for stability operations, long-standing military practice posed a significant if less tangible obstacle. Traditionally, officers responsible for planning conventional

---

[7] The civil affairs forces are discussed in Briefing Slides, JTF-Panama, n.d. [1988].

[8] First and second quotes from Fishel, "Planning for Post-Conflict Panama," p. 171 (see also p. 172). Last quote from Ms, John T. Fishel, War, Peace, and Civil Military Operations: The Role of Civil Affairs in Operations JUST CAUSE, DESERT STORM, and PROVIDE COMFORT, May 1992, copy in author's files.

combat operations have tended to focus almost exclusively on the battles to be fought, while slighting or even ignoring the requirements for stability operations that those battles generate. Such tasks, according to an oft-repeated phrase, were "someone else's job." To be sure, a hard logic has underpinned this well-established approach: should U.S. forces fail to achieve victory, the question of whether or not stability operations and nation building received adequate attention while planning the war would become moot. Added to this has been the fervent desire to devise a plan for defeating the enemy with as few friendly casualties as possible. When Blue Spoon planners took into account the number and diversity of U.S. combat units that would see action in Panama and then factored in the innumerable ways in which war's fog and friction might affect those units adversely, the natural impulse was to spend every available minute working to minimize the inevitable risks and dangers, even if that meant neglecting the Stabilizing Force Operations portion of the combat plan and dodging the requests for coordination emanating from the officers working on the civil-military operations plan.

As late as mid-1989, with the CINCSO version of Blue Spoon being written within his own headquarters and with the JTF-Panama supporting plan being drafted just ten minutes away at Fort Clayton under a general officer answerable directly to him, Woerner was in a position to impose some degree of coordination between the officers drafting Blue Spoon and those drafting Blind Logic. Fishel went so far as to characterize the general's intervention to this end as pivotal. "If General Woerner had not taken such personal interest in postconflict operations, there is no way that the staff elements responsible for *Blue Spoon* would have devoted any time to the necessary coordination with *Blind Logic*."[9] The question remained, however, as to whether Woerner could exact the same level of cooperation from the XVIII Airborne Corps at Fort Bragg when, during Nimrod Dancer, that headquarters began to assert its role as the executive agent for writing an operational-level version of Blue Spoon, one that would supersede the work of JTF-Panama if the corps ended up forming the conventional joint task force for directing combat operations in Panama. Woerner's feelings toward the Army's contingency corps were mixed. On the one hand, he welcomed the staff support it could provide the planning effort; on the other, he perceived the headquarters as composed of "outsiders" unfamiliar with the subtleties and nuances of the situation in Panama. The corps, by its own admission, preferred planning and executing combat operations that incorporated the rapid insertion of U.S. troops and firepower, the use of decisive force to achieve clearly stated objectives, and the quick withdrawal of the troops once the shooting had ended. While Woerner admitted that this approach might be appropriate for certain contingencies, he did not consider it the best prescription for Panama, where the idea of an all-out American invasion ran counter to his own stratagem for a gradual troop buildup and, if necessary as a last resort, U.S. military strikes against a limited number of PDF targets. But the XVIII Airborne Corps had its reputation and way of doing business. Unless he provided it with firm guidance, he anticipated receiving a supporting plan that, among other shortcomings, would pay only

[9] Fishel, "Planning for Post-Conflict Panama," p. 172 (and see also p. 173).

lip service to the civil-military operations contained in the CINCSO version of Blue Spoon.

At first, Woerner enjoyed some success, as, in Fishel's words, the general's "command emphasis forced coordination with the Corps when it was brought on board as execution planner for Blue Spoon." The corps, one could argue, had little choice. Despite its dislike of Woerner's gradual and limited strategy for a possible military showdown in Panama, the Joint Chiefs had designated the Southern Command as the supported headquarters for contingency planning, and the XVIII Airborne Corps as a supporting headquarters. The corps commander and his staff would have to comply with Woerner's enjoinders—"we had to salute the flag," as one of them put it—which included meeting with Fishel and the Blind Logic planning cell.[10] The discussions took place in Panama during the early weeks of Nimrod Dancer after General Stiner had dispatched a three-man team led by Maj. David Huntoon Jr. to help determine how the corps should respond to the intensifying crisis. Fishel and his group, knowing that the corps was now giving serious attention to preparing its supporting plan for Blue Spoon, wanted assurances that Huntoon and the staff officers he represented understood the ramifications flowing from the civil-military operations portion of that plan. Specifically, in a rehash of issues the J–5 planners had at one time discussed with JTF-Panama personnel, Fishel needed to be sure that the corps would make provisions for the timely arrival of military police, engineers, medics, civil affairs units, and reserve personnel essential to the process of restoring law and order and government services during and after hostilities. Command and control arrangements for these forces as well as for U.S. combat troops engaged in stability operations also received careful scrutiny. When the talks concluded, the participants had reached some tentative agreements. Huntoon's team returned to Fort Bragg soon thereafter, leaving the J–5 planners at Quarry Heights confident that the XVIII Airborne Corps now fully understood "the law and order mission and emergency service restoration mission" it would have to conduct if it was activated as the conventional joint task force headquarters executing Blue Spoon.[11]

During this period in mid-1989, Fishel's planners were also engaged in a general review of the CINCSO and the COMCMOTF versions of the Blind Logic operation order. One of the assumptions reflected in drafts from the previous year drew their immediate attention: if war broke out, there would be a presidential call-up of Selected Reserve units trained in civil-military operations. By mid-1989, according to the "political sensing" of one of the J–5 planners, convincing the Joint Chiefs to recommend such a call-up "might well be more difficult than expected." The Blind Logic group therefore devised an alternative that would allow the deployment of twenty-five "pre-selected volunteers," mainly from the 361st Civil Affairs Brigade, to perform the civil-military mission. According to Fishel, "These individuals . . . would have gone on active duty for 31 days under TTAD orders. There was sufficient

---

[10] Ibid., p. 172. The "salute the flag" phrase was said to the author during a conversation with Lt. Col. David Huntoon at Fort Bragg, North Carolina, sometime in 1993.

[11] Fishel, *The Fog of Peace*, pp. 20–24.

redundancy among the volunteers to make certain that a full team would have been activated. This approach would have provided the minimal capability to execute the mission along with in-place forces."[12]

The review of Blind Logic left most other significant parts of the plan unchanged, including the assumption that the SOUTHCOM commander would briefly act as the U.S. military governor of a postinvasion Panama. As things turned out, this assumption proved false. At some point in 1989, President Bush decided that, if Blue Spoon were executed, the three opposition candidates who had clearly won the May presidential election in Panama—the results of which Noriega had annulled—should make good on their victory by taking office immediately, even before U.S. forces began their attacks. The United States would do all it could to support the new government and to get it functioning, but there would be no interim period necessitating the establishment of a CINCSO-led military government. That the president's decision did not percolate down to SOUTHCOM's J–5 and into the revised Blind Logic would lead to problems on the eve of Operation Just Cause. If there was to be no U.S. military government, to whom would the military organizations and personnel involved in stability operations and nation-building activities ultimately be accountable?

Odds were that, on paper at least, these military endeavors would fall under the jurisdiction of the new Panamanian government, with the American ambassador serving as the senior U.S. official involved (a prospect that would require reaching some accommodation with the SOUTHCOM commander, whose chain of command did not run through the State Department). Yet, so long as the assumption persisted at Quarry Heights that there would be a U.S. military government in Panama, operations security ensured that routine requests by Blind Logic planners to coordinate their efforts with the political counselor in the U.S. Embassy in Panama City were denied, in Fishel's words, "on the grounds that the plan was classified and controlled within the JCS channels only and, therefore, could not be coordinated with any non-DOD agencies." The best the planners could obtain was permission to talk with the political counselor "on a hypothetical situation but not to reveal the existence of the plan."[13]

Amid this flurry of drafting and updating plans, the Pentagon announced that Woerner would be retiring as the SOUTHCOM commander, to be replaced by General Thurman at the end of September. The impact of the change in command on Blind Logic was enormous, once Thurman authorized the new concept of operations for Blue Spoon. In addition to calling for the overthrow of the Noriega regime through a surprise attack, what became 90–2 also added the goal of destroying the Panama Defense Forces as an institution. This change and those necessitated by the 3 October coup attempt caused several more PDF units and facilities to be added to the D-day target list. Consequently, combat units that under the original Blue Spoon would have swept through Panama City—thus being in a position to prevent looting and various criminal activities in the capital's business centers and neighborhoods—

[12] Fishel, War, Peace, and Civil Military Operations.

[13] Ibid.

would now be dispersed in and outside the capital so that they could assault targets that the former SOUTHCOM commander had considered peripheral and nonthreatening. As formally approved by the Joint Chiefs on 3 November, JTF-South OPLAN 90–2 did contain the mission statement that U.S. troops should be prepared "to restore law and order, and support the installation of a U.S.-recognized government of Panama."[14] Yet, given the plan's expanded combat operations, there was little in the operational annexes to educate troop commanders as to what specific tasks they might have to perform—and how to perform them—in support of the stabilization mission.

All of this and more worried the Blind Logic planners at Quarry Heights. Further coordination with the XVIII Airborne Corps was essential, they believed, but Thurman did not move to arrange the kind of face-to-face session that his predecessor had mandated during Major Huntoon's Nimrod Dancer visit. Indeed, between the beginning of October and the execution of Just Cause, the general never asked for a briefing on Blind Logic, while the Blind Logic planners, preoccupied with other duties, never prepared one in the absence of a formal request. As for the XVIII Airborne Corps, when Stiner wanted to know who was responsible for "conflict termination," Thurman replied, "Not you. That is the USARSO staff. I want you to focus on combat operations."[15] Stiner complied, to the consternation of Fishel's planning cell, which soon discovered that the agreements they thought they had reached with Huntoon's team months earlier were being either ignored by the corps' planners as a result of Woerner's departure or disregarded as irrelevant because of changes in the war plan.

In addition to the SOUTHCOM planners working on Blind Logic, there was another group of staff officers concerned over how combat and civil-military operations would mesh together in the event of armed conflict in Panama. Under the troop deployment schedule contained in OPLAN 90–2, the two brigades of the 7th Infantry Division still at Fort Ord were not likely to arrive in the country until most of the heavy fighting had subsided. This suggested that the light fighters would be heavily involved in the Stabilizing Force Operations phase of Just Cause. Given that probability, one of the division planners responsible for the 7th's version of Blue Spoon contacted Huntoon at Fort Bragg and recommended that the coordination of combat and civil-military operations be placed on the agenda for the planning session slated to be held in Panama in mid-December. Huntoon readily agreed, but when Lieutenant Paz was killed two days before the conference was to begin, the scheduled session turned into a war council, with last-minute adjustments to the Assault Force Operations portion of the plan dominating the proceedings.

One change in Blue Spoon that Thurman himself had personally initiated upon taking over at Quarry Heights had a direct, if not immediately obvious, impact on plans regarding the Civil-Military Operations Task Force called for in Blind Logic. By removing the Joint Special Operations Task Force

[14] JTF-South OPLAN 90–2, 3 Nov 1989. In a reversal of normal planning procedures, the updated CINCSO version of Blue Spoon, Operation Order 1–90, had been revised to reflect, in more general terms, the contents of the supporting plan, 90–2.

[15] Interv, Lt Col Steven P. Bucci with Gen Carl W. Stiner, U.S. Army (Ret.), 1999, USAWC/USAMHI Senior Officer Oral History Program.

from the operational control of the SOUTHCOM commander and putting it under JTF-South, Thurman had sought to streamline BLUE SPOON command and control arrangements. The point was not lost on the BLIND LOGIC planners. With "all the executing elements" now under the war-fighting joint task force, "unity of command" could be further strengthened by taking the Civil-Military Operations Task Force and having "an entity under the command and control of JTF-South" take charge of it. For several reasons, the "logical candidate" was General Cisneros' U.S. Army, South, a headquarters, the J–5 officers reasoned, that had dealt with the crisis since its inception and, equally important, that would "be without a mission in the event of action in Panama." In the planners' opinion, most duties USARSO was performing while wearing its JTF-Panama "hat" would be assumed by JTF-South, which would also absorb Cisneros and his staff. They would serve as deputies to Stiner's people, who were likely to regard them as excess baggage. So why not call on them to run the Civil-Military Operations Task Force under Stiner? After some preliminary discussions between the USARSO and SOUTHCOM staffs, the SOUTHCOM J–5 director, Air Force Brig. Gen. Benard W. Gann, formally requested that Thurman give Cisneros the civil-military operations mission.[16]

General Gann

Thurman consented, but the proposed transfer quickly stalled. In early December, when Col. Norman Higginbotham, the acting USARSO chief of staff, received a copy of BLIND LOGIC, he asked Maj. Les Knoblock to review the document and provide feedback. A week later, Higginbotham, back in his position as deputy chief of staff for logistics, together with USARSO Chief of Staff Col. Leonard Hardy Jr., met with SOUTHCOM's deputy J–5 director to discuss the transfer. According to Higginbotham, he and Hardy told the J–5 officer that the current version of BLIND LOGIC was "full of crap"; that it contained too many false assumptions, such as the establishment of a U.S. military government; and that, as written, it could not be executed. What was needed, they maintained, was a new operation order, a USARSO plan that would be coordinated with the U.S. Embassy and other American governmental agencies in Panama. Given the problems that the SOUTHCOM J–5 had already experienced in trying to obtain

[16] Fishel, *The Fog of Peace*, pp. 27–28.

Pentagon approval for coordination outside military channels, the redrafting of Blind Logic along the lines Hardy and Higginbotham suggested promised to be a difficult and time-consuming process. The prospect of further delay, however, caused little concern at the time. As Fishel recalled, "There was no great sense of urgency about any of this—the assessment of the probability of execution remained low."[17]

Such was the status of the Prayer Book planning for civil-military operations when President Bush made his decision on 17 December to use armed force in Panama. "At that point," according to Fishel, "the SOUTHCOM staff 'discovered' *Blind Logic*." There followed two days of "frantic activity" in which planners tried to harmonize that operation order with OPLAN 90–2 and with certain political realities. It could not be done. One USARSO officer summoned to Quarry Heights to review the Blind Logic command and control arrangements, as well as the transition from combat to civil-military operations, assessed the J–5 plan as "a disaster waiting to happen." With H-hour looming, the SOUTHCOM J–5 director and staff had little choice but to draft as quickly as possible a shorter version of the command's operation order; attach the still highly relevant annexes from the existing plan; give the on-order mission of commander, Civil-Military Operations Task Force, to the commander, JTF-SOUTH (not the USARSO commander); and, over Thurman's signature on 20 December, send Operation Order 2–90 (Blind Logic) to the Joint Chiefs for approval, several hours after the fighting in Panama had commenced. Despite the last-minute failure to save the original plan, Fishel could take comfort in knowing that much had been salvaged from it. The CINCSO version, he noted, "provided alternative blueprints for force structure and command and control of post-conflict operations. The COMCMOTF plan provided checklists of things that would need to be done by those forces to restore a functioning government to Panama."

The message to the Joint Chiefs spoke of the likely security vacuum and breakdown of law and order, public safety, and some basic infrastructure services resulting from a combination of Panama's preinvasion economic decline and the demise of the Panama Defense Forces that Just Cause would effect. The threat to law and order would come from former PDF members, paramilitary personnel, hostile third-country nationals, and criminals and could take the form of guerrilla warfare, terrorist attacks, sabotage, and criminal activity. Given this threat, Operation Order 2–90 authorized Thurman to establish the Civil-Military Operations Task Force under a military commander to protect American lives and property, ensure law and order, and provide essential services. Simultaneously, the Southern Command would conduct nation-building projects in support of the new government in Panama.

The next day, the Joint Chiefs approved the plan. Looking beyond the combat operations that were winding down, they concurred that the Civil-Military Operations Task Force would come under JTF-Panama upon the inactivation and departure from Panama of JTF-South. In a setback the J–5 planners had anticipated months before, the chiefs refused to ask the White

[17] Second quote from ibid., pp. 27–28. First quote from Interv, author with Col Norman D. Higginbotham, U.S. Army, 20 Jun 1990, Fort Clayton, Panama.

House for the presidential call-up of reserve civil affairs units, including the 361st Civil Affairs Brigade and, under it, the 478th Civil Affairs Company. But they did approve the SOUTHCOM J–5's alternate plan for relying on twenty-five preselected volunteers and additional reservists if they were needed. There was a hitch, though, that invalidated much of what the alternative sought to accomplish. The chiefs directed that the volunteers would have to request tours of 139 days, not 31 days, as BLIND LOGIC recommended. This, according to Fishel, "played havoc with the planning of the 361st, which had teams prepared to go for 31 days but could not, in general, volunteer for the longer 139 days." On 22 December, the Joint Chiefs weighed in again, this time approving the recommendation from the SOUTHCOM J–5 to change the name of BLIND LOGIC to PROMOTE LIBERTY.[18]

Thurman regarded the Joint Chiefs' approval of the revised civil-military operations order as an inconvenient but necessary formality. Faced with the need for immediate action, he had in essence already executed portions of the order on his own authority on the morning of 20 December. Because the handover of the civil-military mission to USARSO had not been accomplished prior to D-day, the SOUTHCOM commander told General Gann to move most of his staff to Panama's Legislative Assembly building, down the hill from Quarry Heights, and to begin helping the Endara government. Thurman also directed Gann to provide U.S. Chargé d'Affaires in Panama John Bushnell "what he needed to assist the newly inaugurated Panamanian government, as well as such additional support that might be required." If Thurman did not formally activate the Civil-Military Operations Task Force at that moment, his instructions giving the J–5 staff its new mission and putting Gann in charge of it had the same effect. Moreover, once established, the task force would be a separate headquarters under Thurman, although it would share a liaison officer with JTF-South. By midmorning of the twentieth, two-thirds of the J–5 staff was en route to the assembly building.[19] By that time, many U.S. troops in Panama were already conducting law and order and stability operations in the field.

THE TRANSITION IN THE FIELD: THE TROOPS AND THE PEOPLE

At Rio Hato, midday on 20 December, 1st Lt. David Haight, a rifle platoon leader with Company B, 2d Battalion, 75th Ranger Regiment, had finished a night of combat—the "easy part," as he would later describe it—and was

[18] Activities surrounding BLIND LOGIC on 17–22 December and the first, second, and fourth quotes are from Fishel, "Planning for Post-Conflict Panama," pp. 173, 175–76. Third quote from Interv, author with Lt Col Robert Pantier, U.S. Army, 13 Jun 1990, Fort Clayton, Panama. Last quote from Fishel, *The Fog of Peace*, pp. 31–32. Msg, CINCSO to JCS, 202055Z Dec 1989, sub: USCINCSO OPORD 2–90 (BLIND LOGIC); Fact Sheet, Maj Robert A. Richardson to Deputy Chief of Staff for Operations (DCSOPS) [USARSO], 15 Mar 1990, sub: Research Reason For Lack of JTF-PM Supporting Order in Panama. This last document, a report to Brig. Gen. Joseph W. Kinzer, the occupant of the newly created position of deputy commanding general, USARSO, contains a chronology of the BLIND LOGIC/PROMOTE LIBERTY messages between the Southern Command and the Joint Chiefs of Staff on 20, 21, and 22 December 1989.

[19] Fishel, *The Fog of Peace*, pp. 32–33.

engaged in mopping-up operations when he learned that he and his men were to be transported to a small town north of Panama City.[20] There, as part of a company-size operation, they were to maintain order and engage in various administrative duties. As he listened to a colonel go over the platoon's new assignment, Haight realized that none of the various tasks being enumerated related to anything his unit had been trained to do. Nevertheless, he and thirty-seven of his men prepared to move out. When they arrived at their destination, they discovered "the town was a mess." "Murder, rape, pillage" numbered among the worst offenses, the result of there being "no one in charge." Over the next three weeks, the Rangers assisted in reorganizing the community and helping the residents cope with the temporary disruptions and uncertainties created by the invasion. In the process, the troops would serve as policemen, politicians, logisticians, sanitation workers, counselors, and more—almost everything but warriors. Consequently, they had to adjust psychologically from the "kill and leave" mind-set their training and professional self-image had instilled in them to the unexpected and unwelcome reality of being ordered to stay and be "polite."

Despite the abruptness of the transition and the perplexity of the troops, there was, in Haight's mind, "no room for mistakes." One of the first orders of business was simply finding out who the Rangers could trust and who they could not, who could best aid them in their mission and who was most likely to obstruct progress. The mayor of the town was still in residence but with little real authority, given that Just Cause had eliminated his police force. After assessing the situation, Haight's soldiers helped restore the mayor's credibility and political legitimacy, two prerequisites to his being able to govern again. As Haight described the process, the Rangers would tell the official what to say and then publicly demonstrate their support for him after he had said it. The success of this approach took time, during which the Rangers had to deal with crime, security issues, trash collection, plumbing problems, and, on occasion, the mediation of personal disputes.

All of this transpired in an urban environment characterized in part by the ubiquitous civilian. Most of the residents were friendly, but that often proved a mixed blessing, since their presence sometimes interfered, unintentionally, with the Rangers carrying out their duties. There were, for example, the curious children, groups of whom seemed to be everywhere. Early on, Haight's men decided it prudent to keep the kids following them around at arm's length. The same approach, however, could not be taken with the town's adults. Their cooperation was essential, yet, they could be the source of too much information or conflicting reports. They could also be overzealous in showing their appreciation for what the Rangers were doing, as when they would organize one social event after another, prompting Haight to warn his men about the consequences of fraternization. Some residents were not so friendly, and on two occasions, members of the platoon came under sniper fire. In response, the lieutenant told the mayor that, if another such incident

[20] The account that follows and the quotes in this and the next three paragraphs are from Presentation, Capt David Haight, author's History of Modern Warfare course, n.d. [early to mid-1990s], U.S. Army Command and General Staff College.

occurred, the Rangers would seize the building from which the fire emanated and kill everyone in it. The threat was a bluff, of course; moral and legal considerations alone precluded such a response, as did the postconflict rules of engagement. Not that the full range of circumstances under which deadly force might be used was all that cut and dried. According to Haight, elements from the battalion's Company C, also located in the town, opened fire on drunk drivers on four different occasions, killing three of the four inebriates.

Over the course of three weeks, U.S. military policemen, civil affairs officers, and psychological operations personnel augmented the ranger force and stayed behind when Haight's platoon pulled out of the town in mid-January prior to redeploying to the United States. The unit, in its leader's opinion, had accomplished its mission. The mayor's authority had been bolstered, and law and order had been restored. But getting those results had not been easy for the rangers. While Haight acknowledged that over time a series of prolonged stabilization missions of the sort he had been given for the town in Panama could lessen his rangers' combat skills, he also asserted that the Army would have served his men better by forgoing at least "a few iterations of assault training" in favor of instruction covering the civil-military-type tasks his platoon had been called on to perform.

From one side of the canal area to the other, thousands of U.S. soldiers and marines in the field shared Haight's sentiment as they began conducting a variety of on-order and follow-on missions in the wake of their combat achievements. In the days and weeks after 20 December, most of these troops would engage in some form of stability operations, with the details of each unit's experience depending on time, place, and circumstances. Diversity prevailed. For example, to help secure and stabilize a location, some units, like Haight's, had to be transported far from where they had engaged the enemy with armed force. Other units, such as those in Colonel Snell's Task Force Bayonet, simply moved outward from their assault targets into adjoining neighborhoods, partly in search of fleeing PDF and Dignity Battalion members and partly to conduct patrols aimed at restoring or maintaining law and order. Some neighborhoods Snell's troops entered had hardly been touched by the fighting; others had not been so fortunate, with burned-out El Chorrillo suffering the worst damage. There were also units, like the company from the 82d at Las Tinajitas, that did not have to leave the target area at all before encountering civilians who, as soon as the shooting stopped, began beseeching the soldiers to provide food and medical treatment, restore electrical power, and perform additional "unfamiliar tasks" until the newly established Endara administration could begin running the country effectively.[21]

---

[21] Interv, Capt John Hollins with Lt Col Billy Ray Fitzgerald, U.S. Army, 20 Jun 1990, Fort Clayton, Panama. Quote from CALL Bull 90–9, Operation Just Cause Lessons Learned, vol. 2, 1990, p. 23. Concerning the elements of the 82d Airborne Division, in General Cisneros' opinion they did not move into Panama City quickly enough to help restore law and order to the areas hit by looting and criminal activity. He cited their unfamiliarity with the area, rather than any pressing civil-military operations at Las Tinajitas or Panama Viejo, as the reason for their slow response. Interv, author with Maj Gen Marc A. Cisneros, U.S. Army, 29 Jun 1990, Fort Clayton, Panama.

If the distance one had to travel before beginning civil-military operations varied from unit to unit, so, too, did the time spent at a given location once the troops arrived. While Haight's platoon stayed three weeks in the town assigned to it, in other cases units received orders to secure an area for just a matter of hours or for a day or two before moving on. During the first several days into the invasion, the frequency of unit relocations was based in part on quickly changing circumstances, with troops being transferred here and there as real or potential security threats became apparent. In these instances, there was often little time to become acquainted with the local population or to acquire the level of information and expertise needed to perform the prescribed duties. Complicating matters further, the tasks themselves changed—or at least the emphasis they received did—as, over a period of days and weeks, nation building became the priority mission, even though concerns over maintaining security persisted. As Brig. Gen. Joseph W. Kinzer of the 82d Airborne Division observed, "The search and clear combat piece vis-à-vis the nation assistance piece began to disconnect. Fewer and fewer combat-type operations were planned and executed, and we got more into nation assistance—those kinds of programs that were conducive to assisting in restructuring or reconstructing the new government."[22] At the tactical level, the impact of this shift could be disruptive. Combat troops, for instance, who had just begun to develop a degree of proficiency in screening traffic at the checkpoint they manned could find themselves discarding those newly acquired policing skills overnight in order to learn the ins and outs of running a successful food distribution program.

As civil-military operations continued throughout late December and into January, the frequent movement of troop units from one location to another often deposited forces from vastly different organizations next to one another without sufficient coordination in advance (a common occurrence during combat operations as well). The risks of friendly fire inherent in such placements, especially if communications between adjacent units had not been well established and if variations in operating procedures between them had not been clearly identified, could be high. An incident on 23 December brought home this point. A hastily drawn boundary line in Colón separated a company of the 82d Airborne Division from elements of the 7th Infantry Division. On the night in question, one of the airborne company's security patrols came under fire from light fighters engaged in a similar activity. No one was hurt, and, afterward, the paratroopers "raised hell about it," receiving the explanation, "Well, we saw you had weapons, so we figured you were enemy." Officers from both headquarters worked to prevent a repetition of the incident, concluding in the process that the "importance of fire control and friendly boundaries can't be understated."[23]

---

[22] Interv, Dolores De Mena with Brig Gen Joseph W. Kinzer, U.S. Army, 20 Apr 1992, Fort Clayton, Panama. At the time of Operation JUST CAUSE, Kinzer was the 82d's assistant division commander for operations. He returned to Panama in February 1990 as U.S. Army, South's deputy commanding general.

[23] Quotes from Briggs, *Operation Just Cause*, p. 109. Interv, author with Lt Col Johnny Brooks, U.S. Army, 11 May 1993, Fort Leavenworth, Kans.

By early January 1990, the movement of troop units conducting stability and nation-building operations, especially into Panama City, began to reflect the redeployment schedules of the stateside forces who had taken part in the invasion. The Rangers were the first to leave—the 1st Battalion, 75th Ranger Regiment—on 3 January, the 2d Battalion a week later. As the 2d Battalion redeployed to Fort Lewis, units from the 82d Airborne Division began their return to Fort Bragg. Before they moved out, they were relieved from their positions in and around Panama City by the two brigades of the 7th Infantry Division located on the Pacific side of the country. For a week, until the 7th's 1st Brigade redeployed to Fort Ord on 17 January, the two headquarters shared responsibility for securing and stabilizing most of the capital. The 2d Brigade remained in Panama until 6 February, where it "expanded operations to the east toward the Colombian border . . . to show a strong U.S. presence, support for the new Panamanian government, and neutralize any remaining PDF elements." With its departure and, on 3 February, that of the division's 3d Brigade, which had constituted the core of Task Force Atlantic, the principal stateside combat units had returned to their home bases, leaving the U.S. conventional and Special Operations Forces stationed in Panama to follow through with ongoing stability and nation-building activities.[24]

While the elements of time, place, and circumstance made the conduct of civil-military operations a unique experience for each combat unit during Operations JUST CAUSE and PROMOTE LIBERTY, there were, in a broader context, certain themes and patterns that emerged. By 21 December 1989, D plus 1, almost all units had become deeply involved in performing basic security and law and order tasks that required continuous patrolling, search and clear operations, and the manning of roadblocks and other checkpoints. On Saturday, 23 December, General Thurman issued his priorities for U.S. military activities in Panama, the first being to establish stability in the cities. By then, many U.S. troops had already spent three days, in the words of one company executive officer, "maintaining the peace, searching out PDF and Dignity Battalion members, arresting drunkards, stopping looters, and settling domestic disputes." Another officer included among his list of immediate concerns the need to "resolve hostage situations; capture weapons and munitions caches; secure US and Panamanian facilities." And there were more. Besides security issues, Thurman's priorities also encompassed a variety of essential services that had to be restored in Panama—"removal of debris, reopening of schools and businesses, distribution of food, establishment of a magistrate system, establishment of combined patrols (US Army Military Police and reorganized Panamanian police forces), and restoration of public utilities." These reflected the shift to nation assistance activities to which Kinzer later referred. Prior to their redeployment to the United States, paratroopers, Rangers, and light fighters were not exempt from these longer-term restoration projects, even though most of the stateside units would not have the capabilities

[24] Dates for the redeployment of U.S.-based troops and quote are from CALL Bull 90–9, Operation Just Cause Lessons Learned, vol. 1, pp. 6–13. On the 82d's turnover of its area of operations in Panama City to elements of the 7th Infantry Division (Light), see Interv, De Mena with Kinzer, 20 Apr 1992.

or resources to become involved in more than one or two such large-scale and time-consuming enterprises.[25]

But whether engaged in on-going security operations or highly specialized restoration programs, almost all U.S. troops shared one common experience: daily contact with the Panamanian people, some operating in an official capacity, but most just wanting to show their appreciation for the service the Americans had rendered in ridding the country of its dictatorial regime. On both sides of the isthmus, the scene was similar. As the Americans entered a town or village, the locals would turn out cheering and waving U.S. and Panamanian flags. More than a few soldiers and marines recalled accounts of World War II, when their predecessors in western Europe and certain areas of the Pacific had been greeted as liberators by crowds behaving in the same jubilant way. Over forty years later in Panama, the assemblages may have been smaller, but the story was the same. For days, perhaps weeks, after a unit arrived at its assigned location, it would experience a near-continuous outpouring of goodwill manifested in handshakes and pats on the shoulder, deliveries of homemade food, offers of beverages of all sorts, invitations to parties, and groups of curious and grateful people just trying to be helpful.

The troops appreciated the show of gratitude, but, as Haight had discovered, too much of a good thing could at times be a nuisance, as when groups of people inadvertently obstructed a routine patrol or when U.S. officers had to spend time discouraging fraternization between their young soldiers and the friendly Panamanian women in their midst. There was, of course, a host of much more serious problems. At least two officers in different locations recounted how children playing with toy guns jumped from hiding places to shoot at passing U.S. soldiers. The discipline of the kids' targets prevented a tragic outcome in each case, but, with split-second decisions being made, the results could easily have been otherwise, especially in an environment that still harbored a variety of dangers. Despite the overwhelming support the Americans enjoyed from the Panamanian people, there remained enemies at large with real weapons waiting for a chance to do harm to those who had vanquished them. The violence unleashed by JUST CAUSE, U.S. troops had to constantly remind themselves, was not necessarily over. True, no other units would find themselves in the kind of firefight that engulfed Task Force Wildcat at the DNTT on 22 December. But for an extended time that lasted well after D-day, sniper fire, drive-by shootings, and attempted ambushes were almost routine occurrences. There was also speculation that remnants of the Panamanian military and Dignity Battalions were planning some form of resistance, a "Black Christmas" counterattack, as some rumors phrased it.[26]

The attack never came, but a serious incident did occur shortly before the holiday that brought home the dangers still facing U.S. forces. On 23 December, paratroopers of Company D, 3d Battalion, 504th Infantry, at Madden Dam stopped a vehicle with five Panamanians inside. Seeing a tear gas grenade on the front seat, the Americans ordered the men out of the vehicle and began

[25] Author's notes, n.d. First quote from Briggs, *Operation Just Cause,* p. 95. Second quote from Bloechl, "Operation 'Just Cause,'" p. 23. Last quote from JTF-South AAR, summary section, p. 8.

[26] Interv, Hollins with Fitzgerald, 20 Jun 1990.

searching them. At that point, one of the suspects lobbed a grenade at his tormentors. Nine paratroopers were wounded. The reaction was immediate. "All hell broke loose," according to one account. "After the shooting stopped, . . . all the vehicle's occupants were presumed dead." When one moved, however, a noncommissioned officer opened fire and killed him. The psychological carryover of the incident may have contributed to the shooting later that night of a Panamanian driver as he tried to flee a U.S. checkpoint at the dam. Reports

Panamanians providing information to U.S. troops

of what happened varied, but one battalion commander in nearby Colón concluded that the paratroopers had probably overreacted to an innocent civilian just trying to get out of harm's way.[27]

All of this served as a caution to the troops not to let down their guard as they went about their daily duties. Here, again, the locals tried to help out, be it in leading the troops to caches of arms or in revealing the hiding places of PDF and Dignity Battalion members still on the loose. And until Christmas Eve, when Noriega sought refuge in the nunciature, there was no shortage of information concerning his whereabouts as well. In the words of battalion commander Lt. Col. Billy Ray Fitzgerald of Task Force Black Devil, "Everybody was reporting that [Noriega] was in their basement." One of Fitzgerald's counterparts on the Atlantic side, Lt. Col. Johnny Brooks, echoed that sentiment: "Everybody knew where Noriega was. They all wanted to get their million dollars." In defense of those offering such information, Fitzgerald maintained that he did not "think the Panamanians were trying to misrepresent the truth. I think that they

[27] Quotes from Briggs, *Operation Just Cause*, pp. 103–04. Operations Log Extract, in After Action Review, Operation Just Cause, 3d Bn (Abn), 504th Inf, n.d.; Interv, author with Brooks, 11 May 1993.

were just so pleased with US forces that they wanted to do something. And if they thought they had some tip that might head us in the right direction, they wanted to give it to us. So the main problem we had with intelligence was the abundance of inaccurate intel."[28]

Fitzgerald and others could have made the more comprehensive point that troops in the field simply had too much information to process, be it accurate or inaccurate, and not just from the people they encountered on the streets but from higher headquarters as well. At H-hour, psychological operations specialists in Panama activated a radio station. At the same time, Special Operations Forces neutralized the PDF-managed national television station, thus clearing the way for personnel aboard Volant Solo aircraft from the 1st Special Operations Group to begin making broadcasts over that channel directly to the Panamanian people. (Despite plans to the contrary, the Southern Command's own radio and television stations remained off-limits for PSYOP use; instead, they confined themselves strictly to English-language broadcasts aimed at American citizens in Panama.) In the days that followed, elements from the 1st Psychological Operations Battalion and, soon thereafter, its parent 4th Psychological Operations Group churned out an assortment of posters and leaflets to supplement radio and television announcements in which Panamanians were encouraged to provide American forces with specific information concerning weapons caches and fugitives, two items listed in almost every U.S. unit's priority intelligence requirement at the time. The telephone numbers of key U.S. staff directorates usually accompanied the broadcasts and printed material. In the first day or two of Operation Just Cause, the handful of officers and employees who could make it to their offices were quickly overwhelmed by ringing phones, not that the capability to handle the enormous body of incoming information improved much when a near full workforce returned over the succeeding days. What intelligence could be processed at this level frequently found its way to units in the field, but there it only deepened the avalanche of information under which the troops were already being smothered.[29]

In the beginning, as seemingly critical information poured in, units tried to react to every tip they received concerning weapons locations and enemy personnel, but that approach soon became impossible for many commanders and staff. In the words of Task Force Wildcat's Colonel Huff, "We could not respond to them fast enough. . . . We didn't have enough MI people, Spanish speaking/reading MI people." Fitzgerald made a similar point, stating that he soon decided that his men would not act on a tip "unless we got some hard intel from some credible source." In an early assessment of intelligence problems affecting Just Cause, the Center for Army Lessons Learned at Fort Leavenworth, Kansas, concluded that in light of "the amount of reports generated by the local populace, maneuver elements found that they were

[28] First and third quotes from Interv, Hollins with Fitzgerald, 20 Jun 1990. Second quote from Interv, author with Brooks, 11 May 1993.

[29] Directorate of Psychological Operations and Civil Affairs, J–9, *Psychological Operations in Panama During Operations Just Cause and Promote Liberty* (MacDill Air Force Base, Fla.: U.S. Special Operations Command, March 1994); Encl 8 (Psychological Operations), to JTF-South AAR, p. 3.

quickly overwhelmed by *battlefield* reporting. From D to D + 10, battlefield reports were so numerous that most units were only able to look at each report once. At times, this led to improper evaluations and prioritization of information which may have delayed operations and/or presented a skewed view of the battlefield." There was another problem as well. In trying to determine the accuracy of information and the credibility of its source, U.S. troops on both sides of the country quickly learned that much of the poor or inaccurate information they received originated not just from honest mistakes by well-intentioned people, but also from deliberate attempts to deceive the Americans. Supporters of the Noriega regime, for example, might lie about the location of an arms cache in order to lure a small U.S. unit into an alley or house where it could be ambushed. More often, though, a Panamanian engaged in a personal feud or vendetta with a neighbor would denounce the man as a PDF or Dignity Battalion member in order to have him arrested and sent off to the enemy prisoner-of-war camp. As this tactic became increasingly apparent to the Americans, the concern, according to Fitzgerald, "was for us not to turn into some vigilante organization or some resource that a neighbor could use to get back at another neighbor." No one ever developed a guaranteed method for separating accurate from inaccurate information. Field commanders could urge their troops to "ask those lead-in questions that often showed that the source didn't know what they were talking about," although, in some cases, that fact became obvious without resort to an informal interrogation. In other cases, a corroborating piece of information often bolstered the credibility of the original source, making a quick U.S. response more likely.[30]

In the event a small-unit leader or company commander believed his troops had legitimate intelligence that required action by other units in the area or by higher headquarters, the information went first to a battalion intelligence staff officer, or S–2, for logging, cataloging, and prioritization, and then to brigade for dissemination laterally or up the chain of command. At the highest levels, Huff reflected, "we couldn't get it processed by the MI people." "It was tough getting anything out of 470th MI [Brigade]," he cited as an example, "because they were directed at other priorities." These included "going through the Comandancia, the G–2 offices, even the DNTT and DENI offices that we'd taken. And Buildings 9 and 8 at Amador became sources/storehouses of loads of intelligence, which they were trying to sift through." Colonel Hale, 7th Infantry Division's 1st Brigade commander, went further than Huff in critiquing the value of the information processed and intelligence generated by higher headquarters in Panama. "Intelligence was most effective from the bottom up. It was lousy from the top down," he maintained, rating the latter as only 5 to 6 percent accurate. Exacerbating the problem was the "broken" system that often put pressure on him to ignore valuable information he was receiving from one of his battalions in order to respond "to the things that came from above which were, in effect, dramatically less likely to produce results." Citing one example of how bits of information from above could be

[30] First quote from Interv, Hollins with Col William H. Huff III, U.S. Army, 20 Jun 1990. Second, fourth, and last quotes from Interv, Hollins with Fitzgerald, 20 Jun 1990. Third quote from CALL Bull 90–9, Operation Just Cause Lessons Learned, vol. 3, p. 5. Haight Presentation; Briggs, *Operation Just Cause*, p. 95.

"more of a nuisance and more of a distraction in mission performance than they were assistive," Hale recalled how higher headquarters would use six- to eight-digit grid coordinates in reporting to him the location of a fugitive. On urban terrain, such coordinates often proved useless in that the designated area could contain several structures or, worse, an apartment building housing over two hundred people. Precise addresses, not map coordinates, were what he needed.[31]

The poor communication that sometimes existed among the Southern Command, JTF-South, and other key headquarters located in the Panama City area and the units in the field had an adverse effect on more than the assessment and dissemination of intelligence requirements. Other priorities suffered as well, at least initially, as was the case with the money-for-weapons program. The precise date for the formal beginning of the program has been recorded variously as 20 December, 21 December, 24 December, and 27 December—the discrepancies themselves serving as an indication that a communication problem existed. What seems clear in retrospect is that when the newly installed Panamanian president, Guillermo Endara, addressed his people over the PSYOP-operated radio station on the twentieth, he included in his remarks a promise that anyone turning in a weapon to U.S. troops would receive American dollars in return. In fact, no money-for-weapons program existed in the field at that time, one reason being that half of the equation was missing—the money. Furthermore, no one stationed in Panama had any experience setting up such a program. Once Thurman blessed the endeavor, the job of finding the cash and establishing appropriate procedures fell to Col. William Connolly, U.S. Army, South's deputy chief of staff for resource management, and his people. Connolly, whom the general had already graced with last-minute responsibility for running the refugee camp that had sprung up overnight in the Balboa area, threw himself into his new assignment, agreeing to set up a contingency fund to support the proposed purchases. The first step, he quickly determined, obliged him "to get guidance on how to legally proceed with the expenditure of funds for this." A series of hastily arranged telephone conversations with people in the U.S. Army Finance and Accounting Center and in Army Budget gave him the initial funding he sought, $150,000, together with a promise of more money if needed and advice on how to make the process for spending it legally accountable. On the twenty-first, Connolly's office took the next step and "issued payment procedures for the weapons program."[32]

---

[31] Interv, Hollins with Fitzgerald, 20 Jun 1990. Huff quotes from Interv, Hollins with Huff, 20 Jun 1990. Hale quotes from Video Interv, Army–Air Force Center for Low Intensity Conflict (CLIC) with Col David R. E. Hale, U.S. Army, 1990. Related to a point made by Hale, the Center for Army Lessons Learned, in addressing the issue of processing prisoners, declared that an area requiring specific attention was "the use of six digit grid coordinates instead of street names." The ambiguity of the sentence, unfortunately, left doubts as to which method was preferred or the level of command to which the statement referred. CALL Bull 90–9, Operation Just Cause Lessons Learned, vol. 3, p. 4.

[32] *Psychological Operations in Panama During Operations Just Cause and Promote Liberty*. First and second quotes from Interv, Dolores De Mena with Col William J. Connolly, 20 Feb 1991, Fort Clayton, Panama. Last quote from Encl 9 (Special Staff), to JTF-South AAR, p. 5.

All this was taking place while combat units were still winding down their operations against enemy forces. Some commanders did not receive word about money for weapons until well after the local population had been so informed. Where this was the case, any Panamanian advancing toward a U.S. troop position carrying a weapon he wanted to turn in for money risked being shot. As word about the program gradually trickled down to the units, many did not have the resources on hand—cash, forms, and legally authorized personnel for handling both—or the experience to run such an undertaking. Given these shortcomings, some commanders welcomed the memorandum they received on the subject that did not commit them to a specific date for starting the program in their area of operations but advised them only to begin it "as soon as possible." Yet any delay risked exacerbating existing difficulties. When troop units lacked the cash and forms needed to complete a transaction, soldiers sometimes closed a deal by signing makeshift IOUs—pieces of paper that were easily duplicated by recipients who sought a maximum return on their contribution. Also, unless both U.S. troops and Panamanian civilians were acquainted with accepted procedures for how to carry a weapon when presenting it at a check-in point, the potential for a lethal encounter remained. To minimize this possibility, the psychological operations people arriving in Panama began producing flyers that identified the prescribed series of steps, together with cautions, in printed and illustrated form. The flyers—often accompanied by loudspeaker broadcasts—provided other useful information, such as the location of turn-in points and, once U.S. authorities had agreed on a scale of payments, the price each surrendered weapon would fetch. The initial absence of such a uniform and equitable pricing system created its own problems. A Panamanian citizen, most likely a former PDF or Dignity Battalion member, might pull up to a turn-in point with a pickup truck full of AK47s. Expecting to be paid a generous price for each rifle, he would more than likely have to settle for a much reduced lump sum. (A rumor concerning a related issue held that one Panamanian wanted to turn in an armored personnel carrier for $2,500; Colonel Snell intervened with a take-it-or-leave-it offer of $150. The man took it.) Once a pricing system was in place, however, it often had an unforeseen consequence of its own: the number of reports U.S. units had been receiving on the location of arms caches dropped precipitously as informers realized more money could be made if they themselves turned in the stockpiled weapons.[33]

If the starting date for the money-for-weapons program is not certain, the date on which it ended is—8 January 1990. U.S. forces accepted almost nine thousand weapons—although the antiquated condition of some invited the sarcastic phrase, "money for muskets"—at a price of nearly $811,000. Lessons learned afterward abounded, including reminders that such a program put strains on a force's transportation assets, which would

[33] *Psychological Operations in Panama During Operations Just Cause and Promote Liberty.* Quote from Briggs, *Operation Just Cause*, pp. 11–12. McConnell, *Just Cause*, pp. 267–68; CALL Bull 90–9, Operation Just Cause Lessons Learned, vol. 3, p. 9. For the armored personnel carrier story, see Herbert A. Friedman, U.S. PSYOP in Panama (Operation Just Cause), n.d., http://www.psywarrior.com/PanamaHerb.html.

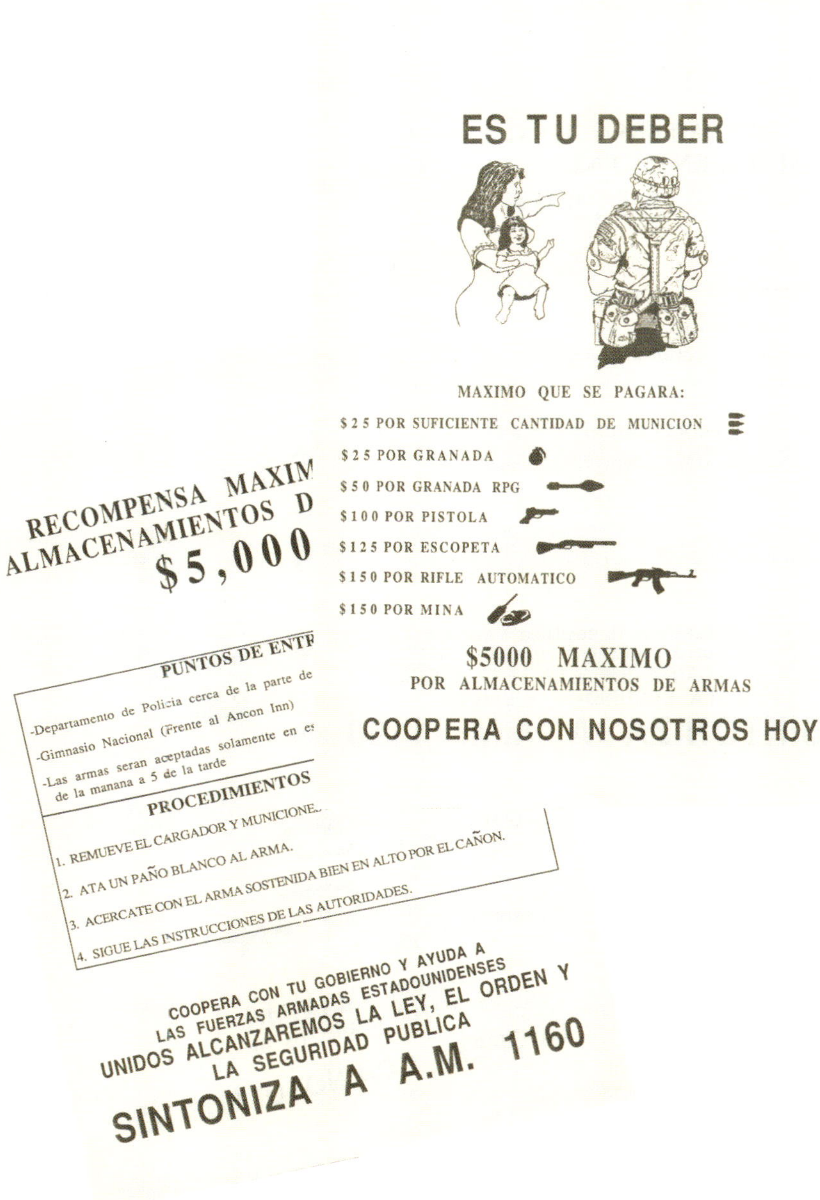

The front and back of a money-for-weapons flyer

have to take the purchased weapons somewhere; and required the presence of authorized people and forms to ensure "that valuable data, such as serial number tracking and weapon location, will not be lost for future analysis." Despite the problems that plagued money for guns, especially in its first several days, the U.S. military viewed it as a success, contributing to the total of 61,000 Panamanian weapons captured or bought during the invasion. Also, in some locations, military intelligence officers benefited from setting up shop near a money-for-guns pay table in order to interview people who had just received their cash and were presumably in a good and, ideally, loquacious mood.[34]

Preinvasion discussions concerning money for weapons did not receive a high priority—if any at all—within planning staffs and troop units earmarked for combat. Thus, the early confusion, risks, and problems associated with the program were to some degree understandable, given its sudden and fitful beginnings, the lack of effective coordination surrounding it, and the unavoidable learn-while-you-earn approach to its procedures. Less explicable were the glitches associated with another program that also required careful preparation, expertise, and close communication and coordination—the detainment of enemy prisoners of war and other captured fugitives. This issue had been addressed in detail during preinvasion planning and preparation sessions, with those involved determining the locations of prisoner collecting points and camps and the procedures for handling the detainees. As for the tactical units who would actually be gathering up the enemy, part of their drill on the eve of JUST CAUSE was to sit through briefings on how their captives were to be treated. Yet, despite all this preparation, difficulties arose.

During combat and mopping-up operations on 20 December, Panamanian defenders surrendered to U.S. troops in significant numbers. As news footage would confirm, almost all of the prisoners were treated well from the time they were flex-cuffed (although some U.S. units apparently ran out of the restraining device), placed on trucks for their ride to the collection points (despite crowds in some locations trying to pull them off the trucks), and incarcerated in a prisoner-of-war camp. The main facility for most enemy personnel was set up at the Empire Range across the canal from Fort Clayton, while a separate detention center at Clayton was used for high-profile political and military prisoners. Visits to the Empire Range camp by Red Cross officials and other humanitarian observers confirmed that the prisoners were being properly fed and cared for.

The initial problem, then, stemmed not from the treatment of the prisoners but from the failure of some tactical forces to find time in the midst of engaging the enemy to process the captives properly. A shortage of Spanish-speaking troops in some units did not help. Thus, too many prisoners were arriving at the collection points and camps without the tags, printed prior to JUST CAUSE, providing such key information as a person's full name (which in Spanish includes the mother's surname after the father's) and the location and circumstances of that person's capture. Of those prisoners who did have tags, the information on them had often been filled out incorrectly. The situation

---

[34] The figures are from Briefing Slides, U.S. Southern Command, Operation Just Cause: "Rebirth of a Nation," n.d. The author received a set of the slides from Lt. Col. James Willbanks in June 1990. Quote from CALL Bull 90–9, Operation Just Cause Lessons Learned, vol. 3, p. 9. Intervs, Robert K. Wright Jr. with Huff and Senior Officers of the 5th Bn, 87th Inf, 193d Inf Bde, 5 Jan 1990, Fort Clayton, Panama; Wright with Lt Gen Carl W. Stiner, U.S. Army, 2, 7, 27 Mar and 11 Jun 1990, Fort Bragg, N.C. Wright is the source for the "money for muskets" reference, a variation he picked up in the course of interviews he conducted after Operation JUST CAUSE.

The detainee camp at the Empire Range

did not improve once the prisoners reached the Empire Range. The military police company in command of the camp itself did not have the capability to process the arrivals and had to rely on the 92d Personnel Service Company at Fort Clayton "to assume the administrative portion of the EPW mission." But even the 92d did not have all the supplies it needed to do the full fourteen-step processing procedure, so it cut the process to eight steps. The ramifications of all this incomplete, inaccurate, and delayed paperwork were felt immediately, as U.S. authorities found it difficult to identify and free the hundreds of prisoners who, in only a matter of days, would be needed to fill the ranks of any new organization being stood up to replace the Panama Defense Forces.[35]

Once a combat unit had dispatched its earliest and generally largest group of prisoners to a collection point, sometimes without escort, the troops still had to round up individual and small groups of PDF and Dignity Battalion members who remained at large. To assist in the search and to facilitate the extraction of better, more reliable information from the Panamanian people, counterintelligence and interrogator teams from the 29th Military Intelligence Battalion made their in-country expertise available to units in the field, while, back at Quarry Heights and Fort Clayton, military intelligence and other officers developed "black-gray-white" (hostile-questionable-friendly) and most-wanted lists containing the names of high-profile targets. As the lists

[35] On the handling of Panamanian forces who surrendered during the first day or two of JUST CAUSE, see CALL Bull 90–9, Operation Just Cause Lessons Learned, vol. 3, p. 4. Quote from Encl 2 (Personnel), to JTF-South AAR, p. 3, and see also Encl 3 (Intelligence), p. 1. Intervs, De Mena with Col Larry L. Raab, U.S. Army, 29 Apr 1991, Fort Clayton, Panama; Hollins with Huff, 20 Jun 1990. Video Interv, CLIC with Hale, 1990. The CALL Bulletin, p. 10, mentions that counterintelligence and interrogator teams from the 29th Military Intelligence Battalion "relied on their in-country expertise in conducting mission area analysis and screening of EPW/displaced persons."

made their way down to the tactical level, so, too, did wanted posters cranked out by PSYOP personnel. Many of the posters displayed a photograph of the fugitive to augment the written physical description and summary of his misdeeds. As a result of these efforts, the trickle of prisoners continued unabated for days after combat operations had wound down.[36]

As U.S. troops pursued fleeing Panamanian hostiles and sought their hiding places in the neighborhoods and built-up areas near the battle zones, talks were already under way in Panama City to determine the kind of organization that would replace the Panama Defense Forces and who would qualify for membership in it. The country's newly sworn-in governing officials discussed the issue with a host of American military and civilian authorities, including Thurman, Cisneros, Gann, Colonel Cornell, and U.S. Chargé d'Affaires Bushnell. All agreed on one point: the need for a decision was urgent.[37] What had become crystal clear to Lieutenant Haight and the other U.S. service personnel conducting patrols in Panamanian cities and towns was not lost upon these senior officials: for the moment, American troops were the principal force responsible for law and order and, to a slightly lesser degree, public services from one side of the canal area to the other. With parts of the capital city destroyed and looted, with criminals and pro-Noriega forces armed and running loose in several urban areas, and with seemingly everything from traffic lights and power plants to grocery stores and the judicial system barely functioning if at all, President Endara, his two vice presidents, and the ministers they had begun appointing to office had to offer more than pledges to lead their country into a new and better era. They had to demonstrate that they were indeed in charge and capable of making democracy in Panama work. At stake was nothing less than the legitimacy and credibility of their government.

As one who would help execute any decisions coming out of the high-level talks, Colonel Fishel in the SOUTHCOM J–5 recalled that, for both the United States and Panama, "the objective was to establish a first-rate, professional police force capable of implementing an effective system of law enforcement as rapidly as possible." Fishel could easily have emphasized the word *police*, in that a consensus was reached early on among the conferees that the PDF's successor would not be a military organization. According to one account, the three top Panamanian leaders regarded their country's "situation as analogous to that of Costa Rica. A standing army was not necessary," although the new security force might require a small paramilitary capability. Furthermore, "to build democracy, it was essential to expunge all vestiges of praetorian rule."

With that difficult decision made, the key question shifted to who would constitute the new police force. Two options presented themselves. The first entailed starting from scratch by recruiting members who had not served in the Panama Defense Forces. The psychological and utilitarian value of this approach required little elaboration. New recruits would not be tainted by the corruption and other abuses of power that had so contaminated many of their

---

[36] CALL Bull 90–9, Operation Just Cause Lessons Learned, vol. 3, p. 10; Bloechl, "Operation 'Just Cause,'" p. 33; *Psychological Operations in Panama During Operations Just Cause and Promote Liberty.*

[37] Fishel, *The Fog of Peace*, pp. 33, 36.

predecessors. With proper training they could be molded into a professional, if initially inexperienced force that, from the outset, would earn the respect of the people they would be sworn to help and protect. Yet, if those aspects of the proposal seemed promising, there were drawbacks. To begin with, the recruiting and training process could take a year or two, during which time U.S. troops would have to police the country as an occupying force. Cisneros, who took part in the talks, held that in "the euphoria of the first month, I know it would have been acceptable to most Panamanians because they supported us and the vast majority of the people still support our presence and operations here. But sooner or later Panamanian nationalism is going [to] resurface." Cisneros usually talked about a "honeymoon period" for U.S. forces of about a hundred days. Fishel agreed. After that time, he surmised, a continuation of the American occupation pending the emergence of an effective police force would likely result in such a negative reaction that the political fallout for Endara "would have measurably degraded the new government's international legitimacy."[38]

The second option was to fill the new force with individuals who had served in the now defunct PDF. Since they were already trained and experienced, they could be on the streets in short order—for the first recruits, in just a matter of days. The pitfalls to this approach, as with the first, were also obvious and numerous. Some retraining would be essential, mainly to break the corrupt and criminal habits many of the prospective recruits had acquired in their old positions. There was another psychological barrier that had to be broken down as well. Most of the recruits would be coming from military units where they had learned to disparage policemen as second-rate functionaries, while glorifying the *machismo* of the true soldier. Would these erstwhile warriors accept shotguns and .38-caliber revolvers in place of the AK47s and .9-mm. semiautomatic pistols to which their military status had entitled them? How would they adapt to such tasks as handing out traffic tickets and calming domestic disputes? Would they be willing to place themselves under civilian control and to swear allegiance to the new government? Few discussing these issues believed the transformation to an enlightened force would be immediate, and that reality raised another, even more critical issue. How would the Panamanian people react to news that the same individuals who had so recently tormented and brutalized them were back on the streets with the blessing of both Panama's new government and the United States?

After careful reflection and further consultations with U.S. officials, Endara, Arias Calderón, and Ford decided in favor of the second option, describing their approach as "evolutionary" instead of "revolutionary" and as one they firmly believed would work, despite its drawbacks. Cisneros later remarked, "There were a lot of salvageable people within the PDF." That

[38] First and last quotes from Fishel and Downie, "Taking Responsibility for Our Actions?" pp. 66–67. Fourth quote from Interv, De Mena with Cisneros, 12 Jun 1990, Fort Clayton, Panama. Second and third quotes from Richard H. Shultz Jr., *In the Aftermath of War: US Support for Reconstruction and Nation-Building in Panama Following JUST CAUSE* (Maxwell Air Force Base, Ala.: Air University Press, 1993), p. 45. What follows on standing up the police force is taken largely from the Fishel and Downie article. During a trip to Panama in January 1990, the author heard General Cisneros refer to the 100-day honeymoon on several occasions.

many members had been highly dissatisfied with the Noriega regime, some to the point of voting against it in the May elections, was well known. As for the issue of corruption, it was an institutional affliction that some members managed to avoid while others, given peer pressure, "went along to get along," but perhaps only in a small way. A key, then, to making the new force work was to vet the former PDF members thoroughly. Logic suggested that the officers who had tried and failed to overthrow Noriega in March 1988 and October 1989 were prospective leaders. Just Cause had rescued from prison those who had survived the attempts, and, before the end of the week, they were brought en masse to Fort Clayton for discussions concerning their future in the new organization.

As for that organization's rank and file, ideally they would have clean or nearly clean records, although determining who had not been corrupt and who had not committed criminal acts—or who had been involved in both but to a tolerable degree—in an organization in which such behavior had been endemic would not be easy. Nevertheless, given the pressure to get policemen on duty as soon as possible, the initial vetting would be a hasty process, rationalized in part by the fallback position that, if further investigation revealed that a new policeman had grossly abused his authority under Noriega, the person could be removed from the force once that information had been verified. The screening got under way within the first week of Just Cause. So, too, did efforts to make the new force palatable to Panamanians. The people would have to be won over through a persuasive public relations campaign and, once the police turned up in their neighborhoods, by the behavior of the cops themselves. Citizens would have to be convinced that the uniformed men they had formerly despised were now deserving of their trust and, ultimately, their respect.[39]

With a decision made on the composition of the successor organization to the Panama Defense Forces, the Endara government and the Southern Command announced over television and radio that recruiting would begin Friday, 22 December, at the Ancon DNTT.[40] Given this short notice, the several hundred people who showed up were mainly PDF members who either had not reported for duty on the twentieth or had eluded the efforts of U.S. troops to capture them during and after the fighting. Despite the shootout that disrupted the proceedings, the recruitment program was under way. Meanwhile, the initial screening process, in Cisneros' words, sought to get "rid of many majors, lieutenant colonels and all the colonels—cleared them out as much as possible." (One source claimed that the vetting "eliminated all the colonels, 83 percent of the lieutenant colonels, and 38 percent of the majors, 31 percent of the captains, and 19 percent of the lieutenants.") The opposite was the case with former PDF enlisted men, almost all of whom were declared suitable for service in the new force. The immediate problem was that many of them were still incarcerated at the Empire Range. Not surprisingly, the commander of the

[39] First quoted words from Interv, De Mena with Col Jack Pryor, U.S. Army, 20 May 1991, Fort Clayton, Panama. Cisneros quote from Interv, De Mena with Cisneros, 12 Jun 1990. Interv, author with Pryor, 21 Jun 1990, Fort Amador, Panama.

[40] In *The Fog of Peace*, p. 36, Fishel suggests that Cisneros may have influenced the decision to send Colonel Cornell and his entourage to the DNTT on the twenty-second, the day of the firefight there. See previous chapter.

prisoner-of-war camp there soon found himself inundated with a stream of memorandums from the Southern Command's intelligence directorate listing the Panamanians to be released and transported to the Ancon DNTT.[41]

The new police force, later named the Panama National Police (PNP), served under what would soon become the Panama Public Force (PPF), an umbrella organization for several subordinate groups—the air service and the maritime service, for example—that filled the vacuum created when Endara announced on 22 December that the Panama Defense Forces would be dissolved. On the recommendation of Cisneros and Cornell, the Panamanian president named Col. Roberto Armijo, the former PDF naval chief, to head the PPF, only to relieve him a few weeks later after unnamed officials, according to one report, "discovered his million dollar bank account." Armijo was replaced by his deputy, Col. Eduardo Herrera Hassan, a twenty-year PDF veteran who had also served as Panama's ambassador to Israel.[42] Cisneros suggested that the DNTT building become the PPF headquarters and directed Colonel Higginbotham to set up an ad hoc advisory group there, the U.S. Forces Liaison Group, to help establish and work with the public force. The initial group was composed of the colonel and five other people. Cisneros also created the Judicial Liaison Group, another ad hoc organization, and he himself soon set up an office in the DNTT, where he worked daily with U.S. advisers (who would be supplemented by civil affairs personnel) and Arias Calderón. The Panamanian vice president, as minister of justice under Panama's system of government, would be in charge of most of the new security force. His discussions with the general centered on ways to organize the public force and how best to allot power within it. (Cisneros remained at the DNTT until he was tapped to head negotiations at the nunciature.)

---

[41] Interv, Sgt Daniel C. Wagner with 2d Lt Paul H. Fredenburgh and 1st Lt Robert M. Mundrell, both U.S. Army, 18 Oct 1990, Fort Davis, Panama. First quote from Interv, De Mena with Cisneros, 12 Jun 1990. Second quote from Shultz, *In the Aftermath of War*, p. 47. Fishel and Downie, "Taking Responsibility for Our Actions?" p. 71; Memos, Brig Gen Michael M. Schneider, Dir of Intelligence, SOUTHCOM, for Cdr, Empire Range Detention Facility; Cdr, Fort Clayton Installation Detention Facility, 1, 2, 6, 12 Jan 1990, sub: Detainee Release from U.S. Control. The text of several memorandums included the sentence "Screening conducted by both US and Panamanian authorities has determined that detention of the individuals listed on the enclosure is no longer required."

[42] As with the PDF, the English name and acronym for the new force will be employed instead of the Spanish because the English usage is found in most of the documents consulted by the author. Quote from CBS Evening News, 3 Jan 1990. Fishel and Downie, "Taking Responsibility for Our Actions?" pp. 67–68; Fishel, *The Fog of Peace,* pp. 36–37. For the State Department plot that was ridiculed in U.S. military circles as "the Rambo option," see Yates, *U.S. Military Intervention in Panama, June 1987–December 1989*, p. 57. Colonel Herrera Hassan was a key figure in the State Department's stillborn plot in 1988 to restore the presidency of Eric Arturo Delvalle, whose removal from office Noriega had engineered in February of that year. In another development affecting the new force, in May 1990, Aristides Valdonedo, the deputy chief of the national police, was arrested for actions he had taken between 1986 and 1988—mainly closing down radio stations and newspapers—when he was a PDF major. That he had been a central conspirator in the March 1988 coup attempt against Noriega and was still in prison at the time of JUST CAUSE did not spare him, although many observers believed that the Endara government's action against him deprived the new police force of a capable leader.

Other issues that demanded immediate attention included uniforms, weapons, supplies, vehicles, and facilities. Many of the police recruits showed up for work in their PDF uniforms, attire guaranteed to elicit a hostile reaction from Panamanians. Designing, purchasing, and awaiting receipt of new uniforms would take weeks, but fortuitous circumstances offered a solution to the problem. U.S. troops had just adopted a new uniform, with stocks of their old one temporarily stored in warehouses in sufficient numbers to issue three sets to each Panamanian policeman until a purchase request could be submitted for obtaining a permanent uniform. Weapons came from captured materiel, but not before Stiner's people intervened with U.S. logistical units preparing to ship the assortment of arms they were in charge of back to the United States. Each recruit received a pistol to carry while on duty, although many of the revolvers were in poor condition and holsters were scarce. It would take several months and U.S. security assistance funds to purchase newer weapons. Meanwhile, short-term fixes also had to be found to supply the vehicles and radios so essential to police work. Police stations, too, required attention. Nearly all of them were in a natural state of disrepair, worsened by the damage to which they had been subjected during the fighting. They could function neither as offices nor jails. U.S. engineers helped with initial repairs, but ultimately the police themselves had to restore their own facilities.[43]

A year would pass before the Panama National Police would outgrow many of its birth pangs. In December 1989, however, the crash measures enacted largely by the U.S. military had succeeded in putting policemen back on the beat. Despite the positive rhetoric designed for public consumption, expectations did not run high at the beginning. Cisneros, for example, recognized that "a major effort is required in their retraining and re-education." "We had to start with basics," he continued, "and teach them that you just don't take a pistol out and shoot somebody because he starts arguing with you." In time, a long-term, professional training program for the police would be institutionalized. In the first days, however, the quick fix consisted of a twenty-hour course taught by U.S. reservists who were policemen in civilian life. Subsequent training came on the job, as Panamanian police joined U.S. troops in patrolling the country's cities and towns.[44]

Around Christmas Eve, liaison officers from the new police force began showing up at the brigade, battalion, and company headquarters of American forces. Their arrival often came as a surprise to U.S. personnel, who in some

---

[43] Cole, *Operation Just Cause,* p. 51; Intervs, author with Higginbotham, 20 Jun 1990, with Cisneros, 29 Jun 1990; Fishel, *The Fog of Peace*, p. 36; idem, War, Peace, and Civil Military Operations; Fishel and Downie, "Taking Responsibility for Our Actions?" pp. 70–75. Concerning the dates on which the new government abolished the Panama Defense Forces and created the Panama Public Force, the initial moves, all are agreed, occurred within days of the beginning of Operation Just Cause, although Endara's cabinet did not formally approve the dissolution decree until 11 February 1990 and the creation of the public force until 17 February. The UN Refugee Agency's *Refworld,* 1 Jun 1990, gives the February dates but notes "the de facto existence of the [public force] very soon after the U.S. invasion." A copy of this issue of *Refworld* can be found at http://www.unhcr.org/refworld/type,QUERYRESPONSE,,PAN,3ae6ab6a50,0.html.

[44] Quotes from Interv, De Mena with Cisneros, 12 Jun 1990. Fishel and Downie, "Taking Responsibility for Our Actions?" p. 73, discuss the quick-fix training course.

cases could barely conceal their displeasure. "I must admit to you that it was a very unusual situation," Colonel Fitzgerald later commented, "that 24/36/48 hours ago we were shooting at a guy that looked very much like this guy in a very similar uniform. And now, all of a sudden, he was in my Tactical Operations Center, and I was supposed to put full confidence in him. So that was a very difficult thing, but we did have that liaison, we did have that contact." Another officer had the same reaction. "Do we want to walk with these guys, because we were shooting at them last week?" The issue was not open to debate, of course, so U.S. officers adjusted to the awkward situation as best they could. At some point early on, Panama City was divided into seven police precincts. Colonel Hale responded by redrawing his boundary lines so that each of his battalion's areas of responsibility would coincide with a precinct. Still, the consternation among U.S. troops at being asked to work side by side with their recent enemy was not easily overcome, making the initial patrols, once more Panamanian policemen began arriving, tense if not outright contentious. A somewhat common sight was to see a Panamanian cop walking slightly in front of his armed U.S. companions, enabling the soldiers to keep a wary eye on him.[45]

Serving as something of a buffer between American combat troops and the new arrivals were U.S. military policemen. As Col. Larry Raab, the USARSO military police commander, related, "Our primary mission was to work with the Panamanian police at every level of their command structure in order to get them back on the streets of Panama as quickly as possible. . . . We had the vehicles, we had the weapons, we had the radios." Without question, the involvement of the U.S. military police was constructive, if not free from criticism. As Colonel Huff crisply stated, "they do not walk." This, in his opinion, was counterproductive. "You cannot glean intelligence or get to meet and befriend the people sitting on top [of] a HUMVEE behind an M-60 machine gun. It does not work." For Colonel Hale, the problem was one of command and control. Hale praised the military police's work with the Panamanian police in his area of operations as being "extremely helpful" but regarded as an unacceptable violation of the principle of unity of command the fact that, despite his protests, they were not under his operational control. "That forced me to do one of two things," he recounted.

> Either to not pay attention to civil-military operations and to turn that over . . . to the Military Police Command . . ., or, the choice I chose, to de facto make them OPCON, to coordinate so closely with them, to be so involved in their operations, to demand that they give me a liaison officer full-time at battalion and at regimental level, to demand that they coordinate all their operations with me, and to demand that I know precisely what patrols they were conducting . . . that, in effect, I made them OPCON.[46]

---

[45] Fitzgerald quote from Interv, Hollins with Fitzgerald, 20 Jun 1990. Other quote from Interv, Wright with Huff and Senior Officers of the 5th Bn, 87th Inf, 193d Inf Bde, 5 Jan 1990. Interv, De Mena with Raab, 29 Apr 1991.

[46] Raab quote from Interv, De Mena with Raab, 29 Apr 1991. Huff quotes from Interv, Wright with Huff and Senior Officers of the 5th Bn, 87th Inf, 193d Inf Bde, 5 Jan 1990. Hale quotes from Video Interv, CLIC with Hale, 1990.

Initially, the reappearance of Panamanian police on the streets of the cities and towns produced the expected hostile reaction. Panamanians were reluctant even to approach the new recruits, much less seek their assistance or reveal to them valuable information. The people clearly preferred to continue dealing with the Americans. By early January, though, this approach had softened somewhat, so that one U.S. operations officer could report that Panamanians "are now consciously going to the DENI with some faith that the new government will take care of them." Hale could also state that the combined patrols in his area of operations were effectively maintaining law and order.[47]

But, despite these signs of progress, the popular distrust and American suspicions would not be easily dispelled. Concerning working relations with the recruits, Hale later echoed Cisneros (and many other observers), "I think the one difficulty that we had was what I would call imparting a new set of values to the members of the police force." He recalled one episode in which the ranking Panamanian officer in his area was walking a beat with one of the 1st Brigade's company commanders. The former PDF member knew the neighborhood well and suggested the two men have lunch in a nearby restaurant. Upon entering the establishment, the Panamanian told the proprietor, "Today, you have the pleasure of donating lunch to my American friend and I." The U.S. officer quickly interrupted to inform the owner that he and his partner would have the pleasure of paying for the lunch, that day and every day thereafter. Other stories like this made the rounds, but not all of them had such satisfactory conclusions. There were instances, word had it, in which some of the new police on patrol with U.S. troops who did not speak Spanish actually tried shaking down Panamanian citizens in front of the Americans, smiling all the while and ordering the victims to do the same. The stories raised an obvious question: what would happen once the U.S. troops redeployed to the United States or, for those based in Panama, returned to their regular peacetime duties, which did not include the combined police patrols? As Hale was well aware, the patrols were designed, in part, to demonstrate the U.S. troops' outward "vote of confidence" in the new police force. Many observers throughout the country, however, seemed more confident that, without the American presence, police work in Panama would quickly return to business as usual, as it had been under the Panama Defense Forces. In May, *La Prensa,* a Panamanian newspaper that had been shut down under Noriega in 1988, joined the chorus expressing this opinion, writing that one could not transform "military personnel humiliated by defeat into policemen that respect the laws and rights of citizens." A week later, another newspaper, *El Siglo,* wrote that 80 percent of the PDF's members were now part of the new police force, adding that "no [PDF] member has turned into a rabbit considering they have had 21 years of indoctrination."[48]

---

[47] Interv, Huff and Senior Officers of the 5th Bn, 87th Inf, 193d Inf Bde, 5 Jan 1990. Quote from Interv, Hollins with Maj John Woods, U.S. Army, 20 Jun 1990, Fort Clayton, Panama. Video Interv, CLIC with Hale, 1990.

[48] All quotes except for last two from Video Interv, CLIC with Hale, 1990. While in Panama in January and June 1990, the author heard stories describing the ways some of the new police engaged in what U.S. officers considered misconduct. The translated quotes from *La Prensa*

These pessimistic evaluations may have worried U.S. troops based in Panama but had little impact on those combat units preparing to redeploy to the United States. Aside from their involvement in toppling Noriega, the latter had little stake and only a moderate interest in the long-term prognosis for the country they had liberated. Still, when dealing with the new police force or in performing the broader range of stability and nation-building operations, they generally welcomed the help of military personnel who had experience in these issues. Providing that assistance, at least in part, were small civil affairs teams that began showing up at various U.S. headquarters on both sides of the canal to lend their expertise in restoring law and order and rebuilding the country. Some of these specialists had arrived in Panama on D-day, as was the case with elements of the only active duty civil affairs unit, the 96th Civil Affairs Battalion based at Fort Bragg. Within the next two days, most of the rest of the battalion had flown in as well. Another four days would pass before individual civil affairs specialists from the reserves began arriving, and another six days after that before a civil affairs support team composed of five functional offices—embassy support teams, public security, public works, government, and humanitarian assistance—joined them. At their peak, civil affairs personnel on the ground would number 152 during JUST CAUSE. Of these, 57 had managed to arrange 139-day temporary tours of active duty; another 75 volunteers came on 31-day tours.[49]

Regardless of when they arrived, the civil affairs specialists quickly discovered that there would be more to their mission than working to get the new government of Panama functioning; they were also to interact with tactical commanders in the field, particularly in providing food and medicine to the general population. The teams assuming these duties usually consisted of an officer and one or two enlisted men or noncommissioned officers. When such a team arrived in a town or village—usually to the great relief of the chaplains, S–1s, and S–5s whose superiors had "volunteered" them to take the lead in civil affairs duties in the absence of trained personnel—its members would assess the situation and suggest ways in which to start a program or to expedite those already under way. An existing program did not always mean an efficient one. Combat units that had set up food distribution points, for example, often confronted chaos and a lack of resources during the first days. Sometimes there was no food to hand out, although any nearby PDF facility might have a food locker that could be raided or, given the season, undistributed Christmas

---

and *El Siglo* are from the UN Refugee Agency's *Refworld,* 1 Jun 1990. Almost a year and a half after JUST CAUSE, the U.S. military police commander in Panama still observed public doubts about the Panamanian police. "The Panamanian people continue to feel as though nobody is in charge and they feel unsafe," Colonel Raab stated. "Therefore, no matter how proficient their police force is they will be perceived to be ineffectual." Interv, De Mena with Raab, 29 Apr 1991.

[49] Stanley Sandler, *Glad to See Them Come and Sorry to See Them Go: A History of U.S. Army Tactical Civil Affairs Military Government, 1775–1991* (Fort Bragg, N.C.: U.S. Army Special Operations Command History and Archives Division, 1993), pp. 379–80, 382; Interv, author with a captain from the 96th Civil Affairs Bn, U.S. Army, 20 Dec 1989, Fort Clayton, Panama. According to the captain, thirteen members of the battalion parachuted with the Rangers at Torrijos-Tocumen; he and a full colonel followed soon thereafter; and another forty-six members of the battalion were scheduled to arrive later on the night of 20 December.

baskets of food. There were also boxes of meals, ready to eat that could be obtained from the Americans' own supplies. With little or no training in riot control, many combat units found it nearly impossible to impose order on the mobs of Panamanians who often overwhelmed several of the food sites. As a result of the confusion and the absence of a systematic way to dispense the food, many recipients quickly learned they could stand in line several times, in the process receiving more than their share. Where such dysfunctional programs existed, the civil affairs teams could help modify, streamline, and oversee the procedures. As for Panamanians in need of medical treatment, civil affairs personnel and the doctors who sometimes arrived with them set up Medical Civic Action Programs (MEDCAPs); ensured that adequate medical supplies were on hand (again, often supplemented from those found in PDF facilities); and, in conjunction with any PSYOP teams present, spread the word as to when and where the clinics would be open. Between 26 December and 3 January, civil affairs teams helped provide 1,660 tons of food and 218 tons of medical supplies to the population.[50]

In addition to the food and medical programs, civil affairs specialists worked with local political leaders to restore essential services. The teams also produced surveys, based on information compiled by American troops in the area, to determine what problems needed fixing, either by the troops, other U.S. government agencies, or the Panamanians themselves. One U.S. officer summarized the breadth of the surveys when he wrote, "We were to report on a region's sewage system, electricity, water and food supply, medical coverage, fire-fighting capability, garbage disposal, and a whole array of systems that make civilization possible." In assessing the effectiveness of the civil affairs teams, the Center for Army Lessons Learned concluded that they "were most successful when allowed by commanders to operate in the field as advisors to units and liaison with the local population. Teams that served as augmentation to the [battalion or brigade] staff were not as effective."[51]

Despite their accomplishments, the reviews civil affairs personnel received were mixed. Among the positive comments, one staff officer rhapsodized that "I just think that we won every heart and mind that we lost with Noriega's propaganda." In a similar vein, a U.S. commander declared, "The best surprise I had, with all the outside agencies, was with the Civil Affairs." More negative assessments stemmed from several causes. In at least one case, the tactical commander did not believe the team that showed up to be large enough or qualified enough to do its job. There was only a captain and a sergeant, he complained, neither of whom spoke Spanish or was "familiar with the way the operation worked in general." Nor did they "interlock or intermix." In response, a lot of civil affairs officers ridiculed what they regarded as the foolishness of many of the conventional forces they were trying to help, criticizing, among other things, "the kindly tendency of the tactical unit

[50] Sandler, *Glad to See Them Come,* p. 381; Briggs, *Operation Just Cause*, p. 125; Intervs, Hollins with Fitzgerald, 20 Jun 1990; with Huff, 20 Jun 1990; Wright with Huff and Senior Officers of the 5th Bn, 87th Inf, 193d Inf Bde, 5 Jan 1990; Cole, *Operation Just Cause,* p. 66.

[51] Sandler, *Glad to See Them Come,* pp. 378, 383. First quote from Briggs, *Operation Just Cause*, p. 124. Second quote from CALL Bull 90–9, Operation Just Cause Lessons Learned, vol. 2, p. 20.

troops to give away MRE rations to civilians long after any immediate need had passed. This frustrated Panamanian food distribution, did the economy no good and fostered an unhealthy dependence on the U.S. military."[52]

Most of the friction was not so personal but derived from procedural and organizational tensions. To begin with, there were not enough civil affairs personnel to cover each combat unit's area of operations; indeed, some civil affairs teams had to scramble to perform their duties over an area that combined the jurisdictions of two or three tactical commanders. Not surprisingly, the overextended teams were not always available or responsive when a particular unit needed them. From that unit's limited perspective, it seemed that the team would show up for a few hours or, at best, a day or two and then disappear. There was also the issue of command and control similar to the one that had plagued Hale with the military police. According to Huff, "You have to understand we are very particular in the way we do operations. You don't come in, and we need to borrow you. When you come to us, we own you, and, therefore, you respond to what we tell you to do, so we can have positive control of the action, and make things go where they should go. That did not happen." As a result, Huff's subordinates believed, resources were not "properly utilized," whether because civil affairs teams had not returned to carry out duties they had promised to perform or because the teams, when they did respond, did not do so quickly enough. The latter had been the case, in Huff's opinion, when food his men had found in the PDF engineer compound rotted before a civil affairs team had arrived with a plan to transport the provisions to the refugee camp at Balboa, where they were sorely needed. Huff's final judgment on the civil affairs assets in his area of operations was reflective: "We wish we could have worked with them a little sooner in training." Fitzgerald had much the same critique to add to what otherwise was his high praise of the civil affairs personnel who, he stated emphatically, should have been in his chain of command, thus allowing him to receive directly the resources and guidance on "how we were to run a Civil Affairs operation." "As infantrymen," he noted, "we didn't necessarily know how to go about doing Civil Affairs correctly."[53]

If the generally productive working relationships between combat troops and the military specialists sent to assist them—military police, civil affairs personnel, psychological operators, and others—occasionally led to friction, the intensity and duration of the resulting criticism rarely exceeded that generated by another source: the rules of engagement. Part of the problem stemmed from the increasingly stringent nature of the rules, although most soldiers and marines understood the need for greater restrictions on the use of force once major combat operations had ceased. Still, the additional limitations seemingly carried with them greater risks, as well as reminders that a change had occurred as "infantrymen, trained primarily 'to close with and destroy the enemy'" suddenly "were expected to act as diplomats and policemen." As

[52] First, third, and fourth quotes from Intervs, Wright with Huff and Senior Officers of the 5th Bn, 87th Inf, 193d Inf Bde, 5 Jan 1990. Second quote from Interv, Hollins with Fitzgerald, 20 Jun 1990. Last quote from Sandler, *Glad to See Them Come,* pp. 378, 381.

[53] First three quotes from Intervs, Wright with Huff and Senior Officers of the 5th Bn, 87th Inf, 193d Inf Bde, 5 Jan 1990. Remaining quotes from Interv, Hollins with Fitzgerald, 20 Jun 1990.

one officer wrote later, "No grunt worth his salt will tell you that his primary function is to prevent kids from stealing tennis shoes, help old ladies cross the street, break up domestic quarrels, or rescue babies from burning buildings— but we found ourselves doing all of these things in Panama."[54]

When combat troops attended a unit briefing in which new and stricter rules of engagement were announced, the news usually signaled that they would be taking on another role for which they had not been trained or prepared. The degree of difficulty experienced in making the necessary adjustment depended on the unit. While documentation on this point is sketchy, what exists suggests that the adverse psychological impact fell hardest on combat forces who, having deployed for JUST CAUSE from outside Panama, were receiving their first exposure to the political, economic, and social considerations and requirements that contingency operations with limited military objectives often entail. These units had engaged and defeated the enemy in a matter of hours, a day or two at most. After some brief follow-on and mopping-up operations, they expected to return home, only to discover that their mission was far from over and involved work for which others, in their opinions, were better suited. This same realization caused less consternation, it seems, within deploying units that had been involved in the crisis before JUST CAUSE. Colonel Hale's 1st Brigade was such a unit, in which two of its battalions had lived with strict rules of engagement and unorthodox taskings during their NIMROD DANCER rotations earlier in the year. In Hale's opinion, the rules of engagement he received upon his return in December were "cautious, but not overly restrictive." Finally, units based in Panama probably felt the least stress from the adjustment; indeed, the major transition for them had been to engage in combat operations after having grown accustomed over the course of two years to exercising restraint so that higher authorities, in a chain of command that reached all the way to Washington, could use the troops' presence to send political signals to the Noriega regime.[55]

If the degree of difficulty varied from unit to unit with respect to the adjustments each had to make as Operation JUST CAUSE passed from the Assault Force Operations phase to the Stabilizing Force Operations phase, one feature of the rules of engagement aggravated nearly everyone equally: besides becoming more restrictive, they changed continuously, sometimes daily. Also, "there were different Rules of Engagement for different areas and different types of situations." Concerning the latter point, the rules for troops around the Cuban Embassy differed from those in effect at a traffic control point. Moreover, at both locations and others, what was considered permissible behavior on one day could be prohibited on the next. Hale, for example, recounted how the rules for house searches changed over a short period, beginning with troops informing occupants their houses were going to

[54] Quotes from Briggs, *Operation Just Cause*, pp. 4, 141. For a dissenting view, one that argues the changes in the rules of engagement presented no significant problems as stability and nation-building operations superseded combat in Panama, see Interv, author with Brooks, 11 May 1993.

[55] Video Interv, CLIC with Hale, 1990. A similar assessment as that applied to Hale's brigade could be used for light fighter units already in Panama under Colonel Kellogg's 3d Brigade, the successor to the 1st Brigade in the NIMROD DANCER rotation.

be searched; then switching to troops having to request permission to search a house; then to requiring soldiers to provide advance notification as well; and, finally, mandating that troops get clearances and provide formal justifications before conducting a search. For a unit frequently moving from one location to another, the need to memorize a new set of rules for each new area and to have those rules change regularly became a source of exasperation and, worse, confusion. "Behavior deemed meritorious under one set of rules could be construed as unacceptable under another set," wrote a company executive officer from the 82d Airborne Division. "It is not difficult to understand how a soldier can become confused when he is praised for an act in one instance but is then reprimanded for a similar act in another. This is especially true in an environment where hesitation or a lapse in judgment could very well kill you or your fellow soldiers. The result was often frustration, tension, and ambivalence that further complicated an already confusing state of affairs." More succinctly, the same officer noted, "Testimonies too numerous to list here focus on the same problem—'I wasn't sure when to use force, or when I could shoot, or what to do IF. . . .'"[56]

If the troops were often confused and frustrated, there was only so much that their superiors could do to remedy a problem arising in part from the decentralized nature of the situation. With small units assigned individual tasks such as guarding a facility or manning a checkpoint or patrolling an area and with each task having its own rules of engagement that, in turn, varied according to location and circumstances, the brigade and battalion commanders found it, in Hale's words, a "very complex thing to keep that type of information flowing down to the individual soldier. . . . We tried to put it in very simple terms . . . what you could do and couldn't do. We tried to ensure that we had written instructions at each one of these specific sites so that subordinate leaders had something like an SOP constantly to refer to." Not helping matters was the fact that most rules of engagement arrived at brigade and battalion headquarters in classified form, generally at the secret level, to be passed along to troops who might have a confidential clearance (one level below secret) or none at all. "So we would have to sanitize it," Fitzgerald said of his battalion, "and give it to a soldier in words that he could understand and apply to his situation."

After higher headquarters had done what they could to disseminate the information they received concerning the rules of engagement, a platoon leader or squad leader became the "man on the spot," according to Hale, responsible for "enforcing and interpreting the situation" and "making the judgments there in terms of what should or should not be done." To ensure that the rank and file understood the instructions they had been given, some brigade and battalion commanders would develop ROE training programs for small-unit leaders to administer to individual soldiers. Also, senior officers, when visiting those units, could themselves quiz the troops, asking them how they would apply the rules in a variety of hypothetical situations. Finally, the colonels and lieutenant colonels could try pulling whatever strings they could to keep a small unit in a given assignment for as long as possible, thus enabling the

[56] First quote from Interv, Hollins with Fitzgerald, 20 Jun 1990. Video Interv, CLIC with Hale, 1990. Remaining quotes from Briggs, *Operation Just Cause*, pp. 4, 104.

troops to become more knowledgeable with the rules in effect at that location, as well as more proficient in implementing them.[57]

As U.S. combat forces engaged in stability and nation-building operations in Panama and performed unfamiliar tasks with a variety of military and civilian specialists, the frustration, resentment, and puzzlement they often experienced was counterbalanced by the deep-seated gratitude the Panamanian people displayed toward them and by the sense of accomplishment the troops themselves often felt at what tangible progress they were making in bringing security, law and order, and stability to the country. If what they were doing seemed confusing at times or if the means to an end were not always clear or resourced, some way could generally be found to get a task done. Many of the troops shrugged off the difficulties and assumed that, at the highest echelons, there was probably a coherent plan and purpose that simply had not been communicated effectively to them. Others assumed that things at the top were probably as screwed up as down in the streets and alleys. Both groups were right to some extent. There was also some validity to a conclusion reached by troops located in areas outside of Panama City that their stability and nation-building initiatives lacked the high priorities assigned to similarly categorized activities taking place in the capital.[58]

PANAMA CITY: FROM ORDERED CHAOS TO THE MILITARY SUPPORT GROUP

Common sense supported the disproportionate attention accorded U.S. civil-military operations in Panama City after 20 December, even if that meant efforts in other parts of the country might not be fully resourced. The capital had witnessed some of the heaviest fighting in Operation JUST CAUSE, had seen one neighborhood—El Chorrillo—completely destroyed and two others heavily damaged, and had suffered (as had Colón) the economic consequences of widespread looting. It was also the seat of the national government, while adjacent to it was the headquarters of the U.S. Southern Command. For these and other reasons, the city became the centerpiece for the most visible stabilization and nation-building programs.

One urgent issue in need of immediate attention concerned the refugee, or displaced persons, camp at Balboa High School. By the end of the day on 20 December, over ten thousand people, mostly from burned-out El Chorrillo, occupied the camp, creating an immediate demand for security, food, sanitation, shelter, and a host of other services that the U.S. military would have to provide. Once Colonel Connolly arrived at the facility to take charge, he quickly realized,

[57] Hale quotes from Video Interv, CLIC with Hale, 1990. Fitzgerald quotes from Interv, Hollins with Fitzgerald, 20 Jun 1990. Fitzgerald, a battalion commander, commented in the aforementioned interview on how his brigade commander, Colonel Snell, handled rules-of-engagement issues effectively as the relevant information made its way down the chain to Snell's battalion commanders.

[58] Lt. Col. Johnny Brooks, referred to for a time in late December and January as "el jefe de Colón," was one officer to make the "stepchild" analogy when assessing the lower priority that his stability and nation-building efforts received compared to those in Panama City.

"There were no ground rules. We had to establish the policies and the procedures as we were going along." On the positive side, Connolly knew that "we had the expertise to do the basic care and feeding and accomplished those kinds of missions, setting up food lines and making sure cots and blankets and that kind of stuff was available." Other services, however, were not secured so easily. With the fighting still winding down, there were not enough U.S. soldiers or military police to stand guard, to deal with troublemakers, and to cull out the PDF and Dignity Battalion members and criminals who had mingled with the horde of civilian families, children, elderly people, and other innocents. Additional manpower and resources would eventually arrive. In the meantime, according to Connolly, "Our biggest problem when we started out was just giving it some semblance of organization." Until that was done, something as seemingly routine as a food line for breakfast could degenerate into a free-for-all. A solution to the organizational crisis soon offered itself. "We had the actual elected mayor of the area of Chorrillo in a tent and we brought him up and told him it was his problem," Connolly recounted. "We told him they were his people and to organize it so that we could have five or six groups each with a group leader." To the surprise of the colonel and other U.S. officers present, the mayor "came back a little later and told us he had worked it out." Within each of the groups the mayor had organized, individual Panamanians accepted certain responsibilities, such as cleaning latrines, which had overflowed on the first day, or passing out food. As a result, "almost over night the camp became more livable."[59]

Security remained a problem, however. "We didn't have any permanently assigned force to protect the site at night," Connolly noted. He tried to obtain help from the combat units, but, even several days after the fighting had stopped, most of them were still overwhelmed with follow-on missions, some known in advance, some not, such as escorting news reporters. Succor for the colonel came in the form of military police units who were arriving in Panama. Several of them, before heading for their preassigned locations, would help guard the camp for a brief time. "I have a tremendous respect for them," Connolly said later. He had the same feeling for several other groups who showed up, including U.S. medics, Panamanian and American civilian volunteers, and the "computer literate" high school students who helped register the refugees. There was also what he called "the most sought [after] resource of every commander," the psychological operations personnel with their loudspeakers. According to one after action report,

> A tactical PSYOP loudspeaker team set up full-time operations in the compound, amplifying the PSYOP radio station broadcasts . . . to put the people at ease and keep them informed. This team, along with other PSYOP soldiers, moved throughout the camp handing out newspapers prepared by the PSYOP [task force], pretesting proposed PSYOP products, chatting with the people, observing their actions, and reporting potential problems, concerns and troublemakers to the camp administrators—members of the 96th Civil Affairs Battalion.[60]

---

[59] Interv, De Mena with Connolly, 20 Feb 1991.

[60] First four quotes from ibid. Last quote from *Psychological Operations in Panama During Operations Just Cause and Promote Liberty*.

Two elements of the civil affairs battalion had parachuted into Torrijos-Tocumen with the Rangers. Once on the ground, several of the men headed for the high school, which was also the site of the battalion's aid station. Within a week, there were at least a dozen civil affairs personnel at the camp. Soon thereafter, Connolly departed to assume other duties. By that time, the facility was running smoothly; the inhabitants numbered around four thousand, less than half of the eleven thousand who had ultimately been registered; and a potential humanitarian disaster had been averted.[61] The civil affairs contingent stuck with the organizational arrangement already in place, coordinating its activities with other civil-military groups, the Panamanian Red Cross, the Corps of Engineers, the U.S. Agency for International Development (USAID), and officials of the Endara administration. The contingent also helped to relocate the camp once word arrived that the U.S. State Department wanted Balboa High School reopened. With the help of military engineers and the Air Force, a new displaced persons center was constructed on nearby Albrook Air Station, with the transfer of the remaining refugees taking place in mid-January. On 19 January, USAID took control of the facility, which was supposed to furnish temporary housing for no more than ninety days. Under contract, the Red Cross assumed responsibility for managing the center on a day-to-day basis, and a platoon of U.S. troops provided security. Given the innovative measures put in place to overcome the early adversity facing homeless Panamanians, few observers would have disputed a statement later praising the handling of the refugees up to that point as "the earliest and most successful CA operation" during JUST CAUSE. Within a year, though, the evaluations would be less flattering, as a failure to build permanent housing for the nearly two thousand displaced civilians still living at Albrook had become a festering scandal for Endara and the U.S. military.[62]

While Connolly and others were occupied with the refugee situation, a host of American officers were busy providing direct help to the new Panamanian government. That the U.S. military would play a role—indeed, the predominant role—in getting the Endara administration stood up and running reflected both understaffing at the U.S. Embassy in Panama and, in stark contrast, the availability of manpower, especially staff officers, and other resources to the Southern Command. After President Bush had recalled Ambassador Arthur Davis in May 1989 following the postelection violence in Panama, the embassy in Panama City was headed by the chargé d'affaires, who on 20 December was John Bushnell. Under Bushnell were forty-five employees, about a third of whom were at their duty stations in

[61] Many of the camp's inhabitants returned to El Chorrillo once it had been cleared of debris. The rebuilding of the neighborhood, however, fell victim to the failure of financial aid promised by Washington to arrive in a timely way, a delay that affected numerous reconstruction programs.

[62] Quote from Sandler, *Glad to See Them Come,* pp. 380–81. Memo for Col William M. Graham, Cdr, Albrook Displaced Persons Camp, 16 Jan 1990, sub: Transfer of Responsibility for the Displaced Persons Center Albrook Hangar; *Resettlement of Panama's Displaced El Chorrillo Residents* (Washington, D.C.: General Accounting Office, 20 December 1990).

the days after hostilities broke out. The State Department immediately dispatched three political advisers to Panama, including Michael Kozak, a veteran of the crisis and the deputy assistant secretary of state for inter-American affairs. Ambassador Davis also returned, but, only briefly, as he would be replaced in early January by Deane R. Hinton. All would take part in the discussions with Endara, Arias Calderón, Ford, and others over what needed to be done in Panama. But while their contributions and those of the country team were invaluable, the U.S. Embassy in late December and well into January simply did not have the wherewithal to help the newly inaugurated officials govern effectively. The U.S. military filled the vacuum by default and by design. Despite the flaws in Blind Logic, the Southern Command had the personnel in its J–5 shop and elsewhere who could get the plan started, and it had the specialists—some already on the ground, some en route—who could sustain the effort and make it work.

On the first day of Just Cause, General Thurman had set the effort in motion when he sent the revised Blind Logic operation order to the Joint Chiefs of Staff for what would become Promote Liberty and when he directed General Gann to command the Civil-Military Operations Task Force. According to Fishel, Gann quickly realized that, given the disarray in the country, the task force "had no choice but lead the Government of Panama, couching its actions in the form of suggestions and recommendations to the 'government' and the Chargé." The same morning that Thurman gave the J–5 director his marching orders, he also telephoned Col. Jack Pryor in the J–5 shop, telling him, "Put your uniforms away, go ahead and get your guayaberas, . . . you're going to work for the new government of Panama. You're going to be my mini-MacArthur of Panama. Go to liaison, work with them, tell us what they need, protect them." Following the conversation, Pryor presented himself to Endara, Arias Calderón, and Ford and set up an office next to theirs. Over the succeeding days and weeks, the colonel served as a bridge between the Endara government, including the ministries that were beginning to take shape, and virtually all outside organizations, agencies, and people involved with assisting that government, including SOUTHCOM, USARSO, the U.S. Embassy, USAID, and a host of others.[63]

Even a brief summary of the U.S. military's work with the Endara government from late December into mid-January covers an extensive list of activities. Pryor recalled that among the first things he did was read the Panamanian constitution, after which he worked with other U.S. officers to devise an organizational chart that they could put forward as a representation of the new government. At first, the diagram only "had three names in it, Endara, Arias, and Ford." More names followed, but many more lines still remained blank. "We had to fill in all the little boxes," Pryor continued, "Minister of Education, minister of this, minister of that. So, the starting point was just to fill that out and blend it with the Constitution of Panama," recognizing that some of the current constitution's dictatorial verbiage

[63] First quote from Fishel, *The Fog of Peace,* pp. 33–36. Second quote from Interv, De Mena with Pryor, 20 May 1991.

introduced by Torrijos and Noriega would have to be ignored until it could be removed by due process. Meanwhile, as Endara, his vice presidents, and other advisers fleshed out the wiring diagram, offices and supplies had to be found for the incoming ministers and their staffs. Also requiring immediate attention were certain housekeeping chores, as two female civilians and a captain from the USARSO protocol office drove through the still dangerous streets of the capital to the presidential palace Wednesday night, 20 December, to vacuum and clean the premises so that a more formal inauguration ceremony for the new government could be held as soon as possible.[64]

Over the ensuing days, stability and nation-building activities gained momentum, as the civil affairs regional and local surveys became available; as shipments from various humanitarian relief agencies began arriving in the country; as U.S. units dealt with refugee, detainees, fugitives, and criminals; as engineers began repairing damaged areas and engaging in several civic action projects; as contractors were hired to help with reconstruction; as the government ministries began functioning; and as business activity resumed, despite the negative impact of the looting. To keep the public informed and to generate and maintain popular support for all these activities, the new government commenced a barrage of television and radio broadcasts, posters and flyers, and other material, produced and, through a highly sensitive arrangement, provided to it by the psychological operations unit from Fort Bragg. After a month of uninterrupted work, Panama by mid-January could boast a functioning, if clearly overwhelmed, government and the restoration of most public services. Furthermore, in Fishel's words, the PDF "had been broken into its component parts, elements had been detached and reorganized, and there were Panamanian policemen on the street working with U.S. MPs."[65]

On a less positive note, Fishel also recorded that "the general impression that one received in Panama during the last 10 days of December 1989 and the first 2 weeks of January 1990 was a kind of ordered chaos." The impression stemmed to some extent from the organizational framework the U.S. military set up to execute Promote Liberty. For the first week, the Civil-Military Operations Task Force formed around General Gann's J–5 staff and bolstered by specialists from the 96th Civil Affairs Battalion and certain military police units took the lead on most civil-military matters. Others were handled by the two ad hoc liaison groups Cisneros had created at the DNTT and by Colonel Pryor in his capacity as an adviser to Endara, Arias Calderón, and Ford. Also weighing in after their arrival in Panama on 26 December were 25 civil affairs reservists, the preselected volunteers (114 more volunteer reservists would enter the country around 1 January). The senior officer among the twenty-five was Col. William Stone from the reserve's 361st Civil Affairs Brigade. Stone had not worked on Blind Logic since 1988 and, thus, having been read off the planning process, had not seen any of the revised versions since

---

[64] Interv, De Mena with Pryor, 20 May 1991. The author was present in the USARSO command group area when the two protocol officers donned Kevlar vests and left with the captain to go vacuum the palace.

[65] Intervs, De Mena with Pryor, 20 May 1991; author with Lt Col John T. Fishel, U.S. Army Reserve, 20 Jun 1990, Panama City, Panama; *Psychological Operations in Panama During Operations Just Cause and Promote Liberty*. Quote from Fishel, *The Fog of Peace*, p. 39.

An example of material prepared by U.S. psychological operations personnel for use by the Endara government

then, including the Joint Chiefs–approved PROMOTE LIBERTY. Once back in Panama, he had the group with him begin setting up headquarters for a civil-military operations task force and, under that, a civil affairs task force, all in accordance with the last draft of BLIND LOGIC to which he had been privy. He assumed that all of this would come under Gann's operational control, but soon "discovered"—no one "told him," he made it clear—that Gann's group belonged to Thurman, while his fell under JTF-South. The time taken to establish each of these headquarters entailed delays, and straightening out the organizational relationships prevented the reservists from getting a quick start on the projects over which they expected to exercise responsibility.[66]

[66] First quote from Interv, author with Fishel, 20 Jun 1990. Remaining quotes from Interv, author with Col William Stone, U.S. Army Reserve [1990], Panama.

As Fishel succinctly described the situation, "We had three headquarters running into each other." And that did not take into account the two liaison groups at the DNTT and Colonel Pryor's activities. Over a couple of weeks, however, the civil-military mishmash began to sort itself out. Stiner quickly realized that he was making little use of Stone's people in JTF-South, so he transferred control of the group to the SOUTHCOM J–5, under whose direction Stone believed he should have been working in the first place. Also, something in the way of a division of labor began to emerge. The staff of Gann's task force broke up into ministry teams, with each helping a government department with the problems and issues unique to it. In the meantime, the task force Stone had formed focused on the country surveys and took control of the displaced persons camp, while the U.S. Forces Liaison Group under JTF-South, with Fishel serving as a representative from Gann's task force, continued its work with the public force. On a wiring diagram, the command and control arrangements in effect by early January appeared functional and efficient (*Chart 3*). That said, in the day-to-day activities of these various groups there was still, at one extreme, a good deal of overlap and duplicated functions, and, at the other, a number of operational voids. Daily coordination meetings helped reduce the disarray, but the perception of "ordered chaos" remained.[67]

As each of the elements responsible for civil-military operations was sorting through the organizational turmoil of late December and early January, a larger effort was under way to overhaul and rationalize the whole process. The initiative came from the commander in chief of the Special Operations Command at MacDill Air Force Base, General James J. Lindsay, who expressed his concerns about the stability and nation-building phases in Panama even before Colonel Stone's group had arrived in the country. Lindsay had lobbied vigorously and successfully to have civil affairs and psychological operations units brought into the special operations fold. He now bore ultimate responsibility for the performance of those units and did not want to see them fail in Panama, especially as a result of being misused by conventional force commanders who did not understand their unique capabilities

General Lindsay

[67] Interv, author with Fishel, 20 Jun 1990.

Chart 3—Civil-Military Operations Organizations

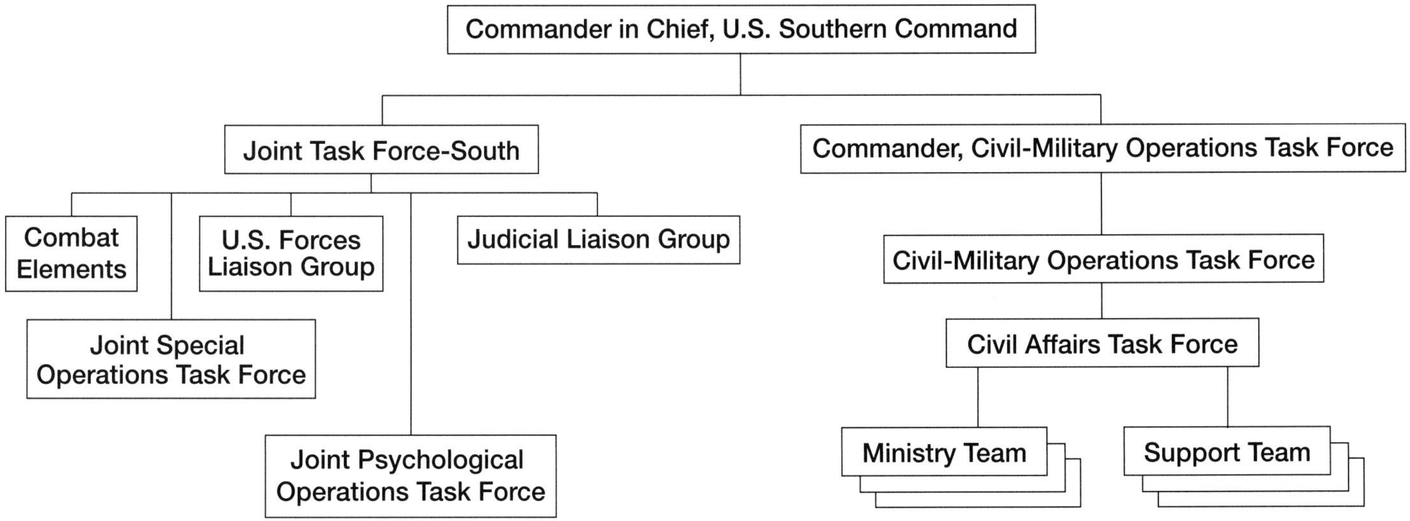

Source: John T. Fishel, *The Fog of Peace: Planning and Executing the Restoration of Panama* (Carlisle Barracks, Pa.: Strategic Studies Institute, 1992), p. 93.

and, as a result, how to apply them. Stiner, of course, was not a problem; he had spent much of his career in special operations. Yet, as commander of JTF-South, his principal concern would be winning the combat phase of the war. As for other officers on the SOUTHCOM, JTF-South, and USARSO staffs and in the field, Lindsay was less confident in their civil-military expertise or even interest. Reinforcing Lindsay's apprehension were reports from one of his staff who, having interacted at times with Blind Logic planners in Panama, believed the plan to be unworkable.[68]

In the first days of Just Cause operations, Lindsay telephoned Thurman several times to discuss the relationship between Just Cause and Promote Liberty, and then, once JTF-South folded its tents, the transition from the former to the latter. At one point, Lindsay asked permission to send an officer from MacDill to Panama to assess the situation and perhaps suggest an organizational alternative to the one that was taking shape. Thurman, who would later be quoted as saying, "I did not even spend five minutes on Blind Logic," was preoccupied with the nunciature negotiations at the time and willingly accepted the offer. On Christmas Day, Col. Harold Youmans, Lindsay's emissary, left MacDill heading south. During his first two days in Panama, Youmans helped to improve the dissemination of civil affairs surveys to the tactical units that would collect the specified information. After that, he began lobbying for a single civil-military organization under which most of the others could be subsumed. In developing a concept for what this organization would look like, he talked to the SOUTHCOM J–3 and J–5 staffs; the commander, Special Operations Command, South; the commanders of the 4th Psychological Operations Group and the 96th Civil Affairs Battalion; and the staff officers at U.S. Army, South. To the USARSO chief of staff, Colonel Hardy, Youmans stressed what he saw as the inevitability of SOUTHCOM's largest service component being assigned the civil affairs stability operations mission, just as had been attempted without success earlier in the month. Hardy agreed, provided Youmans an office (thus allowing him to move to Fort Clayton from Quarry Heights where he had not been popular with the J–5 staff), and told the Army staff at Clayton to prepare for the new tasking.[69]

One critical issue surrounding Youmans' proposal concerned how a single, centralized civil-military organization would be structured. The colonel himself initially favored an agency composed of civilians and military personnel working together, an arrangement that Fishel likened to the Civil Operations and Revolutionary Development Support (CORDS) that had enjoyed some success in pacifying Vietnam, although not in time to change the outcome of that war. General Lindsay, however, preferred another approach, that of the Special Action Force (SAF) that had existed from the 1960s into the early 1970s and had been employed in Latin America and Southeast Asia. Unlike CORDS, the SAF was an all-military organization that, according to Fishel, "put CA, PSYOP, Special Operations Forces, and Combat Service Support all in the same command subordinate to the JTF and commanded by

[68] Fishel, *The Fog of Peace,* p. 39; Interv, author with Col Harold W. Youmans, U.S. Army, 13 Aug 1992, MacDill Air Force Base, Fla.

[69] Fishel, *The Fog of Peace,* pp. 39–40; Interv, author with Youmans, 13 Aug 1992. Quote from Sandler, *Glad to See Them Come,* p. 379.

a general officer." Lindsay made his preferences known to Youmans in a series of telephone calls and during a visit to Panama. Youmans, in turn, adapted the SAF concept to the needs of Promote Liberty, discussed the resulting model with others, passed his revised proposal along to General Hartzog, Thurman's director of operations, and then returned to MacDill. In early January, after having further modified the concept, Hartzog briefed it to Thurman, who gave his blessing to what had been labeled the Military Support Group (MSG).[70]

In the ten days between Hartzog's briefing and the activation of the Military Support Group, much had to be done. Colonel Higginbotham and Lt. Col. Michael W. Menser, the liaison officer to U.S. Army, South, from the Army's Training and Doctrine Command, moved to Building 1 at Fort Amador, the support group's designated headquarters, and began work on getting the personnel authorizations needed to staff the organization. At first, these came exclusively through the Department of the Army, although a decision that the Military Support Group would be a joint undertaking nullified much of this effort. The group's designated commander, Col. (P) James Steele, had made the decision regarding jointness and stuck to it. Steele was a Vietnam veteran, an expert on Latin America, and a former U.S. Military Group commander in strife-torn El Salvador. He and Thurman had worked together in the past, and it was the SOUTHCOM commander who personally picked Steele for the new assignment. Steele responded enthusiastically, sending a note to Thurman with an attached memorandum, "Thoughts on Panama." Next to his signature, Steele wrote, "¡Estoy listo! [I'm ready!]." Steele's deputy would be Pryor, whose presence in Panama's presidential palace had become untenable when what political opposition there was to the Endara administration had begun referring to the colonel sarcastically as the country's "real vice president." As for command and control arrangements, the new organization would fall under JTF-Panama, thus fulfilling to some extent Youmans' prediction that U.S. Army, South, would acquire the nation-building mission. As for JTF-Panama, it underwent its own reorganization that gave Cisneros two deputies: the Military Support Group commander and General Kinzer, the commander of the newly established Ground Forces component that would oversee day-to-day operations.[71]

The Military Support Group was activated on 17 January 1990 with the mission to "conduct nation building operations to ensure democracy,

[70] Quote from Fishel, *The Fog of Peace,* pp. 39–40. Interv, author with Youmans, 13 Aug 1992; Memo, John W. Partin for author, 26 Jun 1990, sub: Military Support Group Origins, and attached Memo, Youmans for Commander in Chief, SOUTHCOM, 8 Jan 1990, sub: Organization of Nation Building Forces.

[71] Fishel, *The Fog of Peace,* p. 40; Interv, author with Higginbotham, 20 Jun 1990. Quote related to Pryor's presence at the presidential palace from Interv, author with Pryor, 21 Jun 1990. Interv, author with Col (P) James Steele, U.S. Army, 28 Jun 1990, Fort Amador, Panama; Note and Memo, Steele for Thurman, 3 Jan 1990. According to a fact sheet prepared in March 1990 by the USARSO Operations Division, at no point after receiving the civil-military operations mission did JTF-Panama, either under General Cavezza or General Cisneros, prepare a supporting plan for Promote Liberty. In receiving this news, General Kinzer noted on the report that, if the information was correct, "we're on our ass!" Fact Sheet, USARSO Deputy Chief of Staff for Operations/Plans Division to Deputy Chief of Staff for Operations, 15 Mar 1990, sub: Research Reason for Lack of JTF-PM Supporting Order in Panama.

Building 1 at Fort Amador where the Military Support Group worked

internationally recognized standards of justice, and [that] professional public services are established and institutionalized in Panama." Soon thereafter, Gann's task force was phased out, after having handed off "all ongoing civil-military operations and activities." Over the coming year, the Military Support Group would experience several organizational changes, but initially it absorbed Cisneros' liaison group (renamed the Public Force Liaison Division) and created components for military police, psychological operations, civil affairs, and special operations (*Chart 4*). With the redeployment of General Cavezza who, given his seniority, served briefly as the JTF-Panama commander, Steele reported daily to Cisneros, who had reclaimed that post. Wiring diagrams notwithstanding, Steele and other support group personnel, with Cisneros' blessing, worked directly for whoever needed their services: the new government and its ministries, the Southern Command, and, perhaps most important, the U.S. Embassy under Ambassador Hinton. As the Military Support Group began to function, one assumption was that it would be disbanded at year's end.[72]

On 11 January, six days before the Military Support Group stood up, the Joint Staff in the Pentagon had informed General Thurman that, if he had no

[72] Fishel, *The Fog of Peace,* pp. 43–48.

Chart 4—Military Support Group Command Structure (as of 31 March 1990)

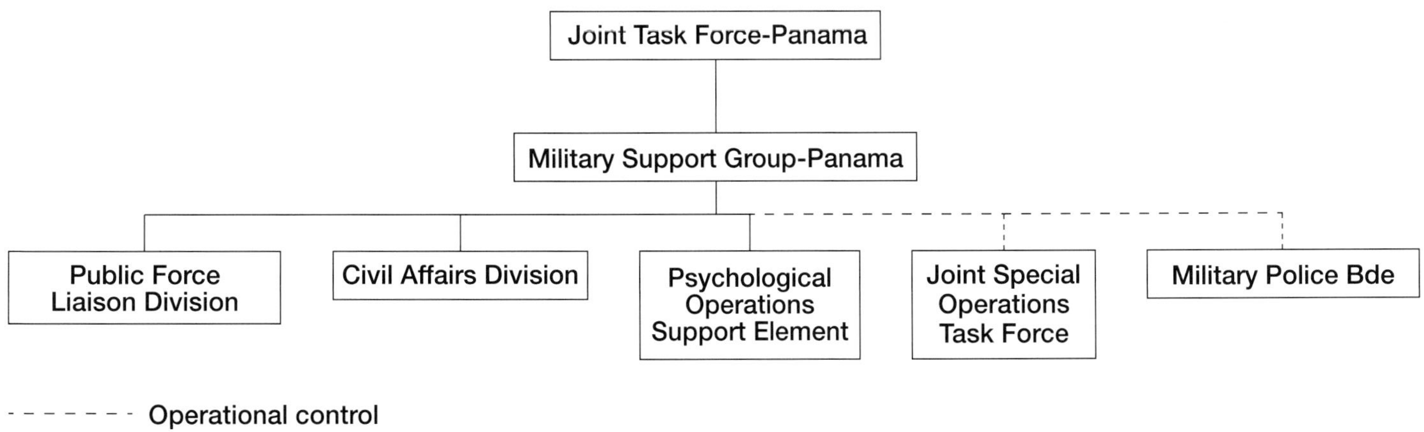

Source: Fishel, *The Fog of Peace*, p. 97.

objection, an announcement would be made that day declaring the formal end to Operation Just Cause. Thurman did not object. Noriega was in custody in the United States. The Panama Defense Forces had been defeated, and the threat of significant, organized armed resistance was over. The Panama Canal was again open, as were the Bridge of the Americas and Torrijos International Airport. A new police force was on the streets, a semblance of law and order was returning, programs for repairing the damage caused by the invasion were under way, and the Military Support Group would soon begin to assist the new Panamanian government as it struggled to become an effective, legitimate, and popular entity. As for the American forces who had flown in from the United States to ensure a victory in Panama, most had redeployed to their home bases or were in the process of doing so. Opinion polls and the public reaction of thousands of Panamanians with whom American troops came into contact indicated that an overwhelming majority of the people approved of the U.S. invasion. Finally rid of a ruthless dictatorship, Panama, with the assistance of the United States, was in the midst of a fitful but promising start toward becoming a stable, functioning democracy.

14

Follow-Through and Assessments

☆ ☆ ☆

The day after Noriega turned himself over to U.S. authorities, he was arraigned in a federal court in Miami on drug-trafficking, racketeering, and money-laundering charges. The trial began over a year and a half later, with the jury delivering its guilty verdict in April 1992. That summer, the judge sentenced Noriega to forty years in prison (reduced to thirty years at the end of the decade). The trial in the United States generated headlines but little public interest. Most Americans had closed the book on the Panama crisis once the widely circulated photograph of Noriega's official mug shot appeared on the front pages of their newspapers.[1]

Public inattention extended as well to the fortunes of the country he had dominated. At the time of the invasion, American officials had talked at length about what needed to be done to transform Panama into a democratic, law-abiding republic. Within a year, however, such discourse was rarely heard, while the aid and assistance the United States had promised to help effect the desired reconstruction proved halting and insufficient. Panama's new leaders complained about the gap between Washington's initial promises and ultimate performance, but, by the end of 1990, their protests were barely heard over the clamor engulfing the Middle East, where war loomed between the United States and Iraq. Given Panama's precipitous demotion on the Bush administration's list of foreign policy priorities, the Endara government, with whatever outside assistance it could muster, proceeded as best it could with the difficult and complex task of rebuilding its country in the wake of the U.S. invasion.

A Busy Year

The growing indifference of the American public toward Panama after Operation Just Cause and the decreasing commitment to nation

[1] In 2010, the former dictator was extradited to France where, in July, a French court found him guilty of laundering drug money and sentenced him to seven years in jail. Then, in September, France granted Noriega a conditional release so that he could be extradited to Panama. Upon his arrival there in December, he was placed in Renacer Prison, a Just Cause target. On 5 February 2012, six days before his seventy-eighth birthday, he was hospitalized for a possible stroke but returned to prison four days later. As of this writing, he is still incarcerated. Associated Press, 12 Dec 2011; *New York Times*, 5 Feb 2012.

building there on the part of the Bush administration caused only mild consternation at the Southern Command where, with the end of hostilities against the Noriega regime, General Thurman declared to a subordinate, "Now, let's get on with the drug war."[2] As a result, most Panama-based U.S. military units, which for over two years had been consumed and distracted by the crisis with the country's dictator, were able to turn their attention back to broader, high-priority issues affecting the hemisphere. Some personnel, to be sure, had to postpone the transition. In Thurman's J–5 directorate, several of the staff remained actively involved in Operation PROMOTE LIBERTY, while at U.S. Army, South, General Cisneros retained responsibility for overseeing what nation-building activities were being enacted by the Military Support Group under Colonel Steele. For these U.S. officers and the people working with them, events in Panama would remain very much in the forefront of their official duties for the remainder of 1990.

Of the urgent concerns confronting the reconstruction and nation-building effort, one of the most potentially disruptive in terms of public support for the new government was crime. While the ruinous looting that occurred during JUST CAUSE had subsided after a few days, the more common forms of crime against property and violent crime against people—murder, assault, and armed robbery—climbed to abnormally high levels during the first three months of 1990, thus undermining official statements that, with the establishment of the new police force, order in the streets was being restored.[3] The causal factors pushing the crime rates higher were not difficult to determine. To begin with, the Panamanian economy had been sliding downward for months before Operation JUST CAUSE, propelled in part by the economic sanctions Washington had imposed on the country in hopes of unseating Noriega without force. Unemployment had reached distressing rates well before December 1989, with the disruptions caused by the invasion only exacerbating the bleak job market and the sense of desperation it generated. Many Panamanians out of work and searching for a way to make ends meet turned to extralegal or illegal means in an effort to support their families and themselves. Hard times determined their course of action. For others, crime was simply a way of life, their chosen profession. For these habitual lawbreakers, the dissolution of the Panama Defense Forces removed many of the prewar restraints on their illicit activities. Compounding the problem were the actions of several PDF officers who, in the early hours of the invasion, had emptied local jails in hopes that the inmates' presence would interfere with U.S. military

[2] Conversation, author with a lieutenant colonel stationed at the Southern Command, 1990.

[3] Briefing Slides, U.S. Military Support Group Panama, n.d. (used during a briefing given to the author on 22 June 1990 at Fort Amador, Panama). Besides the spike in crime, U.S. officials also worried about possible terrorist attacks from recalcitrant hardliners among Noriega's followers. Despite the low probability given such a threat, incidents such as the 2 March 1990 grenade attack at My Place, a *discothèque* in Panama City, in which an American was killed and fifteen U.S. servicemen and twelve Panamanians wounded, kept the issue alive. Eyewitnesses claimed that two men shouting, "Long live Noriega!" were responsible for the attack. *New York Times*, 4, 5 Mar 1990; Donnelly, Roth, and Baker, *Operation Just Cause,* p. 387.

operations. In reality, the scheme accomplished little more than to inflate the number of criminals walking the streets of cities, towns, and villages where the confusion and chaos that prevailed served as an open invitation for lawbreakers to ply their skills. For those who chose to do so, regardless of motive, weapons were readily available for a price, despite the success of the JTF-South money-for-guns program.

The obvious response to the problem of rampant crime was to field a more competent police force and to restore a judicial system capable of handing out justice to lawbreakers who had been arrested. As discussed in the previous chapter, the Endara government made the decision in the first few days of Just Cause to replace the Panamanian military with a police force manned at the outset almost exclusively by former PDF members. The risks to this approach, while deemed to be more acceptable than those inherent in the alternative of starting from scratch with inexperienced recruits, quickly became manifest, afflicting the inchoate program from the outset. The early initiates into the force, formally named the Panama National Police, received only cursory training, a fact that became all-too obvious from the numerous incidents of unprofessional and questionable behavior that were reported once the new cops arrived at their designated headquarters and began walking the streets. The heavy-handedness many PNP officers exhibited in dealing with their countrymen gave rise to increasing resentment among a population already alienated by the cops' PDF lineage. Not that most of the fresh recruits were themselves all that happy, having been compelled to make the abrupt transition from soldiering to police work, with the perceived loss in status and self-esteem associated with the latter. Adding to their discontent was the fact that in the first weeks on the job, many in the Panama National Police lacked uniforms and equipment, as well as any real familiarity with police methods and procedures.

Throughout the remainder of 1990, as U.S. and Panamanian officials sought remedies for each of these problems, measures to redress the shortage of weapons, radios, vehicles, and uniforms yielded the most immediate and visible results. Here, the U.S. Forces Liaison Group—renamed the Public Force Liaison Division after its incorporation into the Military Support Group in late January—played a leading role, searching inventories of captured Panamanian materiel, drawing on U.S. military surplus, and employing U.S. security assistance funds to acquire many of the needed items.[4] The progress achieved in this area, however, was not matched by efforts addressing the more intangible problem of recasting the mind-set of former PDF members who had joined the national police. Training and advisory programs essential for creating an efficient and respected force could take months, even years, to overcome the legacy of the past and instill within PNP recruits a new set of standards and values that would lead to more professional behavior. Time, unfortunately, was in short supply. The purpose of Operation Just Cause,

[4] Fishel, *The Fog of Peace*, p. 43. Briefing Slides, U.S. Military Support Group Panama, n.d., provide a longer list of the Panama National Police's equipment needs and the means by which they would be fulfilled. Besides the methods mentioned in the text, the new government of Panama itself purchased many of the required items and relied on countries other than the United States for additional equipment.

the Bush administration insisted, had been to liberate, not occupy, Panama. Avoiding any impression to the contrary meant, in part, that U.S. military personnel working with the Panama National Police would not enjoy the luxury of an extended timetable for their efforts to succeed. Bureaucratic and legal constraints, together with what often seemed like official indifference in Washington, further impeded attempts to implement long-term training and advisory programs.[5]

The only training the entire force received in its first year was the twenty-hour "quick-fix" transition course developed by reservists and supervised by the Public Force Liaison Division's Maj. Richard Downie, with the help of Colonel Stone's Civil Affairs Task Force. Panamanian officers took the course and then taught it to the rank and file stationed in the cities and countryside. All involved recognized the defects of the quick-fix, which had to cover at least a dozen substantive topics in a very short time: ethics, first aid, basic police techniques, community relations, patrol techniques, crowd control, traffic control, criminal code procedures, the use of force, weapons familiarization, police reporting procedures, and handling problems involving juveniles. Downie's sardonic take on the course: its primary goal was to make the new cops safe to walk the streets.[6]

As the transition course struggled forward, U.S. officials labored to put other measures into place that would ensure graduates received continuous on-the-job training and advice. In late January, U.S. Chargé d'Affaires Bushnell determined that members of the U.S. reserve component on the scene who had been policemen in the United States were eligible to serve as trainers for the Panama Public Force. Thus began what participants dubbed the RC Cop program, in which two reservists would be assigned to every city police precinct and one reservist to every rural precinct. Also in each precinct, whether urban or rural, would be two U.S. Green Berets, there to place their language skills and other capabilities at the disposal of the reservists. In further support of the program would be U.S. military police, who would continue to conduct patrols and share their expertise with the new Panamanian force.[7]

These arrangements for the American military to expedite a systemic training program for the Panama National Police beyond the transition course soon encountered difficulties. Chief among these was a law the U.S. Congress passed in February, the Urgent Assistance to Democracy in Panama Act, that appropriated a meager $43 million for use in reconstructing the country. As the legislation pertained to Operation PROMOTE LIBERTY, Congress, by granting an exemption to Section 660 of the Foreign Assistance Act of 1961, permitted the U.S. military to support the PNP. The exemption, however, did not extend to that portion of Section

[5] According to Fishel, *The Fog of Peace*, p. 43, "The U.S. Embassy had set as its watchword from the beginning that Operation JUST CAUSE was a 'liberation,' not an 'occupation.' Liberation clearly put the onus for success or failure on the Panamanians and severely constrained the U.S. role, including the role of the military."

[6] Ibid., pp. 43–49. The list of topics covered in the twenty-hour course was taken verbatim from Briefing Slides, U.S. Military Support Group Panama. n.d. Downie's comment on the course was mentioned by Fishel in a marginal note on a manuscript copy of this chapter.

[7] Fishel, *The Fog of Peace*, pp. 43–49.

660 prohibiting a military role in *training* foreign police forces. That still left several things American armed forces could do. The Special Forces could help vet the new force and assist zonal police officials in establishing positive relationships with local civilian leaders. Reservists and Green Berets could advise and monitor the new police, and U.S. military police could still set a professional example for their Panamanian counterparts. But any formal training sponsored by the United States would have to come from a nonmilitary source.[8]

Such an entity existed in the form of the International Criminal Investigative Training Assistance Program (ICITAP). Created on 6 January 1986 under the administrative and operational authority of the U.S. Department of Justice, the program in practice received its guidance and funding from the State Department. Drawing on the expertise of available law enforcement personnel and civilian officials in the United States, the training assistance program undertook as one of its principal objectives the development of foreign police forces, mainly in the Caribbean. Prior to 1990, the program had accomplished little in the region as a whole, having devoted most of its limited resources to assisting in the investigation of human rights violations in strife-torn El Salvador. Still, when Congress and the Bush administration sought an alternative to the U.S. military for training the Panama National Police, ICITAP was selected for the assignment. To remove any possible doubts about the legality of the undertaking, the Urgent Assistance to Democracy in Panama Act contained an authorization for ICITAP personnel to "reconstitute" police authority in nations "emerging from instability," including the training of "cops on the beat." Specifically, the training assistance program received the mission of inaugurating a multiyear plan for restoring law enforcement services to Panama. The State Department's Bureau of Inter-American Affairs provided the initial funding in the amount of $13.2 million.[9]

By the end of 1990, this hastily implemented arrangement had not fulfilled all its supporters' expectations. As one analyst has observed, the training assistance program was "wholly unprepared for the tasks thrust upon it." Having completed just its fourth year in operation, the program still lacked adequate funding and consisted of "only a small staff of FBI agents on loan and a few regional language specialists." Ten officers arrived in Panama, their official experience in the Justice Department program having been limited almost exclusively to helping foreign policemen become skilled investigators, not streetwise professionals. In short, the program's personnel lacked the practical knowledge as well as the necessary resources for training cops on the beat. Their work with the Panama National Police over the ensuing months and into the following year reflected those deficiencies as well as other problems. To their credit, ICITAP personnel did design and implement a new three-week

[8] Ibid., pp. 47–50; Anthony Gray and Maxwell Manwaring, "PANAMA: Operation *Just Cause*, in *Policing the New World Disorder: Peace Operations and Public Security*, eds. Robert B. Oakley, Michael J. Dziedzic, and Eliot M. Goldberg (Washington, D.C.: National Defense University Press, 1998), pp. 54–55.

[9] Quotes from Charles T. Call, "Institutional Learning within ICITAP," in Oakley, Dziedzic, and Goldberg, eds., *Policing the New World Disorder*, p. 320. Gray and Manwaring, "PANAMA," pp. 55, 59.

transition course for the former PDF members, establish a police academy, and offer advice on PNP organizational issues. But progress in most areas was glacial, and deadlines and goals were continuously missed. According to one account, "The 3-week transition course . . . was not initiated until May 1990, and the 1991 target of 5,000 course recipients was not met until September 1992." The police academy enjoyed some success because of the positive working relationship established between the PNP major serving as its director and a U.S. official in the Justice Department. Other programs, in which such a level of personal interaction did not exist or was less constructive, often drew criticism. One director general of the Panama National Police dismissed the three-week transition course as useless and the bulk of the ICITAP effort as a waste of U.S. tax dollars. The training assistance program, for its part, conceded its imperfect start but argued that it was making the adjustments and requirements needed to improve its chances for future success. Officials involved in the undertaking also reminded their critics that many of the difficulties being encountered were the result of the decision made well before ICITAP's arrival to rely on former PDF personnel in standing up a new force. Seemingly confirming the problems inherent in this plan, an opinion poll in 1992 indicated that nearly 65 percent of the Panamanian people remained distrustful of the Panama National Police, with only 26 percent recording a positive perception.[10]

ICITAP's fitful approach, initial delays, and staff shortages proved frustrating to all involved, especially those U.S. military personnel who, while providing authorized support to Panama's police force, had to neglect many of their duties not related to the PNP. The inability to effect a workable handoff of the training mission with the training assistance program also impeded the military's attempts to lower its high profile in Panamanian security matters, a move deemed essential for confirming President Bush's assertion that it was not an occupation force. The new U.S. ambassador to Panama, Deane Hinton, was especially concerned about this point. He was the top U.S. civilian official on-the-scene responsible for post–JUST CAUSE reconstruction programs, and, based on his extensive diplomatic experience in other countries, he firmly believed that, in the words of one author, the U.S. military should not be placed "in the forefront of supporting democratization, the creation of civil government, and nation-building." Holding to this conviction, Hinton let Thurman, Cisneros, and Colonel Steele's Military Support Group know that the U.S. Embassy and the State Department should play the leading role in establishing a new police force. Then, having made his point, he bowed to the reality of the situation as it existed throughout 1990. Until the Panama National Police demonstrated greater progress in its development as a professional organization, Hinton could not sever most of the working ties that had been established between the PNP and the Military Support Group. Consequently, the support group's Public Force Liaison Division would continue to interact with ICITAP, while U.S. Special Forces, military police, and reservists would continue to conduct joint patrols with the Panama National Police, support "precinct house

[10] Quotes from Call, "Institutional Learning within ICITAP," pp. 329–33. Fishel, *The Fog of Peace*, pp. 49–51; Gray and Manwaring, "PANAMA," pp. 55–64.

administration" and "other law enforcement functions to maintain order," and monitor the PNP's "adherence to training standards." These activities, as characterized in the Military Support Group's briefing slides, would permit U.S. military personnel to operate with the Panamanian police but not train them.[11]

While Hinton recognized the need for the U.S. military to enjoy some latitude in dealing with the Panama National Police, he faced a larger issue concerning the Military Support Group. Although the new ambassador was formally in charge of the postinvasion reconstruction program in Panama, the Pentagon and the Southern Command had deliberately excluded his predecessor and the U.S. Embassy from the nearly two-year planning process that had culminated in Operation PROMOTE LIBERTY. Complicating matters further, when the Joint Chiefs of Staff ordered execution of the plan on 21 December, most embassy personnel were not even in the country. Not that their presence would have mattered that much. Given the staff's small size and its lack of nation-building expertise, it would not have been able to assume the leading role that diplomatic protocol and tradition expected of it. Nor could these shortfalls be rectified by appeals through official channels back to Washington. The State Department simply did not have a sufficient complement of officials with nation-building experience to commit to Panama, and time constraints precluded hiring volunteers, an option that would have required advertising for the positions, screening the applicants, and then training those selected. Given the scope of all the difficulties he faced, Hinton found himself lamenting, "I had no clue at all, arrived here with a disorganized embassy staff, no [US]AID at all, a few local employees. . . . No resources and a mandate to fix things. . . . I think their [DOD] planning and the way they did things was mistaken in some respects." But mistaken or not, without the assets to address his misgivings along lines more acceptable to him, Hinton acquiesced as the Military Support Group moved to the forefront of the nation-building effort by default. As Colonel Steele succinctly summarized the situation, "The embassy was not functional," so his newly formed organization "played the role of the country team."[12]

As Hinton made his peace with what necessity demanded, he also made his position clear. The Military Support Group would take the organizational lead in reconstructing Panama, but only with him, the ambassador, calling the shots. On paper, it appeared as though this condition would be difficult to meet. The support group's command and control diagram placed the organization squarely within JTF-Panama and answerable directly to its commander, General Cisneros; Cisneros, in turn, reported to General Thurman, who, as a unified commander, answered directly to the National Command Authority, not the State Department or any U.S. Embassy in his area of responsibility. (Thurman later said he made this point forcefully, but amicably, to "my buddy"

[11] Gray and Manwaring, "PANAMA," p. 60; Fishel, *The Fog of Peace*, p. 50. First quote from Shultz, *In the Aftermath of War*, p. 40. Remaining quotes from Briefing Slides, U.S. Military Support Group Panama, n.d. Interv, author with Maj Gen Marc A. Cisneros, U.S. Army, 29 Jun 1990, Fort Clayton, Panama.

[12] Shultz, *In the Aftermath of War*, pp. 40, 63.

Hinton, a "good man, but you've got to straighten him out every now and then.") But Hinton was a tough and seasoned diplomat, and Thurman, eager to shift his command's attention to other issues afflicting Latin America, did not desire a turf battle with the ambassador over nation building in Panama. As a result, the two men hammered out a mutually acceptable arrangement whereby the Military Support Group, while formally remaining within the military's chain of command under Cisneros, acknowledged the primacy of the ambassador's jurisdiction. Cisneros, for his part, accepted the agreement. To be effective, he realized, the support group needed to be in direct and constant communication with virtually any individual or group involved in the rebuilding process or affected by it, including Panamanian officials at the highest levels, American diplomats, and the variety of U.S. agencies contributing to Promote Liberty (*Chart 5*). A daily briefing from Colonel Steele kept the JTF-Panama commander informed of the group's activities and contacts and served as one vehicle for the general to issue directives reflecting what he believed needed to be done and who needed to be doing it. Steele briefed Thurman daily as well, with the military chain in Panama ensuring that Hinton received full accounts of these deliberations.[13]

Once the embassy and the Southern Command reached accord on who would answer to whom for U.S.-sponsored nation-building activities in Panama, civil-military relations for the most part proceeded harmoniously. There were, to be sure, a few exceptions to this concord, one of the most significant of which concerned the role psychological operations would play in Promote Liberty. Once Ambassador Hinton was fully briefed on the Military Support Group, he learned of the PSYOP program, started during the first days of Just Cause and now under Steele, to produce informational material and engage in projects that would help the new government of Panama establish its legitimacy. Of necessity, the task involved the Army's psychological operations personnel in activities directly linked to Panamanian politics, a connection that, if made public, could prove extremely embarrassing to both President Bush and President Endara. Hinton was understandably uncomfortable with the arrangement.

Well before the new ambassador's arrival in Panama, the Psychological Operations Task Force set up under JTF-South was composed primarily of people and materiel from the 1st Psychological Operations Battalion and, later, its parent headquarters, the 4th Psychological Operations Group, both based

[13] Ibid., p. 49; Fishel, *The Fog of Peace,* pp. 39–40; Briefing Slides, U.S. Military Support Group Panama, n.d. Quote from Briefing, Gen Maxwell R. Thurman at SAMS, 23 Sep 1992. Intervs, author with Cisneros, 29 Jun 1990; with Col (P) James Steele, U.S. Army, 28 Jun 1990, Fort Amador, Panama; with Col Jack Pryor, U.S. Army, 21 Jun 1990, Fort Amador, Panama. During a postinvasion trip to Panama in June 1990, the author attended Colonel Steele's briefings to General Cisneros. Thurman, in his SAMS briefing, said that Hinton initially thought of him as the U.S. Military Group commander in Panama. Thurman corrected that misperception, while also laying out his chain of command, which did not include the ambassador. Thurman's "straighten him out" phrase should be viewed as nothing more than the kind of dramatic talk for which the general was known and not as his actual statement or an insinuation that he in any way was brusque with the ambassador. Thurman, in fact, repeatedly mentioned their cordial interaction and daily tennis matches.

Chart 5—Military Support Group Interactions, Access, and Influence

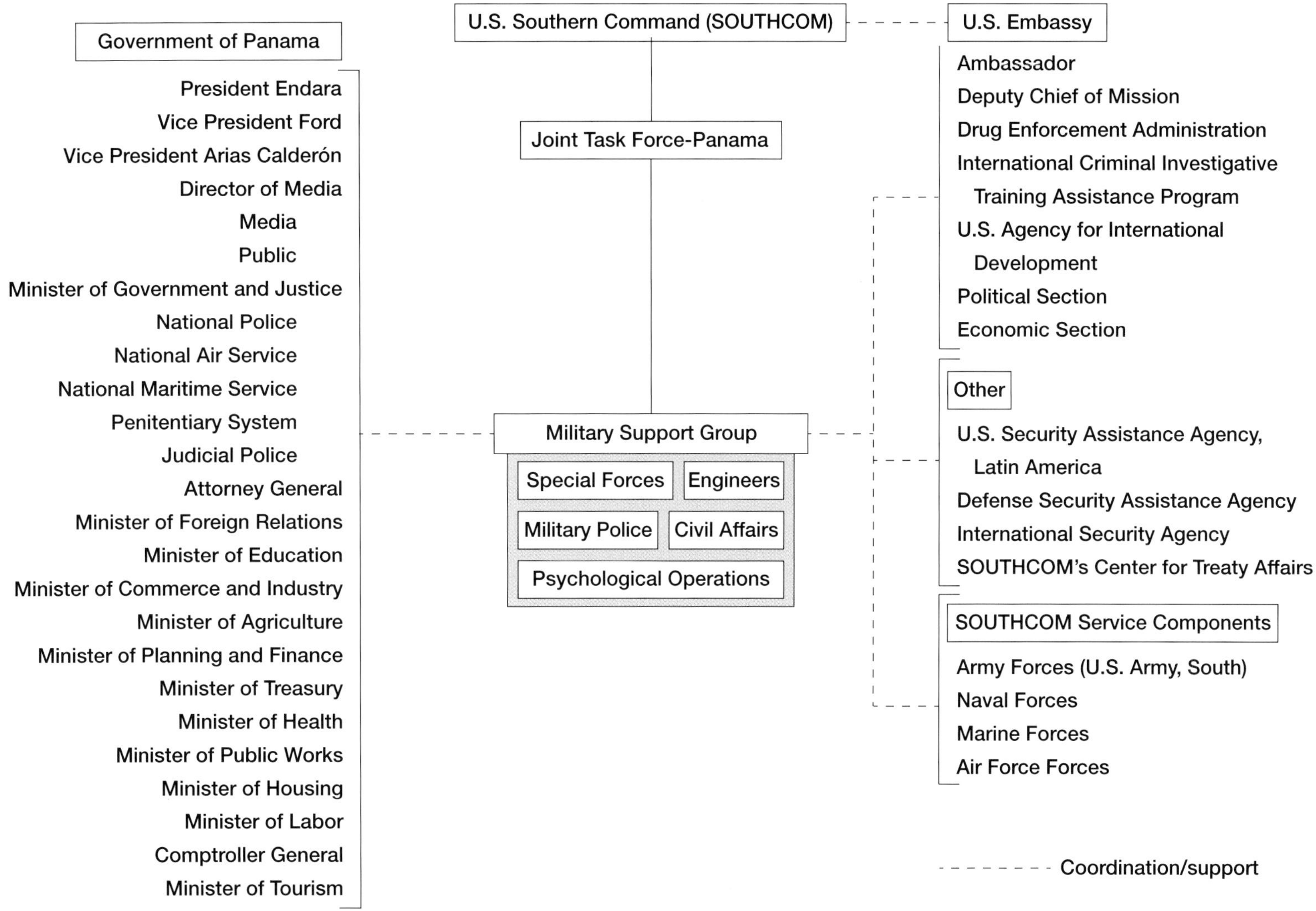

Source: Briefing Slides, U.S. Military Support Group Panama, n.d.; Fishel, *The Fog of Peace*, p. 88.

at Fort Bragg. Until early January, the unit had concentrated most of its effort on explaining and legitimizing the U.S. invasion, soliciting information from the Panamanian people concerning the whereabouts of weapons caches and fugitives, providing information on such topics as displaced persons camps and U.S. military medical services, and getting formerly regime-operated television and radio stations up and running for use by the new government. At the time the Military Support Group was activated, the task force had been cut back to a 48-person Psychological Operations Support Element designed for Steele to incorporate into his organization. By then, the situation in Panama had reached a point where the PSYOP element could shift its priorities from supporting U.S. operations in the field to the task that so troubled Hinton: assisting the Endara government in consolidating and legitimizing its power. Toward that end, psychological operations specialists manufactured a number of audio-visual and printed products—posters, pamphlets, leaflets, television news footage, and radio scripts, among them—that could be used by President Endara, Vice Presidents Arias Calderón and Ford, and the ministers serving under them.[14]

The entire undertaking was highly sensitive and classified and not without reason. Even in the euphoria immediately following the downfall of the Noriega regime, some Panamanian leaders were already criticizing the Endara administration for being a puppet of the United States, a charge that had pushed Col. Jack Pryor out of his advisory position inside Panama's presidential palace and into the Military Support Group as Steele's deputy. If, under these circumstances, the fact became known that U.S. PSYOP experts were secretly involved behind-the-scenes in assisting the new government, the critics' credibility would be strengthened at a time when the American-installed triumvirate was very vulnerable. Any public exposure of the relationship would also raise questions concerning whether the U.S. Army was engaged in what could be construed not just as political activities, which were often unavoidable in stability operations, but as partisan politics, the institutional prohibition against which has been a time-honored American military tradition. For these reasons, security measures surrounding the Psychological Operations Support Element's activities were rigorous.

Security, of course, provided no guarantee that the material produced by the PSYOP element would be free from errors caused by the impossibility on short notice of mastering all aspects and nuances of Panama's political culture. For example, one PSYOP task was to produce a poster showing Panama's national seal "surmounted by an eagle." When completed, the poster was "beautifully done," according to the psychological operations after action report. Unfortunately, from the support group's perspective, the bird depicted atop the Panama crest was not the eagle traditionally found in that country but "a North American bald eagle—not exactly the kind of thing which would make us popular with the population!" The element's operations officer caught the mistake before the poster was distributed—a close call "clearly

[14] *Psychological Operations in Panama During Operations Just Cause and Promote Liberty*; Shultz, *In the Aftermath of War*, pp. 57–61.

demonstrating the complexity of PSYOP product development during JUST CAUSE, and later during PROMOTE LIBERTY."[15]

In light of these glitches and other risks that could result in scandal, unwanted publicity and a counterproductive backlash, Hinton told Steele at the outset of their collaboration, "I do not like PSYOPS." Bushnell shared this view, and pressure from the embassy led to the Psychological Operations Support Element being cut in April 1990 from forty-eight to eighteen people. According to one analyst, the decision crippled the element's "crucial communication and information support" without making any attempt "to establish what was and what was not within the boundaries of military PSYOPS, or to follow up after the groundwork was laid." Whatever the validity of this critique, it was not compelling enough to reverse the new policy.[16]

The retrenchment in psychological operations was just one setback the Military Support Group encountered in performing its mission. Another was lack of sufficient funds with which to conduct a full-scale nation-building program. President Bush had spoken of $1 billion in foreign aid for Panama, but the funding bill that passed Congress in May allocated less than half of that, $420 million, and, of that sum, only $100 million had reached its destination by early 1991. With just one-tenth of the money that had been anticipated, Steele's group and those working with it had to scale back dramatically what they hoped to accomplish. As one author described the adjustment, nation building gave way to "infrastructure reconstruction," especially in the countryside. Translated into specific projects, this meant undertaking repairs on the country's dilapidated schools, hospitals, clinics, roads, bridges, and other facilities in rural areas. To work on these enterprises, the Military Support Group needed military engineers, which Steele had no authority to summon. The colonel was thus appreciative when General Thurman intervened to arrange for the 536th Engineer Battalion and various National Guard engineering assets to come under the support group's operational control (*Chart 6*).[17]

Of overriding importance was that credit for the reconstruction projects go to the new government and that the president and vice presidents and their ministers be observed cooperating with provincial and local officials in whose districts the work would be performed. Yet, these seemingly basic connections took time to arrange. During the Noriega years, Endara and his colleagues had been in the opposition; they had no experience in running a country. Moreover, the dictatorial system dominated by the Panama Defense Forces had stifled coordination between government officials at the national level and those scattered throughout the provinces, cities, towns, and villages.

---

[15] Shultz, *In the Aftermath of War,* pp. 57–61. Quotes from *Psychological Operations in Panama During Operations Just Cause and Promote Liberty.*

[16] Quotes from Shultz, *In the Aftermath of War*, pp. 57–58, 60–61. Most of the eighteen soldiers remaining after the personnel cuts in the support element were replacements from the 8th Psychological Operations Battalion, assisted by a few experts from the 1st Psychological Operations Battalion and the 4th Psychological Operations Group. *Psychological Operations in Panama During Operations Just Cause and Promote Liberty.*

[17] Shultz, *In the Aftermath of War*, pp. 55–56.

Chart 6—Organization of the Military Support Group (Effective 15 May 1990)

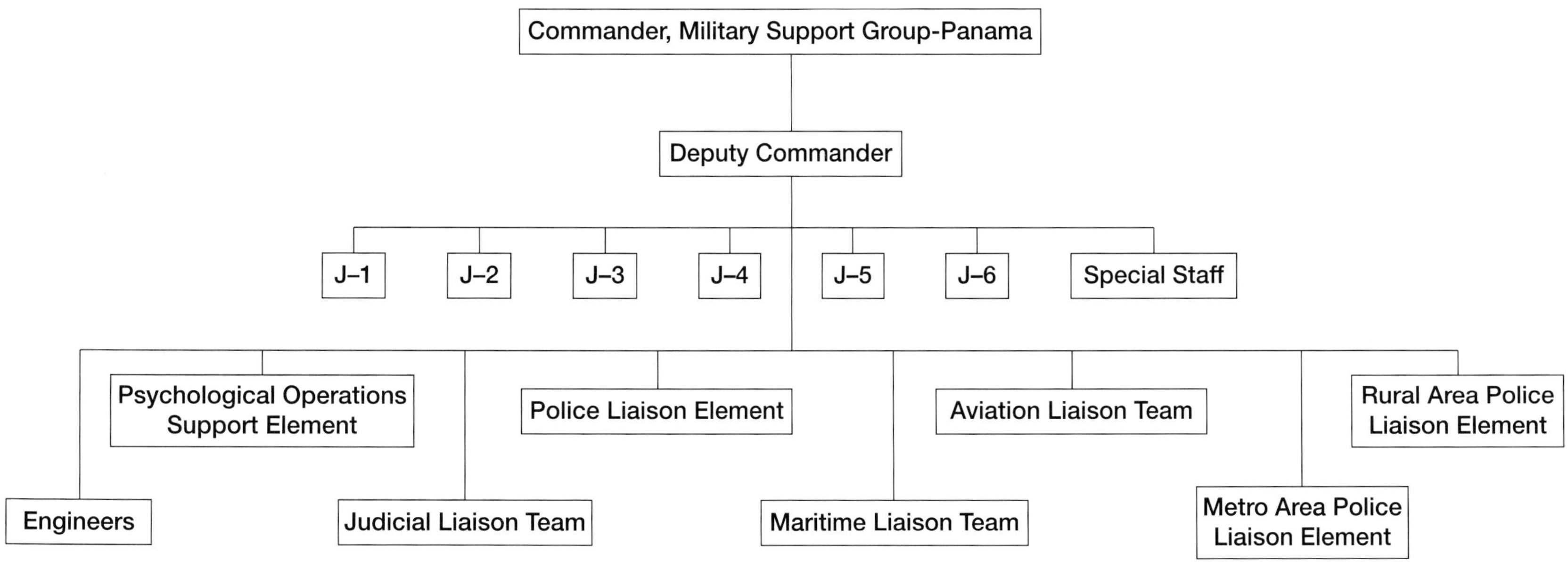

Source: Briefing Slides, U.S. Military Support Group Panama, n.d.

To establish the essential network of contacts among the various levels, the Military Support Group, according to one source,

> encouraged coordination by requiring that every reconstruction project be coordinated between the MSG and the appropriate [government of Panama] ministry. For example, if the Ministry of Education presented a list of schools it would like the MSG to repair, the MSG identified the resources required while ministry representatives took the proposal to the appropriate province officials and then to the local level for their input. When consensus was reached, the project would be initiated.[18]

The local reconstruction projects, while less grandiose than the scope of nation-building activities originally envisaged by U.S. officers planning civil-military operations for postcombat Panama, generally received favorable publicity that redounded to the benefit of the Endara administration. The sense of relief that pervaded the country following Noriega's downfall also enhanced the new government's image. But with the accolades also came the brickbats, as a number of high-profile problems plagued the administration during its first year or so in office. To begin with, Endara had campaigned for the presidency in 1989 as leader of a broad-based four-party coalition that, once in power, fell victim to factional political bickering and in-fighting that frustrated American officials and exasperated many Panamanians. Then, there was the fact of the U.S. invasion with which to contend. In the process of deposing the dictator, Operation JUST CAUSE had taken a psychological and physical toll on the Panamanian people and their country. As previously noted, Cisneros repeatedly warned in early 1990 that the United States and the Endara administration could expect a honeymoon of about a hundred days, after which both governments would be subjected to criticism for any remaining problems stemming from the invasion. Of these problems, perhaps the most salient was the plight of the people displaced by JUST CAUSE, especially in Panama City, where, over two years after being burned out of El Chorrillo and other neighborhoods, victims of the fire saw the construction of new homes slowed by financial and political disruptions or found themselves still living in the "tin city" erected for them at Albrook Air Station.[19] The trickle of promised U.S. aid provided no corrective, and, while Endara at one point conducted a public hunger strike to draw attention to the displaced persons (and his country's poverty level in general), his inability after a meeting with President Bush to obtain a substantial increase in U.S. funding raised questions concerning his clout in Washington, not to mention the White House's interest in the welfare of a post-Noriega Panama.

Another issue proved a two-edged political sword for Endara. While he moved quickly to replace the Panama Defense Forces with a police-oriented organization answerable to civilian officials, the Panama National Police's

[18] Ibid., p. 56.

[19] *The Panama Deception*, a 1992 Oscar-winning "documentary" on JUST CAUSE, offered viewers little more than a well-made piece of propaganda. One of the film's critiques, however, its depiction of the displaced persons' plight at Albrook, raised a valid issue and highlighted a scandalous situation, albeit in a sensationalized way.

glacial adoption of high professional standards, the problems affecting its training program, and its initial inability to curtail crime resulted in a gap between public expectations and actual progress that worked against the new government. Nor was the administration's image helped when, in December 1990, around a hundred members of that new force staged what was variously described as a protest, an uprising, and a coup attempt against the government. U.S. troops stationed in Panama crushed the movement, but that only left Endara exposed as a man who still seemed to need American armed force to help run his country. One result of the December unpleasantness was that Ambassador Hinton asked Steele to postpone inactivation of the Military Support Group, scheduled for that month, until 17 January 1991, the anniversary of its creation.[20]

Operation PROMOTE LIBERTY did not end with the support group's subsequent dismantling. Steele, who became U.S. Army, South's deputy commander for support, used JTF-Panama's J–5 directorate to continue a variety of assistance programs, and, as part of that process, he and Ambassador Hinton met with the Panamanian president once a week for breakfast. Still, on 17 January 1991, a milestone had been reached, as PROMOTE LIBERTY entered a new, even less dynamic phase. Looking back over the preceding year, the balance sheet on nation building and reconstruction had been mixed. In the most general terms, a civilian-led government had replaced a military dictatorship, and the U.S. military had done what it could to make both the new government and that government's new police organization work. In December 1990, news media in the United States ran several broadcasts along the lines of "Panama: One Year Later." Most noted areas in which progress had been made but concluded that poverty, corruption, and drugs remained serious problems in the "liberated" country. By that time, of course, the American people were concentrating on the buildup of U.S. troops in the Persian Gulf in what would be the prelude to Operation DESERT STORM. If the situation in Panama had not improved to the extent that policymakers had hoped as the result of President Bush ordering Operation JUST CAUSE, few Americans cared by the first anniversary of that occasion. Nor did they concern themselves in 1994 with the results of the first Panamanian national elections to be held after the invasion, even though the candidate of a reformed and revitalized Partido Revolucionario Democrático, one of Noriega's principal political organs, won the presidency. U.S. officials assessing the outcome emphasized its positive side; the political structure in Panama had proved itself viable in the toughest test a new democracy could face, the peaceful transfer of power to the opposition through the electoral process. Five years later, another milestone passed, again without notice or fanfare on the part of the American people: on 31 December 1999, the United States, in adherence to the Panama Canal treaties, vacated its remaining military bases in Panama and turned complete control of the waterway over to the government of that country, thus confounding the critics who had charged that the U.S. invasion ten years earlier had been a naked power play, motivated by Washington's desire to reacquire absolute control over the canal's operation.

[20] Fishel, *The Fog of Peace*, p. 52.

CONCLUSIONS

Within months—in some cases weeks—of the cessation of hostilities in Panama, the Pentagon, the Southern Command, the XVIII Airborne Corps, and the numerous other U.S. military headquarters involved in Operation JUST CAUSE began conducting briefings and distributing after action reports—some classified, some not—touting their respective roles. Somewhere near the beginning of practically each of these assessments was the mission statement that the featured headquarters had received, while the information contained in the slides and pages that followed reflected an across-the-board consensus: U.S. units and personnel had accomplished their missions in what had been a highly successful operation. The dictatorship in Panama had fallen, and its leader was in the United States awaiting trial. The Panama Defense Forces had been destroyed as an organization and replaced by a police force subject to civilian control. After a rough but short transition from combat to stability operations, law and order had been restored to most of the country, and a government that had been legally elected but denied power was finally in place and functioning. In the process, U.S. property and other interests including the canal had been secured, and, with three tragic exceptions, the lives of American civilians living in Panama protected.[21]

American service personnel had paid a price in lives and limbs to achieve these goals, but the casualties were considered low in light of the worst-case scenarios some planners had predicted. The U.S. military had lost 23 killed in action and 324 wounded. An exact count of enemy dead was difficult to obtain, largely because many defenders fought wearing civilian clothes, making it impossible in some cases to distinguish between them and innocent bystanders who were killed. Well before the end of JUST CAUSE and quickly thereafter, several U.S. and Panamanian agencies conducted rigorous investigations to determine as best they could the number and identities of the Panamanian dead. Most conclusions indicated that somewhat over three hundred Panamanian military, paramilitary, and innocent civilians had died in the invasion. Wounded and injured Panamanians numbered around sixteen hundred.[22]

---

[21] Preinvasion assertions by the SOUTHCOM and JTF-South commanders that the death of one U.S. civilian would mean Operation JUST CAUSE had failed were ignored or explained for what they were—an emphasis on the importance of protecting American lives. As noted in the text, two of the three Americans—the schoolteacher and the college student—were killed (the first by the PDF, the second by friendly fire) the night of the invasion. The body of the third American was found a few days later.

[22] Casualty figures are from Briefing Slides, U.S. Southern Command, Operation Just Cause: "Rebirth of a Nation," n.d.; Cole, *Operation Just Cause,* pp. 65–66; and the Web site http://www.globalsecurity.org/military/ops/just_cause.htm. In commenting on this chapter, John Fishel indicated that he thought the most accurate figure was 302, which was provided by the Roman Catholic Church after an investigation. A list of the figures drawn up by various agencies are at http://william_h_ormsbee.tripod.com/jc_casualties_panamanian_p01.htm. Immediately after JUST CAUSE and for several years to follow, critics of the invasion claimed that the actual number of Panamanians killed was in the thousands, many of them executed by U.S. forces and buried in secret mass graves. Intensive official investigations revealed no evidence to support such figures, while the largest mass grave in Panama City was front-page news at the

It was only natural that the success of Operation JUST CAUSE and the speed with which victory was achieved would call forth a good deal of hyperbole. One boast made by a number of senior military officers was that U.S. forces had attacked twenty-seven targets simultaneously at H-hour—a catchy and evocative, albeit not quite accurate, characterization of the rapid deployment of American firepower. Pronouncements also heralded how each one of the some twenty-seven thousand U.S. service personnel involved in JUST CAUSE had performed superbly, the result of the high standards set for military recruiting, training, and leadership. This was an obvious exaggeration not meant to be taken literally but rather as a generalized tribute to a praiseworthy performance by the post-Vietnam armed forces of the United States. Another embellishment, duly circulated by the news media, came in a comment made to reporters by General Stiner, in which the JTF-South commander stated that "in my judgment, there were no lessons learned on this operation. I don't think that I or my commanders or our armed forces learned a single lesson. . . . But we did validate a lot of things." The statement had an immediate impact on at least one group, a six-man team of officers arriving in Panama from the recently created Center for Army Lessons Learned located at Fort Leavenworth, Kansas. Given Stiner's position, the team's mission changed from finding lessons to compiling "first impressions." Consequently, the center published a largely anecdotal analysis of the operation, a series of vignettes contained in three very thin but insightful pamphlets.[23]

A decade later, Stiner still stood behind his initial assessment. As he elaborated to author Tom Clancy, "I can't say we really learned any lessons. In my opinion, you only learn a lesson as a result of a big mistake, or when you have failed to anticipate an event somewhere along the line—training readiness, plan development, or the like—that would affect the mission." Turning to what the U.S. military had "validated" during JUST CAUSE, the general proceeded to enumerate "some principles and procedures that contributed to our success in Panama, which also apply to future operations." The nine points on the list he supplied Clancy could be found, with much additional material, in the corps' own 1990 after action report, which hails the "many valuable observations

---

time it was dug for the purpose of providing a temporary burial place until the city's morgues and hospitals could care for the bodies properly.

[23] For the simultaneous targets claim and variations on it, see Briefing Slides, U.S. Southern Command, Operation Just Cause: "Rebirth of a Nation," n.d.; Daniel Bolger, *Death Ground: Today's American Infantry in Battle*, paperback ed. (New York: Ballantine Books, 2003), p. 71, and see also p. 70. (The hardback edition of Bolger's book was published by Presidio Press in 1999.) A doctrinal source is Joint Publication 3–0, *Doctrine for Joint Operations,* September 2001, p. IV-11. First quote from "Inside the Invasion," *Newsweek,* 25 Jun 1990, and is also discussed in the Bolger volume and in Brian McAllister Linn, *The Echo of Battle: The Army's Way of War* (Cambridge, Mass.: Harvard University Press, 2007), p. 220. One example of the somewhat exaggerated depiction of individual performance in JUST CAUSE came in a briefing by a high-ranking Army officer at Fort Leavenworth on 9 January 1990. As for the changed mission of the CALL team to Panama, the experience (and the second quote) was related to the author by Col. Frank Akers, the leader of the team. As for CALL's instructive but meager output on Operation JUST CAUSE, compare the three pamphlets—all cited in previous chapters of this volume—with the larger, more thorough, and comprehensive studies the organization produced for later U.S. operations in Somalia, Haiti, Bosnia, Afghanistan, and the Persian Gulf.

and insights relating to the conduct of joint contingency operations." The item topping the lists contained in both sources, whether identified as one of the principles and procedures or observations and insights, was "integrated planning and execution." As phrased in the after action report, "The first key to success was a good planning and execution sequence focused on joint campaign objectives. For Operation Just Cause there was guidance from the NCA and the JCS, properly developed and approved plans at all levels, and the execution with proper advice, but without undue interference."[24]

With respect to the planning process that began in February 1988, the series of strategic-level contingency plans available throughout the crisis had been drafted by joint staffs answerable to the Southern Command. Each iteration of the operation orders had been coordinated up the chain of command with the Joint Chiefs of Staff and downward with the appropriate joint task forces and their components. Laterally, coordination occurred between headquarters for special operations and conventional forces. So refined and flexible was the planning process that it could readily incorporate the input of an additional headquarters—the XVIII Airborne Corps—midway into the enterprise. It could also adjust quickly to a major change in the concept of operations, namely that which occurred when Thurman succeeded Woerner as the SOUTHCOM commander. Finally, under both Woerner and Thurman, the Southern Command determined that, if force had to be used to resolve the crisis in Panama, the U.S. troops involved would be employed in accordance with a broad campaign plan. The Prayer Book served as proof of this intent, containing operation orders that ranged from defensive measures the U.S. military might take in Panama, including a troop buildup (Post Time) and an evacuation of American civilians (Klondike Key), to offensive combat operations (Blue Spoon, 1–90, and 90–2) and civil-military operations, involving nation building and stability and reconstruction activities (Blind Logic). In short, should war come, the U.S. military intended—on paper at least—to be involved across a spectrum of operations beginning with the defense of U.S. interests, moving on to the elimination of a dictatorial regime and its military arm, and finishing up with actions to support a new, democratic government.[25]

As the Prayer Book took shape, U.S. units based in Panama and those deployed there as part of Washington's response to the crisis were able to incorporate the various contingency plans into exercises and training events scheduled by Joint Task Force-Panama. In many cases, these activities amounted to rehearsals carried out under the rubric of the Purple Storms established by General Loeffke during his tenure as the JTF-Panama commander or the Sand Fleas and contingency readiness exercises General Cisneros added after he succeeded Loeffke. In the aftermath of the failed October coup against Noriega, the tempo of such rehearsals increased dramatically and reached out to encompass units that, while based in the United States, were slated for

[24] First and second quotes from Clancy with Stiner and Koltz, *Shadow Warriors*, p. 392. Remaining quotes from Cover memo and Encl 1, to AAR, Operation Just Cause, 20 Dec 1989–12 Jan 1990, JTF-South, 19 Mar 1990, p. 9.

[25] The summary of the planning process should be supplemented by Yates, *U.S. Military Intervention in Panama, June 1987–December 1989*. See also Cole, *Operation Just Cause*.

action in Panama should BLUE SPOON be executed. Stiner would later praise the training regimen during this period: "There is no substitute for live-fire training, under the most realistic conditions, as you expect to fight." In many cases, especially those involving U.S.-based units, the realistic conditions included mock-ups of the actual targets listed in the battle plans.[26]

When the order did come to execute BLUE SPOON, most of these well-prepared combat forces—half of which were already located in Panama, just minutes from their targets—achieved their objectives quickly, an accomplishment attributed by senior officers not just to the troops' rigorous and realistic training and exemplary leadership but also to the manner in which they entered the area of operations: rapidly, violently, and in overwhelming force. Stiner had argued for this approach once the XVIII Airborne Corps became involved in the planning process. Thurman had signaled his support for the concept even before he took command at Quarry Heights. And most officers involved in JUST CAUSE applauded the results as representing a decisive reversal and repudiation of the kind of piecemeal troop commitments that, in their minds, had contributed to the American military defeat in Vietnam. Put another way, Panama seemed to have validated the anti-incrementalist approach to war espoused in the Weinberger-Powell doctrine first enunciated in the mid-1980s. In what was regarded as a further contrast with the war in Indochina, the tactical units and operational headquarters in JUST CAUSE were able to carry out their assignments without unwanted and unwarranted interference from higher authorities far removed from the battlefield, namely the National Command Authority and the Pentagon in Washington. This freedom from micromanagement, when combined with the multiservice composition of the force employed in Panama, caused several officers to declare that JUST CAUSE had also validated the command and control arrangements and the emphasis on jointness found in the Goldwater-Nichols Department of Defense Reorganization Act of 1986. Not until after Operation DESERT STORM in 1991 would President Bush announce, "By God, we've kicked the Vietnam syndrome once and for all!"[27] Some military analysts would have argued that the U.S. military and the American people as a whole had taken a major step in that direction well over a year earlier in the invasion of Panama.

Having showered effusive praise on the overall planning, preparation, and execution of JUST CAUSE, those assessing the relevance of the operation to future contingencies proceeded throughout 1990 to offer much more detailed

---

[26] Quotes from Clancy with Stiner and Koltz, *Shadow Warriors*, p. 392.

[27] President Bush's statement was widely quoted at the time and has been since. One source is *Washington Post*, 9 Jan 2005. General Thurman, in a comment about the provision in the Goldwater-Nichols Act that had the chain of command running from the president through the secretary of defense directly to the unified combatant commanders, noted characteristically that he had a "very streamlined command and control operation," one in which he did not have to "take any crap off subordinate commanders," especially the Air Force component commander. Thurman SAMS briefing, 23 Sep 1992. General Stiner echoed Thurman's opinion on the efficacy of the command and control arrangements under Goldwater-Nichols. As he told the Senate Armed Services Committee in 1990, "There were no problems with ambiguous arrangements or with units receiving guidance from multiple sources." This quote from Stetson M. Siler, "Operation Just Cause: An Air Power Perspective," *Air Power History* 55, no. 4 (Winter 2008).

analyses of the factors they perceived as underlying the victory. Given the dozens of briefings, reviews, and interviews generated at the time, hundreds of observations poured forth, some so specialized or highly technical as to be of interest to but a very select audience. The importance of others often did not survive changes in doctrine, procedures, equipment, or weapon systems that were in progress well before December 1989 but not completed until after the invasion. Still, even with these anomalies, there remained sufficient grounds for categorizing the crisis and the hostilities in Panama as an important milestone in the 200-year saga of the American military. As many spokespersons reminded their audiences, JUST CAUSE was "the most complex contingency operation undertaken by US forces since World War II," or, to quote General Stiner once more, it was "the largest, most sophisticated contingency operation conducted over the longest distances in the history of the U.S. Armed Forces."[28]

The complexity and sophistication of the operation derived in part from the composition of the force involved. As the references to Goldwater-Nichols made clear, it was a joint force. Some critics could not refrain from noting the disproportionate number of Army units in an operation planned by an Army-heavy unified command, the four-star Army commander of which had called on his service's own contingency corps at Fort Bragg to be the war-fighting component. This observation, while correct, did not change the fact that no service had been left out of the undertaking. Soldiers and marines handled most of the land combat, while the Air Force provided timely troop transports, resupply, and close air support. The Navy guarded the air and sea deployment routes from the United States and also participated in the logistical effort. One feature of the joint endeavor that received special attention was the way it integrated conventional units, both heavy and light, with Special Operations Forces to execute a plan designed to capitalize on the strengths of each (although, again, some observers criticized the use of Navy SEALs instead of Rangers at Paitilla and questioned the need for 82d paratroopers instead of light fighters at Torrijos-Tocumen). As a result of this integration, small Special Operations Forces units carried out rescue missions, conducted raids, seized critical terrain, and sabotaged key targets. Rangers and paratroopers put their airborne capability to good use, thus reducing the time needed to insert combat troops into certain critical areas. And, in a more traditional vein, conventional infantry and mechanized units went after a number of well-fortified fixed targets, with the mechanized-heavy Task Force Gator converging on the Comandancia, the most formidable of the PDF's defensive positions. Throughout the fighting, Joint Task Force-South provided the operational control, coordination, and joint communications needed to make the operation's many "moving parts" mesh as well as possible, a task complicated by the fact that the first several hours of combat took place at night, with most targets being located in or near densely populated, built-up urban areas.

For years preceding Operation JUST CAUSE, both the Army and the Marine Corps had largely neglected the study of urban combat. The Army especially, given its post-Vietnam preoccupation with being prepared to fight Soviet forces

---

[28] First quote from Cover memo, to JTF-South AAR. Second quote from Briefing Slides, U.S. Southern Command, Operation Just Cause: "Rebirth of a Nation," n.d.

on the plains of central Europe, adhered to a doctrine of bypassing enemy-held cities encountered along any line of advance. Time constraints and supply issues, senior officers believed, would not permit lengthy and resource-consuming sieges or street-by-street fighting. The Panama experience by itself did not cause a sudden reversal of this position, but it did suggest that a reassessment was overdue. As the JTF-South after action report bluntly states, the "Army must acknowledge the growing significance of MOUT." Military developments throughout the world in the 1980s and 1990s, including the sieges of Beirut and Sarajevo, and the involvement of U.S. forces in such places as Mogadishu, Port-au-Prince, and the Balkans reinforced this conclusion. In the mid-1990s, the marines took the lead in addressing the need to prepare for urban combat. By the turn of the century, the Army had joined in as a full partner, with both services elevating city-fighting to a much higher priority in their doctrinal deliberations and training. One source of insight on which the reorientation relied was an analysis of past urban operations. The lists of case studies are lengthy but manageable and generally contain much overlap, with something like the World War II battle for Stalingrad representing a worst-case nonnuclear scenario. In comparison, Panama was not close in terms of duration, ferocity, death, and destruction. Regardless, the JUST CAUSE experience offered much to study about clearing buildings (from the bottom up when necessary), clearing rooms (the danger of using fragmentary grenades when construction material was flimsy), minimizing collateral damage under certain circumstances (the requirement for restrictive rules of engagement, nighttime operations, and selected weaponry), and interacting with a large civilian population (the need for realistic training in a more inclusive environment than one characterized simply by force on force).[29]

Just as most of JUST CAUSE took place in urban areas, most of the D-day assaults began at night. U.S. officers at the outset of the planning effort in 1988 had designated 0100 as H-hour for this decisive phase of the invasion because of both the reduced number of civilians likely to be on the streets or at the airport during a nocturnal attack and the superior night-fighting capabilities U.S. forces possessed as a consequence of their training and equipment. The first concern, the desire to minimize civilian casualties as well as property damage, produced rules of engagement that placed some restrictions on what U.S. forces could do to achieve victory. The purpose of such rules was to balance the troops' need to defend themselves and deliver a quick knockout blow to the enemy with the requirement to keep civilians and their possessions out of harm's way. That innocent bystanders would be killed or wounded was seen as inevitable, given the congestion found throughout many sections of Panama's two major cities. But the rules of engagement, U.S. officials concluded soon after the shooting

[29] Quote from Encl 4, to JTF-South AAR, p. 13. In September 1999, the author attended an urban operations conference at Fort Leavenworth, during which it became clear that the Army was preparing to join the Marine Corps in the study of military operations on urbanized terrain (MOUT). In January 2001, the commander of the Army's Training and Doctrine Command sent an e-mail to the Command and General Staff College and the Combat Studies Institute (CSI) at Fort Leavenworth formally directing both to address the issue through teaching and research. One result of that directive was an anthology of MOUT case studies, *Block by Block: The Challenges of Urban Operations,* ed. William G. Robertson and Lawrence A. Yates (Fort Leavenworth, Kans.: U.S. Army Command and General Staff College Press, 2003).

stopped, had helped minimize civilian casualties as well as property damage. For combat troops who habitually trained to fight with minimal constraints, their most negative comments about the imposition of restrictive ROE in JUST CAUSE pertained not so much to the D-day firefights, as one might have anticipated, but to the transition from combat to stability operations, a time when rapid and continuous changes to the rules of engagement, together with the variations based on time and location, generated a good deal of confusion and frustration on the ground.

The second reason for choosing 0100 as H-hour—the planners' desire to utilize surprise and the "night-fighting advantages" training and equipment gave U.S. forces—was seen as "a key contributing factor in the success of the operation." The JTF-South after action report highlights the importance of night-vision devices, which, it states, were employed by "individual soldiers, attack and lift helicopter crews, mechanized infantry crews, and US Air Force crews." The report also singles out the AC–130 Spectre gunship as "an effective close support weapons system, particularly in terms of delivery of accurate fires at night."[30] That support proved essential at the Comandancia, where despite the friendly-fire incident involving a mechanized platoon—the one instance in which U.S. troops had cause to rue the aircraft's pinpoint accuracy—the PDF headquarters received a pounding from the air while civilian areas surrounding the target remained largely unscathed. The fact that the AC–130 proved less helpful to Task Force Wildcat at the PDF engineer compound did not find its way into the after action reports of higher headquarters.

Besides serving as an effective weapons platform for nocturnal warfare, the AC–130 also helped compensate for the undesirable force ratio that U.S. troops experienced at certain targets. Normally in urban operations, defenders are perceived to hold the advantage when the attackers do not have at least a two-to-one ratio of forces in their favor (with some experts putting the desired ratio as high as six to one). In Panama, often the friendly-to-enemy ratio was at best only one to one, including at the Comandancia. Historically, under such conditions, the U.S. military has sought to compensate for manpower deficits by relying on massive firepower. But in Panama, concerns about collateral damage precluded the use of indirect fire weapons such as field artillery (on hand only as a backup system or a direct-fire weapon) and certain other heavy weapons such as tanks (the M551 Sheridan used at the Comandancia being considered a light tank or, officially, an armored reconnaissance assault vehicle). Still, what remained in the U.S. inventory even after these exclusions was sufficient to shift the advantage to the attackers. In addition to the AC–130s, the AH–64 Apache and AH–1 Cobra helicopters and the special operations AH–6 Little Birds delivered lethal and effective fire from the air at night. On the ground, the weapons possessed by most troop units generated more than enough firepower, with the 90-mm. recoilless rifle receiving special recognition for quickly demonstrating "our fire superiority" in Panama's urban terrain, largely because of the weapon's "immediate 'psychological effect' due to

[30] Quotes from Encl 4, to JTF-South AAR, with first, second, and third on p. 16, and fourth on p. 1.

[the] size of the warhead and subsequent explosion and the debris producing and penetration capability of the system."[31] Other weapons and weapons platforms found worthy of praise were the Marine LAV–25s, the F–117A fighter aircraft (despite the targeting controversy), the AT4 rockets, and the .50-caliber machine guns mounted on HMMWVs and M113s.

Among the host of other factors singled out as facilitating the victory in Just Cause, the use of psychological operations broadcasts also came in for special mention. At some targets, PSYOP activities helped reduce casualties by convincing enemy forces to surrender or causing them to flee. In other cases, the broadcasts and various means of disseminating information kept the civilian population abreast of what was expected of them, particularly during stability operations, and explained to them the objectives and actions of U.S. forces in their midst. With the exception of a brief period just prior to H-hour, the Noriega regime and the Panama Defense Forces were denied a similar capability, thanks to JTF-South's "use of electronic warfare, specifically US Air Force systems" such as Volant Solo and Compass Call. Operating aboard six Air Force EC–130s, the systems overrode enemy media broadcasts while transmitting pre-prepared radio and television messages stating, "We are the Americans, your friends. We are here to give your freedom back. . . . Stay in your homes and no one will be harmed." These broadcasts started at H-hour, as did the jamming of PDF radar and tactical communications by two Air Force EF–111s.[32]

In general, then, U.S. officers involved in Operation Just Cause used their after action reports, briefings, and interviews to tout the joint composition and rigorous preparation of the U.S. force; command, control, and communications arrangements that integrated Special Operations Forces and conventional operations and that gave operational and tactical commanders the latitude to make critical decisions necessitated by the fluid situation; the nighttime launch of the invasion with overwhelming but selective firepower that employed a variety of high-tech devices the enemy could not counter or match; and a logistical system that provided timely refueling and resupply.

In contrast to the hyperbolic flourishes used in recounting how these and other factors contributed to the swift removal of the Noriega regime and its military support, the rhetoric used to describe the numerous difficulties encountered in achieving that decisive outcome was much more subdued. The list of problems was lengthy, with a representative sample usually including the ice storm that delayed the arrival of many paratroopers from Fort Bragg; the news reports and other breaches in operations security that warned the enemy well before H-hour that U.S. troops were going to attack; intelligence lapses that allowed Noriega to evade capture for several days; friendly fire at the Comandancia and elsewhere; the ineffectiveness of the Sheridans deployed on

---

[31] Ibid., p. 13. The report went on to note that the antipersonnel and high-explosive rounds available for the recoilless rifle in Panama were not the ideal rounds for penetrating "concrete structures commonly found in cities." The Army therefore needed to "provide suitable munitions that can be effectively employed against buildings."

[32] First quote from ibid., p. 19. Second quote from Clancy with Stiner and Koltz, *Shadow Warriors*, p. 322. William C. Martel, *Victory in War: Foundations of Modern Military Policy* (New York: Cambridge University Press, 2007), p. 172.

Quarry Heights; and the loss of several vehicles in the airdrop over Torrijos-Tocumen. Other problems making the list were the thousands of refugees from El Chorrillo who fled through the battle lines and overwhelmed the makeshift displaced persons camp in Balboa; the controversy surrounding the 2,000-pound bombs at Rio Hato; the several hostage situations that developed, despite the rapid infusion of U.S. troops designed to prevent such incidents; the controversial raid on the Nicaraguan ambassador's residence; the looting that remained unchecked for several days; and the often rough transition from combat to stability operations. Given longstanding military etiquette, no official assessment addressed another negative: the impact of personality conflicts on the operation, not that there was much on which to comment. At the top of the chain of command in Panama during JUST CAUSE, Thurman and Stiner had always been in sync concerning how the invasion should be conducted. On the other hand, the friction between Thurman and Cisneros that had surfaced soon after the change in command at Quarry Heights in late September carried over into the invasion, but, unlike the abortive coup of 3 October, in which the mistrust between the two may have cost the United States an opportunity to remove Noriega from power, in JUST CAUSE the main consequence was felt at the nunciature, where Thurman forced a change to Cisneros' negotiating strategy for getting Noriega and his entourage into U.S. custody. The change, while disruptive and most likely unnecessary, did not alter the outcome of the talks.

Many of the problems listed above and countless others encountered during JUST CAUSE were attributed to the fog and friction of warfare. In that context, Stiner's claim that no lessons were learned seemed less incredulous in that, beginning with their precommissioning education, all officers are taught the maxim that no plan survives first contact with the enemy. There are going to be gaps in the information or intelligence that friendly forces require, and there is going to be a plethora of unexpected, undesirable, and uncontrollable developments. That JUST CAUSE had its fair share of these was to be expected and did not, in the eyes of the U.S. military, detract from the undeniable success of the operation and its value as a model of a military contingency operation.

That being said, there were several cautionary conclusions of a general nature that did occasion comment at the time, at least two of which pertained to the planning process. By the time President Bush ordered the execution of BLUE SPOON, U.S. military headquarters from the Southern Command, through the XVIII Airborne Corps (Joint Task Force-South), down to the various task forces operating at the tactical level had copies of the operation orders relevant to their echelon. There were still last-minute revisions being made at all levels, but the basic plans were in place weeks—and, for some portions of them, months or over a year—before D-day. In fact, the planning itself had been under way for almost two years, an anomaly when compared to most of the Cold War contingencies preceding JUST CAUSE and those that would arise in the post–Cold War decade following it. In U.S. military interventions in the Dominican Republic in 1965, in Grenada in 1983, in Somalia in 1992, in Haiti in 1994, and, after that, in Bosnia, the planning staffs involved in each had a few days or at best a few weeks to fashion the strategy, tactics, forces, transportation, and resources for accomplishing American objectives. Even after General Thurman took over the Southern Command and approved a change in the concept of operations from a gradual buildup of U.S. troops to a swift strike that employed overwhelming combat power, his "warfighters" in the XVIII Airborne Corps at Fort Bragg still had over two months to fine-tune and disseminate the new operation plan they had finished drafting in September 1989, parts of which relied on the contents of General Woerner's version of Blue Spoon.

Given the generous, if atypical, amount of time available for contingency planning on the Panama crisis, U.S. staff officers, as previously noted, sought to integrate the Prayer Book series of operation orders into a campaign plan. On paper, they succeeded. In reality, the words they generated did not translate into the necessary command emphasis, staff coordination, and unit training that would have facilitated an effective synchronization between combat and stability operations, especially in the transition from the former to the latter. True, the mission statement contained in each iteration of Blue Spoon and, after 3 November 1989, OPLAN 90–2 alerted combat units to be prepared to restore law and order and to support the installation of a U.S.-recognized government in Panama. Yet, most of the combat troops participating in the operation, especially those deploying from the United States without previous experience in the Panama crisis, believed that their work would be all but over once the shooting stopped. If each plan's reference to follow-on stability operations generated any expectations among the warriors, it was mostly that the mission would be simple, short, and largely handled by specialists in such matters.

Officers in SOUTHCOM's J–5 directorate responsible for planning the civil-military operations contained in Blind Logic blamed the XVIII Airborne Corps for deliberately ignoring the stability operations with which JTF-South would be charged under Blue Spoon and OPLAN 90–2. General Stiner thought otherwise. Yes, the "conflict termination part" of Just Cause was "not done properly," he later concurred, but, as for the reason, he stated the following:

> Now let me address that. I asked General Thurman from the beginning who was responsible for that. He said, "Not you. That is the USARSO [U.S. Army, South] staff. I want you to focus on the combat operations." I said, "Sir, this is a seamless process, it has to be, because when the fighting stops you have to get right into stability operations. You have to have refugees and you have to get into the nation building. These other agencies of government that do this sort of thing are going to have to be brought in on this." He said, "My J5 will take care of that." I said, "I need to be briefed on it while we are planning the actual combat part, Just Cause, so that it is a seamless process." He said, "The next time you come down on a planning mission, you will be briefed." But the next time I would go down, the J5 was not present to brief. He [Thurman] promised that two or three times and it did not occur. On the night that we struck, he said, "I have been let down. This part has not been planned adequately and I want you and Bill Hartzog to work on it." That part was not done well. Or perhaps the transition process could have gone quicker or smoother and we would not have had some of the glitches that we did in the

nation building, conflict termination part. But other than that I think it was a great operation.[33]

Stiner's account omitted the early coordination between his people and Woerner's J–5 and the breakdown of the agreements reached, but his "nonlesson" was clear enough. The planning for military action in Panama had not ensured a seamless process for a realistic transition from combat to stability operations. But even if it had, there was no guarantee at the time that the affected combat units would have been any better prepared to make that transition. As Lt. Col. Mike Menser, the Training and Doctrine Command liaison officer in Panama, asked in April 1990,

> Where do we train an infantry or artillery battalion to run a city, take care of prisoner/refugees, feed and police the populace, and operate the public utilities? Superb effective execution of METLs [mission-essential task lists] was the norm for all units involved in Just Cause. But they had to chart new ground as they faced real challenges in conducting foreign internal defense (FID), civil affairs (CA), civic action, and psychological operations (PSYOP). We had combat (direct action) units working in the political, economic, and social (or indirect) arenas. When, where, and how do we prepare conventional Army forces to do this? They knew what they were sent here to do: protect U.S. lives and property as effectively as possible. They did. They were, however, given no warning order for a follow-on nationbuilding mission. If they had been, how would they have prepared? What's the METL?[34]

Over a decade after Just Cause, Operation Iraqi Freedom demonstrated that answers to Menser's questions remained elusive, while the institutional obstacles to Stiner's seamless process had yet to be removed, as the military's elaborate master plan for reconstructing post-Saddam Iraq was jettisoned by U.S. military and political authorities at higher levels and, thus, never integrated into the projected campaign plan.

Despite the disconnect between combat and stability operations and despite the fog and friction and other problems U.S. forces encountered in accomplishing their missions in Panama, the planning and preparation for Operation Just Cause, together with the way in which it was executed, deserved all the accolades the undertaking received. Moreover, in the initial penchant for official hyperbole, no one tried to pass the victory off as the one thing that it was not: the defeat of a first-rate enemy force. The outcome of Just Cause had never been in doubt. The Panama Defense Forces would be destroyed as an institution, and the Noriega regime would fall. What was not so certain was the price that victory would extract. The worst scenario envisaged Panamanian defenders knocking U.S. troop transports out of the sky with shoulder-fired, surface-to-air missiles, taking large numbers of American citizens hostage, engaging in tenacious and bloody street fighting in Panama City and Colón,

[33] On Stiner's comment that Thurman failed to involve the XVIII Airborne Corps in planning for stability operations, see Interv, Lt Col Steven P. Bucci with Gen Carl W. Stiner, U.S. Army (Ret.), 1999, USAWC/USAMHI Senior Officer Oral History Program.

[34] Memo, Lt Col Mike W. Menser for Distribution List, 1 Apr 1990.

and carrying the combat into neighborhoods full of Panamanian civilians.[35] Each of these prospects bedeviled U.S. planners and commanders, even though more realistic projections held that the PDF did not have the numbers, the leadership, the training and discipline, the technology and equipment, and the determination to fight the Americans head-on. Nor would they provoke America's wrath by committing a prolonged series of calculated atrocities. This assessment, in turn, led analysts at Quarry Heights, Fort Clayton, Fort Bragg, and the Pentagon to predict that many, if not most, enemy units would flee or surrender once hostilities commenced—a prediction that proved prophetic, except for some specific PDF elements and DigBats that fought tenaciously. Some battles, such as at the Comandancia, were indeed ferocious. Yet, with a one-to-one force ratio, U.S. troops, with their advantage in virtually every other category of military power, eliminated most of the opposition within hours. The worst damage enemy forces inflicted was on their own countrymen, as when PDF and Dignity Battalion members deliberately set fire to El Chorrillo. In the wake of JUST CAUSE, some commentators opined on what would have happened had the PDF been a highly professional and determined opponent. The obvious response was that there would have been a different set of contingency plans taking those facts into account. That being said, any remaining discussion became pure conjecture.

Immediately after JUST CAUSE, each service looked beyond Panama to determine what the operation signified for its future. The Army, as the largest service component to take part in the conflict, drew the most far-reaching conclusions, as initially presented in early January 1990 by its deputy chief of staff for operations, Lt. Gen. Gordon R. Sullivan. According to Sullivan, JUST CAUSE had shown low-intensity conflict to be a "growth industry" that recognized "the ascendancy of contingency forces" in the Army's priorities. It was important, he stated, to have the capability to project light, heavy, special operations, or any combination of forces from bases in the United States to trouble spots overseas, often in support of forward-deployed units.[36] To this end, the Army would reassess its current doctrine, force structure, equipment, and training and education programs, using JUST CAUSE as a model. That process, however, had just begun to show results when Saddam Hussein's forces invaded Kuwait in August 1990. Operations DESERT SHIELD and DESERT STORM soon followed, with some observers suggesting that, in DESERT STORM, the U.S. military finally was able to fight the kind of war for which it had been preparing for four decades: a conventional force-on-force conflict in which firepower and maneuver would prove decisive. In the meantime, the prominence that had been accorded the U.S. intervention in Panama dropped off precipitously. Within a year, it had lost its status as a beacon a future fighting force would follow and had become simply another example of the successful if limited exercise of U.S. power in a dangerous world. Such references made their way

[35] On 19 June 1990, General Cisneros conducted a review, or "hot wash," with a PDF officer who had taken part in the defense of Torrijos-Tocumen. At one point, the Panamanian stated that had his men possessed more sophisticated weapons, they would have "created havoc" with the U.S. Rangers dropping onto the air complex. They were "easy targets," he declared. Author's notes, Hot Wash, 19 Jun 1990, Fort Clayton, Panama.

[36] Briefing, Lt Gen Gordon Sullivan, 9 Jan 1990, Fort Leavenworth, Kans.

into the 2008 version of the Army's Field Manual 3–0, *Operations*, which lists Just Cause as an example of a coup de main (a swift, surprise attack), and into the 2011 version of Joint Publication 3–0, *Joint Operations*.[37] Besides listing Just Cause as a coup de main, 3–0 also lists it as an example of forced entry and nonlinear operations. Still, for many who studied the manuals, the U.S. invasion of Panama had become little more than a distant memory.

This assertion raises a familiar conundrum for the military professional. American society, countless observers have pointed out, suffers from historical amnesia. In an equally repeated truism, the U.S. military reflects the society from which it springs. For many officers, an operation occurring twenty-five years ago has little relevance for the present, much less the future. A historian would argue otherwise, of course, maintaining that there is much to learn from past military undertakings, even those that go back a quarter century (as well as centuries and millennia). If Operation Just Cause did not live up to its instant billing as the principal model a U.S. Army of the future would emulate, it still contains many insights for today's officers. As an enhancement to an officer's professional education and training, Just Cause serves as a superb case study for better understanding contingency operations, urban combat, stability operations, nonlinear warfare, and campaign planning. And one should not forget that the actual event was a joint undertaking throughout, as well as an exercise of the operational art that produced a quick and decisive victory. As such, the operation was a major confidence builder at a time when a rejuvenated U.S. Army needed it the most. For this and for the reasons enumerated throughout this volume, the historical verdict on Just Cause is a positive one.

[37] DA FM 3–0, *Operations*, 27 Feb 2008, pp. 2-13, 8-4 to 8-5; Joint Publication 3–0, *Joint Operations,* 11 Aug 2011, pp. V-14, V-46, V-51.

Bibliography

☆ ☆ ☆

Primary Sources

Research for this project began in March 1989, just months before Operation Just Cause ended the two-year crisis in U.S.-Panamanian relations. At the time, the documents available for me to examine were located almost exclusively in Panama, either at Quarry Heights, home of the U.S. Southern Command, or ten minutes north on Gaillard Highway at the headquarters of U.S. Army, South, the organization that sponsored my first six research trips to the country. The agenda for each trip was similar. An assortment of written sources would be copied, including daily and weekly reports, operation orders, messages, memorandums, organizational directives, training and exercise projections and assessments, and policy statements and discussions. Also on the schedule were frequent meetings, especially of JTF-Panama, and a host of interviews with U.S. officers and civilians involved in the crisis. During the fourth research visit, which began on 11 December 1989, the routine was broken by the onset of Just Cause, with the JTF-South operations center at Fort Clayton affording me a vantage point from which to witness the first hours of the operation.

With that event, the scope of my project and the research it required expanded to include contingency planning for the crisis (I was allowed to see the plans the night before the invasion), the execution of the plans, and the impact of Just Cause on both the situation in Panama and the U.S. military, particularly the Army. The geographical area covered in subsequent research trips also expanded after Just Cause, as documents became available to me in military and other governmental collections scattered across the United States: the I Corps headquarters at Fort Lewis, Washington; the Combined Arms Research Library and various organizations located at Fort Leavenworth, Kansas; the Operations Division and the Joint History Office in the Pentagon; the U.S. Army Center of Military History, now situated at Fort McNair, D.C.; the U.S. Army Special Operations Command and the XVIII Airborne Corps headquarters at Fort Bragg, North Carolina; the U.S. Army War College and U.S. Army Military History Institute at Carlisle Barracks, Pennsylvania; the U.S. Marine Corps History Division at Quantico, Virginia; and the U.S. Special Operations Command at MacDill

Air Force Base, Florida. At these locations and from my desk in the Combat Studies Institute, I also continued to conduct interviews with JUST CAUSE participants and to acquire interview transcripts from individuals and organizations engaged in the same activity. Among those who promoted the use of oral history in analyzing the Panama crisis were officers and historians at the XVIII Airborne Corps; U.S. Army, South, now at Fort Sam Houston, Texas; the U.S. Army Military History Institute; the U.S. Army Special Operations Command; the U.S. Marine Corps History Division; U.S. Southern Command, now in Dade County, Florida; and the U.S. Special Operations Command. My position at the Command and General Staff College also provided an ideal position from which to discuss JUST CAUSE with student officers, faculty members, and other personnel on post who had served in Panama. Many of these individuals also shared personal records they had kept or accounts they had written (or, in some cases, would write for one of my classes).

My most intensive research in primary sources took place within the first five years or so after JUST CAUSE. This point deserves emphasis, in that almost all of the documents I collected during that period came from folders lodged in file cabinets still being used for daily business. Today, for the most part, those collections do not exist intact. Presumably, some of the files were formally retired to the National Archives or to a military service's history division, such as the Army's Center of Military History, or to the history office of a subordinate headquarters, such as U.S. Army, South. Some collections were relocated within a given headquarters, while other collections simply vanished, not infrequently the victims of "shredding parties" held for the purpose of freeing cabinet space for files dealing with more immediate and pressing issues. In short, the original documents that I examined and copied during the first few years of my research are rarely to be found in the same locations, if they can be found at all. Still, a researcher looking into the Panama crisis would do well to contact the organizations cited in this bibliography to ascertain the nature and status of any relevant material they might hold.

Another approach would be to seek access to the discrete collection of primary sources I assembled in preparing this volume and its predecessor. Even here, though, the undertaking will not be easy, at least not for some time. As the collection grew, it was housed on the secured floor of the Combined Arms Research Library. As the first volume neared publication and work progressed on the second, that floor was closed for extensive repairs. I removed many of the public records, after which the remaining material was put under plastic wrap for a number of years. At some future point, my intent, and that of the U.S. Army Center of Military History, is to reunite the two portions into one collection that will be made available to researchers.

UNPUBLISHED MATERIAL

Many of the student papers contained in the list below were submitted in classes I taught at the Command and General Staff College. The authors were participants in Operation JUST CAUSE.

Akers, Col. Frank, comp. Operation JUST CAUSE, *The Warriors*. Collected vignettes released by the U.S. Southern Command.

Beech, Michael F. 'Quasi-War': Training Infantry Small Units for Operations Other Than War. Student paper, School of Advanced Military Studies, Fort Leavenworth, Kans., Academic Year 1995–1996.

Collins, James W. "Blue and Purple: Optimizing the Command and Control of Forward Deployed Naval Special Warfare." Master's thesis, U.S. Army Command and General Staff College, 1997.

Fishel, John T. War, Peace, and Civil Military Operations: The Role of Civil Affairs in Operations JUST CAUSE, DESERT STORM, and PROVIDE COMFORT. May 1992.

Haycock, Maj. Robert. NEO—A Platoon Leader's Experience: Operation Just Cause/Marriott Hotel NEO. Student paper, U.S. Army Command and General Staff College, Fort Leavenworth, Kans., 27 May 1999.

Hedegaard, Maj. Michael L. Task Force Wolf in Operation JUST CAUSE: First AH–64 Apache Helicopters in Combat. Student paper, U.S. Army Command and General Staff College, Fort Leavenworth, Kans., 2 March 2001.

Madrigan, Maj. Brod N. How I Spent the Holidays. Unpublished article, n.d.

Marcella, Gabriel, and General Frederick F. Woerner Jr., U.S. Army (Ret.). The Road to War: The U.S.-Panamanian Crisis, 1987–1989. Draft article, 6 May 1991.

Myers, Houston. A First Lieutenant Navigator's Experience in Panama. Student paper, U.S. Army Command and General Staff College, Fort Leavenworth, Kans.

Reeves, Brian W. "Navy SEALs: Theory and Reality." Master's thesis, Naval Postgraduate School, 1997.

Rosello, Victor M. An MI Paratrooper's Memory of Operation JUST CAUSE. n.d.

Ryan, Michael. C-645 essay [on securing France Field]. Student paper, U.S. Army Command and General Staff College, Fort Leavenworth, Kans.

BOOKS, PAMPHLETS, AND ARTICLES

Of the published sources used in this project, three works must be singled out for special mention. First, Thomas Donnelly, Margaret Roth, and Caleb Baker's *Operation Just Cause: The Storming of Panama* (New York: Lexington Books, 1991) is a superb overview of the invasion that relies on interviews and other primary sources. That book was used extensively to supplement my own sources, and it is cited frequently in the footnotes. With near equal enthusiasm, the same sentiments apply to Edward M. Flanagan Jr.'s excellent *Battle for Panama: Inside Operation Just Cause* (New York: Brassey's, 1993). Finally, John T. Fishel's *The Fog of Peace: Planning and Executing the Restoration of Panama* (Carlisle Barracks, Pa.: Strategic Studies Institute, 1992) is one of the best analyses available on planning for a democratic and functioning Panama and, once the Noriega regime fell, the conduct of stability operations and nation-building activities in pursuit of those goals.

Adams, Thomas K. *US Special Operations Forces in Action: The Challenge of Unconventional Warfare*. New York: Routledge, 1998.

Adkin, Mark. *URGENT FURY: The Battle for Grenada*. Lexington, Mass.: D. C. Heath, 1989.

Bahmanyar, Mir. *Shadow Warriors: A History of the US Army Rangers*. Oxford, U.K.: Osprey Publishing, 2005.

———, with Chris Osman. *SEALs: The US Navy's Elite Fighting Force*. Oxford, U.K.: Osprey Publishing, 2007.

Baker, James A., III, with Thomas M. DeFrank. *The Politics of Diplomacy: Revolution, War, and Peace, 1989–1992*. New York: G. P. Putnam's Sons, 1995.

Bloechl, Timothy D. "Operation 'Just Cause': Low Intensity Conflict Stability Operations and Issues of Tactics." *Low Intensity Conflict & Law Enforcement* 2, no. 1 (Summer 1993).

Bolger, Daniel. *Death Ground: Today's American Infantry in Battle*. Paperback ed. New York: Ballantine Books, 2003.

Briggs, Clarence E., III. *Operation Just Cause, Panama, December 1989: A Soldier's Eyewitness Account*. Harrisburg, Pa.: Stackpole Books, 1990.

Buckley, Kevin. *Panama: The Whole Story*. New York: Simon & Schuster, 1991.

Butler, F. K., Jr. "Closed-circuit oxygen diving in the U.S. Navy." *Undersea and Hyperbaric Medicine* 31, no. 1 (2004).

Call, Charles T. "Institutional Learning within ICITAP." In *Policing the New World Disorder: Peace Operations and Public Security*. Edited by Robert B. Oakley, Michael J. Dziedzic, and Eliot M. Goldberg. Washington, D.C.: National Defense University Press, 1998.

Clancy, Tom, with General Carl Stiner (Ret.) and Tony Koltz. *Shadow Warriors: Inside the Special Forces*. New York: G. P. Putnam's Sons, 2002.

Cole, Ronald H. *Operation Just Cause: The Planning and Execution of Joint Operations in Panama, February 1988–January 1990*. Washington, D.C.: Joint History Office, Office of the Chairman of the Joint Chiefs of Staff, 1995.

———. *Operation URGENT FURY: Grenada*. Washington, D.C.: Joint History Office, Office of the Chairman of the Joint Chiefs of Staff, 1997.

Combelles, Pascale M. "Operation JUST CAUSE: A Military-Media Fiasco." *Military Review* 75 (May-June 1995).

Department of the Army. Field Manual (FM) 3–0, *Operations*. February 2008.

———. FM 7–10, *The Infantry Rifle Company*. December 1990.

———. FM 90–10–1, *An Infantryman's Guide to Urban Combat*. September 1982.

DeTreux, 1st Lt. Kenneth M. "Infantry in Low-Intensity Conflict: Observations From Operation JUST CAUSE." *Marine Corps Gazette* 74 (September 1990).

Dinges, John. *Our Man in Panama: How General Noriega Used the United States and Made Millions in Drugs and Arms*. New York: Random House, 1990.

———. *Our Man in Panama: The Shrewd Rise and Brutal Fall of Manuel Noriega*. Rev. ed. New York: Times Books, 1991.

Durant, Michael J., and Steven Hartov, with Lt. Col. Robert L. Johnson (Ret.). *The Night Stalkers*. New York: G. P. Putnam's Sons, 2006.

Falerios, Kenton J. *"Give Me Something I Can't Do": The History of the 82nd Military Police Company from WWI to the Iraq War.* Bloomington, Ind.: Authorhouse, 2007.

Fishel, John T. "Planning for Post-Conflict Panama: What It Tells Us About Phase IV Operations." In *Turning Victory Into Success: Military Operations After the Campaign*. Edited by Brian M. De Toy. Fort Leavenworth, Kans.: Combat Studies Institute Press, 2005.

———, and Richard D. Downie. "Taking Responsibility for Our Actions? Establishing Order and Stability in Panama." *Military Review* 72 (April 1992).

Gray, Anthony, and Maxwell Manwaring. "PANAMA: Operation *Just Cause*." In *Policing the New World Disorder: Peace Operations and Public Security*. Edited by Robert B. Oakley, Michael J. Dziedzic, and Eliot M. Goldberg. Washington, D.C.: National Defense University Press, 1998.

Hammond, Capt. Kevin J., and Capt. Frank Sherman. "Sheridans in Panama." *Armor* 99 (March-April 1990).

Huelfer, Evan A. "The Battle for Coco Solo Panama 1989." *Infantry* 90 (January-April 2000).

Kelly, Orr. *Brave Men, Dark Waters: The Untold Story of the Navy SEALs*. New York: Pocket Books, 1992.

Kempe, Frederick. *Divorcing the Dictator: America's Bungled Affair with Noriega*. New York: G. P. Putnam's Sons, 1990.

LaFeber, Walter. *The Panama Canal: The Crisis in Historical Perspective*. New York: Oxford University Press, 1978.

Len-Rios, Maria. "Operation Just Cause: An Eyewitness Account." *Focal Point* (Winter 1990).

Linn, Brian McAllister. *The Echo of Battle: The Army's Way of War*. Cambridge, Mass.: Harvard University Press, 2007.

Madrigan, Maj. Brod N. "The Attack on the DNTT Station." *Marine Corps Gazette* 74 (April 1990).

Martel, William C. *Victory in War: Foundations of Modern Military Policy*. New York: Cambridge University Press, 2007.

Mauskapf, Col. Robert P., and Maj. Earl W. Powers. "LAVs in Action." *Marine Corps Gazette* 74 (September 1990).

McConnell, Malcolm. *Just Cause*. Paperback ed. New York: St. Martin's Press, 1991.

Meditz, Sandra W., and Dennis M. Hanratty, eds. *Panama: A Country Study*. Washington, D.C.: Headquarters, Department of the Army, 1989.

Military Airlift Command. *Anything, Anywhere, Anytime: An Illustrated History of the Military Airlift Command, 1941–1991*. Scott Air Force Base, Ill.: Office of History, Military Airlift Command, 1991.

Mucciarone, Lt. Cdr. James J., Col. Craig H. Llewellyn, and Lt. Col. John M. Wightman. "Tactical Combat Casualty Care in the Assault on Punta Paitilla Airfield." *Journal of Special Operations Medicine* 8, no.1 (Winter 2008).

Muse, Kurt, and John Gilstrap. *Six Minutes to Freedom*. New York: Citadel Press, 2006.

Nemmers, Capt. Joseph M., et al. *United States Army South Staff Ride: Operation Just Cause, 20 December 1989–31 January 1990*. Rev. ed. Fort Clayton, Panama: Historical Office, Headquarters, U.S. Army, South, 1998.

Raines, Edgar F., Jr. *The Rucksack War: U.S. Army Operational Logistics in Grenada, 1983*. Washington, D.C.: U.S. Army Center of Military History, 2010.

Reynolds, Lt. Col. Nicholas E. *Just Cause: Marine Operations in Panama, 1988–1990*. Washington, D.C.: U.S. Marine Corps History and Museums Division, 1996.

Robinson, Linda. *Masters of Chaos: The Secret History of the Special Forces*. New York: Public Affairs, 2004.

Rottman, Gordon L., and Ron Volstad. *Panama 1989–90*. London: Osprey Publishing, 1991.

Sandler, Stanley. *Glad to See Them Come and Sorry to See Them Go: A History of U.S. Army Tactical Civil Affairs Military Government, 1775–1991*. Fort Bragg, N.C.: U.S. Army Special Operations Command History and Archives Division, 1993.

Shultz, Richard H., Jr. *In the Aftermath of War: US Support for Reconstruction and Nation-Building in Panama Following JUST CAUSE*. Maxwell Air Force Base, Ala.: Air University Press, 1993.

Siler, Stetson M. "Operation Just Cause: An Air Power Perspective." *Air Power History* 55, no. 4 (Winter 2008).

Soldiers in Panama: Stories of Operation Just Cause. Washington, D.C.: Office of the Chief of Public Affairs, Department of the Army, 1990.

Thigpen, Jerry L. *The Praetorian STARShip: The Untold Story of the Combat Talon*. Maxwell Air Force Base, Ala.: Air University Press, 2001.

U.S. Special Operations Command. *History, 1987–2007*. 20th anniversary ed. MacDill Air Force Base, Fla.: U.S. Special Operations Command, n.d.

Woodward, Bob. *The Commanders*. New York: Simon & Schuster, 1991.

Yates, Lawrence A. "Operation JUST CAUSE in Panama City, December 1989." In *Block by Block: The Challenges of Urban Operations*. Edited by William G. Robertson and Lawrence A. Yates. Fort Leavenworth, Kans.: U.S. Army Command and General Staff College Press, 2003.

———. "Panama, 1988–1990: The Disconnect between Combat and Stability Operations." *Military Review* 85 (May-June 2005).

———. *The U.S. Military Intervention in Panama: Origins, Planning, and Crisis Management, June 1987–December 1989*. Washington, D.C.: U.S. Army Center of Military History, 2008.

Internet

Since retirement in 2005, I have limited my research in primary documents to just a few archival holdings, such as those at the U.S. Army Center of Military History and the U.S. Special Operations Command History Office. One relatively new resource, though, has proved particularly helpful—the Internet. A small sample of useful material from various Internet sites is listed below. The problem associated to this asset, however, will be well known to those who surf the Web: what is there today may not be there tomorrow. Copies of the following reside in my background files for this book.

Arvidson, Cheryl. "Press advocates: Cheney kept a tight rein on media during military conflicts." *Freedom Forum Online*. http://www.freedomforum.org/templates/document.asp?documentID=3869.

Bush, George H. W. Speech. http://bushlibrary.tamu.edu/research/public_papers.php?id=1356&year=1989&month=12.

Casualty figures for Operation JUST CAUSE. http://william_h_ormsbee.tripod.com/jc_casualties_panamanian_p01.htm.

Evans, Daniel E. Actions at Fort Espinar, Republic of Panama, 20 December 1989. n.d. Copy at http://www.benning.army.mil/library/content/Virtual/Donovanpapers/other/STUP5/EvansDaniel%20E.%20CPT.pdf.

Friedman, Herbert A. U.S. PSYOP in Panama (Operation Just Cause). n.d. http://www.psywarrior.com/PanamaHerb.html.

Govern, Lt. Col. Kevin H. Sorting the Wolves from the Sheep. [Detainee camps and U.S. military police during Operation JUST CAUSE.] http://www.wood.army.mil/mpbulletin/pdfs/oct%2004%20pdfs/govern.pdf.

Haight, Maj. David B. Operation JUST CAUSE: Foreshadowing Example of Joint Vision. Student paper, Naval War College, Newport, R.I., 2010. http://www.globalsecurity.org/military/library/report/1998/juscause_T18.htm.

Hunter, Thomas B. "Operation Acid Gambit: The Rescue of Kurt Muse." http://www.specialoperations.com/Operations/Just_Cause/Acid_Gambit/default2.html.

Rangers operations at Rio Hato. http://www.suasponte.com/m_riohato.htm.

Reeves, Brian W. Navy SEALs: Theory vs. Reality. Thesis, Naval Postgraduate School, Monterey, Calif., 1997. http://www.dtic.mil/cgi-bin/GetTRDoc?AD=ADA342802.

The SEAL assault on Patilla [*sic*] Airfield during Operation Just Cause. n.d. http://www.specwarnet.net/miscinfo/patilla.htm.

U.S. Army. 7th Infantry Division (Light). At http://www.globalsecurity.org/military/library/report/1990/90-9/9091ch1.htm.

U.S. Army Center for Army Lessons Learned. "Renacer Prison Raid." In Newsletter 91–1, *Rehearsals* (1991). At http://www.globalsecurity.org/military/library/report/call/call_91-1_sec6.htm.

U.S. Army Center of Military History. Web site on Operation JUST CAUSE. http://www.army.mil/cmh-pg/documents/panama/taskorg.htm.

To U.S. civilians who died during Operation JUST CAUSE. http://william_ormsbee.tripod.com/Pages/dedications_p4.html.

U.S. International Development Cooperation Agency. Panama Project Paper. AID/LAC/P-520. Washington, D.C.: Agency for International Development, 3 February 1990. http://pdf.usaid.gov/pdf_docs/PDABB803.pdf.

U.S. Southern Command. Public Affairs After Action Report. Operation JUST CAUSE. http://www.dod.mil/pubs/foi/International_security_affairs/latinAmerica/3.pdf.

Newspapers, Bulletins, and Magazines

Army Times, Boston Globe, Chicago Sun-Times, El Siglo, La Prensa, Los Angeles Times, New York Times, Sarasota Herald-Tribune, Seattle Post-Intelligencer, St. Petersburg Times, Tropic Times, and *Washington Post.*

Department of State Bulletin and *Refworld.*

Newsweek and *Time.*

Abbreviations

☆ ☆ ☆

AAR	after action report
ADA	air defense artillery
AFB	Air Force Base
AFFOR	Air Force Forces
AO	area of operations
AUTODIN	Automated Digital Network
CA	civil affairs
CALL	Center for Army Lessons Learned
CEV	combat engineer vehicle
CG	commanding general
CIA	Central Intelligence Agency
CINCFOR	commander in chief, U.S. Forces Command
CINCMAC	commander in chief, Military Airlift Command
CINCSAC	commander in chief, Strategic Air Command
CINCSO	commander in chief, Southern Command
CJCS	chairman, Joint Chiefs of Staff
CMOTF	Civil-Military Operations Task Force
COMCMOTF	commander, Civil-Military Operations Task Force
CORDS	Civil Operations and Revolutionary Development Support
COSCOM	Corps Support Command
CRE	contingency readiness exercise
DENI	Spanish acronym for PDF National Investigation Department
DNTT	Spanish acronym for Directorate of Traffic and Transportation
DoD	Department of Defense

EDRE	emergency deployment readiness exercise
EPW	enemy prisoner of war
EXORD	execute order
FARP	Forward Area Rearming and Refueling Point
FAST	Fleet Anti-Terrorist Security Team
FORSCOM	U.S. Army Forces Command
FRAGO	fragmentary order
FUFEM	*fuerzas feminines* (female PDF soldiers)
HB	heavy broadcast
HF	high frequency
HHC	Headquarters and Headquarters Company
HMMWV	high mobility, multipurpose wheeled vehicle
ICITAP	International Criminal Investigative Training Assistance Program
JCS	Joint Chiefs of Staff
jg	junior grade
JOTC	Jungle Operations Training Center
JSOTF	Joint Special Operations Task Force
JTF	Joint Task Force
JTF-Panama	Joint Task Force-Panama
JTF-South	Joint Task Force-South
LAV	light armored vehicle
LAW	light antiarmor weapon
LCM	landing craft, mechanized
LZ	landing zone
MAC	Military Airlift Command
MEDCAP	Medical Civic Action Program
MEDEVAC	medical evacuation
MOUT	military operations on urbanized terrain
MP	Military Police
MRE	meal, ready to eat
MSG	Military Support Group
NAVFOR	Naval Forces
NCA	National Command Authority
NSA	National Security Agency
NSD	national security directive

OPLAN	operation plan
OPORD	operation order
OPTEMPO	operational tempo
PCC	Panama Canal Commission
PDF	Panama Defense Forces
PML	Personnel Movement Limitation
PNP	Panama National Police
POV	privately owned vehicle
PPF	Panama Public Force
PSYOP	psychological operations
PZ	pickup zone
QRF	quick reaction force
RC	reserve component
ROE	rules of engagement
RPG	rocket-propelled grenade
R & S	reconnaissance and surveillance
SAF	Special Action Force
SAM	surface-to-air missile
SAMS	School for Advanced Military Studies
SAW	squad automatic weapon
SEAL	Navy Sea-Air-Land element
SKE	station-keeping equipment
SOCOM	U.S. Special Operations Command
SOCSOUTH	Special Operations Command, South
SOF	U.S. Special Operations Forces
SOP	standing operating procedure
SOUTHCOM	U.S. Southern Command
TF	Task Force
TOE	table of organization and equipment
TOW	tube-launched, optically tracked, wire command-link-guided
TTAD	temporary tour of active duty
TU	Task Unit
TEWT	tactical exercise without troops
UESAT	Spanish acronym for the PDF's special operations unit
USAID	U.S. Agency for International Development
USAMHI	U.S. Army Military History Institute
USARSO	U.S. Army, South
USAWC	U.S. Army War College
USCINCLANT	commander in chief, U.S. Atlantic Command

USCINCSOC	commander in chief, U.S. Special Operations Command
VHF–FM	very high frequency–frequency modulated
YMCA	Young Men's Christian Association

Index